S0-BLI-297

Texts and Monographs in Physics

Philippe Blanchard Erwin Brüning

Variational Methods in Mathematical Physics

A Unified Approach

With 7 Figures

Springer-Verlag
Berlin Heidelberg New York
London Paris Tokyo
Hong Kong Barcelona
Budapest

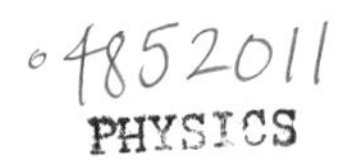

Professor Dr. Philippe Blanchard

Theoretische Physik, Fakultät für Physik, Universität Bielefeld
Postfach 8640, W-4800 Bielefeld, FRG

Dr. Erwin Brüning

Department of Mathematics, University of Cape Town
Private Bag, Rondebosch 7700, South Africa

Translator

Dr. Gillian M. Hayes

Department of Artificial Intelligence, University of Edinburgh
5 Forrest Hill, Edinburgh EH1 2QL, Scotland

Title of the original German edition:
Philippe Blanchard, Erwin Brüning: *Direkte Methoden der Variationsrechnung – Ein Lehrbuch*
© Springer-Verlag Wien 1982

ISBN 3-540-16190-2 Springer-Verlag Berlin Heidelberg New York
ISBN 0-387-16190-2 Springer-Verlag New York Berlin Heidelberg

Library of Congress Cataloging-in-Publication Data. Blanchard, Philippe [Direkte Methoden der Variationsrechnung. English] Variational methods in mathematical physics: a unified approach/ Philippe Blanchard, Erwin Brüning. p. cm. – (Texts and monographs in physics). Translation of: Direkte Methoden der Variationsrechnung: Includes bibliographical references and index. ISBN 0-387-16190-2 1.Variational principles. 2. Mathematical physics. I. Brüning, Erwin. II. Title. III. Series. QC174.17.V35B58 1992 530.1'5564–dc20 92–3355 CIP

Typesetting: Data conversion by Springer-Verlag
55/3140 – 5 4 3 2 1 0 – Printed on acid-free paper

Preface

The first edition (in German) had the prevailing character of a textbook owing to the choice of material and the manner of its presentation. This second (translated, revised, and extended) edition, however, includes in its new parts considerably more recent and advanced results and thus goes partially beyond the textbook level.

We should emphasize here that the primary intentions of this book are to provide (so far as possible given the restrictions of space) a selfcontained presentation of some modern developments in the direct methods of the calculus of variations in applied mathematics and mathematical physics from a unified point of view and to link it to the traditional approach. These modern developments are, according to our background and interests:

(i) Thomas-Fermi theory and related theories, and
(ii) global systems of semilinear elliptic partial-differential equations and the existence of weak solutions and their regularity.

Although the direct method in the calculus of variations can naturally be considered part of nonlinear functional analysis, we have not tried to present our material in this way. Some recent books on nonlinear functional analysis in this spirit are those by K. Deimling (*Nonlinear Functional Analysis*, Springer, Berlin Heidelberg 1985) and E. Zeidler (*Nonlinear Functional Analysis and Its Applications*, Vols. 1–4; Springer, New York 1986–1990).

The reader with some background in the calculus of variations will certainly miss many important aspects and results, for instance (locally) convex analysis and duality theory, minimal surfaces, various recent results in critical point theory and various classical methods and results.

To help to get a fairly complete picture of most of the fascinating aspects of the calculus of variations we would like to give some references covering many important topics not treated in this book.

Two recent books on (locally) convex analysis and duality in the calculus of variations are

I. Ekeland, R. Temam: *Convex Analysis and Variational Problems*, North Holland, Amsterdam 1976, and

V. M. Alekseev, V. M. Tikhimirov, S. V. Fomin: *Optimal Control*, Plenum, New York 1987.

Problems of minimal surfaces have played a prominent role in the development of the calculus of variations and of mathematics itself. A beautiful nontechnical presentation of this subject has been given by S. Hildebrandt and A. Tromba in the book *Mathematics and Optimal Form*, Freeman, New York 1985. Further sources of information are, for instance, E. Gusti: *Minimal Surfaces and Functions of Bounded Variation*, Birkhäuser, Basel 1984, and S. Hildebrandt: Calculus of Variation Today, presented at the Oberwolfach Meetings, in *Perspectives in Mathematics*, ed. by W. Jäger, J. Moser, R. Remmert, Birkhäuser, Basel, 1983, pp. 321–336.

Finally, concerning critical point theory, one should consult books on nonlinear functional analysis and the recent book by J. Mawhin and M. Willem: *Critical Point Theory and Hamiltonian Systems*, Applied Mathematical Sciences *74*, Springer, Berlin Heidelberg, 1989.

The first German and this second edition differ by the following changes and additions. The remark at the end of Section V.6 about the "principle of symmetric criticality" as formulated and proved by R. S. Palais has been replaced by a new Section V.7 where this principle is discussed in some detail.

Section VIII.3 of the first edition about global nonlinear elliptic scalar-field equations on $\mathbb{R}^d$, $d \geq 3$, has been omitted. Now a new and long Chapter IX contains various recent results on global solutions of systems of semilinear elliptic partial-differential equations obtained by variational mathods. In particular we treat such equations on $\mathbb{R}^2$. Furthermore, one section presents the basic results of the "elliptic regularity theory" for this type of equation. In addition some recent results for local problems of this type involving "critical Sobolev exponents" are discussed.

The former Chapter IX on Thomas-Fermi theory has now become Chapter X and has been supplemented by various remarks and comments mainly concerning extensions of the results originally presented.

Finally, to Appendix 4 we have added some further basic facts about weak derivatives, and there is a new technical Appendix 5, which has enabled us to simplify some proofs in Chapter IX.

As in the first edition our lists of references only represent our sources of information or suggest publications for further reading; we have not tried to be complete.

However, we have tried to eliminate all the errors, omissions, and misprints of the first edition, in particular those we learned about from colleagues and students, especially T. Herb and N. Jakobowsky.

Moreover, various useful suggestions and criticism from readers of the first edition have been taken into account. In this connection we would like to thank first of all E. H. Lieb, who not only gave us many valuable comments on our first version of Thomas-Fermi theory (now Chapter X) but also provided helpful explanations of, and comments on, various important results presented in Chapter IX.

Unfortunately, in the German edition, in the long formulae occurring in Section V.6 on Noether's theorem in classical field theory some important terms were omitted. This was brought to our attention by J. Messer. We would like to thank him for this and several other comments on Thomas-Fermi theory.

And last but not least we would like to thank Dipl.-Math. U. Alfes who found an error in the original German formulation of the hypotheses of Theorem III.4.

The new part of the second edition have been written directly in English. Special thanks are due to G. M. Hayes for her corrections of our poor English. Last but not least and with great pleasure we thank W. Beiglböck who managed the realisation of this book in spite of many complications.

Bielefeld and Cape Town, *Ph. Blanchard*
January 1992 *E. Brüning*

Contents

Some Remarks on the History and Objectives of the Calculus of Variations 1

1. Direct Methods of the Calculus of Variations 15

1.1 The Fundamental Theorem of the Calculus of Variations 15
1.2 Applying the Fundamental Theorem in Banach Spaces 20
1.2.1 Sequentially Lower Semicontinuous Functionals 22
1.3 Minimising Special Classes of Functions 25
1.3.1 Quadratic Functionals 28
1.4 Some Remarks on Linear Optimisation 30
1.5 Ritz's Approximation Method 31

2. Differential Calculus in Banach Spaces 35

2.1 General Remarks 35
2.2 The Fréchet Derivative 36
2.2.1 Higher Derivatives 43
2.2.2 Some Properties of Fréchet Derivatives 44
2.3 The Gâteaux Derivative 46
2.4 nth Variation 49
2.5 The Assumptions of the Fundamental Theorem of Variational Calculus 51
2.6 Convexity of f and Monotonicity of f' 52

3. Extrema of Differentiable Functions 54

3.1 Extrema and Critical Values 54
3.2 Necessary Conditions for an Extremum 55
3.3 Sufficient Conditions for an Extremum 60

4. Constrained Minimisation Problems (Method of Lagrange Multipliers) 63

4.1 Geometrical Interpretation of Constrained Minimisation Problems 63
4.2 Ljusternik's Theorems 66

4.3 Necessary and Sufficient Conditions for Extrema Subject to Constraints 72
4.4 A Special Case 75

5. Classical Variational Problems 77

5.1 General Remarks 77
5.2 Hamilton's Principle in Classical Mechanics 80
5.2.1 Systems with One Degree of Freedom 81
5.2.2 Systems with Several Degrees of Freedom 95
5.2.3 An Example from Classical Mechanics 105
5.3 Symmetries and Conservation Laws in Classical Mechanics .. 107
5.3.1 Hamiltonian Formulation of Classical Mechanics 107
5.3.2 Coordinate Transformations and Integrals of Motion 109
5.4 The Brachystochrone Problem 113
5.5 Systems with Infinitely Many Degrees of Freedom: Field Theory 116
5.5.1 Hamilton's Principle in Local Field Theory 117
5.5.2 Examples of Local Classical Field Theories 122
5.6 Noether's Theorem in Classical Field Theory 124
5.7 The Principle of Symmetric Criticality 130

6. The Variational Approach to Linear Boundary and Eigenvalue Problems 142

6.1 The Spectral Theorem for Compact Self-Adjoint Operators. Courant's Classical Minimax Principle. Projection Theorem . 142
6.2 Differential Operators and Forms 148
6.3 The Theorem of Lax-Milgram and Some Generalisations 152
6.4 The Spectrum of Elliptic Differential Operators in a Bounded Domain. Some Problems from Classical Potential Theory 156
6.5 Variational Solution of Parabolic Differential Equations. The Heat Conduction Equation. The Stokes Equations 159
6.5.1 A General Framework for the Variational Solution of Parabolic Problems 161
6.5.2 The Heat Conduction Equation 166
6.5.3 The Stokes Equations in Hydrodynamics 167

7. Nonlinear Elliptic Boundary Value Problems and Monotonic Operators 171

7.1 Forms and Operators – Boundary Value Problems 171
7.2 Surjectivity of Coercive Monotonic Operators. Theorems of Browder and Minty 173
7.3 Nonlinear Elliptic Boundary Value Problems. A Variational Solution 178

8. Nonlinear Elliptic Eigenvalue Problems 192

8.1 Introduction 192
8.2 Determination of the Ground State in Nonlinear Elliptic Eigenvalue Problems 195
8.2.1 Abstract Versions of Some Existence Theorems 195
8.2.2 Determining the Ground State Solution for Nonlinear Elliptic Eigenvalue Problems 203
8.3 Ljusternik-Schnirelman Theory for Compact Manifolds 205
8.3.1 The Topological Basis of the Generalised Minimax Principle . 205
8.3.2 The Deformation Theorem 207
8.3.3 The Ljusternik-Schnirelman Category and the Genus of a Set 210
8.3.4 Minimax Characterisation of Critical Values of Ljusternik-Schnirelman 215
8.4 The Existence of Infinitely Many Solutions of Nonlinear Elliptic Eigenvalue Problems 217
8.4.1 Sphere-Like Constraints 217
8.4.2 Galerkin Approximation for Nonlinear Eigenvalue Problems in Separable Banach Spaces 220
8.4.3 The Existence of Infinitely Many Critical Points as Solutions of Abstract Eigenvalue Problems in Separable Banach Spaces 225
8.4.4 The Existence of Infinitely Many Solutions of Nonlinear Eigenvalue Problems 228

9. Semilinear Elliptic Differential Equations. Some Recent Results on Global Solutions 241

9.1 Introduction 241
9.2 Technical Preliminaries 247
9.2.1 Some Function Spaces and Their Properties 247
9.2.2 Some Continuity Results for Niemytski Operators 252
9.2.3 Some Results on Concentration of Function Sequences 256
9.2.4. A One-dimensional Variational Problem 262
9.3 Some Properties of Weak Solutions of Semilinear Elliptic Equations 266
9.3.1 Regularity of Weak Solutions 266
9.3.2 Pohozaev's Identities 278
9.4 Best Constant in Sobolev Inequality 283
9.5 The Local Case with Critical Sobolev Exponent 287
9.6 The Constrained Minimisation Method Under Scale Covariance 294
9.7 Existence of a Minimiser I: Some General Results 302
9.7.1 Symmetries 302
9.7.2. Necessary and Sufficient Conditions 304
9.7.3 The Concentration Condition 305
9.7.4 Minimising Subsets 308

9.7.5 Growth Restrictions on the Potential 310
9.8 Existence of a Minimiser II: Some Examples 312
9.8.1 Some Non-translation-invariant Cases 313
9.8.2 Spherically Symmetric Cases 316
9.8.3 The Translation-invariant Case Without Spherical Symmetry 319
9.9 Nonlinear Field Equations in Two Dimensions 322
9.9.1 Some Properties of Niemytski Operators on E_q 323
9.9.2 Solution of Some Two-Dimensional Vector Field Equations .. 326
9.10 Conclusion and Comments 332
9.10.1 Conclusion 332
9.10.2 Generalisations 334
9.10.3 Comments 335
9.11 Complementary Remarks 337

10. Thomas-Fermi Theory 340

10.1 General Remarks 340
10.2 Some Results from the Theory of L^p Spaces ($1 \leq p \leq \infty$) 342
10.3 Minimisation of the Thomas-Fermi Energy Functional 344
10.4 Thomas-Fermi Equations and the Minimisation Problem for the TF Functional 351
10.5 Solution of TF Equations for Potentials of the Form $V(x) = \sum_{j=1}^{k} \frac{z_j}{|x-x_j|}$ 357
10.6 Remarks on Recent Developments in Thomas-Fermi and Related Theories 361

Appendix A. Banach Spaces 363

Appendix B. Continuity and Semicontinuity 371

Appendix C. Compactness in Banach Spaces 373

Appendix D. The Sobolev Spaces $W^{m,p}(\Omega)$ 380

D.1 Definition and Properties 380
D.2 Poincaré's Inequality 385
D.3 Continuous Embeddings of Sobolev Spaces 386
D.4 Compact Embeddings of Sobolev Spaces 388

Appendix E 391

E.1 Bessel Potentials 391
E.2 Some Properties of Weakly Differentiable Functions 392
E.3 Proof of Theorem 9.2.3 393

References 395

Index of Names 405

Subject Index 407

Some Remarks on the History and Objectives of the Calculus of Variations

The expression "calculus of variations" was first coined by Leonhard Euler in the year 1756. He used it to describe a new method which J. L. Lagrange had developed in the previous year, and whose formal consequences he was now engaged in unravelling. Nowadays, the expression "calculus of variations", or "variational calculus", as it is often called, is used in a much wider sense. The subject matter of variational calculus is the mathematical formulation of the ideas of

(a) maximising,
(b) minimising and
(c) "criticalising".

In other words, one wants to find the minima, maxima and critical points of a function

$$f : M \to \mathbb{R}, \quad \mathbb{R} = \{\text{real numbers}\}.$$

In most applications, M is a set of numbers, functions, paths, curves, surfaces, fields, and so on.

Around the end of the 17th century, mathematicians were showing great interest in standard extreme value problems. Leibniz' first publication on differential calculus appeared in 1684, with the characteristic title "Nova methodus pro maximis et minimis itemque tangentibus".

The *critical points* of a differentiable function f are the points at which the derivative $f'(x)$ of the function $f(x)$ vanishes, i.e. those points where $f'(x) = 0$. The function $f(x)$ does not need to be a local maximum or minimum at every critical point; it could also be a *point of inflexion* with a horizontal tangent. The critical points, therefore, are simply those points which are *possibly* extrema, since the condition $f'(x) = 0$ is merely necessary in their definition.

In the 18th century one referred to "variation" rather than "differentiation", which is why the expression "calculus of variations" arose. To put this another way, the calculus of variations is a very natural topic of study for all those who think that only the best is good enough. Problems of maxima and minima, or the best and the worst, crop up all the time in everyday life – one need look no further than the minimisation of costs and the maximisation of profit. Many practical problems can be represented in this form. For example, how should one choose the shape of an aeroplane or an automobile in order that its air resistance be as small as possible? Extreme value problems play a large part in industry – transport and warehousing problems, the optimal arrangement of production processes – and in control engineering. In the theory of strategic games, one needs to weigh up many different factors and compute the optimal solution.

Variational calculus is also, in some sense, very old. The ancient Egyptians knew that the shortest path between two points is a straight line. They also knew, if only empirically, that the circle is the geometrical figure with the largest enclosed area inside a given perimeter.

The birth of modern mathematics took place in ancient Greece. There, mathematicians did not only ask *how*, but also *why*: they were no longer content simply with finding rules. Mathematics for the Egyptians was not a scientific undertaking, but rather an aid in administration and commerce. But the Greeks succeeded in solving many extreme value problems – they *proved* that the shortest path joining the points A and B is a straight line. In fact, Archimedes (?287–212 BC) used this fact to define the straight line.

Sometime between 200 BC and 100 AD, Zenodoros proved that the area of a polygon is always smaller than the area of a circle with the same perimeter. At this time it was also known that the cells of a honeycomb have certain maximum-minimum properties. The Greeks also succeeded in determining the geometrical figure with the greatest area for a given perimeter, thereby solving the first *isoperimetric problem.* There is evidence to suggest that Queen Dido made use of this solution in founding the city of Carthage. The legend of her resourcefulness in winning an enormous piece of land from the Numidian King Hierbas is well known. Queen Dido asked him for as much land as could be encompassed by a bull's hide; cutting the hide into thread-like strips, she enclosed the entire area of Hadachat (meaning "new capital"). This event is alleged to have taken place in 814 BC.

Another, very important, Greek discovery is attributed to the Alexandrian scholar Heron, sometime during the first century. It was common knowledge that a light beam from a point P incident on a plane mirror at a point R will be reflected to a point Q such that the angles between the mirror and the lines PR and QR are equal. If R' is some other, arbitrary, point on the mirror, then the total path $PR' + R'Q$ is longer than the path $PR + RQ$. Heron was able to characterise the path PRQ that the light actually takes as the shortest path from P to Q which includes reflection at the mirror. Heron's discovery is thus one of the starting points for geometrical optics.

In 1685, Newton (1643–1727) investigated the problem of finding the least resistance offered to a solid of revolution. He believed that "this theorem would not be completely without use in the construction of ships". His approach, however, did not coincide with later discoveries in hydrodynamics. Christian Huyghens (1629–1695) too wrote about methods for solving variational problems in his book on light.

Notwithstanding, the birth of modern variational calculus is usually given as that day in July 1696, when *Johann Bernoulli* presented the *brachystochrone problem* in the Acta Eruditorum Lipsiae. This problem can be formulated as follows. A point mass glides without friction along a curve joining a point A with a lower point B. Which curve gives the shortest time of travel if the point mass is moving under the influence of gravity only? The motion of the point mass is described by Newton's law. The resultant force acting on the mass is composed of gravity and the force constraining the mass to motion along a fixed curve. The resultant force is in the same direction as the tangent to the curve. It is obvious that the mass will take different lengths of time to travel along different curves, and that the solution is neither a straight line, nor the arc of a circle, even though the former is the shortest path between the two points.

In his "Dialoghi", Galileo poses the question of two equal spheres which start to roll at the same time, the one along a circular arc, the other along this chord. Which of the spheres reaches the bottom of the cord first? From experiment, we know that is the sphere travelling along the arc. A body moving along a straight line accelerates relatively slowly. However, if the curve is steeper near the starting point, then its length increases but a greater proportion is traversed at a higher speed.

Johann Bernoulli starts his treatment of the brachystochrone problem in a very confident tone: "Johann Bernoulli, Professor of Mathematics, sends greetings to the most astute mathematicians of the whole world". His boast was that he had found an elegant solution to this problem; however, he did not intend to make it available publicly, preferring first to challenge his contemporaries to study the problem. This challenge was particularly aimed at his brother and teacher, Jakob, thirteen years older than him, and his bitter enemy.

The brachystochrone problem is concerned with the time it takes to travel along the whole curve, and is thus significantly different from simple problems which depend on only a finite number of variables. It would appear that it was not known at that time that the isoperimetric problem was the same kind of problem. Over the years, many mathematicians gave solutions, including Newton, Leibniz (1646–1716), and l'Hôpital (1661–1704). *Jakob Bernoulli* found a solution, but it was only at the instigation of Leibniz, with whom he had had a lifelong friendship and scientific correspondence, that he sent it to his brother. The solution appeared in May 1697. Ostensibly, Newton sent his solution to a friend, just one day after the solution appeared! To the fascination of the mathematicians, the solution turned out to be a cycloid, a curve which had only recently been discovered. The cycloid is a simple transcendental curve which can be generated mechanically: it is the path taken by a point on the circumference of a circle moving in a straight line along a frictionless surface.

We have already noted that Heron of Alexandria recognised that the reflection of a light beam from a plane mirror could be described by means of a minimum principle. *Fermat* (1601–1665), in the 17th century, noted that the law of refraction could also be expressed in the form of a minimum principle. He proved that light always travels along that path for which the time taken is a minimum. He generalised the predictions of this law to arbitrary optical systems in which the velocity of light is a function of position (as is the case in, for instance, the atmosphere). Fermat's principle of geometrical optics (1662) states that the path of light through an inhomogeneous medium is that for which the time of travel is a minimum. This principle is not only of great theoretical significance but also forms the basis for the practical design of lens systems. Fermat's principle was put to clever use in the design of optical instruments by the physicist Ernst Abbe, one of the founders of the Zeiss firm in Jena. Later, we shall see that minimum principles play a dominant role in other areas of physics. In fact, *Maupertuis* (1698–1759), a man inclined to philosophical and mystical speculations, even spoke of "God's intention of regulating physical phenomena according to the principle of greatest perfection". In 1750, when he was president of the Berlin Academy, he found himself in a hefty argument with the Swiss mathematician, Samuel König (1712–1757), over the principle of least action in mechanics. The term "action" was coined by Maupertuis, "since God's creation takes place with the minimum of effort". Since then, a quantity with the dimension [energy $\times$ time] has been known as an action. Just like Fermat and, later, Einstein (1874–1955), Maupertuis was searching for a universal principle which would draw together the laws of the universe. Maupertuis' formulation was not transparent, a fact he took as proof of the existence of God. The argument reached its climax with a work by Voltaire (1694–1778) in 1752, called "Diatribe du Docteur Akakia, Médecin du Pape", whose acerbic humour ridiculed the unfortunate president.

The various methods used in solving the brachystochrone problem were ad hoc. But the next step in the development of variational calculus was a general theory enabling the solution of general problems. The names *Lagrange* (1736–1813) and *Euler* (1707–1783) are inextricably linked with this step. Shortly before 1732, Euler began a systematic study of extreme value problems. It is rather unlikely that Euler was taught about variational calculus in his early studies with Johann Bernoulli in Basel, since it would otherwise be difficult to explain why Bernoulli pointed Euler towards the problem of geodesics in a letter to Petersburg as late as 1728 or 1729. The method that Euler developed is influenced more by the work of Jakob Bernoulli than that of Johann. It is, however, very different from those of his predecessors, who dealt only with particular problems. Euler's calculations produced a real theory, raising the status of variational calculus to that of a mathematical discipline in its own right. After the argument between König and Maupertuis, Euler put the principle of least action on firm foundations without resorting to the metaphysical views of the latter. This principle was then utilised by Lagrange and, later, by Hamilton (1805–1865). The key role played by the Hamiltonian function in modern mathematical physics makes very clear the fundamental nature of

Euler's contribution to this topic. The term "minimum principle", still in use today, is actually an historical leftover from the 18th century. In fact, only the condition that the first variation must vanish is of immediate significance for the description of the laws of nature.

A geodesic is the shortest path between two given points on a surface. On a sphere these lines are arcs of great circles. In the curved world of a spherical surface, the geodesics play the same role as lines in a planar world. Einstein's general theory of relativity (1915) is a theory of the space-time continuum and gravitation, i.e. it is the modern version of Newton's theory. The core of Einstein's theory was his idea that gravitational effects can be caused by the geometry of the space-time manifold. He explained the gravitational force via the curvature of space-time, which is how the influence of mass is expressed in general relativity. Thus, a particle moving in a gravitational field moves along a geodesic. Since four-dimensional space is curved in the presence of a gravitational field, these extremals are not lines and the actual movement of a particle through space is neither uniform nor rectilinear.

For a special class of functionals, i.e. functions whose arguments are themselves functions, Euler discovered the first necessary condition that a minimising function must satisfy. Nowadays this is known as the Euler-Lagrange equation, and for functionals it corresponds to the condition $f'(x) = 0$ for functions. Euler considered functionals of the form

$$\int_a^b F(t, y, y')\,dt$$

and looked for a function that minimises this functional. At that time, the approach to differential and integral calculus was to calculate with infinitely small quantities. The rigorous formulation in terms of limits is due to Cauchy. Euler's method consists in dividing the interval of integration $[a, b]$ into smaller intervals $[t_{i-1}, t_i]$ using points $a \le t_1 < t_2 < \ldots < t_k < b$, and then in replacing the curve with a polygon. The corners of this polygon are the points (t_i, y_i) where $y_i = y(t_i)$. He also replaced the derivative $y'(t_1)$ by $\Delta y_i / \Delta t_i$, thereby obtaining instead of an integral, a function of the k variables $y_1, \ldots, y_k$. Letting the integral of this function with respect to y_i vanish and the number of points k tend towards infinity subject to the condition that the equation for y still hold, he obtained the well-known "Euler condition"

$$\frac{\partial F}{\partial y} - \frac{d}{dt}\frac{\partial F}{\partial y'} = 0$$

for the unknown function y. In general, Euler's equation is not easy to solve, since second-order differential equations are only integrable by elementary functions in exceptional cases.

Euler's method was later considerably simplified by Lagrange, who noticed that it "does not quite possess that simplicity which would be desirable in a constituent of the field of pure analysis". Instead of making a polygonal approximation, Lagrange let the curve itself vary and replaced $y(t)$ by a function $y(t, \tau)$ with an extra parameter τ such that $y(t, 0) = y(t)$. This idea was later extended to give the concept of homotopy.

The variation is then the function $\delta y(t) = (\partial y / \partial \tau)(t, 0)$, and $\delta(y') = (\delta y')$. Thus, it is sufficient to replace $y(t)$ by $y(t, \tau)$ in $\int_a^b F(t, y, y')\, dt$ and to require that the derivative of this expression with respect to τ vanishes at $\tau = 0$. Partial integration then gives Euler's condition and, additionally, boundary conditions which Euler almost always neglected. Lagrange added, without further justification, that it follows directly from the vanishing of

$$\int_a^b \left[\frac{\partial F}{\partial y} - \frac{d}{dt} \frac{\partial F}{\partial y'} \right] \delta y(t)\, dt$$

that

$$\frac{\partial F}{\partial y} - \frac{d}{dt} \frac{\partial F}{\partial y'} = 0\,.$$

Euler immediately rejoined that a proof was necessary for this. Lagrange wrote to Euler with an attempt at a proof which Euler, however, rejected, with justification. The necessary lemma was not proved until 1879, by the mathematician Du Bois-Reymond from Tübingen, and it was later shown to be the fundamental lemma of variational calculus. To this end Du Bois-Reymond used the notion of continuity which was first introduced by Cauchy in 1820.

Towards the end of the 19th century, the concept of differentiation was generalised by Volterra, Hadamard and his two pupils *Fréchet*, *Gâteaux* and *Hilbert*, and others, to infinite-dimensional spaces, thus providing variational calculus with a solid basis.

The coming together of ideas and methods from algebra, geometry, topology and analysis produced a new branch of mathematics, *functional analysis*, which encompassed a generalisation of all the concepts of classical analysis (limits, convergence, continuity, differential, etc.) to the case of infinite-dimensional spaces. Variational calculus, or the differential calculus of functions, must be regarded as the oldest part of functional analysis.

Lagrange (1736–1813) developed an analytical theory of variational calculus by generalising the method for multiple integrals, and starting to investigate minimal surfaces, i.e. the smallest possible surfaces bounded by a given curve. This problem was also the subject of experimental investigations by the physicist Plateau (1801–1883), and thus goes by the name of Plateau's problem or the soap bubble problem. Mathematically, it involves the solution of a partial differential equation or a system of such equations.

"Mécanique analytique", written by Lagrange in 1788, almost a hundred years after Newton's "Principia", is probably his most valued work. The "Principia", which appeared in 1687, was written in the language of classical Greek geometry. Using the language of variational calculus, Lagrange was able to unify statics and dynamics, which had up to this time been separate. Newton's geometrical viewpoint was dropped completely. This was the triumph of pure analysis, and Lagrange was the first pure analyst. As Hamilton put it, Lagrange raised mechanics to a form of "scientific poetry". In his foreword, Lagrange emphasised that "in this book one will find no figures, but only algebraic operations". The interaction between mathematics and mechanics is characteristic of the 18th century: mathematics was developed primarily as a tool of the natural sciences (geodesy, astronomy and mechanics).

The difference between the various stationary points, maxima, minima and points of inflexion, remained empirical until *Legendre* (1752–1833) started his investigations into "second variation". With the aid of this method, he succeeded in 1786 in deriving the second necessary condition for the existence of an extremum, although Lagrange noticed immediately that Legendre's derivation was not complete.

The methods of variational calculus were applied successfully to mechanics. The crowning achievement was produced by *Jacobi* (1804–1851) in his "Theory of Variational Calculus and Differential Equations" in 1837. He succeeded in producing a perfect derivation of Legendre's condition and proved the third necessary condition for the appearance of a minimum. As the ancient Greeks knew, the geodesic line is that arc of the great circle joining two points on the surface of a sphere. This curve is a minimum as long as the points are not diametrically opposed. If they are, for example the north and south poles, then there is a whole family of semicircular arcs and the points are said to be conjugate. Thus, there is a whole family of curves joining two conjugate points: each member of the family is a minimising or maximising arc but it is not possible to distinguish between them as regards their extremal properties. Jacobi's necessary condition rules out the appearance of conjugate points.

However, as we have already pointed out, the existence of a solution to any extreme value problem still requires a particular proof, and that is usually one of the main difficulties of variational calculus. For long into the 19th century, the greatest mathematicians, such as *Gauß* (1777–1855), *Dirichlet* (1805–1859) and *Riemann* (1826–1866), took the extistence of solutions to extreme value problems for granted. Gauß, in 1839 in his investigations into electrostatics, and Riemann, in 1851 in his doctoral dissertation on the foundational theory of functions, both carried out investigations into the solution of Dirichlet's problem. This problem is to solve the potential equation

$$\Delta u = \frac{\partial^2 u}{\partial x^2} + \frac{\partial^2 u}{\partial y^2} = 0 \quad \text{in } G\,,$$

$$u \restriction_{\partial G} = f\,,$$

for given values on the boundary ∂G of a given area G of the plane $\mathbb{R}^2$. This partial differential equation is the Euler-Lagrange equation of the Dirichlet integral

$$D(u) = \iint_G \left[\left(\frac{\partial u}{\partial x}\right)^2 + \left(\frac{\partial u}{\partial y}\right)^2 \right] dx\, dy\,.$$

This problem arises particularly in electrostatics, where u is the electric potential. The system is in stable equilibrium when the energy $D(u)$ is a minimum. The particular significance of this problem for physics was first recognised by Thomson (1824–1907).

The integrand in the Dirichlet integral is nonnegative and so it has a lower bound which is greater than or equal to zero. This led Riemann, Dirichlet's successor in Göttingen, to the conclusion that there had to be a function u_0 which minimises the integral and thus solves the Dirichlet problem. Riemann therefore declared the "Dirichlet principle" to be the fact that such variational problems always have a solution, without even trying to prove this claim mathematically.

Weierstrass (1815–1897) criticised this claim in 1870 and showed that the a priori existence of a minimising function in a variational problem is by no means assured and that, in the general case, its existence cannot be assumed. He gave many examples of cases in which the lower bound cannot be reached. Weierstrass' critical investigations were instrumental in preparing the way for Hilbert, who was later mainly responsible for providing an existence proof, and thus for establishing the direct methods of the calculus of variations. Hilbert wrote of Weierstrass in 1926: "If complete security reigns in the field of analysis today, then it is largely due to the scientific efforts of Weierstrass". And indeed, the criticisms of Weierstrass and his school are of fundamental significance in the development of the whole of analysis.

A simple geometrical example of a variational problem without a solution is shown in Figure 1. Two points A and B on a straight line are joined by a continuous short curve C, which, in addition, at the points A and B, is perpendicular to the straight line.

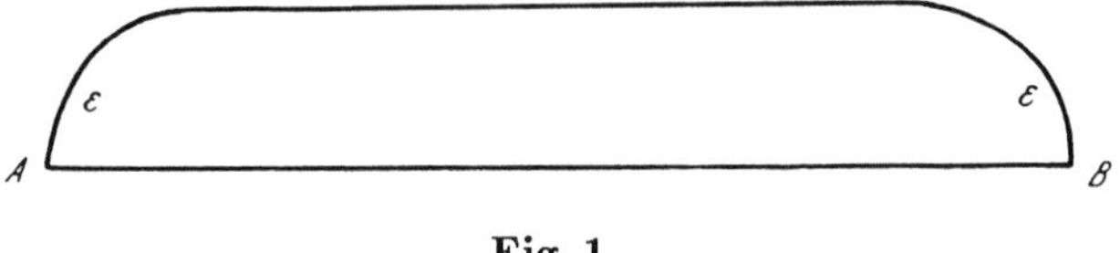

Fig. 1

This problem has no solution because such a curve is always longer than the straight line AB, although its length can be arbitrarily closely approximated to that of the straight line. There is thus a lower bound, but no minimum, for a curve of this particular family. For C can be divided into two small 90-degree arcs of radius ε whose centres lie on AB, and a straight line. The length of C is $l(C) = \varepsilon\pi + a - 2\varepsilon = a + (\pi - 2)\varepsilon$ which can be made arbitrarily close to the distance a between the points A and B. Thus, this quite well formulated problem has no solution. The central difficulty of variational calculus is indeed to determine whether a solution exists and sufficient conditions for this to happen.

Fifty years after the publication of Riemann's results, *David Hilbert* (1862–1943) rescued the Dirichlet principle in his famous lecture "On the Dirichlet Principle" at a conference of the German Mathematical Society, DMV, in 1901. What Riemann and Dirichlet had guessed at, Hilbert proved. This rehabilitation of the Dirichlet principle was one of the greatest triumphs in the history of analysis and mathematical physics. Hilbert's work, and that of *Lebesgue* (1875–1941) on the same topic in 1907, heralded the arrival of modern abstract variational calculus. At the Second International Mathematical Congress in Paris in 1900, Hilbert presented 23 mathematical problems, the 20th problem concerning the existence of a minimum for all regular functions with reasonable boundary conditions. Hilbert ended his presentation with a call for further development of variational calculus. He had tried to describe all the different strands of mathematical research that had taken place in the previous decades and to outline future productive work. He was convinced that proof of the existence theorem must be possible, even though it might be necessary to make some generalisations, e.g. about the concept of derivative or solution. This hunch turned out to be mistaken as far as the minimal surface functional was concerned, although this functional is regular in Hilbert's sense. In 1912, Bernstein (1880–1966) produced a proof of the nonexistence of a solution for nonconvex regions of the plane. However, it says much for Hilbert's mathematical instincts that these 23 problems inspired mathematicians and led to the founding of new disciplines.

As we have already emphasised, variational calculus has always inspired mathematicians to think about new concepts in analysis: continuity, semicontinuity, equicontinuity, neighborhood of a function, and compactness are all worthy of mention. Legendre thought about whether a solution y of Euler's equation gave a relative extremum of the integral $\int_a^b F(t,y,y')\,dt$ or not. His first attempt of a solution was to consider a linear bound by substituting $y+\varepsilon u$ for y and requiring this function of ε to possess a relative extremum for $\varepsilon = 0$, for arbitrary variation $\delta y = \varepsilon u$. One soon realises, however, that this is not sufficient to guarantee that the value of the integral is indeed larger (or smaller) than the value one gets by substituting $y+\delta y$ for y for a "small" variation δy. The crucial point here is the specification of "small". It was assumed for rather a long time that, for a small variation δy, the variation $\delta y'$ of the derivative is also small, until Weierstrass and his students noticed that this property in fact constitutes an extra condition. This led at the end of the 19th century to a distinction being made between a strong and a weak extremum. For a strong extremum, $|y - y_0|$ must be small in a neighborhood of y_0, and for a weak extremum, additionally, $|y' - y_0'|$ must also be small. This notion of different sorts of neighborhood was expressed very clearly in 1906 by *Fréchet* (1878–1973), who introduced the concept of distance in an arbitrary set.

This generalisation of concepts usually used in $\mathbb{R}^n$, such as bounds, continuity, neighborhood, compactness and connectivity, to this extremely general situation is remarkable. Fréchet emphasised, in particular, that the idea of compactness is implicit in the direct methods of variational calculus that Hilbert had developed in 1900. For example, Hilbert proved that a geodesic line exists on every compact, simply-connected, smooth surface in $\mathbb{R}^3$. Fréchet noticed that it is possible to define a distance on the set of curves connecting two points A and B on such a surface, and that for this distance, it turns out that the set of arcs whose length is bounded by a positive number is a compact set. Hilbert's result then follows from a general theorem of *Baire* (1874–1932), who proved that for functions on $\mathbb{R}^n$, a lower semicontinuous function attains its minimum on a compact set. The length introduced by Fréchet is indeed lower semicontinuous on the set of arcs joining A and B. The significance that semicontinuity has for variational calculus was especially emphasised by *Tonelli* (1855–1946).

The name *Carathéodory* (1873–1950) is inextricably linked with variational calculus. Carathéodory was described by Pringsheim as the "Isopérimaître incomparable". His approach to variational calculus is described in his book "Variational Calculus and First-order Partial Differential Equations" (1935).

The relationship between variational calculus and the theory of elliptical partial differential equations has been investigated, as has the regularity of solutions of variational problems. New theories have been developed, closely connected with variational calculus, e.g. the theory of variational inequalities (*Lions*, *Stampacchia*), the theory of monotone operators (*Browder*), and minimax theorems. Variation-theoretical approximation methods have been applied to physical problems, and the work of Rayleigh, Ritz, Galerkin, Courant and Hellinger is particularly noteworthy. The key ideas of variational calculus are today still an important and fruitful source of developments in mathematics, in particular in functional analysis, differential geometry, algebraic topology, and the fields in which they are used, such as physics, economics, engineering, biology, an so on.

Variation-theoretical methods have always played an important role in theoretical mechanics. For example, Hamilton's principle is frequently regarded as an alternative to Newton's equations of motion; we shall often have cause to consider this principle later on. Of course, the significance of a variation-theoretical formulation of the laws of physics goes far beyond a simple alternative description. Its practical advantage lies in the possibility it offers of incorporating in a single functional all the characteristic properties of a problem, such as boundary conditions, initial conditions and constraints. What is also very important is the fact that these formulations provide the best available approximation methods, and in many cases one can use them to construct upper and lower bounds for a solution.

We have already mentioned Hamilton's principle and we shall now discuss it briefly. It is well known that geometrical optics can be formulated in two different ways, one based on Fermat's principle and the other based on Huyghens' principle. Hamilton considered the whole bundle of light beams leaving a point source and was able to explain the relationship between these two approaches. The wavefronts of Huyghens' principle are just those surfaces of equal time of travel. In other words, the light rays are the extremals of the variational problem of finding the shortest time of travel, and the wavefronts are the corresponding transversals. Carrying these results over to mechanics leads to the formulation of Hamilton's principle, which has proved many times to be an efficacious method for handling physical problems, most recently in Schrödinger's quantum theory. The Hamilton formulation is one of the pillars holding up quantum theory. It is also indispensable in statistical mechanics. Moreover, for the last 400 years mechanics has proved to be the nucleus of theoretical physics.

Schrödinger (1887–1961) investigated the analogy that Hamilton pointed out between the mechanics of a point mass in a force field and geometrical optics in an inhomogeneous medium. Inspired by de Broglie, Schrödinger extended this analogy and produced his wave theory of matter.

In many important cases, Hamilton's principle is completely equivalent to Newton's equations of motion. In classical field theory, i.e. in systems with infinitely many degrees of freedom as is the case, for example in Maxwell's theory, the system is described by a Lagrangian which one gets by integrating over a Lagrangian density. The corresponding Euler equations give the field equations. Formulating Newton's equations of motion and the field equations as variational problems has many advantages. The different sorts of constraints can be taken into account in a simple, systematic way, and, moreover, the conservation theorems can easily be derived from the symmetry properties of the system, as *Noether* (1882–1935) showed in 1918.

Because of the historical contribution of variational calculus to the development of mathematics and the possibilities it offers for the solution of mathematical problems in physics, engineering and economics, variational calculus certainly enjoys special status. Newer developments in pure and applied mathematics, physics (relativity and quantum theory), mathematical economics, control theory, biology and engineering science only serve to confirm this, and offer numerous examples of the versatility and power of variation-theoretical methods.

We now leave the history of variational calculus and go on to discuss the basic ideas of its direct methods.

As we outlined above, the task of variational calculus is to determine the extrema and critical points of a function

$$f : M \to \mathbb{R}$$

under the most general conditions possible. This generality is required both of the set M and the function f. If, for example, M is an interval in $\mathbb{R}$ or a subset of $\mathbb{R}^n$, then it is possible to solve the problem using only the methods of classical analysis. On the one hand, Weierstrass' fundamental theorem states that every continuous function $f : K \to \mathbb{R}$ on a compact set K of $\mathbb{R}^n$ is bounded and attains a maximum and a minimum. On the other hand, Heine-Borel's theorem states that a subset K in $\mathbb{R}^n$ is compact if, and only if, it is closed and bounded. We now define the notion of semicontinuity, as introduced in the theory of real functions by Baire. A function $f(x)$ is said to be *lower semicontinuous at a point* x_0, if, for every $\varepsilon > 0$, there exists a $\delta > 0$, such that for all $|x - x_0| < \delta$, we have $f(x) - f(x_0) > -\varepsilon$. If, in Weierstrass' theorem, we limit ourselves to proving the existence of a minimum, then we can see immediately that we need only assume lower semicontinuity, rather than continuity.

In many important applications of variational calculus, however, M is not a subset of a finite-dimensional vector space. In the very old problem of finding the shortest path between two different points, M must be the set of all paths joining these two points. In this example and in applications to physics, M is a subset of an infinite-dimensional function space.

The following thus appears to be a suitable delimitation of "abstract variational calculus".

Variational calculus investigates conditions on a topological space X, a subset $M \subseteq X$ and a function $f : M \to \mathbb{R}$ which guarantee that a point x_0 exists in M such that

$$f(x_0) = \inf_{x \in M} f(x)$$

or

$$f(x_0) = \sup_{x \in M} f(x) .$$

The point x_0 is called the *minimising* or *maximising point.* In its abstract version, then, variational calculus aims to set up and prove an analogue of Weierstrass' theorem. Over a hundred years ago, *Bolzano* (1781–1848) proved that every bounded, infinite point set of the real line $\mathbb{R}$ possesses at least one accumulation point. He pointed to the importance of this theorem for the field of real analysis. The idea of choosing a converging sequence out of infinite sets that do not consist of numbers or points in $\mathbb{R}^n$ but of functions or curves, was and is often used in variational calculus existence theorems. In general, we say that a subset of a topological space is *sequentially compact* if every sequence of elements from this subset contains a converging subsequence. This leads to the general definition of compactness of a subset in an abstract topological space. A subset is said to be *compact* if for every open covering there exists a finite subcovering. Compactness and sequential compactness are equivalent in normed spaces.

Many practical optimisation problems have the following in common. Let there exist a continuous linear functional f on a Banach space X. Then our aim is to find a point x_0 in X with $\|x_0\| = 1$ such that

$$f(x_0) = \|f\| = \sup_{\substack{x \in X \\ \|x\|=1}} |f(x)| .$$

We have thus to find the conditions which guarantee the existence of the point x_0. If the closed unit ball

$$\overline{K}_1 = \{x \in X \mid \|x\| \leq 1\}$$

were compact, then the existence of x_0 would follow immediately from the continuity of f. Unfortunately, a theorem proved by Riesz (1880–1956) in 1918 states that unit balls are compact only in finite-dimensional Banach spaces, i.e. in $\mathbb{R}^n$ or $\mathbb{C}^n$. So although $\overline{K}_1$ is closed and bounded, it is not compact in an infinite-dimensional Banach space. In such spaces one needs an extra condition, in addition to closedness and boundedness, in order to characterise compact subsets. We shall give these extra conditions for various Banach spaces. It is easy to convince oneself of the result of Riesz' theorem by considering the following counterexample. It is impossible to choose a convergent subsequence from a sequence $\{x_n\}_{n\in\mathbb{N}}$ in an infinite-dimensional Hilbert space $\mathcal{H}$, where $\langle x_n, x_m\rangle = \delta_{nm}$ (in other words $\{x_n\}_{n\in\mathbb{N}}$ is an orthonormal system in $\mathcal{H}$), because

$$\|x_n - x_m\| = \langle x_n - x_m, x_n - x_m\rangle^{1/2} = \sqrt{2} \quad \forall n \neq m\,.$$

An obvious way out would seem to be to introduce a metric in the Banach space X such that, on the one hand, the unit ball $\overline{K}_1$ is compact in the new metric and, on the other hand, the continuous linear functionals are also continuous in the new metric. Unfortunately, it is not possible to realise this idea in the general case. Thus, we have no choice than to consider whether we should pass to a more general concept of space than a metric space. This brings in the *weak topology*: a sequence $\{x_n\}_{n\in\mathbb{N}}$ converges weakly towards x_0, if, for every continuous linear functional f on X (i.e. $f \in X'$), we have $\lim_n f(x_n) = f(x_0)$. If it is possible to choose from every sequence of elements of a subset, a weakly converging subsequence, then this subsequence is called *weakly sequentially compact.* The unit ball $\overline{K}_1$ (and also every bounded, convex, closed subset in X) often has a very useful property: it is weakly sequentially compact. These compactness properties are of great importance in the various formulations of the fundamental theorem of variational calculus. As in the case of classical analysis, the second ingredient of these theorems is a natural weakening of the concept of continuity for functions on an infinite-dimensional space, i.e. the concept of semicontinuity. Notice that in, for example, a Banach space, continuity with respect to the norm topology is not continuity with respect to the coarser weak topology. For convex functions one can at least prove weak lower semicontinuity.

A complete solution of the abstract variational problem encompasses:

(A) existence
(B) uniqueness
(C) calculation

of the minimising point $x_0 \in M$. Problems (A) and (B) can be simply solved to a fair degree of generality subject to reasonable conditions. We shall cover this topic in Chap. 1. The problem of explicitly calculating the minimising point is considerably more complicated and requires much more supporting structure. To this end, we shall discuss in Chap. 2 differential and integral calculus in Banach spaces, and look particularly at the concept of the Fréchet derivative $f'(x_0)$ of a function $f : M \to \mathbb{R}$, where M is an open subset of a Banach space X. We can then obtain a solution to problem (C) if we can find the minimising point in the set of critical points of the function f, i.e. in $\{x \in M \mid f'(x) = 0\}$.

Finally, we shall discuss numerous applications which will illustrate the practical versatility and effectiveness of these rather abstract results. The technique of solving mathematical problems by means of variational methods is useful in many fields, and often results in simplification of these problems. However, we shall restrict ourselves to analysis and mathematical physics.

The necessary mathematical tools, such as Banach spaces, the concepts of semicontinuity and compactness, etc. are introduced and discussed in the appendices.

The source literature for the history of variational calculus can be very instructive, and we give here some examples.

Carathéodory, C.: Der Beginn der Forschung in der Variationsrechnung. Gesammelte Math. Schriften, vol. II, 1937

Stäckel, P.: Variationsrechnung. Wissenschaftliche Buchgesellschaft, Darmstadt 1976, with a commentary

Goldstine, H. H.: A history of the Calculus of Variations from the 17th through the 19th Century. Springer, Berlin Heidelberg 1980

Woodhouse, R.: A history of the calculus of variations in the eighteenth century. Chelsea, reprint of the 1810 edition

Todhunter, I.: History of the calculus of variations. Chelsea, reprint of the 1861 edition

Dieudonné, J.: History of functional analysis. North-Holland, Amsterdam 1981

Monna, A. F.: Dirichlet's principle: a mathematical comedy of errors and its influence on the development of analysis. Oosthoede, Scheltema and Holkema, Utrecht 1975

Yourgrau, W., Mandelstam, S.: The significance of variational principles in natural philosophy. In: Variational principles in dynamics and quantum theory. Pitman, London 1968

Helmholtz, H.: Zur Geschichte des Prinzips der kleinsten Action. In: Wissenschaftliche Abhandlungen, Band III, 1895

Planck, M.: Das Prinzip der kleinsten Wirkung. In: Kultur der Gegenwart, Teil III, Abt. III, 692–702 (P. Hinneberg (ed.)). Teubner, Leipzig 1915

1. Direct Methods of the Calculus of Variations

1.1 The Fundamental Theorem of the Calculus of Variations

In our previous chapter we showed that the determination of the extrema, such as the minima, of a function, is central to the calculus of variations. An *extreme value problem* in its most general form can be understood to be the following.

Let M be a set and f a real function on M. Determine the *minimising (maximising) points of f on M*, in other words, those points $x_0 \in M$ for which

$$f(x_0) = \inf_{x \in M} f(x), \quad \left(f(x_0) = \sup_{x \in M} f(x) \right)$$

holds. We would like to obtain a result analogous to the Bolzano-Weierstrass theorem. A complete solution to this problem will, naturally, not be restricted just to proving the existence of minimising points but, if they do exist, will also prove their uniqueness and/or will provide an algorithm for calculating the minimising point(s). Of course, we do not expect to find a useful starting point for the solution in this general formulation of the problem; we will have to make precise assumptions about the set M and the function f.

In many "classical" applications and particularly in, for instance, the Lagrangian formulation of classical mechanics (see Chap. 5), the set M is a submanifold of a Banach space of differentiable paths $q : I = [t_1, t_2] \to \mathbb{R}^n$, $t_i \in \mathbb{R}$, $i = 1, 2$, and the function f then has the particular form

$$f(q) = \int_I L(t, q(t), \dot{q}(t)) \, dt$$

with a "Lagrange function" $L : I \times \mathbb{R}^n \times \mathbb{R}^n \to \mathbb{R}$.

Since we want to tackle the extreme value problem in a general way so that we will be able to apply it in many practical situations, we will formulate the fundamental theorem under the weakest possible assumptions. It will be clear from the following explanation and later applications that this generality is quite appropriate and it will demonstrate the essential content of all minimum problems.

First of all, however, let us recall a few mathematical concepts which will be important to us. The theorems cited below are proved in the appendices, where we also illustrate the concepts we introduce with examples and references to the literature.

Let M be a Hausdorff space. A function

$$f : M \to \mathbb{R} \cup \{+\infty\}$$

is said to be *lower semicontinuous* (l.s.c.) *at a point* $x_0 \in M$ if, for every $\varepsilon > 0$, x_0 is an interior point of $\{x \in M | f(x) > f(x_0) - \varepsilon\}$. The function f is said to be *lower semicontinuous on* M if it is lower semicontinuous at all points of M.

Analogously, a function f is said to be *upper semicontinuous (u.s.c.)* if $-f$ is lower semicontinuous. An example of a function which is not continuous at 0, but which is lower semicontinuous, is the function

$$f(x) = \begin{cases} (1+x^2)\sin(1/x)\,, & x \neq 0\,, \\ -1\,, & x = 0\,. \end{cases}$$

Corresponding to the well-known sequence criterion for continuity, one can also characterise lower semicontinuity with the help of sequences, under a restrictive condition for the topology of M.

Lemma 1.1.1. *Let M be a Hausdorff space and $f : M \to \mathbb{R} \cup \{+\infty\}$ a function on M. Then the following holds:*

(a) *If f is* l.s.c. *at $x_0 \in M$, then for every sequence $\{x_n\}_{n\in\mathbb{N}} \subset M$ converging to x_0,*

$$f(x_0) \leq \liminf_{n\to\infty} f(x_n)\,.$$

(b) *If M satisfies the first axiom of countability (that is to say, if every point of M possesses a countable neighborhood basis) then the converse of* (a) *also holds.*

Proof. See Appendix 2. □

Remark 1.1.1. We recall the definition of lim inf: $a \in [-\infty, +\infty]$ is called an accumulation point of a sequence $\{a_n\}_{n\in\mathbb{N}}$ if and only if there exists a subsequence $\{a_{n_j}\}_{j\in\mathbb{N}}$ which converges to a. $\liminf_{n\to\infty} a_n$ is the *smallest accumulation point*, which always exists in $[-\infty, +\infty]$, of the sequence $\{a_n\}_{n\in\mathbb{N}}$.

A further concept which is crucial to the fundamental theorem of the calculus of variations is that of a sequentially compact set. A subset K of a Hausdorff space M is said to be *sequentially compact* if and only if every infinite sequence in K possesses a subsequence which converges to a point in K.

The relationship between this concept of compactness and other concepts of compactness which are also important in application is examined in detail in Appendix 3.

The task of determining the maximum of a function f is identical to the task of determining the minimum of the function $-f$. Thus we can now formulate and prove the fundamental theorem of the calculus of variations, on the existence of minimising points.

Theorem 1.1.2. Existence of Minimising Points. *Let M be a Hausdorff space and $f : M \to \mathbb{R} \cup \{+\infty\}$ a lower semicontinuous function on M. Let there be a real number λ such that*

(i) $M_{f,\lambda} = \{x \in M | f(x) \leq \lambda\} \neq \emptyset$ *and*
(ii) $M_{f,\lambda}$ *is sequentially compact.*

Then f possesses a minimising point x_0 in M:

$$f(x_0) = \inf_{x \in M} f(x) .$$

Proof. (a) First of all, we show indirectly that f is bounded from below. Supposing f were not bounded from below, then there would exist a sequence $\{x_n\}_{n \in \mathbb{N}}$ in M for which $f(x_n) < -n$ for all $n \in \mathbb{N}$. If n is sufficiently large, $n > n_0(\lambda)$, then $x_n \in M_{f,\lambda}$. Since $M_{f,\lambda}$ is sequentially compact, there exists a subsequence $\{x_{n_j}\}_{j \in \mathbb{N}}$ which converges to a point $y \in M_{f,\lambda}$. Since f is lower semicontinuous, it follows that

$$-\infty < f(y) \leq \liminf_{j \to \infty} f(x_{n_j}) ,$$

which is a contradiction, because by construction of the x_{n_j}, we have

$$f(x_{n_j}) \leq -n_j .$$

Therefore, f is bounded from below and thus has a finite greatest lower bound

$$-\infty < m(f) = \inf_{x \in M} f(x) \leq \lambda .$$

(b) By definition of the number $m(f)$, a sequence $\{x_n\}_{n \in \mathbb{N}} \subset M$ exists such that $f(x_{n+1}) \leq f(x_n)$ and

$$0 \leq f(x_n) - m(f) < \frac{1}{n} , \quad \forall n \in \mathbb{N} .$$

Further, $n_0(\lambda) \in \mathbb{N}$ exists, such that $x_n \in M_{f,\lambda}$ for all $n \geq n_0(\lambda)$. The sequential compactness of $M_{f,\lambda}$ again ensures the existence of a subsequence $\{x_{n_j}\}_{j \in \mathbb{N}}$, which converges to a point $x_0 \in M_{f,\lambda}$. The fact that f is lower semicontinuous implies

$$f(x_0) \le \liminf_{j\to\infty} f(x_{n_j}).$$

On the other hand, by construction we have $\liminf_{j\to\infty} f(x_{n_j}) \le m(f)$. Obviously we also have $m(f) \le f(x_0)$, and thus

$$f(x_0) = m(f) = \inf_{x\in M} f(x) = \lim_{j\to\infty} f(x_{n_j}).$$

Consequently, f has a minimising point. □

Remark 1.1.2. The theorem above should be regarded as a generalisation of Weierstrass's classical theorem (a real continuous function attains its minimum and maximum on a compact interval), so one could also refer to Theorem 1.1.2 as Weierstrass's Variational Calculus Theorem.

Remark 1.1.3. It may seem to be a natural weakening of this existence theorem when we require that f, the function to be minimised, be continuous. The following example shows that this weakened version is too weak even for simple applications, so that the extension of this theorem to l.s.c. functions will be essential.

Let $S^2 = \{x \in \mathbb{R}^3 |\ \|x\| = 1\}$ be the unit sphere in $\mathbb{R}^3$ and let A and B be two points on S^2. Now consider the set of all smooth paths from A to B on S^2.

$$M = \{\sigma : [0,1] \to S^2 | \sigma \in C^{(2)}, \sigma(0) = A, \sigma(1) = B\},$$

and the length functional on this set M,

$$L : M \to \mathbb{R}, \quad L(\sigma) = \left(\int_0^1 \|\dot{\sigma}(t)\|^2\, dt\right)^{1/2}.$$

It is known that on the unit sphere S^2 there is a shortest path between two points A and B, i.e. the geodesic line σ_0 which connects them (see Chap. 5):

$$L(\sigma_0) = \inf\{L(\sigma) | \sigma \in M\} \le \liminf_{n\to\infty} L(\sigma_n)$$

for every sequence of paths $\{\sigma_n\}_{n\in\mathbb{N}} \subset M$ which converges to σ_0. Thus, L is l.s.c. at σ_0. However, L is not continuous σ_0, since in every neighborhood of σ_0 there are paths which are arbitrarily long, i.e.

$$\limsup_{n\to\infty} L(\sigma_n) = +\infty > L(\sigma_0).$$

H. Lebesgue showed in 1902 [1.1] that even a functional as simple as arc length is only semicontinuous, and not continuous. If one embeds a curve C within a narrow, strip-like region, then it is obvious that there are winding curves within this region which join some points A and B and which are of arbitrary length.

The question of uniqueness in minimum problems can easily be answered in geometrical terms using the graph of the function f as illustration.

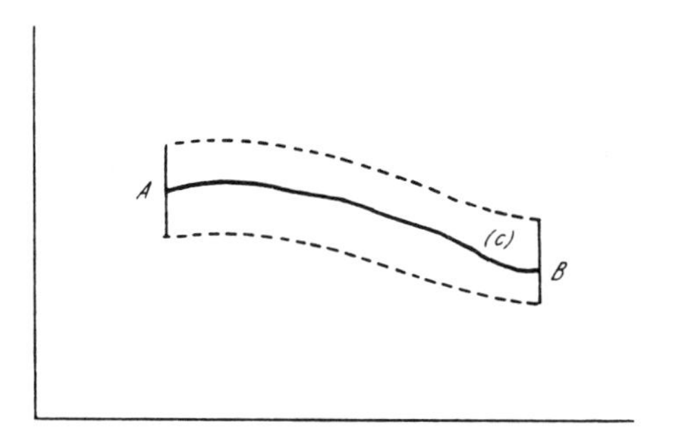

Fig. 2

Theorem 1.1.3. Uniqueness. *Let M be a convex set in a linear space (i.e. $x_j \in M$, $j = 1, 2, 0 \leq \lambda \leq 1 \Rightarrow \lambda x_1 + (1-\lambda)x_2 \in M$) and $f : M \to \mathbb{R}$ a strictly convex function on M (i.e. $x_1, x_2 \in M$, $x_1 \neq x_2$, $0 < \lambda < 1 \Rightarrow f(\lambda x_1 + (1-\lambda)x_2) < \lambda f(x_1) + (1-\lambda)f(x_2)$). Then f has at most one minimising point x_0 in M.*

Proof. If there were two different minimising points $x_0, y_0 \in M$, then we would have for all $\lambda \in (0,1)$

$$f(x_0) = f(y_0) \leq f(\lambda x_0 + (1-\lambda)y_0) < \lambda f(x_0) + (1-\lambda)f(y_0) = f(x_0),$$

which is a contradiction. Under the given assumptions, there exists at most one minimising point. □

Clearly, the combination of Theorems 1.1.2 and 1.1.3 gives an existence and uniqueness theorem. The difficulty in applying this theorem stems from the fact that it is often rather difficult to satisfy simultaneously the hypotheses:

(1) M convex,
(2) $f : M \to \mathbb{R}$ strictly convex,
(3) f lower semicontinuous, and
(4) $\exists \lambda \in \mathbb{R}$ with $M_{f,\lambda} \neq \emptyset$ and $M_{f,\lambda}$ sequentially compact.

In order to satisfy the hypotheses (3) and (4) at the same time in an application, it is necessary to be very careful in choosing the topology on M, since (3) and (4) represent requirements on the topology of M not easily compatible with one another. We recall the following:

(i) the finer topology on M yields more lower semicontinuous functions;
(ii) the coarser topology on M yields more sequentially compact sets.

In applications one can assume that the "function" f will be given in the form of a particular algorithm or some formula, so that careful choice of the domain M will enable one to satisfy the hypotheses (1) and (2).

1.2 Applying the Fundamental Theorem in Banach Spaces

We want to treat the difficulties connected with fulfilling the hypotheses (3) and (4) in a case which is very important for applications, namely, when M is a subset of a Banach space X with norm $\|\cdot\|$. First of all, we state without proof a few important results from the theory of Banach spaces. For basic concepts of this theory, we refer the reader to Appendices 1 and 3, and the literature cited there.

The following theorem shows, in accordance with the comment (ii) above, that the norm topology yields too few compact sets for our requirements as soon as the Banach space becomes infinite-dimensional.

Theorem 1.2.1 (Riesz). *A Banach space is finite-dimensional if and only if the closed unit ball $\{x \in X | \, \|x\| \leq 1\}$ (or every bounded closed set) is compact.*

Proof. See Appendix 3. □

We recall that, for the norm topology, the concepts of compactness and sequential compactness coincide (Appendix 3).

Remark 1.2.1. In his investigations of the Dirichlet Principle, Riemann assumed automatically that the closed unit ball would also be compact in an infinite-dimensional Banach space.

Riesz's result implies, for example, that a compact subset in an infinite-dimensional Banach space contains no interior points. In particular, it follows from Theorem 1.2.1 that in an infinite-dimensional Banach space X, a sequence $\{x_n\}_{n\in\mathbb{N}} \subset \{x \in X | \, \|x\| \leq 1\}$ and an $\varepsilon > 0$ always exist, such that $\|x_n - x_m\| > \varepsilon$, $\forall n \neq m$. In the special case of a Hilbert space X, for example, for every orthonormal system one has $\|x_n - x_m\| = \sqrt{2}$, $\forall n \neq m$.

Luckily, a topology on a Banach space is available which

(a) is coarser than the norm topology on X, and
(b) in the sense of the comment (ii) above, has sufficiently many compact sets.

This is the so-called *weak topology* $\sigma(X, X')$ on X (see Appendix 2), by definition the coarsest locally convex topology on X with respect to which all elements f of the dual space X' of X are continuous. Thus, $\sigma(X, X')$ is generated by the system $\{q_f(\cdot) | f \in X'\}$ of seminorms on X:

$$x \mapsto q_f(x) := |f(x)|, \quad x \in X.$$

Accordingly, the weak topology $\sigma(X, X')$ on a Banach space X possesses, of all the locally convex topologies on X with the same dual space X', the most compact sets, and it is therefore particularly suited for the purposes of variational calculus.

There is, however, one complication. The weak topology on X does not necessarily satisfy the first axiom of countability, so that we have to distinguish between compact and sequentially compact sets. In this context, it is necessary to know of sufficient conditions as to when subsets of a Banach space X are weakly sequentially compact (that is to say, sequentially compact with respect to the weak topology) or weakly compact. The following theorem serves this purpose.

Theorem 1.2.2 (Eberlein). (a) *Every weakly countable compact subset of a Banach space X is weakly compact and weakly sequentially compact.*

(b) *In a reflexive Banach space X, every bounded subset is relatively compact with respect to the weak topology $\sigma(X, X')$.*

For the proof of (a) see [Ref. 1.2, p. 185], where a subset K of X is called *weakly countably compact* when every sequence in K has an accumulation point in K with respect to the weak topology (or, equivalently, when every countable covering of K by weakly open sets possesses a finite subcovering). Part (b) will be proved in Appendix 3.

Remark 1.2.2. One of the nicest properties of the weak topology is seen in the case of a reflexive Banach space X (Appendix 3). The weak topology $\sigma(X', X) = \sigma(X', X'')$ on the dual space X' of X admits a characterisation of compact sets which is just as simple as in the well-known case of finite-dimensional Banach spaces: a subset $K \subset X'$ is $\sigma(X', X)$-compact if and only if it is $\sigma(X', X)$-closed and bounded with respect to the norm.

Therefore, every bounded sequence in an infinite-dimensional reflexive Banach space X has a weakly convergent subsequence, i.e.

$$\begin{gathered} \{x_n\}_{n \in \mathbb{N}} \subset X, \quad \|x_n\| \leq c < \infty, \quad \forall n \in \mathbb{N} \\ \Rightarrow \exists \text{ subsequence } \{x_{n_j}\}_{j \in \mathbb{N}} \text{ and } x_0 \in X, \end{gathered}$$

so that

$$f(x_0) = \lim_{j \to \infty} f(x_{n_j}) \quad \text{for all } f \in X'.$$

In this particular version, Eberlein's theorem is highly applicable to most variational calculus problems (cf. the proof of Theorem 1.1.2). It remains only to ensure that the function which is to be minimised is sequentially l.s.c. with respect to the weak topology. We will discuss this point now.

1.2.1 Sequentially Lower Semicontinuous Functionals

Let X be a Banach space and M a subset of X. A function $f : M \to \mathbb{R}$ is called *weakly sequentially lower semicontinuous at a point* $x_0 \in M$ if and only if for every sequence $\{x_n\}_{n\in\mathbb{N}} \subset M$, with $x_0 = w - \lim_{n\to\infty} x_n$, the inequality

$$f(x_0) \leq \liminf_{n\to\infty} f(x_n)$$

holds. If this property is true for every $x_0 \in M$, then f is called *weakly sequentially lower semicontinuous on* M. A weakly sequentially upper semicontinuous functional is defined accordingly.

Example 1.2.3. The norm $\|\cdot\| : X \to \mathbb{R}$ of a Banach space is, naturally, continuous with respect to the norm topology on X. In the case of an infinite-dimensional Banach space X, however, the norm is no longer weakly sequentially continuous, but is, as we now show, weakly sequentially lower semicontinuous.

Suppose that the sequence $\{x_n\}_{n\in\mathbb{N}} \subset X$ converges weakly to a point $x_0 \in X$. Then

$$\|x_0\| \leq \liminf_{n\to\infty} \|x_n\| .$$

As proof, note that, according to Appendix 1,

$$\|x_0\| = \sup_{f\in S_1(X')} |f(x_0)| , \quad S_1(X') = \{f \in X' | \; \|f\|' = 1\} .$$

Now $x_0 = w - \lim_{n\to\infty} x_n$ implies that $f(x_0) = \lim_{n\to\infty} f(x_n)$, and thus $|f(x_0)| = \lim_{n\to\infty} |f(x_n)|$.

However, $|f(x_n)| \leq \|f\|' \, \|x_n\| = \|x_n\|$, so that

$$|f(x_0)| \leq \liminf_{n\to\infty} \|x_n\|$$

for all $f \in S_1(X')$. Therefore, $\sup_{f\in S_1(X')} |f(x_0)| \leq \liminf_{n\to\infty} \|x_n\|$ and the assertion follows.

In order to be able to apply the fundamental theorem of the calculus of variations effectively in Banach spaces, it is obviously very useful to have at our disposal criteria which are specific to Banach spaces and which yield the hypotheses of the fundamental theorem. We will discuss the following points for this reason.

- When is a function weakly (sequentially) l.s.c. ?
- The relationship between convexity and semicontinuity.
- The concept of coerciveness.

In Chap. 3, we will be able to give sufficient extra conditions with the aid of differential calculus in Banach spaces. This has proved to be a particularly useful tool in variational calculus, since it reduces the problem of determining extreme values of a functional to the problem of solving a particular differential equation. The solution of many theoretical variational problems was treated in this way in the 18th century, though with no particular mathematical justification.

Let X be a Banach space, $M \subset X$ a subset and $f : M \to \mathbb{R}$ a function on M. The next lemma characterises the weak (sequential) lower semicontinuity of f in terms of the set $M_{f,\lambda}$ of all points which lie in or below the λ-level $f^{-1}(\lambda) \subset M$ of f.

Lemma 1.2.3. *Let X be a Banach space and $M \subset X$ a weakly (sequentially) closed subset. A function $f : M \to \mathbb{R}$ is weakly (sequentially)* l.s.c. *on M if the sets*

$$M_{f,\lambda} = \{x \in M | f(x) \leq \lambda\}$$

are weakly (sequentially) closed for every $\lambda \in \mathbb{R}$.

Proof. We shall prove this lemma explicitly for sequential convergence and sequential closure. The general case follows analogously if one uses weakly convergent nets [1.2].

First let f be weakly sequentially l.s.c. For arbitrary $\lambda \in \mathbb{R}$, let $\{x_n\}_{n\in\mathbb{N}}$ be a weakly convergent sequence in $M_{f,\lambda}$; then $w - \lim_{n\to\infty} x_n = x \in M$, since M is weakly sequentially closed. If follows from Lemma 1.1.1 that

$$f(x) \leq \liminf_{n\to\infty} f(x_n) \leq \lambda\,,$$

and therefore $x \in M_{f,\lambda}$, i.e. $M_{f,\lambda}$ is weakly sequentially closed.

Conversely, let all sets $M_{f,\lambda}$, $\lambda \in \mathbb{R}$, be weakly sequentially closed. If f were not weakly sequentially l.s.c. on M, then a point $x \in M$ and a sequence $\{x_n\}_{n\in\mathbb{N}} \subset M$ would exist, with

(i) $x = w - \lim\limits_{n\to\infty} x_n$,
(ii) $\liminf\limits_{n\to\infty} f(x_n) < f(x)$.

Then there is also a number $\lambda \in \mathbb{R}$ with

$$\liminf_{n\to\infty} f(x_n) < \lambda < f(x)$$

and, moreover, a subsequence $\{x_{n_j}\}_{j\in\mathbb{N}}$ in $M_{f,\lambda}$, i.e. $x_{n_j} \in M_{f,\lambda}$, $\forall j \in \mathbb{N}$.

Condition (i) implies $x = w - \lim_{j\to\infty} x_{n_j}$. Since $M_{f,\lambda}$ was assumed to be weakly sequentially closed, the weak limit point x of the sequence $\{x_{n_j}\}_{j\in\mathbb{N}} \subset M_{f,\lambda}$ is also contained in $M_{f,\lambda}$, i.e. $f(x) \leq \lambda$. But this is in contradiction to $\lambda < f(x)$. Therefore, f is weakly sequentially l.s.c. on M. □

As a first application of this lemma, we shall show how from its continuity, one can infer that a convex function $f : M \to \mathbb{R}$ is weakly l.s.c.,

Lemma 1.2.4. *Let X be a Banach space, $M \subset X$ a closed convex subset and $f : M \to \mathbb{R}$ a continuous convex function on M. Then f is weakly* l.s.c.

Proof. Let $\lambda \in \mathbb{R}$ be arbitrary. Because f is continuous, $M_{f,\lambda} = \{x \in M | f(x) \leq \lambda\}$ is a closed subset of X. Since f is convex (i.e. $x_1, x_2 \in M$, $0 \leq t \leq 1 \Rightarrow f(tx_1 + (1-t)x_2) \leq tf(x_1) + (1-t)f(x_2)$), $M_{f,\lambda}$ is a convex subset.

As is known (Appendix 1), a convex set of a Banach space is closed if and only if it is weakly closed. $M_{f,\lambda}$ is therefore weakly closed, and applying Lemma 1.2.3, the result follows. □

It remains for us to realise the last of the crucial hypotheses of the fundamental theorem, the sequential compactness of the sets $M_{f,\lambda}$, $\lambda \in \mathbb{R}$. We have already explained that one can effectively use the fundamental theorem in Banach spaces for the weak topology. Consequently, we shall now discuss a condition on the function $f : M \to \mathbb{R}$ which implies the weak sequential compactness of $M_{f,\lambda}$, $\lambda \in \mathbb{R}$.

Let X be a Banach space, $M \subset X$ a subset and $f : M \to \mathbb{R}$ a function on M. Then f is called *coercive* on M if $x \in M$ and if $\|x\| \to +\infty$ always implies that $f(x) \to +\infty$.

Remark 1.2.4. The coerciveness condition will often be made more stringent in later applications to take into account a specification of the growth of $f(x)$ for $\|x\| \to \infty$. Furthermore, we shall discuss a few sufficient conditions for coerciveness. It is worth emphasising that very many solutions to minimisation problems can be achieved via a realisation of the coerciveness condition.

If $f : M \to \mathbb{R}$ is a coercive function, then all the sets $M_{f,\lambda}$, $\lambda \in \mathbb{R}$ are bounded. If X is now a reflexive Banach space, which is fortunately the case in most applications, then the sets $M_{f,\lambda}$, $\lambda \in \mathbb{R}$ are weakly relatively compact, due to Eberlein's Theorem. This fact brings us to the following variation of the fundamental theorem of variational calculus.

Theorem 1.2.5. *Let X be a reflexive Banach space and $M \subset X$ a weakly (sequentially) closed subset. Let $f : M \to \mathbb{R}$ be a coercive weakly (sequentially)* l.s.c. *function on M. Then $m(f) = \inf_{x \in M} f(x)$ is finite and is attained at a point $x_0 \in M$, i.e. $m(f) = f(x_0)$.*

Proof. $M_{f,\lambda}$ is not empty for suitable $\lambda \in \mathbb{R}$. By virtue of Lemma 1.2.3, $M_{f,\lambda}$ is weakly (sequentially) closed. The coerciveness of f implies that $M_{f,\lambda}$ is bounded and therefore weakly (sequentially) compact. Theorem 1.1.2 completes the proof. □

1.3 Minimising Special Classes of Functions

The choice of the classes of functions for whose elements we want to establish the hypotheses of the fundamental theorem, and to which we actually want to apply the theorem, is determined by the selection of applications we shall be discussing later (Chap. 5–10). Further applications may be found in the literature [1.3–5].

Once again, our starting point is a reflexive Banach space X and a function $f : M \to \mathbb{R}$ defined on a subset M of X. We immediately have from Lemma 1.2.4:

Corollary 1.3.1. *Let $M \subset X$ be convex and closed; let $f : M \to \mathbb{R}$ have a representation of the form $f = f_1 + f_2$ with*

(i) *$f_1 : M \to \mathbb{R}$ convex and continuous,*
(ii) *$f_2 : M \to \mathbb{R}$ weakly (sequentially) continuous.*

Then f is weakly (sequentially) lower semicontinuous.

Proof. By Lemma 1.2.4 f_1 is weakly l.s.c. and the sum of two weakly l.s.c. functions is also weakly l.s.c. □

The following lemma generalises this result.

Lemma 1.3.2. *Let X be a reflexive Banach space and $F : X \times X \to \mathbb{R}$ a function having the following properties:*

(i) *for all $y \in X$,*

$$x \mapsto F(x,y)$$

is a continuous convex function on X.

(ii) *For all bounded subsets $B \subset X$,*

$$\{y \mapsto F(x,y) | x \in B\}$$

is a family of functions which is weakly (sequentially) equicontinuous on bounded subsets of X.

Then

$$x \mapsto f(x) := F(x,x)$$

is weakly (sequentially) l.s.c.

Proof. (a) Because of Lemma 1.2.4, the hypothesis (i) implies that

$$x \mapsto F(x,y)$$

is a weakly l.s.c. function on X for every $y \in X$. Now let $x \in X$ be the weak limit of a sequence $\{x_n\}_{n\in\mathbb{N}} \subset X$. We show that

$$f(x) \le \liminf_{n\to\infty} f(x_n)$$

by considering

$$f(x_n) = F(x_n, x) + \{F(x_n, x_n) - F(x_n, x)\}$$

and showing that

$$\lim_{n\to\infty} \{F(x_n, x_n) - F(x_n, x)\} = 0 \, . \tag{$*$}$$

Having shown this, it follows that

$$\liminf_{n\to\infty} f(x_n) = \liminf_{n\to\infty} F(x_n, x) \ge F(x, x) = f(x) \, .$$

(b) Since $\{x_n\}_{n\in\mathbb{N}} \subset X$ converges weakly, $\{x_n | n \in \mathbb{N}\}$ is a bounded set in X (Appendix 1). It then follows by virtue of the hypothesis (ii) that $\{y \mapsto F(x_n, y) | n \in \mathbb{N}\}$ is a family of functions on X that is weakly (sequentially) equicontinuous on bounded sets, that is, for all weakly converging sequences

$$\{y_j\}_{j\in\mathbb{N}} \text{ in } X \, , \quad y = w - \lim_{j\to\infty} y_j \, ,$$

the following holds:

$$\bigwedge_{\varepsilon>0} \ \bigvee_{j_0 = j_0(\varepsilon)} \ \bigwedge_{j \ge j_0} \ \bigwedge_{n\in\mathbb{N}} |F(x_n, y) - F(x_n, y_j)| < \varepsilon \, .$$

Specialising to the sequence $\{x_j\}_{j\in\mathbb{N}}$ easily leads to statement ($*$). □

Functionals of the type $f(x) = F(x, x)$, which we considered in Lemma 1.3.2, appear quite often in applications. We should mention here that in Chaps. 6, 7 and 8 we shall discuss the furthest reaching applications, in the form of F. Browder's results on nonlinear elliptic boundary value and eigenvalue problems. In these applications, F has the form

$$F(u, v) = \int \sum_{j=0}^{n} \overline{A_j(x, u(x), Du(x))} \, D_j v(x) \, dx \, ,$$

and with suitable hypotheses about the A_j it is possible to show that F satisfies the hypotheses of Lemma 1.3.2.

For the class of functions that satisfy the hypotheses of Lemma 1.3.2 and that are sometimes [1.4] also called *semiconvex*, it is easy to show the following version of the fundamental theorem of the calculus of variations.

Theorem 1.3.3. *Let X be a reflexive Banach space and $F : X \times X \to \mathbb{R}$ a function which satisfies the hypotheses* (i) *and* (ii) *in Lemma* 1.3.2.

Then the function $f(x) = F(x,x)$ is bounded from below on every weakly closed bounded subset K of X, and attains its minimum at a point $x_0 \in K$:

$$\inf_{x \in K} f(x) = f(x_0)\,.$$

Proof. According to Theorem 1.2.2, K is weakly compact, and according to Lemma 1.3.2, f is weakly l.s.c. on K; the conclusion then follows from the fundamental theorem of the calculus of variations. □

There is a simple corollary for coercive functions.

Corollary 1.3.4. *Let X and F be as in Theorem 1.3.3 and $f(x) = F(x,x)$ be coercive on X. Then the function f attains its minimum on X:*

$$f(x_0) = \inf_{x \in X} f(x)\,.$$

Proof. $K_R = \{x \in X|\ \|x\| \le R\}$ is a closed, convex, bounded subset of X and thus weakly compact. According to Theorem 1.3.3 the function f attains its minimum on K_R. Moreover, since f is coercive, then for sufficiently large $R > 0$:

$$\|x\| > R \Rightarrow f(x) > f(0) \ge \inf_{x \in K_R} f(x)\,,$$

and we conclude immediately. □

A version of Theorem 1.3.3 which is often used in applications to nonlinear eigenvalue problems (Sect. 8.4) is formulated in the following corollary.

Corollary 1.3.5. *Let X and F be as in Theorem 1.3.3, and $f(x) = F(x,x)$ be coercive on X. Moreover, let $g : X \to \mathbb{R}$ be a function which is weakly continuous on bounded sets of X. Then f attains a minimum on every level surface*

$$M = g^{-1}(x) = \{x \in X | g(x) = c\} \ne \emptyset\,, \quad c \in \mathbb{R}$$

of g.

Proof. Since f is coercive, there exists an $R > 0$, such that for all $x \in X$ with $\|x\| > R$

$$f(x) > \inf_{y \in M} f(y)\,.$$

It follows that

$$\inf_{x \in M} f(x) = \inf_{x \in M_R} f(x)\,,$$

when we put $M_R = \{x \in M|\ \|x\| \le R\}$. Since g is weakly continuous, M_R is a weakly closed, bounded subset of X. Theorem 1.3.3 can then be applied to f and M_R. □

1.3.1 Quadratic Functionals

Now we come to a special class of functionals of the sort we looked at in Lemma 1.3.2. We assume that F is a Hermitian sesquilinear form on the reflexive Banach space X. Then

$$x \mapsto f(x) := F(x, x)$$

is called a *quadratic functional* on X. The study of quadratic functionals on a Hilbert space is one of the oldest and best known areas of variational calculus. Interest in these functions has its origins in the classical "theory of second variation" (Chap. 3) and in the treatment of Dirichlet's problem in potential theory (Chap. 6). Here we shall discuss a simple abstract version which will be the basis for numerous applications (Chaps. 6, 7, 8).

Theorem 1.3.6. *Let X be a reflexive Banach space and Q a Hermitian sesquilinear form on X having the following properties:*

(i) *Q is strictly coercive, i.e. there exists a constant $c > 0$ such that*

$$Q(x, x) \geq c\|x\|^2$$

holds for all $x \in X$.

(ii) *Q is weakly continuous in the second variable, which means that for arbitrary but fixed $x_0 \in X$,*

$$x \mapsto Q(x_0, x)$$

is a weakly continuous linear form on X.

The the following holds: for every $l \in X'$ and for every $r > 0$, there is exactly one $x_0 = x_0(l, r) \in \overline{B}_r = \{x \in X \mid \|x\| \leq r\}$ such that the function

$$x \mapsto f(x) := Q(x, x) - \operatorname{Re} l(x)$$

attains its minimum (on the closed ball $\overline{B}_r$) at the point x_0:

$$f(x_0) = \inf_{x \in \overline{B}_r} f(x) .$$

Proof. Since the balls $\overline{B}_r$, $r > 0$ are bounded convex subsets of the reflexive Banach space X, they are weakly compact (Theorem 1.2.2). The assertion follows from Theorems 1.1.2 and 1.1.3 if we show that f is

(i) strictly convex and
(ii) weakly (sequentially) l.s.c.

A simple calculation shows that for $x_1, x_2 \in X$ and $0 < t < 1$,

$$f((1-t)x_1 + tx_2) = (1-t)f(x_1) + tf(x_2) - t(1-t)Q(x_1 - x_2, x_1 - x_2).$$

As a result of the coerciveness of Q the following holds:

$$x_1 \neq x_2\,, \quad 0 < t < 1 \Rightarrow -t(1-t)Q(x_1 - x_2, x_1 - x_2) < 0\,.$$

This implies the strict convexity of f.

Another simple calculation shows that

$$Q(x,x) - Q(x_0,x_0) = Q(x - x_0, x - x_0) + Q(x - x_0, x_0) + Q(x_0, x - x_0).$$

If $(x_\alpha)_{\alpha \in A}$ is now a net in X (a sequence $\{x_n\}_{n \in \mathbb{N}}$) which converges weakly to x_0, then it follows that

$$Q(x_\alpha, x_\alpha) \geq Q(x_0, x_0) + Q(x_\alpha - x_0, x_0) + Q(x_0, x_\alpha - x_0)$$

and thus

$$\liminf_\alpha Q(x_\alpha, x_\alpha) \geq Q(x_0, x_0)\,,$$

since

$$\lim_{\alpha \in A} Q(x_\alpha - x_0, x_0) = \lim_{\alpha \in A} Q(x_0, x_\alpha - x_0) = 0$$

because of the hypothesis (ii). Therefore, $x \mapsto Q(x,x)$ is weakly l.s.c. Thus

$$x \mapsto f(x) = Q(x,x) - \mathrm{Re}\; l(x)$$

is also weakly l.s.c. for every continuous linear form l on X. □

Remark 1.3.1. In the context of the *Dirichlet principle* (see Chap. 6), we will introduce the following bilinear form on the Sobolev space $H_0^1(\Omega)$, where $\Omega \subset \mathbb{R}^n$ is bounded and has a smooth boundary (see Appendix 4):

$$Q(u,v) = \int_\Omega \nabla u \cdot \nabla v \, dx\,.$$

The coerciveness of Q will follow from the Poincaré inequality, i.e.

$$\|v\|_2 \leq C(\Omega) \|\nabla v\|_2\,, \quad \forall v \in H_0^1(\Omega)\,.$$

Corollary 1.3.7 is a somewhat more concrete version of this theorem.

Corollary 1.3.7. *Let A be a bounded Hermitian operator on a Hilbert space $\mathcal{H}$. Let A be strictly positive, that is there exists a constant $c > 0$ such that $\langle x, Ax\rangle \geq c\langle x, x\rangle$ for all $x \in \mathcal{H}$.*

Then for every $y \in \mathcal{H}$, the function

$$x \mapsto f(x) = \langle x, Ax\rangle - \mathrm{Re}\,\langle y, x\rangle$$

attains its unique minimum on every ball $\overline{B}_r = \{x \in \mathcal{H} | \; \|x\| \le r\}$, *i.e. exactly one* $x_0 = x_0(y,r)$ *exists, such that*

$$f(x_0) = \inf_{x \in \overline{B}_r} f(x) .$$

Proof. $Q_A(x,x) = \langle x, Ax \rangle$ is a Hermitian sesquilinear form on $\mathcal{H}$ that is strictly coercive. On the other hand, $x \mapsto Q_A(y,x) = \langle Ay, x \rangle$ is a weakly continuous linear form on $\mathcal{H}$, which is a property of the scalar product in a Hilbert space. Theorem 1.3.6 is then applicable. □

Remark 1.3.2. If $\mathcal{H}$ is an infinite-dimensional Hilbert space and the bounded Hermitian operator A is not necessarily strictly positive, then simple examples show that the corollary above does not necessarily hold.

1.4 Some Remarks on Linear Optimisation

The basic problem in linear optimisation consists, in its "simplest" form, in determining the maximum (or minimum) of a continuous linear function f on a subset K of a Hausdorff locally convex topological vector space X. One readily realises that in the general situation, interesting results are only available for bounded sets K. A fundamental result, not only in this context, is *H. Bauer's maximum principle* (e.g. Theorem 25.9 in [Ref. 1.6]), which assumes the compactness of K but which weakens the hypotheses of f, in the sense of the fundamental theorem of the calculus of variations.

Theorem 1.4.1. *Let* X *be a real Hausdorff locally convex topological vector space and* $K \subset X$ *a (nonempty) convex compact subset. Then every convex upper semicontinuous function* $f : K \to \mathbb{R}$ *(especially every continuous linear function* $f : X \to \mathbb{R}$*) attains its maximum at an extremal point* x_0 *of* K*:*

$$f(x_0) = \sup_{x \in K} f(x) .$$

Remark 1.4.1. (a) An *extremal point* x_0 of a convex set K is a point of K which cannot be represented in the form

$$x_0 = \lambda x_1 + (1-\lambda) x_2 \quad \text{with } x_1, x_2 \in K , \quad x_1 \ne x_2 , \quad 0 < \lambda < 1 .$$

The extremal points of a triangle, for instance, are just its vertices. The extremal points of a closed disc in the plane are all the points on its boundary.

(b) On the basis of the fundamental Theorem 1.1.2, the existence of (at least) one maximising point is certain: an upper semicontinuous function attains its maximum on a compact set ! The difficult part of the proof of Theorem

1.4.1 is showing that all maximising points are necessarily extremal points of K.

(c) The proof of Theorem 1.4.1 is simple for the case when f is assumed to be strictly convex, so that by Theorem 1.1.3 it possesses at most one maximising point. As a result of the remark (b), f possesses exactly one maximising point. We shall show that this maximising point x_0 is necessarily an extremal point of K. We assume the following:

$$x_0 = \lambda x_1 + (1-\lambda)x_2\,, \quad 0 < \lambda < 1\,, \quad x_1, x_2 \in K\,, \quad x_1 \neq x_2\,.$$

The strict convexity of f gives, because $f(x_1) \leq f(x_0)$, $i = 1, 2$,

$$\begin{aligned} f(x_0) &= f(\lambda x_1 + (1-\lambda)x_2) < \lambda f(x_1) + (1-\lambda) f(x_2) \\ &\leq \lambda f(x_0) + (1-\lambda) f(x_0) = f(x_0) \end{aligned}$$

and thus a contradiction.

(d) The assertion of the theorem is very important for the determination of the maximum because it places very strong restrictions on the set of potential extremal points. Particularly in the case of convex polyhedra, which often appears in applications, one need only look for the extrema of a function in the finite set of the extremal points of the polyhedron.

Example 1.4.2. In an optimisation problem, we try to find the maximum (or minimum) of a particular function. This function depends on variables which cannot be freely chosen but which are subject to restrictions in the form of inequalities and, in some cases, equalities. The simplest type of problem is the linear optimisation problem. Our function in this case is a linear function $f(x, y) = ax + by$ and the restrictions are linear inequalities, for example

$$cx + dy \leq e\,, \quad fx + gy \leq h\,, \quad x \geq 0\,, \quad y \geq 0\,.$$

Values of x and y are sought which fulfil the constraints and which give the largest possible value of $f(x, y)$.

1.5 Ritz's Approximation Method

Let $M \subset X$ be a subset of a topological space X and $f : M \to \overline{\mathbb{R}} = \mathbb{R} \cup \{\pm\infty\}$ a function on M. Every sequence $\{x_n\}_{n\in\mathbb{N}} \subset M$, for which

$$\lim_{n\to\infty} f(x_0) = \inf\{f(x) | x \in M\}$$

holds is called a *minimising sequence* (for the problem of minimising the function f on the set M). This definition does not assume that the sequence of the x_n converges.

When does a minimising sequence exist? That is already the case under very weak assumptions, namely, when

(i) there are points $x \in M$ with $f(x) < \infty$;
(ii) f is bounded from below, i.e.

$$\inf\{f(x)|x \in M\} = \mu > -\infty .$$

Then minimising sequences exist (definition of the infimum). However, this does not solve the minimisation problem at all. As a solution to this problem we regard a point $x_0 \in M$ for which

$$f(x_0) = \mu = \inf_{x \in M} f(x)$$

holds. Since the minimising sequences $\{x_n\}_{n \in \mathbb{N}}$, whose existence are assured by hypotheses (i) and (ii), do not necessarily converge, and even if one does converge, for instance $x_n \underset{n\to\infty}{\longrightarrow} y \in M$, then

$$\lim_{n\to\infty} f(x_n) = f\left(\lim_{n\to\infty} x_n\right)$$

need not hold.

Thus three steps are necessary to solve the problem of minimising a function $f : M \to \mathbb{R}$:

(i) Construct a minimising sequence $\{x_n\}_{n \in \mathbb{N}}$.
(ii) Prove that this sequence converges to a point $x_0 \in M$.
(iii) Justify the equation

$$\lim_{n\to\infty} f(x_n) = f\left(\lim_{n\to\infty} x_n\right) .$$

The fundamental theorem of the calculus of variations provides sufficient conditions to ensure that these three steps can be carried out. But this theorem does not provide a practical method of calculating the minimising point or even the minimal value of the function (or at least only an approximate one). However, the Ritz approximation method, in a particular situation, does [1.7]. This method consists, in an infinite-dimensional separable Hilbert space, in constructing for the original problem a series of corresponding problems in finite-dimensional subspaces, solving these finite-dimensional problems, and then showing that the sequence of the solutions of all finite-dimensional subproblems is a minimising sequence.

Various versions of this method have been formulated, including some which show that the sequence of the solutions of the approximating finite-dimensional problems converges to a solution of the original problem [1.8]. We mention here the so-called *Galerkin approximation* [1.9]. We shall discuss in more detail a special version of the Galerkin approximation in a separable reflexive Banach space together with Browder's results on nonlinear elliptical eigenvalue problems. The following is a simple version for separable Hilbert spaces.

Theorem 1.5.1. *Let $\mathcal{H}$ be a separable Hilbert space and $f : \mathcal{H} \to \mathbb{R}$ a continuous weakly* l.s.c. *coercive functional.*

One can then obtain a minimising sequence $\{x_n\}_{n\in\mathbb{N}}$ for f by minimising f on certain finite-dimensional subspaces $\mathcal{H}_n$:

$$f(x_n) = \min_{x\in\mathcal{H}_n} f(x), \quad \text{i.e.} \quad x_n \in \mathcal{H}_n .$$

Here $(\mathcal{H}_n)_{n\in\mathbb{N}}$ is a sequence of subspaces of $\mathcal{H}$ with the following properties:

(i) $\dim \mathcal{H}_n = n$,

(ii) $\mathcal{H}_n \subset \mathcal{H}_{n+1}$,

(iii) $\mathcal{H} = \overline{\bigcup_{n=1}^{\infty} \mathcal{H}_n}$.

Proof. Since f is coercive and weakly l.s.c., then for all $n \in \mathbb{N}$ points $x_n \in \mathcal{H}_n$ exist with

$$f(x_n) = \min_{x\in\mathcal{H}_n} f(x) .$$

and there is a point $x_0 \in \mathcal{H}$ with

$$f(x_0) = \min_{x\in\mathcal{H}} f(x)$$

(Theorem 1.2.5). Now let $P_n : \mathcal{H} \to \mathcal{H}_n$ denote the orthogonal projector onto the subspace $\mathcal{H}_n$. The continuity of f implies that

$$\lim_{n\to\infty} f(P_n x_0) = f\left(\lim_{n\to\infty} P_n x_0\right) = f(x_0)$$

since the hypotheses (ii) and (iii) imply

$$\lim_{n\to\infty} P_n x = x \quad \text{for all} \quad x \in \mathcal{H} .$$

Since $f(P_n x_0) \geq f(x_n) \geq f(x_0)$, then

$$\lim_{n\to\infty} f(x_n) = f(x_0)$$

also holds, i.e. the sequence $\{x_n\}_{n\in\mathbb{N}}$ is indeed a minimising sequence. □

Remark 1.5.1. The theorem above does not assert that the sequence of the x_n converges; it only asserts that the minimising sequence thus constructed provides a good approximation for the minimal value of the function f if the dimension n of the space $\mathcal{H}_n$ on which the function f is approximately minimised is chosen to be large enough. Further, it makes no quantitative prediction as to how large n should be for the approximation to attain any required degree of precision. The size of n necessary for a useful approximation depends first of all on f, but also on the apt (or inept) choice of the sequence $\{\mathcal{H}_n\}$ of the subspace. A "good sense of intuition" about the problem and its possible solutions can be very helpful in the choice of the sequence of the subspaces $\{\mathcal{H}_n\}$ and can lead to a fast convergence of $\{f(x_n)\}_{n\in\mathbb{N}}$!

Remark 1.5.2. Algebraic equations are at our disposal in simple special cases for the determination of the points x_n of the minimal sequence in Ritz's method. We have, for instance,

$$f(x) = \frac{1}{2}Q(x,x) - \langle y, x\rangle\,,$$

with a coercive, continuous, symmetrical bilinear form Q on a Hilbert space $\mathcal{H}$ and a fixed point $y \in \mathcal{H}$. The sequence $\{\mathcal{H}_n\}$ of the subspaces is chosen as in Ritz's theorem, and $v_1, \ldots, v_n$ are an orthonormal basis of $\mathcal{H}_n$. Now

$$f(x_n) = \min_{x\in\mathcal{H}_n} f(x)$$

implies (see Theorem 3.2.1)

$$f'(x_n)(v) = 0 \quad \text{for all} \quad v \in \mathcal{H}_n\,,$$

which is true if and only if

$$f'(x_n)(v_k) = 0 \quad \text{for} \quad k = 1,\ldots,n\,.$$

Now $f'(x_n)(v) = Q(x_n, v) - \langle y, v\rangle$. Thus, for the determination of

$$x_n = \sum_{j=1}^{n} \alpha_j^n v_j$$

we have at our disposal n algebraic (linear) equations:

$$\sum_{j=1}^{n} \alpha_j^n Q(v_j, v_k) = \langle y, v_k\rangle\,, \qquad k = 1,\ldots,n\,.$$

Remark 1.5.3. Ritz's method (1908) is a generalisation of a method used by Lord Rayleigh (see "Theory of Sounds", 1877–1878) for the special case of the determination of eigenvalues. Ritz also used his method to investigate the vibrating plate; he calculated the form of the sound patterns, which showed a very satisfying agreement with experiment.

2. Differential Calculus in Banach Spaces

2.1 General Remarks

The previous chapter contains the fundamental theorems of existence and uniqueness of the calculus of variations in quite general form, suitable for applications. However, these theorems do not provide any practical information on how to calculate the minimising (maximising) point in a particular case.

In order to get an idea which we can expound for this purpose, let us recall the simplest case of a real continuous function f defined on a compact interval $M = [a, b]$. Our existence Theorem 1.1.2 (or Weierstrass's theorem) states that there exists a minimising (and also a maximising) point $x_0 \in M$. If one knows something more about the function $f : M \to \mathbb{R}$, then one also knows how this minimising point can be calculated, if it is an interior point.

If f is continuously differentiable on M, then x_0 belongs to the set of zeros of the derivative of f, or, in other words, to the set of *critical points* of f,

$$x_0 \in \operatorname{Ker} f' = \{x \in M | f'(x) = 0\} .$$

As we have already pointed out, a critical point does not need to be a local maximum or minimum. It can also be a point of inflexion with horizontal tangent.

If f is sufficiently smooth, then one can conveniently study its behaviour in a suitable neighborhood of x_0 with the aid of its Taylor expansion. For example, if $f'(x_0) = 0$, $f''(x_0) \neq 0$, then

$$f(x) = f(x_0) + f''(x_0)\frac{(x - x_0)^2}{2!} + \{x - x_0)^2 R_2(x_0, x)$$

with $\lim_{x \to x_0} R_2(x_0, x) = 0$.

If f is twice continuously differentiable on M, then one has not only necessary conditions, but also sufficient conditions for a point x_0 to be a local minimising (maximising) point:

$$f'(x_0) = 0 \quad \text{and} \quad f''(x_0) > 0 \quad (f''(x_0) < 0) .$$

If, for example, $f'(x_0) = f''(x_0) = 0$, $f'''(x_0) \neq 0$, then

$$f(x) = f(x_0) + f'''(x_0)\frac{(x - x_0)^3}{3!} + (x - x_0)^3 R_3(x, x_0)$$

with $\lim_{x \to x_0} R_3(x, x_0) = 0$.

Thus x_0 is a point of inflexion with horizontal tangent, since f behaves locally like the third-order polynomial $f(x) - (x - x_0)^3 R_3$.

Remark 2.1.1. Investigation of the behaviour of functionals $x \to f(x)$ on a Banach space E can be reduced to the investigation of real functions $t \to f_1(t)$ of a real variable t by setting $f_1(t) = f(x(t))$, where $t \to x(t)$ is a smooth curve in E.

2.2 The Fréchet Derivative

We shall see that the well-known results above can be generalised in a natural way to the case when f is a real function on a suitable set M of a Banach space E. For this purpose, we need only find an appropriate generalisation of the concept of derivation of a function $f : M \to \mathbb{R}$, $M \subset E$.

In the case of a differentiable function $f : U \to \mathbb{R}$, where $U \subset \mathbb{R}$ is open, the derivative $f'(x_0)$ of f at the point $x_0 \in U$ is usually interpreted as the slope of the tangent to the graph of f at the point x_0. The equation of this tangent T_{f,x_0} is

$$\mathbb{R} \ni x \to T_{f,x_0}(x) = f(x_0) + (x - x_0)f'(x_0)\,.$$

By definition, the tangent is the straight line through the point $(x_0, f(x_0))$ which is the "best" approximation to the graph of f in a neighborhood of the point x_0.

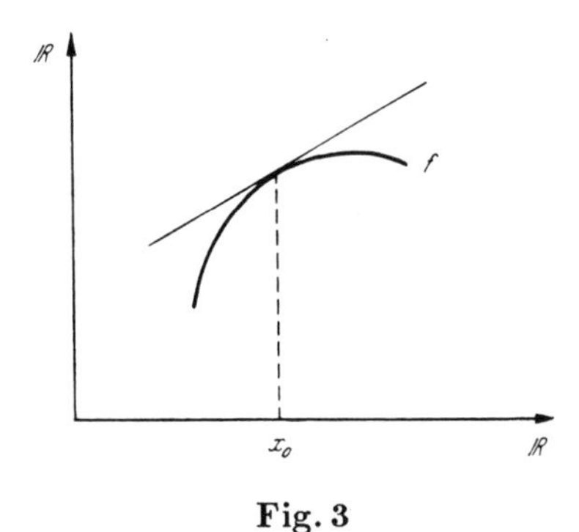

Fig. 3

The idea behind this approximation is expressed precisely by the following equation:

$$f(x_0 + h) - T_{f,x_0}(x_0 + h) = o(x_0, h)$$

where h denotes the distance of the point x from x_0. Here o is a function which specifies this approximation, i.e.

(i) $o(x_0, 0) = 0$,
(ii) $\lim_{|h| \to 0} |o(x_0, h)|/|h|$ exists and is equal to zero.

This idea, which can easily be generalised to the case of functions on Banach spaces, consists simply in regarding the multiplication of $(x - x_0)$ by the derivative of f at the point x_0, that is, by

$$D_{x_0}(f) \equiv (f'(x_0) \in \mathbb{R},$$

as a linear mapping of the Banach space $\mathbb{R}$ into the Banach space $\mathbb{R}$. With this interpretation there are no difficulties to carry over the above definition of a derivative to the case of functions on Banach spaces.

Definition 2.2.1. Let $(E_1, \| \ \|_1)$ and $(E_2, \| \ \|_2)$ be two Banach spaces and $M \subset E_1$ an open nonempty subset. Then a function

$$F : M \to E_2$$

is said to be *differentiable at a point* $x_0 \in M$ if and only if a continuous linear mapping $l : E_1 \to E_2$ exists, such that

$$f(x_0 + h) - [f(x_0) + l(h)] = o(x_0, h) \tag{2.2.1}$$

holds for all $h \in E_1$, $x_0 + h \in M$.

The function $E_1 \ni h \to o(x_0, h) \in E_2$ has the following properties:

(i) $o(x_0, 0) = 0$

(ii)
$$\lim_{h \to 0} \frac{\|o(x_0, h)\|_2}{\|h\|_1} = 0\,. \tag{2.2.2}$$

If f is differentiable at $x_0 \in M$, then the continuous linear mapping $l \in \mathcal{L}(E_1, E_2)$ is called the *derivative of the function* f *at the point* $x_0 \in M$, and one writes accordingly

$$l \equiv D_{x_0}(f) \equiv f'(x_0) \equiv Df(x_0)\,. \tag{2.2.3}$$

If f is differentiable at all points in M, then f is said to be *differentiable on* M: in this case, the mapping

$$Df : M \to \mathcal{L}(E_1, E_2)\,, \tag{2.2.4}$$

which associates with every point $x \in M$ the derivative of f at the point x, is called the *derivative of* f. If $Df : M \to \mathcal{L}(E_1, E_2)$ is a continuous mapping, then f is said to be *continuously differentiable on* M or alternatively *of class* C^1 (i.e. $f \in C^1(M, E_2)$). Df is known as the *Fréchet derivative of* f.

If $E_2 = \mathbb{R}$, then the Fréchet derivative $Df(\cdot)$ is an element in $\mathcal{L}(E_1, \mathbb{R})$, that is in the dual space E_1' of E_1. This situation occurs often in variational calculus.

The proof that Definition 2.2.1 is reasonable follows from the next lemma.

Lemma 2.2.2. *There is at most one continuous linear mapping $l \in \mathcal{L}(E_1, E_2)$ such that Eq.* (2.2.1) *holds.*

Proof. For $l_j \in \mathcal{L}(E_1, E_2)$, $j = 1, 2$, let

$$f(x_0 + h) = f(x_0) + l_j(h) + o_j(x_0, h)$$

for all $h \in B_r = \{x \in E_1, \|x\|_1 < r\}$, $0 < r$, such that $x_0 + B_r \subset M$. Then, for $l = l_1 - l_2$,

$$l(h) = o(x_0, h) \equiv o_1(x_0, h) - o_2(x_0, h), \quad \forall h \in B_r .$$

By virtue of (2.2.2), there exists for every $\varepsilon > 0$ a $\delta(\varepsilon) > 0$, such that

$$\|o(x_0, h)\|_2 \leq \varepsilon \|h\|_1$$

for all $h \in B_r$, $\|h\| < \delta(\varepsilon)$. Taking into account that $\rho B_1 = B_\rho$, $\rho = \inf(r, \delta(\varepsilon))$, the norm of l can be calculated as follows:

$$\begin{aligned} \|l\| &= \sup_{h \in B_1} \|l(x)\|_2 = \sup_{x \in \rho^{-1} B_\rho} \|l(x)\|_2 \\ &= \sup_{h \in B_\rho} \left\| l\left(\frac{1}{\rho} h\right) \right\|_2 = \frac{1}{\rho} \sup_{h \in B_\rho} \|o(x_0, h)\|_2 \\ &\leq \frac{1}{\rho} \sup_{h \in B_\rho} \varepsilon \|h\|_1 = \varepsilon . \end{aligned}$$

Since $\varepsilon > 0$ was arbitrary, it follows that $\|l\| = 0$; thus $l = 0$, i.e. $l_1 = l_2$. □

Remark 2.2.1 One can specialise the definition above to the case of the finite-dimensional Banach spaces

$$E_1 = \mathbb{R}^n , \quad E_2 = \mathbb{R}^m .$$

Let us convince ourselves now that this specialisation gives just the concept of differentiability and derivation that is normally treated in introductory courses on differential and integral calculus (see, for example [2.1]). In this case, the derivative of f at a point $x_0 \in M \subset \mathbb{R}^n$, or in other words the linear mapping $A = Df(x_0) \in \mathcal{L}(\mathbb{R}^n, \mathbb{R}^m)$ with respect to the canonical bases of $\mathbb{R}^n$ and $\mathbb{R}^m$, can be represented by an $m \times n$ matrix:

$$A = \begin{pmatrix} a_{11} & a_{12} & \dots & a_{1n} \\ \vdots & & & \\ a_{m1} & a_{m2} & \dots & a_{mn} \end{pmatrix}.$$

If we consider f in the form

$$f = \begin{pmatrix} f_1 \\ \vdots \\ f_m \end{pmatrix}, \quad f_j : M \to \mathbb{R},$$

it follows that

$$a_{ji} = \frac{\partial f_j}{\partial x_i}(x_0) \text{ is the partial derivative of } f_j \text{ with respect to } x_i \text{ at the point } x_0 .$$

In this case, then, $Df(x_0)$ can be identified in a canonical manner with the Jacobi matrix of the mapping f at the point x_0:

$$Df(x_0) = \left(\frac{\partial f_j}{\partial x_i}(x_0)_{\substack{j=1\dots m \\ i=1\dots n}} \right).$$

Example 2.2.2. In the remark above, we have seen that Definition 2.2.1 indeed represents a generalisation of the concepts of differentiability and derivation for the mappings

$$f : M \to \mathbb{R}^m, \quad M \subset \mathbb{R}^n \text{ open}, \quad n, m \in \mathbb{N} \text{ arbitrary}.$$

Here we have the first large class of examples.

Example 2.2.3. Let $A \in \mathcal{L}(E_1, E_2)$ be a continuous linear mapping of the Banach space E_1 into the Banach space E_2. Then for all $x_0 \in E_1$,

$$Ax = Ax_0 + A(x - x_0).$$

It therefore follows that $x \to Ax$ is differentiable at x_0 with the derivative

$$DA(x_0) = A.$$

A is therefore continuously differentiable on E_1 with the constant derivative

$$DA = A.$$

Example 2.2.4. Let $\mathcal{H} = E_1$ be a real Hilbert space with the scalar product $\langle\, , \rangle$ and let A be a bounded linear operator on $\mathcal{H}$. Consider the function f:

$$f : \mathcal{H} \to \mathbb{R},$$

$$x \to f(x) = \langle x, Ax \rangle .$$

For arbitrary $x_0 \in \mathcal{H}$, a simple calculation shows that

$$f(x_0 + h) = f(x_0) + \langle A^* x_0 + A x_0, h \rangle + \langle h, Ah \rangle .$$

Therefore, f is differentiable at x_0 with the derivative $Df(x_0) \in \mathcal{L}(\mathcal{H}, \mathrm{I\!R}) \cong \mathcal{H}$:

$$Df(x_0)(h) = \langle A^* x_0 + A x_0, h \rangle , \quad \forall h \in \mathcal{H} ,$$

since $o(x_0, h) = \langle h, Ah \rangle$ indeed has the required property

$$|o(x_0, h)| \leq \|A\| \, \|h\|^2 .$$

Consequently, f is continuously differentiable on $\mathcal{H}$. For the special case when

$$A = \mathbb{1} , \quad \text{i.e.} \quad f(x) = \langle x, x \rangle = \|x\|^2 ,$$

one has

$$Df(x_0)(h) = 2\langle x_0, h \rangle .$$

Example 2.2.5. In applications, we very often have real-valued functions on function spaces; then one likes to use the term *functionals*. These functionals are often given by an integral over a kernel K. Let us consider a simple case.

Let E_1 be the Banach space of continuous real functions on a compact interval $I = [a, b]$, with the norm

$$\|x\|_I = \sup_{t \in I} |x(t)| .$$

Let $K : \mathrm{I\!R}^2 \to \mathrm{I\!R}$ be a twice continuously differentiable function. The following is then valid: for all

$$x \in E_1 = (\mathcal{C}(I), \| \cdot \|_I)$$

the function

$$f(x) = \int_I K(t, x(t)) \, dt$$

is a well-defined real function on E_1. We would like to show that f is Fréchet-differentiable and to determine the Fréchet derivative. Let $x \in E_1$ be given and let $h \in E_1$ fulfil the condition $\|h\|_I < r$. Consider the expression

$$f(x + h) - f(x) = \int_I [K(t, x(t) + h(t)) - K(t, x(t))] \, dt .$$

Taylor's formula states

$$K(t, x(t) + h(t)) - K(t, x(t)) = K_x(t, x(t))h(t) + \int_0^1 (1 - \tau) K_{xx}(t, x(t) + \tau h(t)) h(t)^2 d\tau .$$

It follows that

$$f(x+h)-f(x)=\int_I K_x(t,x(t))h(t)\,dt+o(x,h)\,,$$

$$o(x,h)=\int_I\int_0^1(1-\tau)K_{xx}(t,x(t)+\tau h(t))h(t)^2dt\,dt\,,$$

and thus

$$|o(x,h)|\le C_K\|h\|_I^2$$

where

$$C_K=|I|\sup_{t\in I}\sup_{|y|\le r}|K_{xx}(t,x(t)+y)|<+\infty\,;$$

thus

$$\lim_{h\to 0}\frac{o(x,h)}{\|h\|_1}=0\,.$$

Consequently, f is differentiable with the derivative

$$(Df)(x)(h)=\int_I K_x(t,x(t))h(t)\,dt\,,\quad \forall h\in E_1\,.$$

In physics books particularly, one likes to speak of "variational derivative" in this context, and one writes

$$Df(x)(h)=\int_I\frac{\delta f(x)}{\delta x(t)}h(t)\,dt\,;$$

this means that

$$\frac{\delta f(x)}{\delta x(t)}=K_x(t,x(t))\text{ is the \textit{variational derivative of the functional} }f\,.$$

Example 2.2.6. In applications, especially to physics (see the treatment of the Hamiltonian variational principle below), it is often necessary to investigate the differentiability of functions of a form given below.

First, E_1 is the Banach space of all continuously differentiable real functions on the compact interval $I=[a,b]$ with the norm

$$\|x(t)\|_{I,1}=\sup_{t\in I}\sup_{0\le\nu\le 1}|x^{(\nu)}(t)|\,,$$

$$E_1(\mathcal{C}_{\mathbb{R}}(I),\|\cdot\|_{I,1})\,,$$

where $x^{(\nu)}$ is the νth derivative of x. Then we are given a twice continuously differentiable real function $F:\mathbb{R}^3\to\mathbb{R}$, and we want to investigate the differentiability of the function

$$f:E_1\to\mathbb{R}\,,$$

which is defined by

$$f(x) = \int_I F(t, x(t), \dot{x}(t))\, dt .$$

We would like to show that under these conditions f is differentiable at all points $x_0 \in E_1$, and to calculate the Fréchet derivative.

To this end, let $x_0 \in E_1$ be fixed; we want to investigate $f(x_0+h) - f(x_0)$ for all $h \in E_1$, $\|h\|_{I,1} < \varepsilon$.

Taylor's formula states for F that

$$\begin{aligned} F(t, x_0(t) + h(t), \dot{x}_0(t) + \dot{h}_0(t)) \\ &= F(t, x_0(t), \dot{x}_0(t)) + h(t) F_x(t, x_0(t), \dot{x}_0(t)) \\ &\quad + \dot{h}(t) F_{\dot{x}}(t, x_0(t), \dot{x}_0(t)) + R(x_0, t), \quad \forall t \in I, \end{aligned}$$

with the remainder

$$\begin{aligned} R(x_0, t) = \int_0^1 (1-\tau)[\widehat{F}_{xx}(t,\tau)(h(t))^2 + 2\widehat{F}_{x\dot{x}}(t,\tau) h(t) \dot{h}(t) \\ + \widehat{F}_{\dot{x}\dot{x}}(t,\tau)(\dot{h}(t))^2]\, dt , \end{aligned}$$

where we have used the abbreviation

$$\widehat{F}_{xx}(t,\tau) = F_{xx}(t, x_0(t) + \tau h(t), \dot{x}_0(t) + \tau \dot{h}(t))$$

with similar expression for $\widehat{F}_{\dot{x}\dot{x}}$ and $F_{x\dot{x}}$. It follows that

$$\begin{aligned} f(x_0 + h) - f(x_0) = \int_I [F_x(t, x_0(t), \dot{x}_0(t)) h(t) \\ + F_{\dot{x}}(t, x_0(t), \dot{x}_0(t)) \dot{h}(t)]\, dt + o(x_0, h) \end{aligned}$$

where

$$o(x_0, h) = \int_I R(x_0, t)\, dt .$$

A rough estimate shows that

$$|o(x_0), h| \le C_F \|h\|_{I,1}^2$$

with

$$\begin{aligned} C_F = |I| \sup_{\substack{|y_1| \le \varepsilon \\ |y_2| \le \varepsilon}} \sup_{t \in I} [|F_{xx}(t, x_0(t) + y_1, \dot{x}_0(t) + y_2)| \\ + 2|F_{x\dot{x}}(t, x_0(t) + y_1, \dot{x}_0(t) + y_2)| + |F_{\dot{x}\dot{x}}(t, x_0(t) + y_1, \dot{x}_0(t) + y_2)|] . \end{aligned}$$

Therefore f is differentiable at x_0 with the derivative

$$Df(x_0) \in \mathcal{L}(E_1, \mathbb{R}) ,$$

given by

$$Df(x_0)(h) = \int_I [F_x(t, x_0(t), \dot{x}_0(t))h(t) + F_{\dot{x}}(t, x_0(t), \dot{x}_0(t))\dot{h}(t)]\, dt\,.$$

If the second term of the integrand can be integrated by parts, one gets

$$Df(x_0)(h) = h(t)F_{\dot{x}}(t, x_0(t), \dot{x}_0(t))|_a^b$$
$$+ \int_a^b [F_x(t, x_0(t), \dot{x}_0(t)) - \frac{d}{dt}F_{\dot{x}}(t, x_0(t), \dot{x}_0(t))]h(t)\, dt\,.$$

If we now restrict f to a subset $M \subset E_1$,

$$M = \{x \in E_1 | x(a) = x_1, x(b) = x_2\}\,,$$

in which the function $x : I \to \mathbb{R}$ has specified values at the initial and end points of the interval $I = [a, b]$, then we obtain for the Fréchet derivative of the function $f : M \to \mathbb{R}$ at a point $x_0 \in M$:

$$Df(x_0)(h) = \int_a^b [F_x(t, x_0(t), \dot{x}_0(t)) - \frac{d}{dt}F_{\dot{x}}(t, x_0(t), \dot{x}_0(t))]h(t)\, dt\,,$$

since it follows from $x_0 \in M$ and $x_0 + h \in M$ that $h(a) = h(b) = 0$.

2.2.1 Higher Derivatives

Now we come to a discussion of the higher derivatives of a function $f : M \to E_2$, where $M \subset E_1$ is open and E_1 and E_2 are Banach spaces. By Definition 2.2.1, the derivative Df of a continuously differentiable function $f : M \to E_2$ is a continuous mapping from M into $\mathcal{L}(E_1, E_2)$, which is the Banach space of all continous linear mappings from E_1 into E_2. Thus, we are again in the situation mentioned in Definition 2.2.1, where now the Banach space $\mathcal{L}(E_1, E_2)$ takes on the role of E_2. Correspondingly, we say that a function $f : M \to E_2$ is *twice differentiable* if the derivative

$$Df : M \to \mathcal{L}(E_1, E_2)$$

is differentiable in the sense of Definition 2.2.1 (i.e. at a point, on M, continuously), and we speak of the Fréchet derivative of the function Df as the second derivative of f:

$$D^2 f = D(Df)\,.$$

By virtue of this definition, the second derivative $D^2 f$ of f is a function

$$D^2 f : M \to \mathcal{L}(E_1, \mathcal{L}(E_1, E_2))\,.$$

that is, a mapping from M into the Banach space of the continuous linear mappings from E_1 into the Banach space $\mathcal{L}(E_1, E_2)$.

Now the Banach space $\mathcal{L}(E_1, \mathcal{L}(E_1, E_2))$ is naturally isomorphic to the Banach space $\mathcal{L}(E_1 \times E_1, E_2) \equiv \mathcal{L}(E_1^{\times 2}, E_2)$ of the continuous bilinear mappings of the Banach space $E_1^{\times 2} = E_1 \times E_1$ into the Banach space E_2. This isomorphism can be defined as follows.

For $A \in \mathcal{L}(E_1, \mathcal{L}(E_1, E_2))$, let $\widetilde{A} \in \mathcal{L}(E_1^{\times 2}, E_2)$ be defined by

$$\widetilde{A}(x_1, x_2) = A(x_2)(x_1)\,, \quad x_j \in E_1\,.$$

For $B \in \mathcal{L}(E_1^{\times 2}, E_2)$ define $\check{B} \in \mathcal{L}(E_1, \mathcal{L}(E_1, E_2))$ by

$$\check{B}(x_2) = B(\cdot, x_2)\,,$$

$$\check{B}(x_2)(x_1) = B(x_1, x_2)\,.$$

It follows that $\widetilde{\check{B}} = B$. Thus $A \mapsto \widetilde{A}$ is bijective and $B \to \check{B}$ is the corresponding inverse mapping. It can be shown (as an exercise) that $A \mapsto \widetilde{A}$ is an isometric isomorphism of the Banach spaces above.

Thus, for the second derivative of a function $f : M \to E_2$, where $M \subset E_1$, we have

$$D^2 f : M \to \mathcal{L}(E_1^{\times 2}, E_2)\,.$$

The nth derivative is defined correspondingly as the derivative of the $(n-1)$th derivative:

$$D^n f \equiv D(D^{n-1} f) \in \mathcal{L}(E_1; \mathcal{L}(E_1^{\times(n-1)}, E_2))$$

where

$$D^{n-1} f : M \to \mathcal{L}(E_1^{\times(n-1)}, E_2)\,.$$

The natural isomorphism follows here too:

$$\mathcal{L}(E_1, \mathcal{L}(E_1^{\times(n-1)}, E_2)) \cong \mathcal{L}(E_1^{\times n}, E_2)$$

and therefore

$$D^n f : M \to \mathcal{L}(E_1^{\times n}, E_2)\,.$$

In other words, the *nth derivative* $D^n f$ of a function $f : M \to E_2$ is a mapping of M into the Banach space of the continuous n-linear functions from E_1 into E_2.

2.2.2 Some Properties of Fréchet Derivatives

We will conclude our discussion of the Fréchet derivative with a short look at their most important properties.

(1) The Fréchet derivative is a *linear* function, i.e. if $f_j : M \to E_2$, $j = 1, 2$, where $M \subset E_1$ is open, are two (continuously) differentiable functions on M and if E_2 is a $\mathbb{K}(= \mathbb{R}$ or $\mathbb{C})$ Banach space, then for arbitrary $\lambda_j \in \mathbb{K}$, $\lambda_1 f_1 + \lambda_2 f_2$ is also a (continuously) differentiable function on M, and

$$D(\lambda_1 f_1 + \lambda_2 f_2) = \lambda_1 D f_1 + \lambda_2 D f_2\,.$$

The proof is simple. Let $x \in M$, $r > 0$, such that $x + B_r \subset M$, with

$$B_r = \{y \in E_1 | \, \|y\|_1 < r\}\,.$$

Then, $\forall h \in B_r$,

$$f_j(x+h) - f_j(x) = Df_j(x)(h) + o_j(x,h)\,,$$

and thus

$$\begin{aligned}(\lambda_1 f_1 + \lambda_2 f_2)&(x+h) - (\lambda_1 f_1 + \lambda_2 f_2)(x)\\ &= \lambda_1(f_1(x+h) - f_1(x)) + \lambda_2(f_2(x+h) - f_2(x))\\ &= \lambda_1 Df_1(x)(h) + \lambda_1 o_1(x,h) + \lambda_2 Df_2(x)(h) + \lambda_2 o_2(x,h)\,.\end{aligned}$$

Now o_1 and o_2 have the property (2.2.2), and thus so does $h \mapsto \lambda_1 o_1(x,h) + \lambda_2 o_2(x,h)$; the differentiability of $\lambda_1 f_1 + \lambda_2 f_2$ at the point x therefore follows, and the value of the derivative is

$$D(\lambda_1 f_1 + \lambda_2 f_2)(x) = \lambda_1 Df_1(x) + \lambda_2 Df_2(x)\,.$$

(2) The Fréchet derivative fulfils the *chain rule*: let E_1, E_2 and E_3 be three Banach spaces, let $U_j \subset E_j$ be open, nonempty sets, and let $f : U_1 \to U_2$ and $g : U_2 \to U_3$ be (continuously) differentiable. Then

$$g \circ f : U_1 \to U_3$$

is also (continuously) differentiable, and for all $x \in U_1$:

$$D(g \circ f)(x) = Dg(f(x)) \circ Df(x)\,.$$

Proof. "f is differentiable at x" means

$$f(x+h) - f(x) = Df(x)(h) + o_1(x,h)\,, \quad \forall h \in B^1_{r_1}\,, \quad x + B^1_{r_1} \subset U_1$$

and correspondingly, since g is differentiable at $y = f(x) \in U_2$,

$$g(y+k) - g(y) = Dg(y)(k) + o_2(y,k)\,, \quad \forall k \in B^2_{r_2}\,, \quad y + B^2_{r_2} \subset U_2\,.$$

Since f, particularly, is continuous, then for given $r_2 > 0$, we can choose $r_1 > 0$ in such a way that the first expression holds and also $f(B^1_{r_1}) \subseteq B^2_{r_2}$. Then for all $h \in B^1_{r_1}$ we get

$$\begin{aligned}g \circ f(x+h) - g \circ f(x) &= g[f(x+h)] - g[f(x)]\\ &= g[f(x) + Df(x)(h) + o_1(x,h)] - g[f(x)]\\ &= (Dg)(f(x)) \circ Df(x)(h) + o_1(x,h))\\ &\quad + o_2(f(x), Df(x)(h) + o_1(x,h))\\ &= (Dg)(f(x)) \circ Df(x)(h) + (Dg)(f(x) \circ o_1(x,h))\\ &\quad + o_2(f(x), Df(x)(h) + o_1(x,h))\,.\end{aligned}$$

A simple calculation shows that

$$o(x,h) = Dg(f(x)) \circ o_1(x,h) + o_2(f(x), Df(x)(h) + o_1(x,h))$$

does indeed have the property (2.2.2), since o_1 and o_2 have this property, and since $Dg(f(x))$ and $Df(x)$ are continuous linear mappings.

Thus, the equation above proves the differentiability of $g \circ f$ and determines the value of the derivative to be

$$D(g \circ f)(x) = Dg(f(x)) \circ Df(x) \in \mathcal{L}(E_1, E_3)\,.$$

The remaining statements now follow without difficulty. □

Remark 2.2.7. The notation for the Fréchet derivative has been so chosen that, with a little care, the classical rules of differentiation can be applied.

2.3 The Gâteaux Derivative

The generalisation of the concept of partial derivative of a function $f : \mathbb{R}^n \to \mathbb{R}$ leads to the concept of the Gâteaux differential $\delta f(x_0, h)$, just as the generalisation of the total differential led to the concept of the Fréchet derivative.

Definition 2.3.1. Let $(E_1, \|\cdot\|_1)$ and $(E_2, \|\cdot\|_2)$ be two normed spaces and $f : E_1 \to E_2$ some mapping from E_1 into E_2. If for some point $x_0 \in E_1$ there exists a mapping $\delta f(x_0, \cdot) : E_1 \to E_2$ such that

$$\lim_{t\to 0} \left\| \frac{1}{t}(f(x_0 + th) - f(x_0)) - \delta f(x_0, h) \right\|_2 = 0\,, \quad \forall h \in E_1$$

then one calls $\delta f(x_0, h)$ the *Gâteaux-differential of f at the point x_0 in the direction h.* Note that the mapping $\delta f(x_0, \cdot)$ needs to be neither continuous nor linear.

The Gâteaux differential $\delta f(x_0, h)$ is a homogeneous mapping, in other words,

$$\delta f(x_0, \lambda h) = \lambda \delta f(x_0, h)\,, \forall \lambda \in \mathbb{R}\,.$$

If $\delta f(x_0, h)$ is, moreover, linear and continuous in h, then one writes

$$\delta f(x_0, h) \equiv \delta_{x_0} f(h)$$

and one refers to $\delta_{x_0} f$ as the *Gâteaux derivative of f at the point x_0*. If f is a functional, i.e. $f : E_1 \to \mathbb{R}$, then $\delta_{x_0} f$ is an element of the dual space E_1'.

We have already introduced the concept of the Fréchet derivative $D_{x_0} f$. In finite-dimensional spaces the Fréchet derivatives is often known as the total derivative, and the existence of the total differential follows from the existence of all partial derivatives. The next lemmas show the corresponding relationship in the general case between Gâteaux and Fréchet derivatives.

Lemma 2.3.2. *Let $f : E_1 \to E_2$. If the Gâteaux derivative $\delta_{x_0} f$ exists in a neighborhood U of x_0 and is continuous, then*

$$\delta_{x_0} f = D_{x_0} f .$$

Proof. We put $\omega(x_0, h) = f(x_0 + h) - f(x_0) - \delta_{x_0} f(h)$. There exists for all $y' \in E_2'$ a $\tau \in [0,1]$ such that

$$\langle \omega(x_0, h), y' \rangle = \langle (\delta_{x_0 + \tau h}(f) - \delta_{x_0}(f))(h), y' \rangle .$$

It thus follows that

$$\|\omega(x_0, h)\|_2 \leq \|\delta_{x_0 + \tau h} f - \delta_{x_0} f\| \; \|h\|_1 .$$

Therefore

$$0 \leq \frac{\|\omega(x_0, h)\|_2}{\|h\|_1} \leq \|\delta_{x_0 + \tau h} f - \delta_{x_0} f\| ,$$

and the right-hand side tends towards zero for $\|h\|_1 \to 0$, since the operator $\delta_x f$ has been assumed to be continuous. In other words, a continuous Gâteaux derivative is automatically a Fréchet derivative. □

Lemma 2.3.3. *If f is Fréchet-differentiable at the point x_0, then f is also Gâteaux-differentiable and both differentials are equal.*

Proof. It follows from the definition of the Fréchet differential, for $t \in \mathbb{R}$, $t \neq 0$, that

$$\frac{f(x_0 + th) - f(x_0)}{t} - \frac{(D_{x_0} f)(th)}{t} = \frac{o(x_0, th)}{t} .$$

Since $D_{x_0} f$ is a linear operator it also follows that

$$\lim_{t \to 0} \frac{f(x_0 + th) - f(x_0)}{t} = (D_{x_0} f)(h) .$$

Thus f is Gâteaux-differentiable and

$$\delta_{x_0} f(h) = D_{x_0} f(h)$$

□

Just as in the finite-dimensional case, it is possible to draw some conclusions about the continuity of f; for example, the next lemma holds.

Lemma 2.3.4. *If f is Gâteaux-differentiable at the point x_0, then f is continuous at the point x_0 in every direction h, i.e.*

$$\lim_{t \to 0} \|f(x_0 - th) - f(x_0)\| = 0 .$$

Remark 2.3.1. The following rules hold for $\delta_x f$:

$$\delta_x(\lambda f) = \lambda \delta_x f\,, \quad \forall \lambda \in \mathbb{R}\,,$$

$$\delta_x(f_1 + f_2) = \delta_x f_1 + \delta_x f_2\,,$$

$$\delta_x(g \circ f) = \delta_y g \circ \delta_x f\,, \quad y = f(x)\,,$$

where, in the chain rule for the mappings f, g, we have assumed

$$E_1 \xrightarrow{f} E_2 \xrightarrow{g} E_3\,.$$

Example 2.3.2. As a last example we want to prove the differentiability of the functional

$$F(f) = \|f\|_p^p = \int_{\mathbb{R}^n} |f(x)|^p d^n x$$

on $M = L^p(\mathbb{R}^n)$, $1 < p < 2$. To this end we show that the Gâteaux derivative of F at all points $f \in L^p(\mathbb{R}^n)$ exists, and that it is a continuous linear functional on $L^p(\mathbb{R}^n)$. The differentiability of F then follows with the aid of Lemma 2.3.2.

Let $h, f \in L^p(\mathbb{R}^n)$, $f \neq 0$, be given. We have to consider

$$\frac{1}{t}\{F(f + th) - F(f)\}$$

for $t \neq 0$. Let us look at the real function φ,

$$\varphi(\xi) = (|1 + \xi|^p - 1 - p\xi)|\xi|^{-p}\,.$$

The boundedness of φ follows from the fact that $\lim_{|\xi| \to \infty} \varphi(\xi) = 1$ and $\lim_{\xi \to 0} \varphi(\xi) = 0$; therefore

$$C_1|\xi|^p \leq |1 + \xi|^p - 1 - p\xi \leq C_2|\xi|^p\,.$$

Substituting $\xi = t(h(x)/f(x))$ in this inequality at all points $x \in \mathbb{R}^n$ with $f(x) \neq 0$, and multiplying by $|f(x)|^p$, we obtain

$$\begin{aligned} C_1|th(x)|^p &\leq |f(x) + th(x)|^p - |f(x)^p - p\,th(x)|f(x)|^{p-1}\operatorname{sign} f(x) \\ &\leq C_2|th(x)|^p\,. \end{aligned}$$

Integration of the inequality gives

$$\begin{aligned} C_1 t^p F(h) &\leq F(f + th) - F(f) - pt \int h(x)|f(x)|^{p-1}\operatorname{sign} f(x) d^n x \\ &\leq C_2 t^p F(h)\,. \end{aligned}$$

Dividing by t, it follows that

$$\lim_{t \to 0} \frac{1}{t}(F(f + th) - F(f)) = \delta F(f, h) \equiv p \int h(x)|f(x)|^{p-1}\operatorname{sign} f(x) d^n x\,.$$

Since

$$|f(\cdot)|^{p-1}\operatorname{sign} f(\cdot) \in L^{p'}(\mathbb{R}^n), \quad \frac{1}{p'} + \frac{1}{p} = 1,$$

then, on the basis of Hölder's inequality, $h \mapsto \delta F(f, h)$ is a continuous linear functional on $L^p(\mathbb{R}^n)$. It follows that

$$DF_f(h) = p \int_{\mathbb{R}^n} |f(x)|^{p-1}\operatorname{sign} f(x) h(x) d^n x,$$

since $f \mapsto |f(\cdot)|^{p-1}\operatorname{sign} f$ is a continuous map $L^p \to L^{p'}$.

2.4 *n*th Variation

A convenient method of investigating a real fuction f on a Banach space E_1 consists in studying the functions

$$F_h(t) = f(u_0 + th)$$

of a real variable t for arbitrary, but fixed, $u_0, h \in E_1$. Information about the behaviour of $F_h(t)$ in the neighborhood of $t = 0$ can be obtained from Taylor's theorem:

$$F_h(t) = F_h(0) + \sum_{n=1}^{N} \frac{t^n}{n!} F_h^{(n)}(0) + R_N, \quad \forall t \in (-t_0, t_0), \quad t_0 > 0.$$

If F_h is N times differentiable on $(-t_0, +t_0)$, then the following holds for the remainder R_N:

$$R_N = \frac{t^N}{N!}\left[F_n^{(N)}(\delta(t,h)t) - F_n^{(N)}(0)\right] = o(t^N), \quad 0 < \delta < 1,$$

i.e.

$$\frac{R_N}{t^N} \to 0 \text{ for } t \to 0.$$

Definition 2.4.1. Let $f : M \subset E_1 \to \mathbb{R}$ be given, where M is open. The nth *variation of f at the point $x_0 \in M$ in the direction $h \in E_1$* is, by definition, equal to

$$\Delta^n f(x_0; h) = F_h^{(n)}(0) \equiv \left.\frac{d^n f(u_0 + th)}{dt^n}\right|_{t=0},$$

if this derivative exists.

The next lemma describes the relationship between the first variation and the Gâteaux derivative.

Lemma 2.4.2. *Let E_1 be a Banach space, M an open subset and f a real function on M. Then the Gâteaux derivative $\delta_{x_0} f$ of f at a point $x_0 \in M$ exists if and only if the first variation $\Delta f(x_0; h)$ exists for all $h \in E_1$ and if $h \to \Delta f(x_0; h)$ is a continuous linear functional on E_1. In this case,*

$$\Delta f(x_0; h) = \delta_{x_0} f(h) .$$

Proof. It is sufficient to compare Definitions 2.4.1 and 2.3.1. □

Remark 2.4.1. If f possesses, for example, a local minimum at x_0, that is

$$f(x) \geq f(x_0) , \quad \forall x \in U(x_0) ,$$

then obviously F_h too has a local minimum at $t = 0$, that is

$$F_h'(0) = 0 , \quad F_h''(0) \geq 0 ,$$

as long as these derivatives exist. These considerations show that investigation of $F_h^{(n)}(0)$, $h \in E_1$, can play a decisive role in the study of extreme value problems. This justifies our introduction of the concept of the nth variation.

Remark 2.4.2. We have seen that the following chain of implications exists:

Existence of the Fréchet derivative ⇒ Existence of the Gâteaux derivative ⇒ Existence of the first variation.

In many applications, the existence of the nth variation can often be more conveniently verified than, for example, the existence of the nth Fréchet derivative. It is therefore an advantage if one has conditions for local extrema which can be expressed in terms of the nth variation (see Chap. 3).

Remark 2.4.3. There are more than 20 concepts of differentiability on topological spaces. For a discussion and classification of all these concepts, see, for example, the work of V. Averbukh and O. Smolanov [2.2].

In applications, it often happens that the domain M of the functional f being considered is neither a subset nor an open neighborhood. In this situation, the class of curves $x(t)$ for which one is investigating the functional in the form $f(x(t))$ must be restricted correspondingly. For example, if M is convex, one normally chooses $x(t) = tx_0 + (1-t)x_1$ where $x_0, x_1 \in M$ and $t \in [0,1]$.

2.5 The Assumptions of the Fundamental Theorem of Variational Calculus

In Chap. 1 we have already given criteria which tell us when the assumptions of the fundamental theorem of variational calculus are fulfilled. With the help of differential calculus, we are now in a position to derive new criteria.

We shall discuss a sufficient condition for the weak lower semicontinuity of a functional $f : x \to \mathbb{R}$. Let us assume that the equation

$$f(x) - f(x_0) - Df_{x_0}(x - x_0) \geq 0$$

holds $\forall (x - x_0) \in \overline{B}_r = \{y \in X \mid \|y\| \leq r\}$. Clearly, this implies for any sequence $\{x_n\}_{n\in\mathbb{N}} \subset x_0 + \overline{B}_r$ which converges weakly to x_0

$$f(x_n) - f(x_0) \geq f'(x_0)(x_n - x_0).$$

It follows that

$$\liminf f(x_n) \geq f(x_0).$$

In other words, f is weakly l.s.c.. We can now show that the inequality asserted above follows from the condition

$$D^2 f(x)(h,h) \geq 0.$$

Indeed, for suitable τ and $\tau' \in [0,1]$, we have

$$\begin{aligned} f(x) - f(x_0) &= Df(x_0 + \tau(x - x_0))(x - x_0) \\ &= Df(x_0)(x - x_0) + [Df(x_0 + \tau(x - x_0))(x - x_0) \\ &\quad - Df(x_0)(x - x_0)] \\ &= Df(x_0)(x - x_0) + \tau D^2 f(x_0 + \tau'(x - x_0))(x - x_0, x - x_0). \end{aligned}$$

This is shown by the following lemma.

Lemma 2.5.1. *Let X be a Banach space and $f : X \to \mathbb{R}$ a C^2 functional on $\overline{B}_r = \{x \in X \mid \|x\| \leq r\}$, such that*

$$D^2 f(x)(h,h) \geq 0, \quad \forall x \in \overline{B}_r, \quad \forall h \in X.$$

Then f is weakly l.s.c. *on $\overline{B}_r$.*

With the aid of differential calculus, one can also find coerciveness criteria. For example, the following holds.

Lemma 2.5.2. *Let f be a C^1 functional on a Banach space X. There exists a continuous function $g : (0,\infty) \to \mathbb{R}$ with*

(i) $\int_{r_0}^{\infty} \frac{g(t)}{t}\, dt = +\infty$ *for some* $r_0 > 0$,

(ii) $\langle f'(x), x \rangle = (Df)(x)(x) \geq g(\|x\|)$.

Furthermore, let f be bounded from below on $S_{r_0}(X) = \{x \in X \mid \|x\| = r_0\}$. Then f is coercive on X.

Proof. The following holds uniformly in $y \in X$, $\|y\| = 1$, for $s > r_0$:

$$f(sy) - f(r_0 y) = \int_{r_0}^{s} \langle f'(ty); y \rangle dt \geq \int_{r_0}^{s} dt \frac{g(t)}{t} .$$

It follows for $\|x\| \to +\infty$, taking (i) into account, that

$$f(x) \geq f\Big(r_0 \frac{1}{\|x\|} x\Big) + \int_{r_0}^{\|x\|} dt \frac{g(t)}{t} \underset{\|x\| \to +\infty}{\longrightarrow} +\infty .$$

Thus f is coercive on X. □

Next we note a result for weak l.s.c. functionals which corresponds to Rolle's theorem in classical analysis. The following holds:

Lemma 2.5.3. *Let X be a reflexive Banach space and $M \subset X$ a bounded open subset. Let $\overline{M}^w$ denote the weak closure of M and $\partial M = \overline{M}^w \setminus M$ its boundary. Then, let $f : X \to \mathbb{R}$ be a weakly* l.s.c. *functional on $\overline{M}^w$ for which*

$$f(x) \geq f(x_0), \quad \forall x \in \partial M \quad \textit{and for some } x_0 \in M .$$

Then f has a critical point in M.

Proof. It follows from the fundamental theorem of variational calculus that f attains its minimum on $\overline{M}^w$, and one can conclude from the fact that $f(x) \geq f(x_0)$ that x_0 is a minimising point of f. Thus, it must be true that $f'(x_0) = 0$, which means that x_0 is a critical point of f. □

2.6 Convexity of f and Monotonicity of f'

Let X be a real Banach space, and $T : X \to X'$ a mapping from X into the dual space X' that satisfies the following inequality for all $x, y \in X$:

$$(Tx - Ty)(x - y) \geq 0 .$$

An operator which has this property is called *monotone*. (If "$>$" holds for $x \neq y$ instead of "$\geq$", the operator is called *strictly monotone*.)

Now we want to prove a criterion for the convexity of a functional

$$f : X \to \mathbb{R} .$$

Theorem 2.6.1. *Let X be a real Banach space and f a real C^1 function on X. The following assertions are equivalent:*

(i) *f is (strictly) convex on X;*
(ii) *f' is (strictly) monotone on X.*

Proof. Let f be convex. For all x and y in X and for λ, $0 < \lambda \leq 1$, the following holds:

$$f(y + \lambda(x - y)) - f(y) \leq \lambda[f(x) - f(y)].$$

It follows that

$$\langle f'_y, x - y\rangle \leq f(x) - f(y)$$

and analogously

$$\langle f'_x, y - x\rangle \leq f(y) - f(x).$$

By adding these two inequalities we obtain

$$\langle f'_y - f'_x, x - y\rangle \leq 0,$$

and thus f' is monotone.

Now let f' be monotone. For x and y in X, consider the function

$$p(\lambda) = f(\lambda x + (1 - \lambda)y) - \lambda f(x) - (1 - \lambda)f(y)$$

with $\lambda \in [0, 1]$. To show that f is convex, it is sufficient to show that $p(\lambda) \leq 0$ on $[0, 1]$. If $p(\lambda)$ were neither negative nor zero, then p would have a maximum, and there would exist a point $\lambda_0 \in (0, 1)$ with $p'(\lambda_0) = 0$.

Let $\lambda \in [0, 1]$ with $\lambda_0 < \lambda$. Then

$$\begin{aligned}
p'(\lambda) - p'(\lambda_0) &= \langle f'(\lambda x + (1 - \lambda)y) - f'(\lambda_0 x + (1 - \lambda_0)), x - y\rangle \\
&= (\lambda - \lambda_0)^{-1}\langle f'(\lambda x + (1 - \lambda)y) \\
&\quad - f'(\lambda_0 x + (1 - \lambda_0)y), \{\lambda x + (1 - \lambda)y\} - \{\lambda_0 x + (1 - \lambda_0)y\}\rangle \\
&\geq 0.
\end{aligned}$$

In other words, p cannot decrease for $\lambda > \lambda_0$. It follows that $p(\lambda_0) \leq p(1) = 0$ and $p(\lambda) \leq 0$, $\forall\lambda \in [0, 1]$. This means f is convex. □

3. Extrema of Differentiable Functions

In this chapter we would like to investigate how one can calculate, for a variational problem, a minimising point whose existence and possibly also whose uniqueness (in the sense of Chap. 1) have been established. It is not possible, with the theorems we proved in Chap. 1 for the existence of an extremum of a functional $f : M \to \mathbb{R}$ (where M is an open subset of a Banach space E), to find those points at which this functional attains, for example, its minimum. However, it is possible to do this with the aid of differential calculus in Banach spaces, which we introduced in Chap. 2, taking the approach used in the well-known case of differentiable functions on the real axis.

3.1 Extrema and Critical Values

We shall now introduce, in analogy with the simple case of differentiable functions on the real axis, those concepts which play corresponding roles in variational calculus. Let E be a Banach space with norm $\|\cdot\|$ and let $M \subset E$ be an open non-empty subset. A point $x_0 \in M$ is called a *local (relative) extremal point* of a function $f : M \to \mathbb{R}$ if and only if there exists an open ball $K_r(x_0) = \{x \in E| \; \|x - x_0\| < r\} \subset M$ with centre x_0 such that the following holds:

(i) local maximum: $f(x) \le f(x_0)$, $\forall x \in K_r(x_0)$,
(ii) local minimum: $f(x) \ge f(x_0)$, $\forall x \in K_r(x_0)$.

If it is true that $f(x) > f(x_0)$, $\forall x \in K_r(x_0)$, $x \ne x_0$, one talks about a *strict local minimum.*

Note that it is possible to convert every maximum problem into a minimum problem (and vice versa) by changing f to $-f$ – the function f has a relative maximum at x_0 if and only if $-f$ has a relative minimum at x_0. From now on we shall consider only the case of a minimum, and leave it to the reader to carry over the results to maximum problems.

If the inequality $f(x) \ge f(x_0)$ is true for all $x \in M$ then one talks about a *global minimum.*

We shall make use of the following definitions (although we shall not use them in this very general form until Chap. 4).

Definition 3.1.1. Let $M_j \subset E_j$ be open subsets of Banach spaces E_j, $j = 1, 2$, and let $\varphi : M_1 \to M_2$ be a differentiable mapping of M_1 into M_2. Then:

(i) $x_0 \in M_1$ is called a *critical point* of the mapping φ if and only if the Fréchet derivative $D\varphi(x_0)$ of φ at x_0 is not surjective;

(ii) $x_0 \in M_1$ is called a *regular point* of the mapping φ if and only if $D\varphi(x_0) : E_1 \to E_2$ is surjective.

Remark 3.1.1. This definition is indeed an appropriate generalisation of the equivalent expressions in the simplest situation, in which $E_2 = \mathbb{R}$, since then obviously $x_0 \in M_1$ is a critical (regular) point of the mapping $\varphi : M_1 \to \mathbb{R}$ if and only if $D\varphi(x_0) = 0$ ($D\varphi(x_0) \neq 0$). If $E_1 = \mathbb{R}$, then $x_0 \in (a, b)$ is called a critical point of f if and only if $f'(x_0) = 0$. The graph of the function f has a horizontal tangent at a critical point x_0, which means there is either a local extremum or a point of inflexion at x_0.

Example 3.1.2. Let $E_1 = \mathcal{H}$ be a real Hilbert space and let B be a symmetric, bounded operator on $\mathcal{H}$ with bounded inverses $B^{-1} \in \mathcal{L}(\mathcal{H}, \mathcal{H})$. Moreover, let $W : \mathcal{H} \to \mathbb{R}$ be differentiable. The Fréchet derivative of the function $\varphi : \mathcal{H} \to \mathbb{R}$ with

$$\varphi(x) = \frac{1}{2}\langle x, Bx\rangle - W(x),$$

where $\langle , \rangle$ is the scalar product in $\mathcal{H}$, is given by

$$D\varphi(x) = Bx - DW(x).$$

This means that $x_0 \in \mathcal{H}$ is a critical point of the mapping φ if and only if x_0 is a solution of the equation

$$x_0 = B^{-1}DW(x_0).$$

In the case when the function W is $W(x) = \frac{1}{2}\lambda\langle x, x\rangle = \frac{1}{2}\lambda\|x\|^2$, then $DW(x) = \lambda x$, which means x_0 is a critical point of the mapping $\varphi(x) = \frac{1}{2}\langle x, Bx\rangle - \frac{1}{2}\lambda\langle x, x\rangle$ if and only if it is an eigenvector of the operator B with eigenvalue λ.

3.2 Necessary Conditions for an Extremum

Our aim now is to calculate the extremal points of particular mappings. Theorem 3.2.1 establishes the fact that for differentiable functionals the extremal points are necessarily contained in the set of critical points of the mapping.

Theorem 3.2.1. Necessary Condition of Euler-Lagrange. *Let M be an open subset of a Banach space E and $f : M \to \mathbb{R}$ a real function which is Fréchet-differentiable on M. Then every extremal point of f is a critical point.*

Proof. Let $x_0 \in M$ be an extremal point of f. There exists an $r > 0$ such that the following assertions are true:

(i) $x_0 + B_r(0) \subset M$,
(ii) $f \restriction B_r(x_0)$ is extremal at x_0.

Now let $h \in E$ be arbitrary, $h \neq 0$, and $\|h\|\delta < r$. Then a real function

$$F_h(t) = f(x_0 + th)\,, \quad t \in I_\delta\,.$$

is well defined on $I_\delta = (-\delta, +\delta)$. By the chain rule, this function is differentiable with respect to t, and it has the derivative

$$\frac{d}{dt}F_h(t) = Df(x_0 + th)(h)\,.$$

Since x_0 is an extremal point of f on $B_r(x_0)$, then $t = 0$ is an extremal point of $F_h(t)$ on I_δ. Using a well-known result from elementary analysis, it can be shown that

$$0 = \left.\frac{dF_h(t)}{dt}\right|_{t=0} = Df(x_0)(h)\,.$$

Since $h \in E$ was chosen to be arbitrary, it follows that $Df(x_0) = 0$, i.e. x_0 is a critical point of f. □

Example 3.2.1. In Chap. 2 Example 2.2.6 we calculated the Fréchet derivative of the functional

$$S(q) = \int_I \mathcal{L}(t, q(t), \dot{q}(t))dt\,.$$

If S is restricted to the subset $M \subset E_1$,

$$M = \{q \in E_1 = C^1(I = [a,b], \mathbb{R}) \mid q(a) = x_1, q(b) = x_2\}\,,$$

of functions $q : I \to \mathbb{R}$ taking given values at the boundary points of the interval I, then for the Fréchet derivative one has

$$DS(q_0)(h) = \int_I h(t)\left[\frac{\partial \mathcal{L}}{\partial q}(t, q_0, \dot{q}_0) - \frac{d}{dt}\frac{\partial \mathcal{L}}{\partial \dot{q}}(t, q_0, \dot{q}_0)\right]dt$$

for all $h \in C^1(I, \mathbb{R})$, $h(a) = 0 = h(b)$.

In Chap. 5 we shall show (Du Bois-Reymond lemma) that $DS(q_0)(h) = 0$, $\forall h$, implies

$$\frac{\partial}{\partial q}\mathcal{L}(t, q_0(t), \dot{q}_0(t)) - \frac{d}{dt}\frac{\partial}{\partial \dot{q}}\mathcal{L}(t, q_0(t), \dot{q}_0(t)) = 0\,.$$

Here we recognise the Euler-Lagrange equations from classical mechanics.

Remark 3.2.2. It follows from the calculation above that if f has a local minimum at x_0, then the necessary condition

$$\Delta f(u_0; h) = 0, \quad \forall h \in E$$

holds for the first variation. An equivalent relation holds for the Gâteaux derivative.

Remark 3.2.3. Let us consider two nondiametrically opposed points P and Q on the sphere. Then there are two arcs of the great circle through P and Q which connect these two points. The shorter arc γ_{PQ} is the geodesic line (i.e. the shortest arc between two points on the surface). The longer arc Γ_{PQ}, although a "critical point", is neither the shortest nor the longest connecting path between P and Q. This very simple geometrical example illustrates the fact that the necessary condition of Euler-Lagrange is not sufficient to show that a minimum exists.

Remark 3.2.4. It is often possible with the aid of variational calculus to solve "operator equations" of the form

$$T(x) = 0$$

where T is in general a nonlinear function $T : E \to \mathbb{R}$ on a Banach space E: if a function $f \in \mathcal{C}^1(E, \mathbb{R})$, whose Fréchet derivative is equal to T, $f' = T$, exists, then the solutions of this equation are just the critical points of the function f. Determining the extrema of f and thus particular solutions of $T(x) = 0$ is, in some cases, relatively simple. Moreover, methods, essentially topological, have been developed which often provide at least a lower bound on the number of critical points of f and thus on the number of solutions of $T(x) = 0$. We shall discuss in Chap. 8 a modern version of these methods, which go back to Ljusternik and Schnirelman.

In this context, one says that an "operator" $T : E \to \mathbb{R}$ on a Banach space E, which is the Fréchet derivative of a real $\mathcal{C}^1$ function f on E, is a *potential operator* or a *gradient operator*, f is then known as the *potential* of T. Just as in the case of "operators" on finite-dimensional Banach spaces, we also know of criteria for the general case, saying when an "operator" T is a potential operator. There are integral criteria and differential criteria. These criteria are generalisations of the classical criteria which assure, for example, that a force field on $\mathbb{R}^3$ has a potential (see [3.1]). This is the case if

(a) the work associated with this force field is independent of the path along which the work is calculated (integral criterion),
(b) the curl of the force field F is zero (differential criterion): $\operatorname{curl} F = 0$.

Before we come to discuss the conditions which are sufficient to show the existence of extremal points, let us look at the corresponding necessary conditions for twice continuously Fréchet-differentiable functions $f : M \to \mathbb{R}$. First of all, let us consider a simple form of the Taylor expansion of a $\mathcal{C}^2$ function.

Lemma 3.2.2. *Let E be a Banach space, $M \subset E$ an open subset and $f \in \mathcal{C}^2(M, \mathbb{R})$. Then, for all $x_0 \in M$, there exists an $r = r(x_0) > 0$ such that $B_r(x_0) = \{x \in E | \, \|x - x_0\| < r\} \subset M$, and for all $h \in B_r(0)$*

$$f(x_0 + h) = f(x_0) + Df(x_0)(h) + \frac{1}{2} D^2 f(x_0)(h, h) + R_2(x_0, h)(h, h).$$

Here the function $R_2(x_0, \cdot)$ on B_r fulfils the condition

$$\lim_{\|h\| \to 0} R(x_0, h) = 0$$

in $\mathcal{L}(E \times E, \mathbb{R})$.

Proof. Choose r to be as in the lemma and let $h \in B_r$ be arbitrary but fixed. Then

$$F_h(t) = f(x_0 + th) \in \mathcal{C}^2(I, \mathbb{R}), \quad I = [-1, +1].$$

We want to assume that the Taylor expansion for a twice-differentiable function is known and has the following form:

$$F_h(t_0 + \tau) = F_h(t_0) + f_h'(t_0)\tau + \frac{1}{2!} F_h''(t_0)\tau^2 + r_2(t_0, \tau)\tau^2$$

where

$$r_2(t_0, \tau) = \int_0^1 (1 - \xi)[F_h''(t_0 + \xi\tau) - F_h''(t_0)] d\xi.$$

See [3.2] for the proof. Applying this formula for $t_0 = 0$ and $\tau = 1$, we obtain

$$F_h(1) = F_h(0) + F_h'(0) + \frac{1}{2!} F_h''(0) + r_2(0, 1).$$

We can calculate the derivatives using the chain rule:

$$F_h'(0) = Df(x_0)(h), \quad F_h''(t) = D^2 f(x_0 + th)(h, h).$$

Substituting, we have

$$f(x_0 + h) = f(x_0) + Df(x_0)(h) + \frac{1}{2!} D^2 f(x_0)(h, h) + R_2(x_0, h)(h, h)$$

since

$$\begin{aligned} r_2(0, 1) &= \int_0^1 (1 - \xi)[\{D^2 f(x_0 + \xi h) - D^2 f(x_0)\}(h, h)] d\xi \\ &\equiv R_2(x_0, h)(h, h). \end{aligned}$$

As $D^2 f : M \to \mathcal{L}(E \times E, \mathbb{R})$ is continuous, R_2 indeed has the property $R_2(x_0, h) \to 0$ for $h \to 0$ in $\mathcal{L}(E \times E, \mathbb{R})$. □

The first, simple consequence of this lemma is the following theorem.

Theorem 3.2.3. *Let $M \subset E$ be an open subset of a Banach space E and $f \in \mathcal{C}^2(M, \mathbb{R})$. Then if f has a relative minimum (maximum) at a point $x_0 \in M$, the following holds:*

(i) $Df(x_0) = 0$.
(ii) $D^2 f(x_0) \in \mathcal{L}(E \times E, \mathbb{R})$ *is not negative (not positive), that is* $\forall h \in E$, $D^2 f(x_0)(h,h) \geq 0$ $(D^2 f(x_0)(h,h) \leq 0)$.

Proof. Let $x_0 \in M$ be, for example, a relative minimum. Then there exists an $r > 0$ with the properties

(i) $B_r(x_0) = x_0 + B_r(0) \subseteq M$.
(ii) $f(x_0) \leq f(x)$, $\forall x \in B_r(x_0)$.

Now let us apply Lemma 3.2.2. For all $h \in B_r$,

$$f(x_0) \leq f(x_0 + h) = f(x_0) + Df(x_0)(h) + \frac{1}{2} D^2 f(x_0)(h,h) + R_2(x_0,h)(h,h)$$

holds. By Theorem 3.2.1, $Df(x_0) = 0$. It follows that

$$0 \leq \frac{1}{2} D^2 f(x_0)(h,h) + R_2(x_0,h)(h,h)\,, \forall h \in B_r\,.$$

This is true in particular for all εh where $0 < \varepsilon \leq 1$, $h \in B_r$:

$$0 \leq \frac{1}{2} D^2 f(x_0)(\varepsilon h, \varepsilon h) + R_2(x_0, \varepsilon h)(\varepsilon h, \varepsilon h)\,.$$

Thus

$$0 \leq \varepsilon^2 \left[\frac{1}{2} D^2 f(x_0)(h,h) + R_2(x_0, \varepsilon h)(h,h) \right].$$

This means that

$$0 \leq D^2 f(x_0)(h,h) + 2R_2(x_0, \varepsilon h)(h,h)\,, \forall \varepsilon \in (0,1]\,, \quad \forall h \in B_r\,.$$

If we take into account that $\lim_{\varepsilon \to 0} R_2(x_0, \varepsilon h) = 0$, it follows that

$$0 \leq D^2 f(x_0)(h,h)\,, \forall h \in B_r\,.$$

Now if $h \in E$, it follows for suitable $\lambda > 0$ that $h \in \lambda B_r$; thus

$$0 \leq D^2 f(x_0)\Big(\frac{1}{\lambda} h, \frac{1}{\lambda} h\Big) = \lambda^{-2} D^2 f(x_0)(h,h)\,.$$

Therefore $D^2 f(x_0)(h,h) \geq 0$, $\forall h \in E$, which we wanted to show. □

Remark 3.2.5. If the functional $f : M \to \mathbb{R}$ is not twice Fréchet-differentiable on M, necessary conditions can still be derived, for example with the aid of first and second variation

If $f : M \to \mathbb{R}$ has a local minimum at x_0, then

$$\Delta f(x_0, h) = 0\,, \quad \forall h \in E\,,$$

$$\Delta^2 f(x_0, h) \geq 0\,, \quad \forall h \in E\,,$$

if $\Delta^2 f(x_0, h)$ exists for all $h \in E$. As we have already pointed out, it is often easier to check the existence of the nth variation than the existence of the nth Fréchet derivative.

3.3 Sufficient Conditions for an Extremum

After this short discussion of the necessary conditions we now come to some sufficient conditions. Theorem 3.2.3 almost has an inverse, Theorem 3.3.1.

Theorem 3.3.1. *Let E be a Banach space, $M \subset E$ an open subset and $f \in C^2(M, \mathbb{R})$. Let the following hold for a point $x_0 \in M$:*

(i) $Df(x_0) = 0$,
(ii) $D^2 f(x_0)$ *is strictly positive, that is* $\inf\{D^2 f(x_0)(h,h) \mid h \in E, \|h\| = 1\} > 0$.

The function f then has a strict relative minimum at the point x_0, i.e. $f(x_0) < f(x_0 + h)$, $\forall h \in B_r$, $h \neq 0$, for a suitable $r > 0$.

Proof. For $x_0 \in M$ choose $r_1 > 0$ as in Lemma 3.2.2. Then for all $h \in B_{r_1}$

$$f(x_0 + h) = f(x_0) + \frac{1}{2} D^2 f(x_0)(h,h) + R_2(x_0, h)(h,h) .$$

If we scale with $0 < \varepsilon \leq 1$, then it follows for all $h \in B_{r_1}$ that

$$f(x_0 + \varepsilon h) - f(x_0) = \varepsilon^2 \left[\frac{1}{2} D^2 f(x_0)(h,h) + R_2(x_0, \varepsilon h)(h,h) \right] .$$

For sufficiently small $\varepsilon > 0$, $R_2(x_0, \varepsilon h)(h,h)$ is negligible compared to $\frac{1}{2} D^2 f(x_0)(h,h)$ by virtue of the fact that $\lim_{\|h\| \to 0} R_2(x_0, h) = 0$ and condition (ii). Thus the sign of $f(x_0 + \varepsilon h) - f(x_0)$, $h \in B_{r_1}$, $0 < \varepsilon < 1$, is determined by the sign of $\frac{1}{2} D^2 f(x_0)(h,h)$. The assertion follows. □

Corollary 3.3.2. *By changing f to $-f$, Theorem* 3.3.1 *also contains sufficient conditions for f to have a relative maximum at x_0.*

Remark 3.3.1. It is not difficult to use first and second variation to express equivalent sufficient conditions.

As we have already pointed out, particulary strong assertions hold if f is convex. The following corollary holds.

Corollary 3.3.3. *Let M be an open subset of the Banach space E; moreover, let*

(i) *M be convex,*
(ii) *$f \in C^1(M, \mathbb{R})$ be convex.*

Then every critical point of f is also a minimising point.

Proof. Let x_0 be a critical point of f, so that $Df(x_0) = 0$. Consider the auxiliary function on M

$$g(x) = f(x) - f(x_0).$$

This function is also convex and satisfies $g(x_0) = 0$ and $Dg(x_0) = 0$. If f did not have a relative minimum at x_0, there would exist an $x_1 \in B_{r_1}(x_0) \subset M$ such that

$$g(x_1) = -a^2 < 0.$$

Consider another auxiliary function G on $[0,1]$:

$$G(t) = g(x_0 + t(x_1 - x_0)).$$

G is differentiable on $(0,1)$ by the chain rule, with the derivative

$$G'(t) = Dg(x_0 + t(x_1 - x_0))(x_1 - x_0).$$

It follows that

$$G'(0) = \lim_{t \to 0+} G'(t) = Dg(x_0)(x_0)(x_1 - x_0) = 0.$$

But on the other hand,

$$G'(0) = \lim_{t \to 0+} \frac{G(t) - G(0)}{t} = \lim_{t \to 0+} \frac{1}{t} g(x_0 + t(x_1 - x_0)) \leq -A^2 < 0,$$

since the convexity of g implies that

$$g(x_0 + t(x_1 - x_0)) = g(tx_1 + (1-t)x_0) \leq tg(x_1) + (1-t)g(x_0) = tg(x_1).$$

Thus

$$\frac{1}{t} g(x_0 + t(x_1 - x_0)) \leq g(x_1) = -A^2, \quad \forall t \in (0,1).$$ □

Remark 3.3.2. Let $\mathcal{H}$ be a real Hilbert space and $Q : \mathcal{H} \times \mathcal{H} \to \mathbb{R}$ a symmetric coercive bilinear form, i.e. there is some constant $c > 0$ such that

$$Q(x,x) \geq c\|x\|^2, \quad \forall x \in \mathcal{H}.$$

Then the functional

$$f(x) = Q(x,x) - T(x)$$

has, for every $T \in \mathcal{H}'$, a unique minimum x_0 (Corollary 1.3.4). It follows from Theorem 3.2.1 that

$$Df(x_0)(h) = 2Q(x_0, h) - T(h) = 0,$$

which means

$$2Q(x_0, h) = T(h), \quad \forall h \in \mathcal{H}.$$

In the special case when $Q(x, x) = \langle Ax, x \rangle$ and $T(x) = 2\langle x, u \rangle$, the necessary and sufficient condition for the minimising point x_0 is

$$Ax_0 = u.$$

4. Constrained Minimisation Problems (Method of Lagrange Multipliers)

4.1 Geometrical Interpretation of Constrained Minimisation Problems

In many applications of the calculus of variations, what we want to determine is not just the minimum of a function f on an open set U but the minimum of f subject to certain restrictions on the points $x \in U$. A well-known example from classical mechanics can be used to illustrate this type of problem. We want to determine the minimum of the action functional subject to the subsidiary condition that the motion be on a given surface. The restriction in this case is, therefore, that the points $x \in U$ satisfy an equation of the form $g(x) = 0$, i.e. the equation of the surface.

In the general case too we shall only consider those restrictions on the points $x \in U$, subject to which a function $f : U \to \mathbb{R}$ is to be minimised (maximised), which can be expressed as an equation $g(x) = 0$. We will say that we have a *variational problem with subsidiary conditions*, or a *constrained minimisation (maximisation) problem*.

We formulate the variational problem with subsidiary conditions in a general way suitable for application as follows.

Let E_1 and E_2 be two Banach spaces and $U \subset E_1$ an open nonempty subset. A differentiable mapping $\phi : U \to E_2$, a point $y_0 \in \phi(U)$ and a function $f : U \to \mathbb{R}$ are given. We shall refer to the problem of finding extremal points of the function $f \restriction M : M \to \mathbb{R}$ on the level surface $M = \phi^{-1}(y_0) \subset U$ of the mapping ϕ as *the extremal problem for the function $f : U \to \mathbb{R}$ subject to the constraint $\phi(x) = y_0$.*

In many classical applications U is, say, an open subset of $\mathbb{R}^n$ and ϕ is a differentiable mapping from U into $\mathbb{R}^m$. Then the level surface $M = \phi^{-1}(y_0)$ of ϕ at a point $y_0 \in \phi(U) \subset \mathbb{R}^m$ is the "surface" in U which is defined by

$$\{x \in U | \phi_j(x) - y_{0,j} = 0,\ j = 1, \ldots, m\} = M\,.$$

The extreme value problem for a function $f : U \to \mathbb{R}$ subject to the constraint $\phi(x) = y_0$ then becomes the problem of determining the extremal points of the restriction of f to this "surface" M.

In general M will not be an open subset of E_1 so that the theorems of Chap. 3 are not necessarily applicable. However, the idea behind the solution to the problem is very simple: under suitable assumptions about the mapping ϕ it is possible, with the aid of the implicit function theorem, to choose local coordinates on the level surface $M = \phi^{-1}(y_0)$ such that the results of Chap. 3 are applicable in these coordinates.

In order to illustrate the solution of a variational problem with constraints in geometrical terms and to provide the motivation for what follows, let us first of all consider a very simple case and its heuristic solution.

Let U be an open subset of $\mathbb{R}^2$ and $\phi : U \to \mathbb{R}$ a differentiable mapping. We want to illustrate which conditions are necessary for the variational problem for a differentiable function $f : U \to \mathbb{R}$, subject to the constraint $\phi(x) = y_0 \in \phi(U)$, to be soluble. To this end, let us imagine that the level surface $\phi^{-1}(y_0)$ of ϕ has the following form:

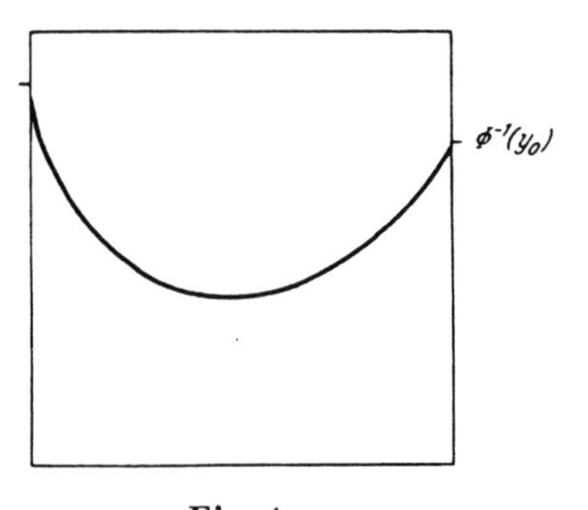

Fig. 4

Further, let us assume that the level surfaces $f^{-1}(c)$ of f in the neighborhood of a point $c_0 \in \mathbb{R}$ depend on c in the following manner (we assume the "monotonicity" $c'' < c' < c_1 < c_0 < c_2$):

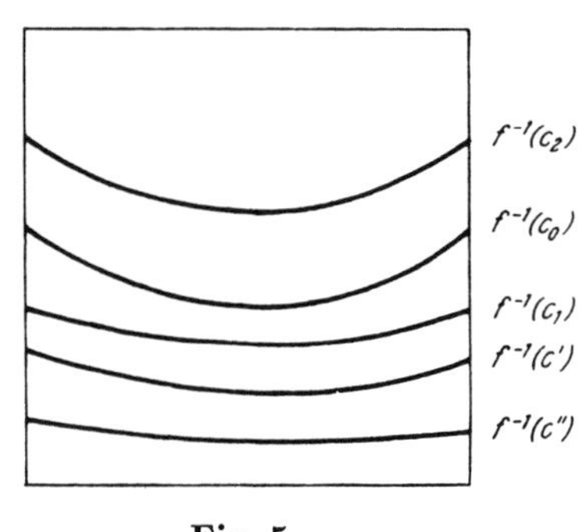

Fig. 5

If we now combine both pictures we can read off an ansatz for the solution:

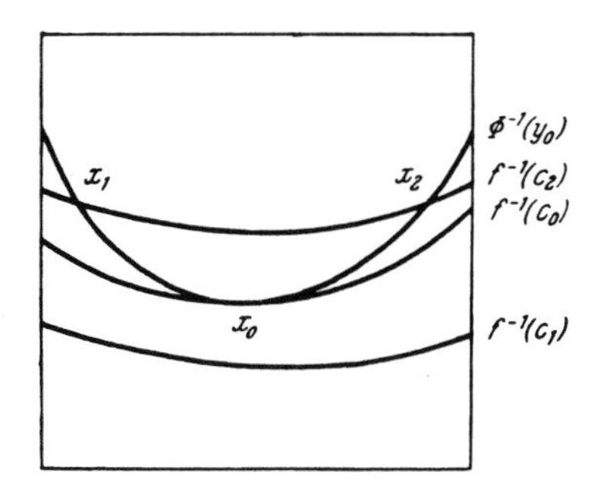

Fig. 6

(1) The subsidiary condition $\phi(x_j) = y_0$ is satisfied at the points x_1 and x_2, but since $c_0 < f(x_j) = c_2$, f is not minimal under the condition $\phi(x) = y_0$.
(2) The level surfaces $\phi^{-1}(y_0)$ and $f^{-1}(c_0)$ touch at a point $x_0 \in U$.
(3) The constraint $\phi(x) = y_0$ is not satisfied at any of the points of the level surface $f^{-1}(c_1)$.

In this example, then, the minimising point is just the point of contact x_0 of the level surfaces $\phi^{-1}(y_0)$ and $f^{-1}(c_0)$, and the value of f at the minimum is $c_0 = f(x_0)$ subject to the constraint $\phi(x) = y_0$.

Now, the "surfaces" $\phi^{-1}(y_0)$ and $f^{-1}(c_0)$ touch at a point x_0 if this point belongs to both surfaces and if the tangents at this point coincide, i.e. if

$$(Df)(x_0) = \lambda(D\phi)(x_0) \qquad (*)$$

for a suitable $\lambda \in \mathbb{R} \cong \mathcal{L}(\mathbb{R}, \mathbb{R})$.

Exercise 4.1.1. Using the implicit function theorem convince yourself that the equation $(*)$ really does imply that the tangents at the point x_0 are equal.

We obtain from these heuristic considerations some hints for the treatment of the general case.

(1) Determine the tangent space $T_{x_0}\phi^{-1}(y_0)$ of the level surfaces of a differentiable mapping $\phi : U \to E_2$, where $U \subset E_1$ is open and E_1 and E_2 are Banach spaces, at a point $x_0 \in U$, $\phi(x_0) = y_0$.

Under certain assumptions, it can be shown for differentiable mappings $\phi : U \to \mathbb{R}^m$, $U \subset \mathbb{R}^n$, that the tangent space of a level surface $\phi^{-1}(y_0)$ at a point $x_0 \in U$, $\phi(x_0) = y_0$, can be described by

$$T_{x_0}\phi^{-1}(y_0) = \{x \in \mathbb{R}^n | \exists h \in \operatorname{Ker} \phi'(x_0) : x = x_0 + h\} = x_0 + \operatorname{Ker} \phi'(x_0).$$

This is also the case here. $\operatorname{Ker} T$, the kernel of T, denotes for a mapping $T : E_1 \to E_2$ those elements of E_1 whose image is the zero element of E_2.

(2) Generalise the condition above for the equality of the tangents to an equivalent relation for the tangent spaces. It will emerge that the appropriate generalisation is

$$T_{x_0}\phi^{-1}(y_0) \subseteq T_{x_0}f^{-1}(x_0).$$

that is with condition (1),

$$\operatorname{Ker}\phi'(x_0) \subseteq \operatorname{Ker} f'(x_0). \qquad (**)$$

Exercise 4.1.2. Convince yourself using simple examples that the equation (**) for tangent spaces is the appropriate generalisation of the equality of the tangents (*) in the very simple example above.

4.2 Ljusternik's Theorems

The following theorems, which provide necessary conditions for a variational problem with constraints, go back to Ljusternik [4.1]. The second theorem contains the necessary conditions we are looking for, while the first theorem gives the above-mentioned characterisation of tangent spaces.

Theorem 4.2.1 *Let E_1 and E_2 be two Banach spaces, $U \subset E_1$ an open nonempty subset and $\phi : U \to E_2$ a differentiable mapping. Then, at a regular point x_0 of the mapping ϕ, at which $\operatorname{Ker}\phi'(x_0)$ has a topological complement in E_1, the set*

$$T_{x_0}\phi^{-1}(y_0) = \{x \in E_1 | \exists h \in \operatorname{Ker}\phi'(x_0) : x = x_0 + h\} = x_0 + \operatorname{Ker}\phi'(x_0)$$

is a "true" tangent space of the level surface $\phi^{-1}(y_0)$ of the mapping ϕ ($y_0 = \phi(x_0)$); that is there exists a homeomorphism χ of a neighborhood U' of x_0 in $T_{x_0}\phi^{-1}(y_0)$ onto a neighborhood V of x_0 in $\phi^{-1}(y_0)$ with the following properties:

(i) $\chi(x_0 + h) = x_0 + h + \varphi(h), \quad \forall x_0 + h \in U'$.
(ii) *φ is continuous and satisfies*

$$\lim_{h\to 0} \frac{1}{\|h\|_1} \|\varphi(h)\|_1 = 0.$$

Remark 4.2.1 We refer the reader to [4.2] for a definition and discussion of the existence of topological complements.

Proof. It is very useful to imagine a two-dimensional version of the problem to illustrate geometrically the following calculations.

(a) $\phi'(x_0)$ is a surjective continuous linear mapping of E_1 onto E_2 at a regular point $x_0 \in U$ of ϕ. The kernel $K = \operatorname{Ker}\phi'(x_0)$ of this mapping is therefore a closed subspace of E_1 and has by assumption a topological

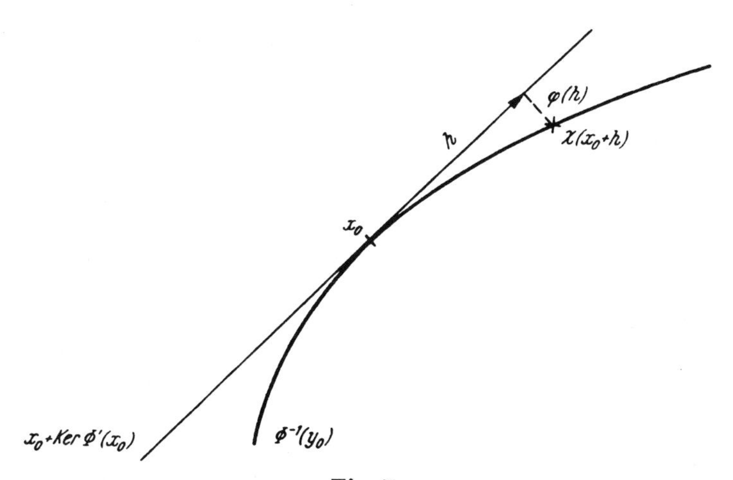

Fig. 7

complement L in E_1, so that the decomposition of E_1 into a direct sum of K and L follows:

$$E_1 = K + L\,.$$

It follows that there are continuous linear mappings p and q of E_1 onto K and L, respectively, which have the following properties:

$$K = \operatorname{Ran} p = \operatorname{Ker} q\,; \quad L = \operatorname{Ker} p = \operatorname{Ran} q\,;$$

$$p^2 = p\,, \quad q^2 = q\,, \quad p + q = id\,.$$

(b) Since U is open, there exists an $r > 0$ such that the open ball

$$B_r = \{x \in E_1 |\ \|x\|_1 < r\}$$

satisfies the relation

$$x_0 + B_r + B_r \subset U\,.$$

Now define a mapping

$$\psi : K \cap B_r \times L \cap B_r \to E_2$$

by

$$\psi(h,g) := \phi(x_0 + h + g)\,, \quad \forall h \in K \cap B_r\,, \quad \forall g \in L \cap B_r\,.$$

ψ is well defined through the choice of r. It has the following properties:

(1) $\psi(0,0) = \phi(x_0) = y_0$.
(2) ψ is continuously differentiable and
 (i) $\psi_{,h}(0,0) = \phi'(x_0) \upharpoonright K = 0 \in \mathcal{L}(K, E_2)$,
 (ii) $\psi_{,g}(0,0) = \phi'(x_0) \upharpoonright L \in \mathcal{L}(L, E_2)$.

Since the mapping $\phi'(x_0)$ is surjective, and injective on the complement L of its kernel K, then $\psi_{,g}(0,0)$ is a bijective continuous linear mapping of the Banach space L onto the Banach space E_2. It follows by the inverse mapping theorem (Appendix A) that

$$\psi_{,g}(0,0)^{-1} : E_2 \to L$$

is also a continuous linear mapping.

(c) The properties of the mapping ψ which we have just established in (b) allow to apply the implicit function theorem (see, for example [4.3]) to the problem $\psi(h,g) = y_0$. This theorem implies that there exist a $\delta \in (0,r)$ and a uniquely determined continuously differentiable function $\varphi : K \cap B_\delta \to L$ such that

$$\begin{aligned} &y_0 = \psi(h, \varphi(h)), \quad \forall h \in K \cap B_\delta, \\ &\varphi(0) = 0, \\ &\varphi'(0) = -\psi_{,g}(0,0)^{-1} \psi_{,h}(0,0). \end{aligned}$$

It follows that $\varphi'(0) = 0$ and thus

$$\lim_{h \to 0} \frac{1}{\|h\|_1} \|\varphi(h)\|_1 = 0.$$

(d) Since φ is continuous, the mapping

$$\chi : x_0 + K \cap B_\delta \to M, \chi(x_0 + h) = x_0 + h + \varphi(h)$$

is continuous too. By construction,

$$y_0 = \psi(h, \varphi(h)) = \phi(x_0 + h + \varphi(h)).$$

so that χ does indeed map into M. Since h and $\varphi(h)$ are in complementary subspaces of E_1, χ is injective; it follows that χ is invertible on

$$V = \{x_0 + h + \varphi(h) \mid h \in K \cap B_\delta\} \subset M,$$

that is

$$\chi^{-1}(x_0 + h + \varphi(h)) = x_0 + h.$$

Since $\operatorname{Ran} p = K$ and $\operatorname{Ker} p = L$, we also have

$$\chi^{-1}(x_0 + h + \varphi(h)) = x_0 + p(h + \varphi(h)),$$

so that χ^{-1} is continuous too. It follows that χ is a homeomorphism of $U' = x_0 + K \cap B_\delta$ onto $V \subset M$.

The properties of φ established in (c) complete the proof. □

Apart from the natural assumption about the regularity of the point x_0 this theorem also contains the "technical" assumption that the kernel $K = \operatorname{Ker} \phi'(x_0)$ of $\phi'(x_0) : E_1 \to E_2$ has a topological complement in E_1. In a general situation it is not so easy to check this assumption. We want to show now that this assumption is quite adequate for the general situation of Theorem 4.2.1 by proving that it is automatically satisfied for three large and important classes of special cases:

Case 1: $E_1 = H_1$ is a Hilbert space.
Case 2: E_2 is a finite-dimensional Banach space.
Case 3: $\phi : U \to E_2$ is a *Fredholm mapping.*

Here, a differentiable mapping $\phi : U \to E_2$ is called a *Fredholm mapping* of U into E_2 if, for all $x \in U$, $D\phi(x) \equiv \phi'(x)$ is a Fredholm operator from E_1 into E_2. By definition, a continuous linear operator A from E_1 into E_2 is a *Fredholm operator* if and only if

(i) $\dim \operatorname{Ker} A < \infty$,
(ii) $\operatorname{Ran} A = A(E_1) \subset E_2$ is closed in E_2,
(iii) $\operatorname{codim} \operatorname{Ran} A = \dim E_2/\operatorname{Ran} A < \infty$.

The most important properties of Fredholm operators are discussed and proved in, for example [Ref. 4.4, pp. 107–112].

If E_1 is a Hilbert space, then $K = \operatorname{Ker} \phi'(x_0)$ is a subspace such that the projection theorem (Appendix A) guarantees the existence of the topological complement $L = K^\perp$; in fact, we then have a decomposition of E_1 into an orthogonal direct sum: $E_1 = K \oplus K^\perp$.

If E_2 is a finite-dimensional Banach space and $\phi'(x_0) : E_1 \to E_2$ is a surjective continuous linear mapping, then there exist linearly independent vectors $e_1, \dots, e_m$ in E_1 such that $\{f_1, \dots, f_m\}$, $f_j = \phi'(x_0)(e_j)$, is a basis of E_2. Let V denote the m-dimensional subspace of E_1 generated by $\{e_1, \dots, e_m\}$. Simple considerations, or a general theorem [Ref. 4.2, p. 107, Theorem 32], show that V has a topological complement in E_1. (For every $x \in E_1$ one has $\phi'(x_0)(x) = \sum_{j=1}^m \lambda_j(x) f_j \in E_2$; define $px := \sum_{j=1}^m \lambda_j(x) e_j$ and $qx := x - px$ and show that p and q have the properties required; we can assume that $f_1, \dots, f_m$ is the standard basis of $\mathbb{R}^m \cong E_2$.)

Corollary 4.2.2. *Let E_1 and E_2 be two Banach spaces, $U \subset E_1$ an open nonempty subset and $\phi : U \to E_2$ a differentiable mapping. In each of the three cases mentioned above, the tangent space of the level surface $\phi^{-1}(y_0)$ at every regular point $x_0 \in \phi^{-1}(y_0)$ is given by*

$$T_{x_0}\phi^{-1}(y_0) = x_0 + \operatorname{Ker} \phi'(x_0) .$$

The following theorem, due to Ljusternik, gives necessary conditions for the solution of the extreme value problem subject to constraints. It is primarily useful for explicit calculation of the extremal points once their existence has been established, say as a consequence of the results of Chap. 1. The theorem also establishes the famous method of Lagrange multipliers and one should thus regard it as a fundamental theorem.

Theorem 4.2.3. *Let E_1 and E_2 be Banach spaces, $U \subset E_1$ an open subset, $\phi : U \to E_2$ a differentiable mapping, and $f : U \to \mathbb{R}$ a differentiable function.*

Let f have a local extremum at the point $x_0 \in U$, subject to the constraint $\phi(x) = y_0 = \phi(x_0) \in E_2$. If x_0 is a regular point of the mapping ϕ and if $\operatorname{Ker} \phi'(x_0)$ *has a topological complement in E_1, then there exists a continuous linear function $l : E_2 \to \mathbb{R}$, such that x_0 is a critical point of the function*

$$F = f - l \circ \phi : U \to \mathbb{R},$$

that is,

$$f'(x_0) = l \circ \phi'(x_0).$$

Proof. (a) Let L be the topological complement of $K = \operatorname{Ker} \phi'(x_0)$ in E_1; then $H := \phi'(x_0) \restriction L$ is an injective continuous linear mapping of the Banach space L onto the Banach space E_2, since x_0 is a regular point of ϕ. The inverse mapping theorem (Appendix A) implies that $H^{-1} : E_2 \to L$ is also linear and continuous.

(b) The hypotheses of this theorem allow Theorem 4.2.1 to be applied. We can thus describe the points x of the level surface $\phi^{-1}(y_0)$ in the neighborhood V of the point $x_0 \in \phi^{-1}(y_0)$ by

$$x = x_0 + h + \varphi(h), \quad h \in K \cap B_\delta$$

(where $\delta > 0$) is as in the proof of Theorem 4.2.1). Let f have a local minimum at, say, x_0 subject to the constraint $\phi(x) = y_0$. Then there exists a $\delta_0 > 0$, $\delta_0 \le \delta$, such that

$$f(x_0) \le f(x_0 + h + \varphi(h)), \quad \forall h \in K \cap B_{\delta_0}.$$

It follows that

$$0 \le f'(x_0)(h) + f'(x_0)(\varphi(h)) + o(x_0, h + \varphi(h)), \quad \forall h \in K \cap B_{\delta_0},$$

and from this we can infer that $f'(x_0)(h) = 0$, $\forall h \in K \cap B_{\delta_0}$, and thus also $f'(x_0)(h) = 0$, $\forall h \in K$, that is

$$K = \operatorname{Ker} \phi'(x_0) \subseteq \operatorname{Ker} f'(x_0). \tag{$*$}$$

(c) If we denote by q the canonical projection from E_1 onto the topological complement L of K, then by virtue of the equation $(*)$ we can define a continuous linear function $\widehat{f'}(x_0) : L \to \mathbb{R}$ by

$$\widehat{f'}(x_0)(qx) = f'(x_0)(x), \quad \forall x \in E_1,$$

since $q(x_1) = q(x_2)$ implies $x_1 - x_2 \in K$ and thus by $(*)$ we have $f'(x_0)(x_1) = f'(x_0)(x_2)$.

Thus, by (a),

$$l : \widehat{f'}(x_0) \circ H^{-1} : E_2 \to \mathbb{R}$$

is a well-defined continuous linear function. If $p : E_1 \to K$ denotes the canonical projection onto K we can write every $x \in E_1$ as $px + qx$ and obtain the following chain of equations:

$$\begin{aligned} l \circ \phi'(x_0)(x) &= l \circ \phi'(x_0)(qx) \\ &= l \circ H(qx) \\ &= \widehat{f}'(x_0)(qx) \\ &= f'(x_0)(x) \end{aligned}$$

so that $l \circ \phi'(x_0) = f'(x_0)$ follows; thus x_0 is indeed a critical point of the function $F = f - l \circ \phi$, and we have proved Theorem 4.2.3. □

If E_2 is an m-dimensional real Banach space, say $E_2 = \mathbb{R}^m$, then every (continuous) linear function $l : E_2 \to \mathbb{R}$ is characterised uniquely by an m-tuple of real numbers $(\lambda_1, \ldots, \lambda_m)$ and Theorem 4.2.3 immediately gives the well-known theorem on the existence of Lagrange Multipliers [4.5].

Theorem 4.2.3′. *Let $U \subset \mathbb{R}^n$ be open and $f : U \to \mathbb{R}$ a differentiable function; furthermore, let $\phi : U \to \mathbb{R}^m$ be a continuously differentiable mapping. Let the function f attain a local extremum at a regular point $x^0 \in U$ of the mapping ϕ ($\operatorname{rg} \phi'(x_0) = m$) subject to the condition $\phi(x) = y^0 \in \mathbb{R}^m$. Then real numbers $\lambda_1, \ldots, \lambda_m$ exist such that*

$$\frac{\partial f}{\partial x_i}(x^0) = \sum_{j=1}^{m} \lambda_j \frac{\partial \phi_j}{\partial x_i}(x^0), \quad i = 1, \ldots, n.$$

This theorem asserts that the set of extremal points of the function f subject to the constraint $\phi(x) = y^0$ is contained in the set of solutions $(\lambda, x^0) \in \mathbb{R}^m \times U$ of the following system of equations:

$$\begin{gathered} \frac{\partial f}{\partial x_1}(x^0) = \lambda_1 \frac{\partial \phi_1}{\partial x_1}(x^0) + \ldots + \lambda_m \frac{\partial \phi_m}{\partial x_1}(x^0), \\ \cdots\cdots\cdots\cdots\cdots\cdots \\ \frac{\partial f}{\partial x_n}(x^0) = \lambda_1 \frac{\partial \phi_1}{\partial x_n}(x^0) + \ldots + \lambda_m \frac{\partial \phi_m}{\partial x_n}(x^0), \\ y_1^0 = \phi_1(x^0), \\ \cdots\cdots \\ y_m^0 = \phi_m(x^0). \end{gathered}$$

There are exactly $n + m$ equations for $n + m$ unknowns.

The theorem on the existence of Lagrange multipliers has, in the form of Theorem 4.2.3′, numerous important applications [4.5] and we mention here a simple example which utilises, at least partially, the generality of Theorem 4.2.3.

Let A be a bounded self-adjoint linear operator in a Hilbert space $\mathcal{H}$ and $x_0 \in \mathcal{H}$, $\|x_0\| = 1$, a minimising point of the function

$$f(x) = \langle x, Ax \rangle$$

$$\phi(x) = \|x\|^2 = 1 .$$

Then Theorem 4.2.3′ asserts the following: there exists a $\lambda \in \mathbb{R}$ such that

$$f'(x_0) = \lambda \phi'(x_0) .$$

Now

$$f'(x_0)(h) = 2 \operatorname{Re}\langle Ax_0, h \rangle \quad \text{and} \quad \phi'(x_0)(h) = 2 \operatorname{Re}\langle x_0, h \rangle , \quad \forall h \in H .$$

This gives

$$2\, Re\langle Ax_0, h \rangle = 2\lambda \operatorname{Re}\langle x_0, h \rangle , \quad \forall h \in \mathcal{H}$$

and thus

$$Ax_0 = \lambda x_0 ,$$

or, in other words, x_0 is an eigenvector of A with the real eigenvalue λ.

Here we have made use of the fact that every point $x \in \mathcal{H}$, $\|x\| = 1$, is a regular point of the mapping $\phi(x) = \|x\|^2$.

This example indicates how one can prove the existence of eigenvalues of certain bounded self-adjoint operators in Hilbert spaces. We shall discuss this in more detail later (Chap. 6) in the proof of the spectral theorem for compact operators.

4.3 Necessary and Sufficient Conditions for Extrema Subject to Constraints

Theorem 4.2.3 establishes those conditions necessary for the existence of a local minimum of a function $f : U \to \mathbb{R}$, subject to the constraint $\phi : U \to E_2$ where $U \subset E_1$ is open, which are most often used in applications. However, it is also important to know of sufficient conditions. These can be obtained relatively easily from Theorem 4.2.1 and its proof.

Let $x_0 \in M_{y_0} = \phi^{-1}(y_0)$ be a regular point of the mapping ϕ, for which $K = \operatorname{Ker} \phi'(x_0)$ has a topological complement L in E_1. According to Theorem 4.2.1 there exists a uniquely determined $\mathcal{C}^1$ function $\varphi : K \cap B_\delta \to L$ which, in a suitable neighborhood V of x_0 in M_{y_0}, provides local coordinates of the level surface M_{y_0}:

$$x \in V \subset M_{y_0} \;\Leftrightarrow\; x = x_0 + h + \varphi(h) , \quad h \in K \cap B_\delta .$$

If we now take ϕ to be a $\mathcal{C}^2$ mapping of U into E_2, it follows that φ too is a $\mathcal{C}^2$ mapping $K \cap B_\delta \to L$. We can easily calculate the first and second derivatives of φ from the equation $y_0 = \phi(x_0 + h + \varphi(h))$, $\forall h \in K \cap B_\delta$. Taking into account the fact that $\varphi(0) = 0$ and $\varphi'(0) = 0$, we get

$$D^2\varphi(0) = -(\phi'(x_0) \restriction L)^{-1} \circ (D^2\phi)(x_0),$$

where, clearly, $D^2\phi(0)$ must be regarded as a continuous bilinear form of K in L.

The extremal properties of a $\mathcal{C}^2$ function $f : U \to \mathbb{R}$ subject to the constraint $\phi(x) = y_0$ are determined by the corresponding properties of the restriction $f_{y_0} = f \restriction M_{y_0}$ of f to the level surface M_{y_0} of ϕ; in other words, in the neighborhood $V \subset M_{y_0}$ of a regular point $x_0 \in M_{y_0}$ of ϕ, at which $K = \operatorname{Ker} \phi'(x_0)$ has a topological complement, these properties are determined by the equivalent extremal properties of the function F from $K \cap B_\delta$ into $\mathbb{R}$ defined by

$$F(h) = f(x_0 + h + \varphi(h)).$$

Now F is a $\mathcal{C}^2$ function on the open subset $K \cap B_\delta = \{h \in K \mid \|h\| < \delta\}$ of the Banach space K for which we can again use our old criteria, Theorems 3.2.3 and 3.3.1. To this end let us calculate the first and second Fréchet derivatives of F at $h = 0$. It follows from $\varphi(h) = \frac{1}{2}(D^2\varphi)(0)(h,h) + o(h,h)$ that

$$\begin{aligned} F(h) &= f(x_0) + Df(x_0)(h + \varphi(h)) + \frac{1}{2!}D^2f(x_0)(h + \varphi(h), h + \varphi(h)) \\ &\quad + o(h + \varphi(h), h + \varphi(h)) \\ &= f(x_0) + Df(x_0)(h) + \frac{1}{2}[D^2f(x_0) + Df(x_0) \circ D^2\varphi(0)](h,h) \\ &\quad + o(h,h), \end{aligned}$$

and thus

$$DF(0) = Df(x_0) \restriction K,$$
$$D^2F(0) = (D^2f(x_0) + Df(x_0) \circ D^2\varphi(0)) \restriction K \times K.$$

If we now substitute the above expression for $D^2\varphi(0)$ we obtain the following theorem from Theorems 3.2.3 and 3.3.1.

Theorem 4.3.1. Necessary and Sufficient Conditions for Extrema Subject to Constraints. *Let E_1 and E_2 be real Banach spaces, $U \subset E_1$ an open nonempty subset and $\phi : U \to E_2$ a $\mathcal{C}^2$ mapping. For a $y_0 \in \phi(U)$, let the level surface $M_{y_0} = \phi^{-1}(y_0) \subset U$ of ϕ consist only of regular points x of ϕ at which the kernel $K_x \subset E_1$ of $D\phi(x) : E_1 \to E_2$ has a topological complement L_x in E_1. Then for every $\mathcal{C}^2$ function $f : U \to \mathbb{R}$ the following holds:*

(a) *if f attains a local minimum at a point $x_0 \in M_{y_0}$ subject to the constraint $\phi(x) = y_0$, then the following conditions hold:*

(i) $Df(x_0) \restriction K_{x_0} = 0$,
(ii) $[D^2f(x_0) - Df(x_0) \circ (D\phi(x_0) \restriction L_{x_0})^{-1} \circ D^2\phi(x_0)](h,h) \geq 0$, $\forall h \in K_{x_0}$.

(b) *Conversely, if* (i) *and* (ii) *hold and if the bilinear form in* (ii) *is strictly positive in the sense of condition* (ii) *in Theorem* 3.3.1, *then the function* f *has a local minimum at* x_0 *subject to the constant* $\phi(x) = y_0$.

Remark 4.3.1. By the proof of Theorem 4.2.3 we know that condition (i) of Theorem 4.3.1 implies the existence of a Lagrange multiplier $l : E_2 \to \mathbb{R}$ satisfying $f'(x_0) = l \circ \phi'(x_0)$. Using this relation, condition (ii) can be reformulated as

(ii′) $(D^2 f(x_0) - l \circ D^2\phi(x_0))(h,h) \geq 0$, $\forall h \in K_{x_0}$.

Theorems 4.2.3 and 4.3.1 are concerned with extremal points of real functions subject to constraints. The proofs of these results indicate that there is an immediate generalisation to "critical points".

Definition 4.3.2. Let E_1, E_2 be Banach spaces, $U \subset E_1$ be open, $\phi : U \to E_2$ a differentiable mapping and $f : U \to \mathbb{R}$ a differentiable function. $x_0 \in U$ is called *a critical point of* f *subject to the constraint* $\phi(x) = y_0 \in \phi(U)$ if and only if

(i) $\phi(x_0) = y_0$ and
(ii) x_0 is a critical point of the function

$$f_{y_0} := f \restriction \phi^{-1}(y_0) : \phi^{-1}(y_0) \to \mathbb{R}.$$

Then Theorem 4.3.3 follows easily.

Theorem 4.3.3. *In the situation described in Definition* 4.3.2, *a regular point* x_0 *of the mapping* ϕ *is a critical point of the function* f *subject to the constraint* $\phi(x) = y_0$ *if and only if*

(a) $f'(x_0) \restriction \operatorname{Ker} \phi'(x_0) = 0$ *or*
(b) $f'(x_0) = l \circ \phi'(x_0)$,

where $l : E_2 \to \mathbb{R}$ *is a continuous linear function. It is assumed here that the kernel* K *of* $\phi'(x) : E_1 \to E_2$ *for* $x \in \phi^{-1}(y_0)$ *has a topological complement.*

Proof. In the proof of Theorem 4.3.1 we showed, among other things, that at a regular point $x \in \phi^{-1}(y_0)$ of the mapping ϕ one has

$$f'_{y_0}(x) = f'(x) \restriction \operatorname{Ker} \phi'(x).$$

Thus condition (a) is necessary and sufficient.

Condition (a) follows immediately from condition (b). Conversely, let (a) be true; this condition asserts that

$$\operatorname{Ker} \phi'(x_0) \subseteq \operatorname{Ker} f'(x_0).$$

We can conclude here, as in the proof of Theorem 4.2.3, that a "Lagrange multiplier" $l \in \mathcal{L}(E_2, \mathbb{R})$ exists, so that condition (b) results. □

4.4 A Special Case

The method of Lagrange multipliers can be applied to good effect in the treatment of nonlinear eigenvalue problems (see Chap. 8). In such applications a C^1 function $\phi : E \to \mathbb{R}$ on a real Banach space E is given, and one wants to investigate a C^1 function $H : E \to \mathbb{R}$ on a level surface $M_c = \phi^{-1}(c)$, $c \in \mathbb{R}$. In particular, one wants to characterise the critical points of the restriction H_c of H to M_c. Since in this case the Banach space E_2 is simply equal to $\mathbb{R}$, we can reformulate our previous results in slightly more detail.

Theorem 4.4.1. *Let $\phi : E \to \mathbb{R}$ be a C^1 function on the real Banach space E and let $H_c = H \restriction M_c$ denote the restriction of the C^1 function $H : E \to \mathbb{R}$ to the level surface $M_c = \phi^{-1}(c)$, $c \in \phi(E) \subset \mathbb{R}$, of ϕ.*

Then for each regular point x of ϕ there exists a point $N(x) = N_\phi(x) \in E$, which is not in the kernel K_x of $D\phi(x) \equiv \phi'(x) \in \mathcal{L}(E, \mathbb{R}) = E'$, such that

$$\langle \phi'(x), N(x) \rangle = 1 .$$

With the help of the function $x \to N(x)$ on the regular points of ϕ, we can establish the following relations between the differentials of H and H_c:

$$DH_c(x) = DH(x) \restriction \operatorname{Ker} \phi'(x) .$$

(a) *If we put*

$$\xi_H(x) = \langle H'(x), N(x) \rangle ,$$

then

$$\|H'(x) - \xi_H(x)\phi'(x)\|_{E'} \leq (1 + \|\phi'(x)\|_{E'} \|N(x)\|_E) \|DH_c(x)\|_{T_x^*(M_c)} .$$

(b) *A regular point x of ϕ is a critical point of H_c if and only if*

$$H'(x) = \xi_H(x)\phi'(x) .$$

Proof. If $x \in M_c$ is a regular point of ϕ, then $\phi'(x) \in \mathcal{L}(E, \mathbb{R}) = E'$ is a nonzero element of the dual space E' of E and thus it uniquely determines a hyperplane $K_x = \operatorname{Ker} \phi'(x)$ in E.

Furthermore, there exists a $y(x) \in E \setminus K_x$, such that $\langle \phi'(x), y(x) \rangle = 1$. The decomposition of E into a topological direct sum

$$E = K_x + \mathbb{R} N(x)$$

results if we set $N(x) = (\mathrm{id} - p(x))y(x)$, where $p(x)$ denotes the projector from E onto K_x. It follows that

$$\langle \phi'(x), N(x) \rangle = 1 \quad \text{and} \quad N(x) \in E \setminus K_x .$$

If we now set $\xi_H(x) = \langle H'(x), N(x)\rangle$, then it follows for all $u \in E$ that

$$\langle \xi_H(x)\phi'(x), u\rangle = \langle H'(x), \beta(u)N(x)\rangle\,, \quad \beta(u) = \langle \phi'(x), u\rangle$$

and thus

$$\begin{aligned}\langle H'(x) - \xi_H(x)\phi'(x), u\rangle &= \langle H'(x), u - \beta(u)N(x)\rangle = \langle H'(x), p(x)u\rangle \\ &= \langle DH_c(x), p(x)u\rangle\,.\end{aligned}$$

Thus, because

$$\|p(x)u\|_{T_x(M_c)} = \|u - \beta(u)N(x)\|_E \le (1 + \|\phi'(x)\|_{E'}\|N(x)\|_E)\|u\|_E$$

we have

$$\|H'(x) - \xi_H(x)\phi'(x)\|_{E'} \le \|DH_c(x)\|_{T_x^*(M_c)}(1 + \|\phi'(x)\|_{E'}\|N(x)\|_E)$$

which is precisely the assertion

If x is a critical point of H_c then by definition $DH_c(x) = 0$, and the relation $H'(x) = \xi_H(x)\phi'(x)$ thus follows from (a).

Conversely, if this relation holds, then taking into account the fact that $\beta(u) = 0$, it follows for all $u \in T_x(M_c) \cong K_x$ that

$$\begin{aligned}\langle DH_c(x), u\rangle &= \langle H'(x), u\rangle = \langle H'(x), u\rangle - \langle H'(x), \beta(u)N(x)\rangle \\ &= \langle H'(x), u\rangle - \langle \xi_H(x)\phi'(x), u\rangle \\ &= \langle H'(x) - \xi_H(x)\phi'(x), u\rangle \\ &= \langle 0, u\rangle = 0\,.\end{aligned}$$

Thus $DH_c(x) = 0$, or in other words x is a critical point of H_c. □

Remark 4.4.1. Part (a) of this theorem will be important when we want to establish sufficient conditions for the so-called Palais-Smale condition (see Chap. 8). The Lagrange multiplier is explicitly determined in part (b).

5. Classical Variational Problems

5.1 General Remarks

In many applications of the calculus of variations we have specific information about the form of the functional f which we want to minimise. In practical terms, this means that the functional f has the form

$$f(\varphi) = \int_I F(t, \varphi(t), \varphi'(t), \ldots, \varphi^{(n)}(t))\, dt\,, \quad \varphi^{(p)}(t) = \frac{d^p \varphi}{dt^p}(t)\,,$$

where we have a certain function $F : I \times \mathbb{R}^{n+1} \to \mathbb{R}$, $\varphi : I \to \mathbb{R}$, and a compact interval $I = [a, b]$ or, alternatively, an equivalent generalisation to functions φ of several variables and to functions φ with values in $\mathbb{R}^p$, $p > 1$.

We do not intend to treat explicitly the most general case in this chapter, since it is technically rather involved and the crucial arguments can in any case be illustrated by simple special cases which have application in physics. We shall start with some classical examples, through which the calculus of variations evolved and which give some indication of the sort of problem which can typically be solved with the aid of variational calculus.

A. *Minimum Surface of Revolution*

Let two points in the plane, A and B, be given, with coordinates (a, a') and (b, b'), $a < b$. The problem of the minimum surface of revolution then consists in finding the curve C given by the equation $y = \varphi(t)$ and passing through A and B, for which the area

$$f(\varphi) = 2\pi \int_a^b \varphi \sqrt{1 + \varphi'^2}\, dt$$

of the surface which results from revolving the curve C about the t-axis is a minimum.

B. *The Brachystochrone*

We have to determine the curve C passing through two points on a plane perpendicular to the earth's surface, such that a point of mass $m = 1$, travelling from A to B along C under the influence of gravity, reaches B in the shortest possible time. If we denote the coordinates of the points A and B by $(0,0)$ and (b, b') (where $b > 0$, $b' > 0$, say), then the magnitude of the velocity at an arbitrary point $(x, q(x))$ of C is

$$v = \frac{ds}{dt} = \sqrt{2gx(t)}\,,$$

if the initial velocity at A is zero. The point mass needs a time

$$T(q) = \int_0^b \frac{ds}{v} = \int_0^b \sqrt{\frac{1 + q'^2(x)}{2gx}}\,dx$$

to travel along the curve C from A to B.

Mathematically speaking, then, the brachystochrone problem proves to be a minimisation problem for the functional $T(q)$. We can recognise the problem in this form as a special case of Fermat's Principle of Least Time which requires in its general case the minimisation of the functional

$$T(q) = \int_0^b \frac{ds}{v}$$

where v is the velocity of light in the medium being investigated.

C. *Geodesics*

The problem of the geodesics consists in determining the shortest curve C connecting two points A and B on the plane or in space. Somewhat more generally, this problem consists in determining the shortest curve between two given points A and B on a given surface Σ in $\mathbb{R}^3$, for instance. This case is obviously a variational problem subject to a constraint.

D. *The Isoperimetric Problem*

This problem requires to determine a curve C through two given points $A = (0,0)$ and $B = (b,0)$ on the plane, such that

(i) the length of the curve C has a given value l, and
(ii) the curve C and the line AB enclose a surface with the largest possible area.

In other words, the functional

$$f(\varphi) = \int_0^b \varphi(t)\, dt$$

is to be maximised, and at the same time the constraint

$$l = \int_0^b \sqrt{1 + \varphi'^2}\, dt$$

must be satisfied. Problems of this kind, where, of two given functionals f_1 and f_2, one has to take on an extreme value and the other a given value, are known as *isoperimetric problems.*

The examples above show that one has to consider functionals of the form

$$f(\varphi) = \int_I F(t, \varphi(t), \dot{\varphi}(t))\, dt\,,$$

$\varphi \in C^1(I)$, $\dot{\varphi} = d\varphi/dt$, with given boundary conditions. As we have already pointed out several times, many problems in variational calculus can be regarded as infinite-dimensional analogues of the extreme value problem for functions of a real variable. However, additional phenomena now appear. Let us consider two examples, the first due to David Hilbert. Let

$$f_1(\varphi) = \int_0^1 t^{2/3} \dot{\varphi}^2(t)\, dt\,, \quad \varphi \in C^1([0,1])\,, \quad \varphi(0) = 0\,, \quad \varphi(1) = 1\,.$$

The Euler-Lagrange equation is

$$\frac{d}{dt}(2t^{2/3}\dot{\varphi}) = 0\,,$$

and its general solution is

$$\varphi(t) = Ct^{1/3} + D\,.$$

The curve $\varphi_0(t) = t^{1/3}$ passes through the given points, but it is not, however, in $C^1([0,1])$. The solution of the Euler-Lagrange equation exists in this example; it is even uniquely determined and gives an absolute minimum, but it is not an element of the underlying function space $C^1([0,1])$.

The second example is due to Karl Weierstraß; he used it as an argument against Riemann's explanation of the Dirichlet Principle. Consider the functional

$$f_2(\varphi) = \int_0^1 t^2 \dot{\varphi}^2(t)\, dt$$

with the boundary conditions $\varphi(0) = 0$, $\varphi(1) = 1$. In this case the Euler-Lagrange equation is

$$\frac{d}{dt}(2t^2\dot{\varphi}) = 0\,,$$

and the general solution is $\varphi(t) = C/t + D$. However, no curve in this family goes through the given points. Thus, no solution of the Euler-Lagrange equation satisfies the given boundary conditions. No absolutely continuous solution of the variational problem exists at all in this case. For absolutely continuous φ where $\varphi \neq 0$, we have $f_2(\varphi) > 0$. The lower limit of the functional f_2 has the value zero. To convince oneself of this, it is sufficient to consider the sequence

$$\varphi_n(t) = \begin{cases} nt\,, & t \in [0, 1/n]\,, \\ 1\,, & t \in [1/n, 1]\,, \end{cases}$$

for then

$$f_2(\varphi_n) = \int_0^{1/n} t^2 n^2 \, dt = \frac{1}{3n}$$

holds, and so $\lim_{n \to +\infty} f_2(\varphi_n) = 0$.

5.2 Hamilton's Principle in Classical Mechanics

In classical mechanics we learn that the dynamical behaviour of a physical system Σ can be described in terms of the *Lagrangian*

$$L = L_\Sigma = T - V$$

(where T is the kinetic energy and V is the potential energy of Σ) assigned to the system. For this, we use *Hamilton's Principle* which states that the actual movement of the system Σ, starting at time $t = a$ from a point q_S and arriving at time $t = b > a$ at a point q_E is such that the *action functional* $W = W_\Sigma$, where

$$W = \int_I L \, dt\,, \quad I = [a, b]\,,$$

is stationary. Consequently, determining the time behaviour of a classical mechanical system turns out to be a typical variational problem. Hamilton's Principle is also known as *the Principle of Least Action*, since there are cases in which the functional W is not only stationary, but attains a minimum. We want to investigate the naturally arising question as to whether, and when, the stationary value is a minimum. Many textbooks more or less maintain that the problem of determining the minimum of a functional is solved when all the critical points have been found, i.e. all stationary values of this functional. There is often a physical or geometrical reason for this mathematical nonsense: from the way in which the problem is put, one knows a priori that only (local) minima are possible as stationary points.

In this section we want to show which assertions can be obtained in this special case from our general theorems. For the sake of a clear and transparent presentation of the basic arguments, we start with a detailed discussion of the case of mechanical systems with one degree of freedom. The more general case with N degrees of freedom can then easily be treated in an analogous fashion, but with more technical effort. We shall clarify later which modifications are necessary for the case $N > 1$.

We shall discuss the problem of minimising action functionals of the form (5.2.1) (and the generalisations to several degrees of freedom and field theory) in great detail, because

(i) in many important applications we are indeed looking for a minimising point of W and not just a critical point, and
(ii) the property of being a minimum point is often essential for the interpretation of the solution.

5.2.1 Systems with One Degree of Freedom

Let $L = L(t, q, \dot{q})$ be the Lagrangian of a physical system with one degree of freedom. To start with, our standard assumption will be only that L is twice continuously differentiable, i.e. $L \in \mathcal{C}^2(I \times \mathbb{R}^2)$. Later, with the aid of the fundamental theorems of variational calculus, we shall obtain further conditions on the Lagrangian which guarantee that the principle of least action can indeed be realised with a particular Lagrangian.

Let $I = [a, b]$ be a compact interval and $\mathcal{C}^1(I) = \mathcal{C}^1(I, \mathbb{R})$ be the Banach space of the continuously differentiable functions $q : I \to \mathbb{R}$ with norm

$$\|q\| = \|q\|_{I,1} = \sup_{t \in I}\{|q(t)|, |\dot{q}(t)|\},$$

where $\dot{q}$ is the derivative of q. Further, let $E_1(I) = E_1(I; q_\mathrm{S}, q_\mathrm{E})$ denote the set of all continuously differentiable curves $q \in \mathcal{C}^1(I)$ from an initial point q_S to an endpoint q_E,

$$E_1(I) = \{q = \mathcal{C}^1(I) | q(a) = q_\mathrm{S}, q(b) = q_\mathrm{E}\}.$$

The elements of $E_1(I)$ can be regarded as potential trajectories of the system Σ. The action function W for the time interval I is then the function

$$W : E_1(I) \to \mathbb{R},$$

$$W(q) = \int_I L(t, q(t), \dot{q}(t))\, dt, \tag{5.2.1}$$

on the subset $E_1(I)$ of the Banach space $\mathcal{C}^1(I)$. If we now put

$$E_1^0(I) = \{h \in \mathcal{C}^1(I) | h(a) = h(b) = 0\},$$

then for arbitrary $q \in E_1(I)$ and for arbitrary $h \in E_1^0(I)$ it is true that $q + h \in E_1(I)$ too. We can thus investigate the action function's dependence on the trajectories.

Now we shall start to discuss those conditions on the Lagrange function L which are necessary for a trajectory $q \in E_1(I)$ to minimise locally the action function W. Moreover, we include a fairly detailed discussion and proof of the necessary regularity conditions. But we shall not repeat these arguments for systems with several degrees of freedom or for field theories (Sect. 5.5). In both cases, we assume where necessary the required regularity for the calculation, and we hope that the discussion which follows, together with the discussion of regularity in Sect. 9.3, will provide enough information so that the reader will be able to fill in the gaps.

If a trajectory $q \in E_1(I)$ minimises the action functional W locally, then by Theorem 3.2.3, if $D^2W(q) \neq 0$,

$$DW(q)(h) = 0\,, \quad \forall h \in E_1^0(I)\,, \tag{5.2.2 a}$$

$$D^2W(q)(h,h) \geq 0\,, \quad \forall h \in E_1^0(I)\,. \tag{5.2.2 b}$$

Remark 5.2.1. One can show that $E_1(I)$ is a differentiable submanifold of the Banach space $C^1(I)$. The tangent space of $E_1(I)$ at the point $q \in E_1(I)$ is $q + E_1^0(I)$ (see Sects. 4.1, 5.2, 5.7). Since $q + E_1^0(I)$ is isomorphic to $E_1^0(I)$, one naturally considers the Fréchet derivative of $W : E_1(I) \to \mathbb{R}$ at $q \in E_1(I)$ as a continuous linear mapping $DW(q) : E_1^0(I) \to \mathbb{R}$.

The Fréchet derivative of W was calculated in Example 2.2.6. Similarly, using the chain rule, the second Fréchet derivative of W can be computed. We then have

$$DW(q)(h) = \int_I [L_q(t)h(t) + L_{\dot{q}}(t)\dot{h}(t)]\, dt \tag{5.2.3 a}$$

and

$$D^2W(q)(h,h) = \int_I [L_{qq}(t)h(t)^2 + 2L_{q\dot{q}}(t)h(t)\dot{h}(t) + L_{\dot{q}\dot{q}}(t)\dot{h}(t)^2]\, dt\,, \tag{5.2.3 b}$$

where we have used the abbreviations

$$L_q(t) = \frac{\partial L}{\partial q}(t, q(t), \dot{q}(t))$$

$$L_{qq}(t) = \frac{\partial^2 L(t,q,\dot{q})}{\partial q^2}$$

with similar expressions for $L_{\dot{q}}$, $L_{q\dot{q}}$ and $L_{\dot{q}\dot{q}}$.

In order to derive from (5.2.2, 3) conditions on q and L which are necessary for a local minimiser of the action functional W, we need a few basic lemmas.

Lemma 5.2.1 (Lagrange. Fundamental Lemma in the Calculus of Variations). *If a continuous function $g : I \to \mathbb{R}$, $I = [a,b]$, satisfies the condition*

$$\int_I g(t)h(t)\, dt = 0$$

for all $h \in E_1^0(I)$, then this function vanishes identically: $g(t) = 0$, $\forall t \in I$.

Proof. For the indirect proof we suppose that there is some $t_0 \in I$ such that $g(t_0) = 2A > 0$. Then, by continuity of g, there is a $\delta_0 > 0$ such that $g(t) \geq A$ for all $t \in I \cap [t_0 - \delta_0, t_0 + \delta_0] = I_1$. Now we construct a $\mathcal{C}^\infty$ function on $\mathbb{R}$ which vanishes outside I_1 and use our hypothesis to derive a contradiction.

There are nonnegative $\mathcal{C}^\infty$ functions f with support in the interval $[-1, +1]$, for example the function f defined by $f(t) = \exp[-(1-t^2)^{-1}]$ for $|t| < 1$ and $f(t) = 0$ for $|t| \geq 1$. Using the centre point t_1 and the width $2\delta_1$ of the interval I_1, we can represent the interval as $I_1 = [t_1 - \delta_1, t_1 + \delta_1]$. Then, for $0 < \varepsilon < \delta_1$, define a function

$$h_\varepsilon(t) = \frac{1}{\varepsilon} f\Big(\frac{t - t_1}{\varepsilon}\Big).$$

This is again a nonnegative $\mathcal{C}^\infty$ function with support in $[t_1 - \varepsilon, t_1 + \varepsilon] \subset I_1$. Thus we get the contradiction

$$\begin{aligned} 0 &= \int_I g(t) h_\varepsilon(t)\, dt = \int_{I_1} g(t) h_\varepsilon(t)\, dt \\ &\geq \int_{I_1} A h_\varepsilon(t)\, dt = A \int f(\tau)\, d\tau > 0 \end{aligned}$$

and it follows that $g(t) \leq 0$ for all $t \in I$. Since our hypothesis is invariant under the transformation $g \to -g$, it follows in the same way that $(-g)(t) \leq 0$ and therefore that $g(t) = 0$ for all $t \in I$. □

Lemma 5.2.1′ (Du Bois-Reymond). *If a continuous function $g : I \to \mathbb{R}$, $I = [a, b]$, satisfies for all $h \in E_1^0(I)$ the condition*

$$\int_I g(t) \dot{h}(t)\, dt = 0,$$

then this function is constant: $g(t) = g(a)$ for all $t \in I$.

Proof. Fix a nonnegative $\mathcal{C}^\infty$ function f on $\mathbb{R}$ with support in I and normalise it by $\int f(t)\, dt = 1$. Then, with every $\mathcal{C}^\infty$ function φ on $\mathbb{R}$ with support in I, associate the functions

$$\chi_\varphi = \varphi - M(\varphi) f, \quad M(\varphi) = \int_I \varphi(t)\, dt,$$

$$h_\varphi(t) = \int_{-\infty}^t \chi_\varphi(\tau)\, d\tau.$$

Since $M(\chi_\varphi) = 0$, it follows that h_φ has its support in I. The function h_φ is easily seen to be $\mathcal{C}^\infty$ on $\mathbb{R}$ and in particular it follows that $\dot{h}_\varphi = \chi_\varphi$, and thus by hypotheses

$$0 = \int_I g(t) \dot{h}_\varphi(t)\, dt = \int_I g(t)[\varphi(t)] - M(\varphi) f(t)]\, dt = \int_I \varphi(t)[g(t) - c]\, dt$$

where

$$c = \int_I g(t)f(t)\,dt\,.$$

Since this equation holds for all $\varphi \in \mathcal{C}^\infty(\mathbb{R})$ with support in I, Lemma 5.2.1 implies that $g(t) - c = 0$ for all $t \in I$, i.e. $g = \text{const.}$ □

Remark 5.2.2. Notice that we have actually proved stronger statements than those of Lemmas 5.2.1 and 5.2.1′ since we have used only the assumption that the hypothesis holds for every $\mathcal{C}^\infty$ function on $\mathbb{R}$ with support in I. These lemmas are used quite often in this form in the theory of distributions on $\mathbb{R}$, and in the language of this theory they say that

(a) if a continuous function vanishes in the sense of distributions, it vanishes as a function;
(b) if the derivative, in the sense of distributions, of a continuous function vanishes, then the function is constant.

We now prepare to investigate the implications of condition (5.2.2b) by proving the following lemma on nonnegative quadratic forms on $E_1^0(I)$.

Lemma 5.2.2. *If the quadratic form*

$$Q[h] = \int_I [F_1(t)h(t)^2 + F_2(t)h(t)\dot{h}(t) + F_3(t)\dot{h}(t)^2]\,dt$$

on $E_1^0(I)$ with continuous functions $F_j : I \to \mathbb{R}$, $I = [a,b]$, is nonnegative, i.e. $Q[h] \geq 0$ for all $h \in E_1^0(I)$, then the function F_3 is nonnegative: $F_3(t) \geq 0$ for all $t \in I$.

Proof. As in the proof of Lemma 5.2.1 we choose a nonnegative $\mathcal{C}^\infty$ function f with support in $[-1,1]$ and introduce for $0 < \varepsilon < \infty$ and $t_0 \in \mathbb{R}$ the functions

$$h_\varepsilon(t) = f\Big(\frac{t - t_0}{\varepsilon}\Big)\,.$$

These are $\mathcal{C}^\infty$ functions on $\mathbb{R}$ with support in $[t_0 - \varepsilon, t_0 + \varepsilon]$. If this interval is contained in I, the change of variables $\tau = (t - t_0)/\varepsilon$ gives

$$\begin{aligned} Q[h_\varepsilon] = \varepsilon \int F_1(t_0 + \varepsilon\tau)f(\tau)^2\,d\tau + \int F_2(t_0 + \varepsilon\tau)f(\tau)\dot{f}(\tau)\,d\tau \\ + \frac{1}{\varepsilon}\int F_3(t_0 + \varepsilon\tau)\dot{f}(\tau)^2\,d\tau\,. \end{aligned}$$

Suppose that there is a point $t_1 \in I$ such that $F_3(t_1) = -2A < 0$. Then by continuity of F_3, a point $t_0 \in I$ and a number $\varepsilon_0 > 0$ exist such that

$$F_3(t) \leq -A \quad \forall t \in [t_0 - \varepsilon_0, t_0 + \varepsilon_0] \subset I\,.$$

We evaluate the quadratic form Q on the functions h_ε constructed above, with this point t_0, for all $0 < \varepsilon < \varepsilon_0$ and get

$$Q[h_\varepsilon] \leq \varepsilon \int F_1(t_0 + \varepsilon\tau) f(\tau)^2 \, d\tau + \int F_2(t_0 + \varepsilon\tau) f(\tau) \dot{f}(\tau) \, d\tau - \frac{A}{\varepsilon} \int f(t)^2 \, dt$$

and therefore $Q[h_\varepsilon] \to -\infty$ for $\varepsilon \searrow 0$. This shows that Q cannot be non-negative on $E_1^0(I)$ if the function F_3 is negative at one point. Thus we conclude. □

Theorem 5.2.3 (Euler's Equation and Legendre's Condition). *For a Lagrangian L which satisfies the standard assumptions, define the action functional W on $E_1(I)$,*

$$W(q) = \int_I L(t, q(t), \dot{q}(t)) \, dt \,,$$

and assume $D^2 W(q) \neq 0$. If a trajectory $q \in E_1(I)$ is a local minimum of W then the following conditions hold:

(a) *the function*

$$L_{\dot{q}}(t) \equiv \frac{\partial L}{\partial \dot{q}}(t, q(t), \dot{q}(t))$$

is continuously differentiable with derivative

$$\frac{d}{dt} L_{\dot{q}}(t) = L_q(t) \equiv \frac{\partial L}{\partial q}(t, q(t), \dot{q}(t)) \tag{5.2.4}$$

for all $t \in I$ (Euler's equation for a path of stationary action).

(b) *For all $t \in I$,*

$$L_{\dot{q}\dot{q}}(t) = \frac{\partial^2 L}{\partial \dot{q}^2}(t, q(t), \dot{q}(t)) \geq 0$$

(Legendre's condition).

Proof. We already know that for a local minimum the conditions (5.2.2) are satisfied. Now we introduce the function

$$\sigma(t) = \int_a^t L_q(\tau) \, d\tau$$

into (5.2.3 a), the equation for the Fréchet derivative $DW(q)$. Then σ is continuously differentiable and

$$\frac{d\sigma}{dt} = L_q \,.$$

Hence integration by parts in 5.2.3) yields

$$DW(q)(h) = \int_I [L_{\dot{q}}(t) - \sigma(t)] \dot{h}(t) \, dt \,,$$

since $\sigma h|_a^b = 0$ for all $h \in E_1^0(I)$. Now we use (5.2.2 a) and Lemma 5.2.1′ to conclude

$$L_{\dot q}(t) - \sigma(t) = \text{const.} \quad \forall t \in I$$

and thus the function $L_{\dot q}$ is continuously differentiable. Euler's equation follows by differentiation. Similarly, we use (5.2.3 b) for the second Fréchet derivative and Lemma 5.2.2 with $F_1 = L_{qq}$, $F_2 = 2L_{q\dot q}$ and $F_3 = F_{\dot q\dot q}$ to deduce Legendre's condition from (5.2.2 b). □

The proof of *regularity of a local minimum* of the action functional W will follow from part (a) of Theorem 5.2.3 and the following elementary lemma.

Lemma 5.2.4. *If the Lagrangian L satisfies our standard assumptions and if $q \in E_1(I)$ is a trajectory for which*

$$L_{\dot q}(t) = \frac{\partial L}{\partial \dot q}(t, q(t), \dot q(t))$$

is continuously differentiable with respect to t, then the function q is twice continuously differentiable at all points $t \in I$ where $L_{\dot q\dot q}(t) \neq 0$.

Proof. By continuity of $L_{\dot q\dot q}$ we know that the set $J = \{t \in I \,|\, L_{\dot q\dot q}(t) \neq 0\}$ is open in I. For $t, t' \in J$ we calculate the difference

$$\begin{aligned}
\Delta(L_{\dot q}) &= L_{\dot q}(t') - L_{\dot q}(t) = L_{\dot q}(t', q(t'), \dot q(t')) - L_{\dot q}(t, q(t), \dot q(t)) \\
&= \int_t^{t'} L_{t\dot q}(\tau, q(t'), \dot q(t'))\, d\tau + \int_{q(t)}^{q(t')} L_{q\dot q}(t, \tau, \dot q(t'))\, d\tau \\
&\quad + \int_{\dot q(t)}^{\dot q(t')} L_{\dot q\dot q}(t, q(t), \tau))\, d\tau \\
&= \Delta(t) I_1(t') + \Delta(q) I_2(t') + \Delta(\dot q) I_3(t')
\end{aligned}$$

where

$$\Delta(t) = t' - t\,, \quad \Delta(q) = q(t') - q(t)\,, \quad \Delta(\dot q) = \dot q(t') - \dot q(t)\,.$$

By continuity of q and $\dot q$ and by the continuity assumption for the second derivatives of L, the integrals $I_j(t')$ have limits for $t' \to t$:

$$\begin{aligned}
I_1(t') &\equiv \int_0^1 L_{t\dot q}(t + s\Delta(t), q(t'), \dot q(t'))ds \underset{t' \to t}{\longrightarrow} L_{t\dot q}(t, q(t), \dot q(t)) \equiv L_{t\dot q}(t)\,, \\
I_2(t') &\equiv \int_0^1 L_{q\dot q}(t, q(t) + s\Delta(q), \dot q(t'))ds \underset{t' \to t}{\longrightarrow} L_{q\dot q}(t)\,, \\
I_3(t') &\equiv \int_0^1 L_{\dot q\dot q}(t, q(t), \dot q(t) + s\Delta(\dot q))ds \underset{t' \to t}{\longrightarrow} L_{\dot q\dot q}(t)\,.
\end{aligned}$$

One also knows that

$$\frac{\Delta(q)}{\Delta(t)} \underset{t' \to t}{\longrightarrow} \dot{q}(t),$$

and by assumption

$$\frac{\Delta(L_{\dot{q}})}{\Delta(t)} \underset{t' \to t}{\longrightarrow} \frac{d}{dt} L_{\dot{q}}(t).$$

The above identity can be written

$$\frac{\Delta(\dot{q})}{\Delta(t)} I_3(t') = \frac{\Delta(L_{\dot{q}})}{\Delta(t)} - I_1(t') - \frac{\Delta(q)}{\Delta(t)} I_2(t') \tag{$*$}$$

and for $t \in J$ we know that $L_{\dot{q}\dot{q}}(t) \neq 0$. Hence, for $t \in J$, there is an $\varepsilon > 0$ such that for $|t' - t| < \varepsilon$, $t' \in J$, $I_3(t') \neq 0$. For these points t', the equation $(*)$ implies

$$\frac{\Delta(\dot{q})}{\Delta(t)} = \frac{1}{I_3(t')} \left\{ \frac{\Delta(L_{\dot{q}})}{\Delta(t)} - I_1(t') - \frac{\Delta(q)}{\Delta(t)} I_2(t') \right\}$$

and according to our statements above, the right-hand side is known to have the limit

$$\frac{1}{L_{\dot{q}\dot{q}}(t)} \left\{ \frac{d}{dt} L_{\dot{q}}(t) - L_{t\dot{q}}(t) - \dot{q}(t) L_{\dot{q}\dot{q}}(t) \right\}$$

for $t' \to t$. Thus, the left-hand side too has such a limit. This proves that $q \in E_1(I)$ is twice differentiable at $t \in J$ and we have the identity

$$\ddot{q}(t) = \frac{1}{L_{\dot{q}\dot{q}}(t)} \left\{ \frac{d}{dt} L_{\dot{q}}(t) - L_{t\dot{q}}(t) - \dot{q}(t) L_{q\dot{q}}(t) \right\} \tag{5.2.5}$$

for the second derivative. The right-hand side of (5.2.5) is a continuous function on J, and hence $\ddot{q}(t)$ is continuous. □

Theorem 5.2.5. *If the Lagrangian L satisfies the standard assumptions and if $q \in E_1(I)$ is a local minimiser of the action functional W associated with L on I such that $D^2W(q) \neq 0$, then q is twice continuously differentiable at all points $t \in I$ for which $L_{\dot{q}\dot{q}}(t) > 0$.*

Proof. By Theorem 5.2.3 we know that, for a local minimiser q of the action functional W, the hypotheses of Lemma 5.2.4 are satisfied. So by this lemma we conclude. □

Theorem 5.2.5 shows how, for a local minimiser, one gains one order of differentiability. Hence, this is a result on the improved *regularity* for a solution of a minimisation problem. Actually, it is the first result of this kind in the calculus of variations.

In many typical application, the Lagrangian L not only belongs to class $\mathcal{C}^2$, but also to class $\mathcal{C}^3$, and satisfies $L_{\dot{q}\dot{q}} \geq a > 0$. Then for a trajectory $q \in E_1(I)$ which is a critical point of the action functional, we know that q actually belongs to the class $\mathcal{C}^2(I)$. Therefore, the function

$$L_{q\dot{q}}(t) = \frac{\partial^2 L}{\partial q\, \partial \dot{q}}(t, q(t), \dot{q}(t))$$

is continuously differentiable. Integration by parts yields for all $h \in E_1^0(I)$

$$\int_I 2L_{q\dot{q}}(t)h(t)\dot{h}(t)\,dt = \int_I L_{q\dot{q}}(t)\frac{d}{dt}h(t)^2\,dt = -\int_I h(t)^2\frac{dL_{q\dot{q}}}{dt}(t)\,dt\,.$$

This allows the second Fréchet derivative of the action functional W to be rewritten in the form

$$D^2W(q)(h,h) = \int_I \left[\left(L_{qq} - \frac{d}{dt}L_{q\dot{q}}\right)h(t)^2 + L_{\dot{q}\dot{q}}\dot{h}(t)^2\right]dt \qquad (5.2.3\,\mathrm{b}')$$

and it is in this form that it is usually studied.

Remark 5.2.3. According to its derivation, Euler's condition (5.2.4) characterises the fact that a trajectory $q \in E_1(I)$ is a stationary or critical point of the action functional, but not necessarily a point of extremal or even minimal action. In order to ensure that q is actually a trajectory of minimal action, at least locally, it is necessary that, in addition, Legendre's condition be satisfied. Thus, in order to realise the *principle of least action*, Euler's condition does not suffice.

Now we come to the analysis of the sufficient conditions, which are considerably more complicated. According to Theorem 3.3.1, the following conditions are sufficient for the curve q in $E_1(I)$ to realise a strict local minimum of the action function W:

$$DW(q) = 0\,; \quad \text{Euler's equation}\,, \qquad (5.2.6\,\mathrm{a})$$

$$D^2W(q) \text{ is } \textit{strictly} \text{ positive on } E_1^0(I) \times E_1^0(I)\,. \qquad (5.2.6\,\mathrm{b})$$

Our next objective is to translate these general conditions into actual requirements on the Lagrangian L. Legendre tried to show that

$$L_{\dot{q}\dot{q}}(t, q(t), \dot{q}(t)) > 0\,, \quad \forall t \in I\,, \qquad (5.2.7)$$

which is a simple and natural strengthening of the necessary condition (b) of Theorem 5.2.3, is also sufficient for obtaining condition (5.2.6 b) as long as Euler's equation (5.2.6 a) is valid. He started by considering a simple rearrangement of $D^2W(q)(h,h)$. According to (5.2.3 b), for all $h \in E_1^0(I)$,

$$J(h) \equiv D^2W(q)(h,h) = \int_I [A(t)h^2(t) + B(t)\dot{h}^2(t)]\,dt \qquad (5.2.8)$$

holds, where

$$A(t) = L_{qq}(t, q(t), \dot{q}(t)) - \frac{d}{dt}L_{q\dot{q}}(t, q(t), \dot{q}(t)) \qquad (5.2.9\,\mathrm{a})$$

$$B(t) = L_{\dot{q}\dot{q}}(t, q(t), \dot{q}(t)) . \tag{5.2.9 b}$$

Legendre's conjecture was that, given the hypothesis (5.2.7), i.e. $B(t) > 0$, $\forall t \in I$, one can always diagonalise the quadratic form J on $E_1^0(I)$; in other words, J can be written as a sum of integrals over I of the squares of functions. Legendre used a second rearrangement of J to this end: for arbitrary $\omega \in C^1(I)$ and for all $h \in E_1^0(I)$

$$0 = \int_I \frac{d}{dt}[\omega h^2]\, dt = \int_I [\dot{\omega} h^2 + 2\omega h \dot{h}]\, dt$$

holds, and therefore

$$J(h) = \int_I \{B(t)\dot{h}^2(t) + 2\omega(t)h(t)\dot{h}(t) + [A(t) + \dot{\omega}(t)]h^2(t)\}\, dt \tag{5.2.8'}$$

too. It follows that, for $B(t) > 0$ on I,

$$J(h) = \int_I B(t)\Big[\Big(\dot{h} + \frac{\omega}{B}h^2\Big)^2 + \Big(\frac{h}{B}\Big)^2 [B(A + \dot{\omega}) - \omega^2]\Big]\, dt . \tag{5.2.10}$$

It follows from this representation that $J(h)$ can be expressed as an integral over a product of nonnegative continuous functions as long as *Ricatti's differential equation*

$$B(A + \dot{\omega}) = \omega^2 \tag{5.2.11}$$

has a solution ω in $C^1(I)$. Although this equation always has a local solution, it does not necessarily have a solution on the whole of I as simple examples show. Extra conditions on the functions A and B are needed in order to guarantee that this nonlinear equation is soluble on the whole of I.

To this end, let us consider the following simple relationship between solutions $\omega \in C^1(I)$ of equation (5.2.11) and solutions u of Euler's equation for the functional J on $E_1^0(I)$, i.e.

$$-\frac{d}{dt}[B\dot{u}] + Au = 0 , \tag{5.2.12}$$

which do not vanish on I. If $\omega \in C^1(I)$ is a solution of (5.2.11), then

$$u(t) = u(a) \exp\left(-\int_a^t \frac{\omega(\tau)}{B(\tau)}\, d\tau\right) , \quad u(a) \neq 0 \tag{5.2.13}$$

is a solution of (5.2.12) which does not vanish on I, and conversely, if u is a solution of (5.2.12) which does not vanish on I then

$$\omega = -\frac{\dot{u}}{u} B \tag{5.2.14}$$

is a solution of (5.2.11). Assuming $B \in C^1(I)$, equation (5.2.12) is a second-order linear differential equation whose coefficients are continuous functions on I. Then the differential equation (5.2.12) always has nontrivial solutions on all of I. In other words, the problem of determining the existence of a solution of Ricatti's nonlinear differential equation (5.2.11) defined on the whole of I can be reduced to the problem of determining the existence of a solution, which vanishes nowhere on I, of the (linear) Euler equation (5.2.12) for the functional $J(h)$. We thus obtain a simple sufficient condition (Theorem 5.2.6) for the strict positivity of $J(h)$.

Theorem 5.2.6. *Let A be a continuous, and B a continuously differentiable, real function on the interval $I = [a, b]$. Let*

(a) $B(t) > 0$, $\forall t \in I$;
(b) *the differential equation* (5.2.12) *have a solution $u \in C^2(I)$, which vanishes nowhere on I (Jacobi's condition).*

Then the functional

$$J(h) = \int_I [A(t)h^2(t) + B(t)\dot{h}^2(t)]\, dt$$

is strictly positive on $E_1^0(I)$.

Proof. Let $u \in C^2(I)$ be a solution of (5.2.12) which vanishes nowhere on I. Then $\omega = -(\dot{u}/u)B$ is a C^1 solution of (5.2.11) on I. We substitue ω into the expression (5.2.10) for $J(h)$ and we obtain

$$J(h) = \int_I B(t)\left[\dot{h}(t) + \frac{\omega(t)}{B(t)}h(t)\right]^2 dt \geq 0$$

for all $h \in E_1^0(I)$, which shows that J is nonnegative on $E_1^0(I)$. Now let $J(h) = 0$ for one $h \in E_1^0(I)$. $B(t) > 0$ implies that $\dot{h} + (\omega/B)h = 0$, so that

$$h(t) = h(a) \exp\left(-\int_a^t \frac{\omega(\tau)}{B(\tau)}\, dt\right).$$

But $h \in E_1^0(I)$ requires that $h(a) = 0$ and thus $h \equiv 0$. Therefore J is strictly positive on $E_1^0(I)$. □

Before we start to discuss the converse of Theorem 5.2.6 we want to explain a particular simplification of Jacobi's condition which Jacobi himself proposed. We shall do this with the aid of a few elementary facts from the theory of ordinary linear second-order differential equations.

Theorem 5.2.7. *If the functions A and B satisfy the hypotheses of Theorem 5.2.6, then the following conditions are equivalent:*

(a) *A and B satisfy Jacobi's condition on* $I = [a, b]$.
(b) *The solution* $u_{0,1}$ *of the differential equation* (5.2.12) *for the inital conditions* $u_{0,1}(a) = 0$, $\dot{u}_{0,1}(a) = +1$, *has no zero in* $(a, b]$.

Proof. If A and B satisfy Jacobi's condition on I, then there exists a solution v of (5.2.12) which has the property

$$0 < \delta = \inf_{t \in I} v(t)\,.$$

Let $u_{1,0}$ denote the solution of (5.2.12) for the initial conditions $u_{1,0}(a) = 1$, $\dot{u}_{1,0}(a) = 0$. The functions $u_{0,1}$ and $u_{1,0}$ are then linearly independent and therefore generate the solution space of the differential equation (5.2.12). Thus there exist real numbers α and β such that

$$v = \alpha u_{1,0} + \beta u_{0,1}\,.$$

It follows that $0 < \delta \leq v(a) = \alpha$.

We shall now show that the assumption that $u_{0,1}$ has a zero at a point $c \in (a, b]$ leads to a contradiction. We can assume that c is the smallest of all numbers $\gamma \in (a, b]$ with $u_{0,1}(\gamma) = 0$. The differential equation (5.2.12) implies that the derivative of

$$B[u_{0,1}\dot{u}_{1,0} - \dot{u}_{0,1}u_{1,0}]$$

vanishes on I. It follows that

$$B[u_{0,1}\dot{u}_{1,0} - \dot{u}_{0,1}u_{1,0}](t) = B[u_{0,1}\dot{u}_{1,0} - \dot{u}_{0,1}u_{1,0}](a) = -B(a)\,,$$

and thus, since $u_{0,1}(c) = 0$, that

$$\dot{u}_{0,1}(c) = \frac{B(a)}{B(c)}\,\frac{1}{u_{1,0}(c)}$$

for $t = c$. On the other hand, $0 < \delta \leq v(c) = \alpha u_{1,0}(c)$, so that $u_{1,0}(c) > 0$, which means $\dot{u}_{0,1}(c) > 0$. But by definition of c as the smallest zero of $u_{0,1}$, we know for some $\varepsilon > 0$ that $u_{0,1}(t) > 0$ for $t \in (a, c)$ and $u_{0,1}(t) < 0$ for $c < t < c + \varepsilon$. This implies $\dot{u}_{0,1}(c) < 0$ and thus a contradiction.

Conversely, assume that the solution $u_{0,1}$ of the differential equation (5.2.12) has no zero on $(a, b]$. Since the solution $u_{1,0}$ is continuous, there exists a $c \in (a, b)$ such that $u_{1,0}(t) \geq 1/2$, $\forall t \in [a, c]$. If we now put

$$\gamma = \inf_{t \in [c,b]} u_{1,0}(t)\,,$$

$$\delta = \inf_{t \in [c,b]} u_{0,1}(t)\,,$$

then it follows that $\delta > 0$. For arbitrary $\alpha > 0$ put $\beta > |\alpha\gamma/\delta|$. It then follows for the solution $v = \alpha u_{1,0} + \beta u_{0,1}$ of (5.2.12), for $t \in [a, c]$, that

$$v(t) = \alpha u_{1,0}(t) + \beta u_{0,1}(t) \geq \alpha/2 > 0\,,$$

and for $t \in [c, b]$ one has

$$v(t) \geq \alpha\beta + \beta\delta > 0 .$$

Thus, v is a solution of the differential equation (5.2.12) which does not vanish on I. This implies that Jacobi's condition holds for the functions A and B on the interval I. □

Remark 5.2.4. We should mention here the following notation which is often used. If the solution $u_{0,1}$ of the differential equation (5.2.12) has a zero at c in $(a, b]$, then c is called a *conjugate point of* a with respect to the differential equation (5.2.12). Lemma 5.2.7 can thus be formulated as follows.

Lemma 5.2.7′. *The following conditions are equivalent:*

(a) *A and B satisfy Jacobi's condition on $I = [a, b]$.*
(b) *There is no conjugate point of a in $(a, b]$ with respect to the differential equation* (5.2.12).

We are now in a position to prove the converse of Theorem 5.2.6.

Theorem 5.2.8. *Suppose two functions $A \in C(I, \mathbb{R})$ and $B \in C^1(I, \mathbb{R})$ where $I = [a, b]$ satisfy*

(a) $B(t) > 0$, $\forall t \in I$.
(b) *The functional $J(h) = \int_I [A(t)h^2(t) + B(t)\dot{h}^2(t)]\, dt$ is strictly positive on $E_1^0(I)$, i.e. $J(h) > 0$, $\forall h \in E_1^0(I)$, $h \neq 0$.*

Then these functions satisfy Jacobi's condition on $I = [a, b]$.

Proof. The functional

$$J_0(h) = \int_I \dot{h}^2(t)\, dt$$

is obviously strictly positive on $E_1^0(I)$ and it follows that the functional J_λ, $\lambda \in [0, 1]$, defined by

$$J_\lambda(h) = \lambda J(h) + (1 - \lambda) J_0(h) = \int_I [A_\lambda h^2 + B_\lambda \dot{h}^2]\, dt ,$$

$$A_\lambda = \lambda A , \quad B_\lambda = 1 - \lambda + \lambda B$$

is also strictly positive on I. The corresponding Euler equation is

$$-\frac{d}{dt}(B_\lambda \dot{u}) + A_\lambda u = 0 .$$

Since the coefficients of this linear second-order differential equation depend analytically on λ, the solution u_λ of this equation for the initial conditions

$$u_\lambda(a) = 0\,, \quad \dot{u}_\lambda(a) = 1$$

depends differentiably on λ. To prove the theorem it is sufficient to show that the solution u_1 has no zero in $(a, b]$. We show this indirectly by demonstrating the relationship between the possible zeros of u_1 and those of the solution u_0. But the solution u_0 is known explicitly, i.e. it is the solution of the differential equation

$$-\frac{d}{dt}(\dot{u}_0) = 0$$

for the initial conditions $u_0(a) = 0$, $\dot{u}_0(a) = +1$, i.e.

$$u_0(t) = t - a\,.$$

The solution u_0 thus has no zeros in $(a, b]$.

Let us assume that $c \in (a, b]$ is a zero of u_1. The strict positivity of $J_1 = J$ on $E_1^0(I)$ rules out the case $c = b$, since in this case $u_1 \in E_1^0(I)$, $u_1 \neq 0$, and

$$J(u_1) = \int_I \Big[-\frac{d}{dt}(B\dot{u}_1) + Au_1 \Big] u_1 \, dt = 0$$

holds. It follows that necessarily $c \in (a, b)$. Thus the set

$$I_0 = \{\lambda \in [0,1] \,|\, \exists t_\lambda \in (a,b), u_\lambda(t_\lambda) = 0\}$$

is not empty. Since u_λ depends continuously on λ, then both $\lambda = 1$ and a suitable neighborhood of $\lambda = 1$ are contained in I_0, which means that there exists a $\lambda_1 \in (0, 1)$ with $[\lambda_1, +1] \subseteq I_0$. Now let $\lambda_0 \in [0,1)$ denote the smallest number for which $[\lambda_0, 1] \subseteq I_0$ is true. If $\lambda_0 > 0$, then the implicit function theorem would produce a contradiction. It is true that $u_{\lambda_0}(t_0) = 0$ for $t_0 = t(\lambda_0) \in (a, b)$ and $\dot{u}_{\lambda_0}(t_0) = (du_{\lambda_0}/dt)(t_0) \neq 0$, since u_{λ_0} is a nonzero solution of a second-order differential equation. We can therefore solve the implicit equation $u_\lambda(t) = 0$ for t on a suitable neighborhood of (λ_0, t_0) in $(0,1) \times (a, b)$ and in this way obtain a continuous function $\lambda \to t(\lambda)$ which satisfies $u_\lambda(t(\lambda)) = 0$. For a suitable $\delta > 0$, it follows that $[\lambda_0 - \delta, 1] \subseteq I_0$ contradicting our assumption of λ_0 being the smallest number with this property. Thus, $I_0 = [0, 1]$, which means u_0 also has a zero, $t_0 = t(0)$, in (a, b). This is a contradiction, so it follows that u_1 has no zero in $(a, b]$. □

Corollary 5.2.9. *Let $A \in C(I, \mathbb{R})$ and $B \in C^1(I, \mathbb{R})$ where $I = [a, b]$, and assume the following properties:*

(a) *$B(t) > 0$, $\forall t \in I$.*
(b) *$J(h)$ is positive on $E_1^0(I)$, which means $J(h) \geq 0$, $\forall h \in E_1^0(I)$.*

Then the functions A and B satisfy Jacobi's condition on the open interval (a, b).

Proof. We consider solutions u_λ, $0 \leq \lambda \leq 1$, just as in the proof of Theorem 5.2.8. Since $J_1 = J$ has now been assumed to be only positive and not strictly positive, we can no longer exclude the possibility that the point $t = b$ is a zero of u_1. The assumption that a zero $c \in (a, b)$ of u_1 exists can then be shown to lead to a contradiction, as it did above. □

In conclusion, let us sum up the results so far. Let $L(t, q, \dot{q})$ be a Lagrangian which satisfies the standard requirements, i.e. $L \in \mathcal{C}^2(I \times \mathbb{R}^2, \mathbb{R})$. The action function associated with L for the time interval $I = [a, b]$ is

$$W(q) = \int_I L(t, q(t), \dot{q}(t))\, dt\,, \forall q \in E_1(I)\,.$$

We can calculate the functions

$$A^{(q)}(t) = \frac{\partial^2 L}{\partial q^2}(t, q(t), \dot{q}(t)) - \frac{d}{dt}\frac{\partial^2 L}{\partial q \partial \dot{q}}(t, q(t), \dot{q}(t))\,,$$

$$B^{(q)}(t) = \frac{\partial^2 L}{\partial q \partial \dot{q}}(t, q(t), \dot{q}(t))$$

with the aid of a "trajectory" $q \in E_1(I)$ and the Lagrangian L. We can formulate the following theorem using this notation and these quantities.

Theorem 3.2.10 (Principle of Least Action). (a) *If $q \in E_1(I)$ is a trajectory for which the action W attains a local minimum, and if the second Fréchet derivative of W does not vanish at the point q, then*

(i) $\frac{d}{dt} L_{\dot{q}}(t, q(t), \dot{q}(t)) = L_q(t, q(t), \dot{q}(t))$; *Euler's equation.*

(ii) $L_{\dot{q}\dot{q}}(t, q(t), \dot{q}(t)) \geq 0$, $\forall t \in I$.

(iii) *Whenever $A^{(q)} \in \mathcal{C}(I, \mathbb{R})$ and $B^{(q)} \in \mathcal{C}^1(I, \mathbb{R})$, these functions satisfy Jacobi's condition on the (semiopen) interval $[a, b)$.*

(b) *Conversely, suppose that a trajectory $q_0 \in E_1(I)$ satisfies the following:*

(i) $\frac{d}{dt} L_{\dot{q}}(t, q_0(t), \dot{q}_0(t)) = L_q(t, q_0(t), \dot{q}_0(t))$,

(ii) $A^{(q_0)} \in \mathcal{C}(I, \mathbb{R})$, $B^{(q_0)} \in \mathcal{C}^1(I, \mathbb{R})$,

(iii) $B^{(q_0)}(t) > 0$, $\forall t \in I$,

(iv) *$A^{(q_0)}$ and $B^{(q_0)}$ satisfy Jacobi's condition on $I = [a, b]$.*

Then the action functional $W(q) = \int_I L(t, q, \dot{q})\, dt$ has a strict local minimum for the trajectory q_0.

5.2.2 Systems with Several Degrees of Freedom

We now want to analyse systems with more than one degree of freedom using essentially the same ideas we have used up to now. The changes which will be necessary are primarily of a technical nature.

Again, let $\mathcal{C}^1(I,\mathbb{R}^n) = \mathcal{C}^1(I,\mathbb{R})^n$ denote, for a time interval $I = [a,b]$, the Banach space of continuously differentiable functions: $q : I \to \mathbb{R}^n$ with norm

$$\|q\| = \|q\|_{I,1} = \max_{j=1,\dots,n} \sup_{t\in I}\{|q_j(t)|, |\dot{q}_j(t)|\} \quad \text{where } \dot{q}_j = \frac{dq_j}{dt}.$$

In general, the Lagrangian of a system with n degrees of freedom has the form $L = L(t,q(t),\dot{q}(t))$ and we again make the standard assumption that $L = \mathcal{C}^2(I \times \mathbb{R}^{2n}, \mathbb{R})$. Then the action function for a fixed starting point $q_S \in \mathbb{R}^n$ and a fixed endpoint $q_E \in \mathbb{R}^n$, for the time interval $I = [a,b]$, is a function on the subset

$$E_1(I)^n = \{q \in \mathcal{C}^1(I,\mathbb{R}^n) | q(a) = q_S, q(b) = q_E\}$$

of the Banach space $\mathcal{C}^1(I,\mathbb{R}^n)$. The action associated with a trajectory $q \in E_1(I)^n$ is

$$W(q) = \int_I L(t,q(t),\dot{q}(t))\,dt\,.$$

The first two Fréchet derivatives of this action function can be easily calculated using Taylor's formula for functions of several variables. This yields for all $h \in E_1^0(I)^n$:

$$E_1^0(I)^n = \{h \in \mathcal{C}^1(I,\mathbb{R}^n) | h(a) = h(b) = 0\}\,,$$

$$DW(q)(h) = \int_I \sum_{j=1}^n \left(L_{q_j} - \frac{d}{dt}L_{\dot{q}_j}\right) h_j\,dt\,, \tag{5.2.15 a}$$

$$D^2W(q)(h,h) = \int_I \sum_{i,j=1}^n \left\{L_{q_iq_j} h_i h_j + 2L_{q_i\dot{q}_j} h_i \dot{h}_j + L_{\dot{q}_i\dot{q}_j} \dot{h}_i \dot{h}_j\right\} dt \tag{5.2.15 b}$$

where the functions

$$L_{q_j} = \frac{\partial L}{\partial q_j}, \quad L_{\dot{q}_j} = \frac{\partial L}{\partial \dot{q}_j}, \quad L_{q_iq_j} = \frac{\partial^2 L}{\partial q_i \partial q_j}, \quad L_{q_i\dot{q}_j} = \frac{\partial^2 L}{\partial q_i \partial \dot{q}_j}$$

and

$$L_{\dot{q}_i\dot{q}_j} = \frac{\partial^2 L}{\partial \dot{q}_i \partial \dot{q}_j}$$

have the argument $(t,q(t),\dot{q}(t))$ while the functions h_i and $\dot{h}_j$ are evaluated at the point t.

We can write this in a more compact and convenient form by using the following notation. Let $L_q : I \to \mathbb{R}^n$ denote the vector-valued function on I whose components at the points $t \in I$ are just $L_{q_j}(t, q(t), \dot{q}(t))$, $j = 1, \ldots, n$. Let $L_{\dot{q}}$ be defined accordingly. Moreover, let

$$L_{qq} : I \to \mathcal{M}_n(\mathbb{R})$$

denote the matrix-valued function on I whose real coefficients at the points $t \in I$ are the real numbers $L_{q_i q_j}(t, q(t), \dot{q}(t))$, $i, j = 1, \ldots, n$. $L_{q\dot{q}}$ and $L_{\dot{q}\dot{q}}$ are defined accordingly.

Note that the set $\mathcal{M}_n(\mathbb{R})$ of all real $n \times n$ matrices is in a canonical way a Banach space and can be regarded as the Banach space $\mathcal{L}(\mathbb{R}^n, \mathbb{R}^n)$ of all (continuous) linear mappings from the real Hilbert space $\mathbb{R}^n$ into the Hilbert space $\mathbb{R}^n$. We denote the scalar product in $\mathbb{R}^n$ by $\langle \cdot, \cdot \rangle$; thus

$$\langle x, y \rangle = \sum_{j=1}^{n} x_j y_j \quad \text{for } x, y \in \mathbb{R}^n .$$

The Fréchet derivatives of the action function can then be rewritten

$$DW(q)(h) = \int_I \langle L_q(t) - \frac{d}{dt} L_{\dot{q}}(t), h(t) \rangle dt , \tag{5.2.16 a}$$

$$\begin{aligned} D^2W(q)(h,h) = \int_I \{ &\langle h(t), L_{qq}(t) h(t) \rangle + 2 \langle h(t), L_{q\dot{q}}(t) \dot{h}(t) \rangle \\ &+ \langle \dot{h}(t), L_{\dot{q}\dot{q}}(t) \dot{h}(t) \rangle \} \, dt . \end{aligned} \tag{5.2.16 b}$$

Taking into account the fact that $h \in E_1^0(I)^n$ and assuming that $L_{q\dot{q}} \in \mathcal{C}^1(I, \mathcal{M}_n)$, we can bring (5.2.16 b) into the form we used earlier. First of all we note that

$$\begin{aligned} 0 &= \int_a^b \frac{d}{dt} \langle h(t), L_{q\dot{q}}(t) h(t) \rangle \, dt \\ &= \int_I dt \Big\{ \langle h(t), \frac{d}{dt} L_{q\dot{q}}(t) h(t) \rangle + \langle h(t), L_{q\dot{q}}(t) \dot{h}(t) \rangle + \langle \dot{h}(t), L_{q\dot{q}}(t) h(t) \rangle \Big\} . \end{aligned}$$

If $L_{q\dot{q}}(t)$ is a symmetric matrix, i.e. $\langle x, L_{q\dot{q}} y \rangle = \langle L_{q\dot{q}} x, y \rangle$, $\forall x, y \in \mathbb{R}^n$, then we have as before

$$D^2W(q)(h,h) = \int_I \{ \langle h(t), A(t) h(t) \rangle + \langle \dot{h}(t), B(t) \dot{h}(t) \rangle \} \, dt \tag{5.2.17 a}$$

where

$$A(t) = A^{(q)}(t) = L_{qq}(t) - \frac{d}{dt} L_{q\dot{q}}(t) , \tag{5.2.17 b}$$

$$B(t) = B^{(q)}(t) = L_{\dot{q}\dot{q}}(t) . \tag{5.2.17 c}$$

These equations serve to illustrate the formal analogy between this case and a system with one degree of freedom.

If $D^2W(q) \neq 0$, then

$$DW(q)(h) = 0\,, \tag{5.2.18 a}$$

$$D^2W(q)(h,h) \geq 0 \tag{5.2.18 b}$$

necessarily hold for all $h \in E_1^0(I)^n$ for a trajectory $q \in E_1(I)^n$ which minimises the action W. If we now substitute

$$h(t) = \chi(t)e_j\,, \quad j = 1, \ldots, n\,,$$

into (5.2.16 a) for $DW(q)(h)$, where $\{e_i\}_{i=1,\ldots,n}$ is the canonical basis of $\mathbb{R}^n$ and $\chi \in E_1^0(I)$, then, with the aid of Lemma 5.2.1, the condition (5.2.18 a) gives

$$L_q(t) - \frac{d}{dt}L_{\dot{q}}(t) = 0\,,$$

which means

$$\frac{d}{dt}L_{\dot{q}_j}(t, q(t), \dot{q}(t)) = L_{q_j}(t, q(t), \dot{q}(t))\,, \quad j = 1, \ldots, n\,. \tag{5.2.19}$$

These are the *Euler equations* (Euler-Lagrange equations) for the action function W.

The analysis of the condition (5.2.18 b) is somewhat more complicated than in the one-dimensional case. We shall start with the simple part which leads to Legendre's necessary condition. To this end we need the following (trivial) generalisation of Lemma 5.2.2.

Lemma 5.2.11. *Let $A, B \in \mathcal{C}(I, \mathcal{M}_n)$. Let the functional*

$$J(h) = \int_I \{\langle h(t), A(t)h(t)\rangle + \langle \dot{h}(t), B(t)\dot{h}(t)\rangle\}\, dt$$

be nonnegative on $E_1^0(I)^n$, or, in other words, $J(h) \geq 0$, $\forall h \in E_1^0(I)^n$. Then $B(t)$ is a positive definite matrix for all $t \in I$, i.e. $\langle y, B(t)y\rangle \geq 0$, $\forall y \in \mathbb{R}^n$ and $\forall t \in I$.

If we apply Lemma 5.2.11 to the functional (5.2.17) then we obtain *Legendre's necessary condition* for a minimising trajectory $q \in E_1(I)^n$. In other words, the matrix $B(t) = L_{\dot{q}\dot{q}}(t)$ must be positive definite for all $t \in I$:

$$\langle q, L_{\dot{q}\dot{q}}(t)q\rangle \geq 0\,, \quad \forall t \in I\,, \quad \forall y \in \mathbb{R}^n\,. \tag{5.2.20}$$

In our investigations of the necessary and sufficient conditions we shall further assume, for the same reason as in the one-dimensional case, the natural strengthening of the Legendre condition (5.2.20), that is, we let $B(t) = L_{\dot{q}\dot{q}}(t)$ be strictly positive definite $\forall t \in I$, so that

$$\langle y, B(t)y\rangle = \langle y, L_{\dot{q}\dot{q}}(t)y\rangle > 0\,, \quad \forall t \in I\,, \quad \forall y \in \mathbb{R}^n \setminus \{0\}\,. \tag{5.2.21}$$

We investigate on $E_1^0(I)^n$ general quadratic functionals of the form

$$J(h) = \int_I \{\langle h(t), A(t)h(t)\rangle + \langle \dot{h}(t), B(t)\dot{h}(t)\rangle\}\, dt \tag{5.2.22 a}$$

where

$$A \in \mathcal{C}(I, \mathcal{M}_n), \quad A(t)^* = A(t), \quad \forall t \in I, \tag{5.2.22 b}$$

and

$$B \in \mathcal{C}(I, \mathcal{M}_n), \quad B(t) \geq b\mathbb{1}_n, \quad b > 0. \tag{5.2.22 c}$$

A clarification of the question as to when a functional such as J on $E_1^0(I)^n$ is positive definite or strictly positive definite will produce not only necessary but also sufficient conditions (Jacobi's conditions) for trajectories which minimise the action W.

The condition $B(t) \geq b\mathbb{1}_n$, $b > 0$, implies that we can define $B^{-1}(t) = (B(t))^{-1}$ and $B^{+1/2}(t) = (B(t))^{1/2}$. Then

$$B^{-1}(t)B(t) = B(t)B^{-1}(t) = B^{-1/2}(t)B^{1/2}(t) = \mathbb{1}_n\,,$$
$$B^{1/2}(t)B^{1/2}(t) = B(t)\,.$$

For an arbitrary function $\omega \in \mathcal{C}^1(I, \mathcal{M}_n)$ and all $h \in E_1^0(I)^n$, the identity

$$\begin{aligned} 0 &= \int_I \frac{d}{dt}\langle h(t), \omega(t)h(t)\rangle\, dt \\ &= \int_I dt\{\langle \dot{h}, \omega h\rangle + \langle h, \dot{\omega} h\rangle + \langle h, \omega\dot{h}\rangle\} \end{aligned}$$

holds. Thus, here too, we can rewrite the functional (5.2.22 a) in the following way:

$$\begin{aligned} J(h) = \int_I dt\{&\langle B^{1/2}\dot{h} + B^{-1/2}\omega h, B^{1/2}\dot{h} + B^{-1/2}\omega h\rangle \\ &+ \langle h, [A + \dot{\omega} - \omega^* B^{-1}\omega]h\rangle\}\,. \end{aligned} \tag{5.2.23}$$

Here ω^* denotes the adjoint matrix of ω.

The positivity properties of the functional $J(h)$ can again be easily explained if we are able to write J as an integral over the square of a real function, i.e. if the differential equation

$$A + \dot{\omega} = \omega^* B^{-1}\omega \tag{5.2.24}$$

for the matrix ω has a solution in $\mathcal{C}^1(I, \mathcal{M}_n)$.

Since $\omega^* B^{-1}\omega$ is symmetric, every solution of (5.2.24) for the symmetric initial condition $\omega(a) = \omega^*(a)$ is itself symmetric, $\omega(t) = \omega^*(t)$, $\forall t \in I$, since a solution ω of the differential equation (5.2.24) satisfies the integral equation

$$\omega(t) = \omega(a) - \int_a^t A(\tau)\, d\tau + \int_a^t (\omega^* B^{-1}\omega)(\tau)\, d\tau\,.$$

As we might expect from our consideration of the one-dimensional case, and as we can indeed show, the existence or nonexistence of a solution $\omega \in \mathcal{C}^1(I, \mathcal{M}_n)$ of the Ricatti-type differential equation (5.2.24) can be characterised with the aid of the concept of the conjugate point for the Euler equation associated with the functional $J(h)$:

$$-\frac{d}{dt}(B\dot{u}) + Au = 0\,, \quad u \in \mathcal{C}^2(I, \mathbb{R}^n)\,. \tag{5.2.25}$$

As a first step, let us discuss a characterisation of conjugate points. Let $A \in \mathcal{C}(I, \mathcal{M}_n)$ and $B \in \mathcal{C}^1(I, \mathcal{M}_n)$ be given, matrix-valued functions with the following properties:

(i) $A(t)^* = A(t)\,, \quad \forall t \in I\,,$
(ii) $B(t) \geq b\mathbb{1}_n\,, \quad b > 0\,, \quad \forall t \in I\,.$

The Euler equation (5.2.25) associated with the functional J is often known as the *Jacobi equation* of the functions A and B. One then refers to a point $c \in (a, b]$ as a *conjugate point of* a with respect to A and B, as long as a nontrivial solution of the Jacobi equation exists which vanishes at the points a and c.

Making these assumptions, the Jacobi equation (5.2.25) has $2n$ linearly independent solutions on I. As in the one-dimensional case, we distinguish the following solutions.

Let $v_{0,1}^j$ be the solution of (5.2.25) with

$$v_{0,1}^j(a) = 0\,, \quad \dot{v}_{0,1}^j(a) = e_j\,, \quad j = 1, \ldots, n\,, \tag{5.2.26 a}$$

and let $v_{1,0}^j$ be the solution of (5.2.25) with

$$v_{1,0}^j(a) = e_j\,, \quad \dot{v}_{1,0}^j(a) = 0\,, \quad j = 1, \ldots, n\,. \tag{5.2.26 b}$$

Here $(e_1, \ldots, e_n)$ denotes the canonical basis of $\mathbb{R}^n$. We now interpret the solutions $v_{0,1}^1, \ldots, v_{0,1}^n$ as the columns of an $n \times n$ matrix V_{01} and, analogously, the solutions $v_{1,0}^1, \ldots, v_{1,0}^n$ as the columns of an $n \times n$ matrix V_{10}. The functions $V_{01} \in \mathcal{C}^2(I, \mathcal{M}_n)$ and $V_{10} \in \mathcal{C}^2(I, \mathcal{M}_n)$ are then solutions of the differential equation

$$-\frac{d}{dt}(B\dot{V}) + AV = 0\,, \quad V \in \mathcal{C}^2(I, \mathcal{M}_n)$$

for the initial conditions

$$V_{01}(a) = 0\,, \quad \dot{V}_{01}(a) = \mathbb{1}_n\,,$$
$$V_{10}(a) = \mathbb{1}_n\,, \quad \dot{V}_{10}(a) = 0\,.$$

The choice of the initial conditions guarantees that the solutions $\{v_{0,1}^j(t), v_{1,0}^j(t)\}$, $j = 1, \ldots, n$, generate the solution space of the Jacobi equation. For any solution u of the differential equation (5.2.25), we thus know that

$$u(t) = \sum_{j=1}^{n} \left[v_{1,0}^{j}(t) u_j(a) + v_{0,1}^{j}(t) \dot{u}_j(a) \right] . \tag{5.2.27}$$

For $2n$ solutions $u^1, \ldots, u^{2n}$, this relation can be conveniently written as a matrix equation:

$$\begin{pmatrix} u^1(a) \cdots u^{2n}(a) \\ u^1(t) \cdots u^{2n}(t) \end{pmatrix} = \begin{pmatrix} v_{10}^1(a) \cdots v_{10}^1(a) & v_{01}^1(a) \cdots v_{01}^n(a) \\ v_{10}^1(t) \cdots v_{10}(t) & v_{01}^1(t) \cdots v_{01}^n(t) \end{pmatrix} \begin{pmatrix} u^1(a) \cdots u^{2n}(a) \\ \dot{u}^1(a) \cdots \dot{u}^{2n}(a) \end{pmatrix}$$

$$= \begin{pmatrix} V_{10}(a) & V_{01}(a) \\ V_{10}(t) & V_{01}(t) \end{pmatrix} \begin{pmatrix} u^1(a) \cdots u^{2n}(a) \\ \dot{u}^1(a) \cdots \dot{u}^{2n}(a) \end{pmatrix} .$$

One now refers to

$$\Delta_{u^1 \ldots u^{2n}}(t) = \det \begin{pmatrix} u^1(a) \cdots u^{2n}(a) \\ u^1(t) \cdots u^{2n}(t) \end{pmatrix}$$

as the *Mayer determinant* of the solutions $u^1, \ldots, u^{2n}$ and

$$W_{u^1 \ldots u^{2n}}(t) = \det \begin{pmatrix} u^1(t) \cdots u^{2n}(t) \\ \dot{u}^1(t) \cdots \dot{u}^{2n}(t) \end{pmatrix}$$

is known as the *Wronskian* (or *Wronski's determinant*) of these solutions.

By virtue of the initial conditions, equation (5.2.27) gives the important relation

$$\Delta_{u^1, \ldots, u^{2n}}(t) = D_{01}(t) W_{u^1, \ldots, u^{2n}}(t) \tag{5.2.28}$$

where

$$D_{01}(t) = \det V_{01}(t) .$$

This relation is the basis of the characterisation of conjugate points. The following lemma holds:

Lemma 5.2.12. *Subject to the assumption above, $c \in (a, b]$ is a point conjugate to a with respect to A and B if, and only if, one of the following equivalent conditions is true:*

(a) $D_{01}(c) = 0$.

(b) *The Mayer determinant $\Delta_{u^1, \ldots, u^{2n}}$ vanishes at the point c for a system of solutions $u^1, \ldots, u^{2n}$ with $W_{u^1, \ldots, u^{2n}} \neq 0$; in other words, $\Delta_{u^1, \ldots, u^{2n}}(c) = 0$.*

Proof. The equivalence of the statements (a) and (b) is justified by equation (5.2.28). If $c \in [a, b]$ is a point conjugate to a, then a nontrivial solution u of the Jacobi equation (5.2.25) exists, such that $u(a) = u(c) = 0$. Also, the nontriviality of u (i.e. $u \neq 0$) requires that $\dot{u}(a) \neq 0$. Equation (5.2.27) implies that

$$u = \sum_{j=1}^{n} \dot{u}_j(a) v_{01}^{j} = V_{01} \dot{u}(a) ,$$

and accordingly, since $\dot{u}(a) \neq 0$, one has

$$0 = u(c) = V_{01}(c) \cdot \dot{u}(a)\,,$$
$$0 = D_{01}(c) = \det V_{01}(c)\,.$$

Thus, (a) follows.

Conversely, if (a) holds, then, because $\det V_{01}(c) = 0$, the equation $V_{01}(c)\lambda = 0$ possesses a nontrivial solution $\lambda \in \mathbb{R}^n$, $\lambda \neq 0$. Put

$$u(\cdot) = V_{01}\lambda = \sum_{j=1}^{n} \lambda_j v_{01}^j(\cdot)\,.$$

It follows that

$$u(a) = V_{01}(a) \cdot \lambda = 0 \cdot \lambda = 0\,,$$
$$u(c) = V_{01}(c) \cdot \lambda = 0\,,$$
$$\dot{u}(a) = \dot{V}_{01}(a) \cdot \lambda = \lambda \neq 0\,.$$

Thus $V_{01}(\cdot)\lambda$ is a nontrivial solution of Jacobi's differential equation (5.2.25) which vanishes at a and c. It follows that c is a point conjugate to a. □

The following theorem characterises the positivity properties of the functional (5.2.22 a) in terms of conjugate points, assuming that (5.2.22 b) and (5.2.22 c) hold.

Theorem 5.2.13. (a) *The functional J defined by (5.2.22 a) on $E_1^0(I)^n$ is strictly positive definite if, and only if, no point conjugate to a with respect to A and B exists in $(a, b]$.*

(b) *If the functional $J(h) = \int_I dt\{\langle h, Ah\rangle + \langle \dot{h}, B\dot{h}\rangle\}$ is positive definite on $E_1^0(I)^n$, then the open interval (a, b) contains no point c conjugate to a with respect to A and B.*

Proof. Let the functional J be (strictly) positive definite on $E_1^0(I)^n$. Then, by contradiction, we want to rule out the existence of a conjugate point. Suppose the assertion to be false. Then there would exist a point c_1 conjugate to a with respect to the functions A and B. If $c_1 = b$, then there would exist a $u^1 \in E_1^0(I)^n$, $u^1 \neq 0$, which was a solution to the Jacobi equation, which implies

$$J(u_1) = \int_I dt\Big\langle u^1, Au^1 - \frac{d}{dt}(B\dot{u}^1)\Big\rangle = 0\,.$$

If $J(h) > 0$, $\forall h \in E_1^0(I)^n$, $h \neq 0$, then $c_1 = b$ is not possible as a conjugate point. From this we can assume that $c_1 \in (a, b)$. We are led to a contradiction, as in the one-dimensional case, by deforming the given functional $J = J_1$ continuously into the strictly positive functional

$$J_0(h) = \int_I dt\langle \dot{h}, \dot{h}\rangle\,.$$

To this end, consider the family of functionals J_λ, $\lambda \in [0,1]$, on $E_1^0(I)^n$:

$$J_\lambda(h) = \int_I \{\langle \dot{h}, B_\lambda \dot{h}\rangle + \langle h, A_\lambda h\rangle\} dt$$

where

$$B_\lambda = \lambda B + (1-\lambda)\mathbb{1}_n\,, \quad A_\lambda = \lambda A\,,$$

and the corresponding family of Jacobi equations

$$-\frac{d}{dt}(B_\lambda \dot{U}) + A_\lambda U = 0\,, \quad U \in \mathcal{C}^2(I, \mathcal{M}_n)\,.$$

By Lemma 5.2.12, the conjugate points with respect to the functions A_λ and B_λ can be characterised by means of the solution U_λ of these equations for the initial conditions

$$U_\lambda(a) = 0\,, \quad \dot{U}(a) = \mathbb{1}\,.$$

These solutions depend continuously on λ. Thus, the assumption that a conjugate point $c_1 \in (a,b)$ exists for $\lambda = +1$ allows us to conclude, as in the proof of Theorem 5.2.8, that a conjugate point $c_0 \in (a,b)$ for the functions $A_0 = 0$ and $B_0 = \mathbb{1}_n$ must also exist for $\lambda = 0$. But this is a contradiction, since the solution U_0 of

$$-\frac{d}{dt}(B_0 \dot{U}) + A_0 U = -\ddot{U} = 0$$

for the initial conditions $U_0(a) = 0$ and $\dot{U}_0(a) = \mathbb{1}_n$ is

$$U_0(t) = (t-a)\mathbb{1}_n\,.$$

Thus (b) follows, as does the first part of assertion (a).

Conversely, let there be no point conjugate to a with respect to A and B in $(a,b]$. Giving our heuristic observations in connection with equation (5.2.23) and (5.2.24) a more precise form, it is possible to show that the functional J is strictly positive definite.

If there is no conjugate point in $(a,b]$, then the solution $V = V_{01}$ can be inverted on $(a,b]$, i.e.

$$t \to V^{-1}(t) = V(t)^{-1}$$

is a function in $\mathcal{C}^1((a,b], \mathcal{M}_n)$. It follows that

$$t \to \omega(t) = -(B\dot{V}V^{-1})(t) \tag{5.2.29}$$

is also an element in $\mathcal{C}^1((a,b], \mathcal{M}_n)$ with derivative

$$\dot{\omega} = -A - \omega\dot{V}V^{-1} = -A + \omega B^{-1}\omega\,;$$

thus

$$\dot{\omega} + A = \omega B^{-1}\omega\,. \tag{5.2.24'}$$

Now we show the symmetry of the solution defined by (5.2.29) on $(a, b]$ of the differential equation (5.2.24′). Note that for arbitrary solutions U, W of the Jacobi equation $(d/dt)(B\dot{U}) = AU$ the following relation holds:

$$U^* B\dot{W} - \dot{U}^* BW = \text{const.} \tag{5.2.30}$$

To prove this, one shows that because of the Jacobi equation and the symmetry of A and B, the derivative of

$$\langle Ux, B\dot{W}y\rangle - \langle \dot{U}x, BWy\rangle$$

vanishes for all $x, y \in \mathbb{R}^n$.

If we now apply (5.2.30) to $W = V = V_{01} = U$, it follows that

$$V^* B\dot{V} - \dot{V}^* BV = (V^* B\dot{V} - \dot{V}^* BV)(a) = 0\,,$$

so that

$$V^* B\dot{V} = \dot{V}^* BV\,,$$

and thus

$$(-\omega)^* = (B\dot{V}V^{-1})^* = -\omega\,.$$

It follows that

$$(\omega(t))^* = \omega(t)\,, \quad \forall t \in (a, b]\,.$$

Thus the function ω defined by (5.2.29) is also a solution of the differential equation (5.2.24). The representation (5.2.23) of the functional yields

$$\begin{aligned}
J(h) &= \lim_{\varepsilon \to 0+} \int_{a+\varepsilon}^{b} dt\{\langle h, Ah\rangle + \langle \dot{h}, B\dot{h}\rangle\} \\
&= \lim_{\varepsilon \to 0+} \int_{a+\varepsilon}^{b} dt\{\langle B^{1/2}\dot{h} + B^{-1/2}\omega h, B^{1/2}\dot{h} + B^{-1/2}\omega h\rangle \\
&\quad + \langle h, [A + \dot{\omega} - \omega^* B^{-1}\omega] h\rangle\} \\
&= \int_a^b dt\langle B^{1/2}\dot{h} + B^{-1/2}\omega h, B^{1/2}\dot{h} + B^{-1/2}\omega h\rangle \geq 0\,,
\end{aligned}$$

because, since $h \in E_1^0(I)^n$, $\omega(t)h(t)$ has a limit value for $t \to a$, namely $-B(a)h(a)$.

The absence of conjugate points in $(a, b]$ implies therefore that the functional J is positive definite on $E_1^0(I)^n$. Now let $h \in E_1^0(I)^n$ and let $J(h) = 0$. the formula above, i.e.

$$J(h) = \int_I dt \| B^{1/2}\dot{h} + B^{-1/2}\omega h\|^2$$

with $\|\cdot\|^2 = \langle\cdot,\cdot\rangle$, shows that h must satisfy the differential equation $B^{1/2}\dot{h} + B^{-1/2}\omega h = 0$. Moreover, $\omega(t)$ is well defined at the point $t = b$. Since $h \in E_1^0(I)^n$, it follows that $h(b) = \dot{h}(b) = 0$. Since h is a solution of the linear differential equation $\dot{h} + B^{-1}\omega h = 0$, h is the trivial solution, i.e. $h = 0$. Thus the functional J is also strictly positive on $E_1^0(I)^n$. □

To summarise, we can formulate the following characterisation of a minimising trajectory $q \in E_1(I)^n$ of a system with n degrees of freedom.

Theorem 5.2.14 (Principle of Least Action). (a) *Necessary conditions. If $q \in E_1(I)^n$ is a trajectory for which the action $W(q) = \int_I L(t, q(t), \dot{q}(t))dt$ attains a local minimum, and if the second Fréchet derivative of W does not vanish at the point q, then the following hold:*

(i) *Euler's equations*

$$\frac{d}{dt}\frac{\partial L}{\partial \dot{q}_j}(t, q(t), \dot{q}(t)) = \frac{\partial L}{\partial q_j}(t, q(t), \dot{q}(t)), \quad j = 1, \ldots, n, \quad \forall t \in I.$$

(ii) *Legendre's condition. The matrix*

$$L_{\dot{q}\dot{q}}(t) = \left(\frac{\partial^2 L}{\partial \dot{q}_i\, \partial \dot{q}_j}(t, q(t), \dot{q}(t))\right)_{i,j=1,\ldots,n}$$

is positive definite for all $t \in I$.

(iii) *Jacobi's condition. There is no point conjugate to a in (a, b) with respect to the functions $A^{(q)}$ and $B^{(q)}$ defined by*

$$A^{(q)}(t) = L_{qq}(t) - \frac{d}{dt}L_{\dot{q}q}(t),$$
$$B^{(q)}(t) = L_{\dot{q}\dot{q}}(t).$$

Here we assume that $A^{(q)} \in C(I, \mathcal{M}_n)$ and $B^{(q)} \in C^1(I, \mathcal{M}_n)$.

(b) *Sufficient conditions. A trajectory $q \in E_1(I)^n$ may satisfy the following conditions.*

(i) *q is a solution of Euler's equations, i.e.*

$$\frac{d}{dt}\frac{\partial L}{\partial q_j}(t, q(t), \dot{q}(t)) = \frac{\partial L}{\partial q_j}(t, q(t), \dot{q}(t)), \quad \forall t \in I, \quad j = 1, \ldots, n.$$

(ii) *For the functions $A^{(q)}$ and $B^{(q)}$ defined in terms of q in (a) (iii), $A^{(q)} \in C(I, \mathcal{M}_n)$ and $B^{(q)} \in C^1(I, \mathcal{M}_n)$ hold.*

(iii) *$B^{(q)}(t)$ is strictly positive definite for all $t \in I$.*

(iv) *There is no point conjugate to a in the interval $(a, b]$ with respect to the functions $A^{(q)}$ and $B^{(q)}$.*

Then the action $W(q) = \int_I L(t, q(t), \dot{q}(t))dt$ has a strict local minimum for the trajectory q.

5.2.3 An Example from Classical Mechanics

For many systems in classical mechanics with n degrees of freedom, the Lagrangian L has the form

$$L(t,q,\dot{q}) = \frac{1}{2}\langle \dot{q}, M\dot{q}\rangle - V(q)\,,$$

where the matrix $M = M^*$ is symmetric, i.e. $M_{ij} = M_{ji}$, $i,j = 1,\ldots,n$, and V is a potential energy which we take to be twice continuously differentiable. It follows then that

$$L_q = -V_q\,, \quad L_{\dot{q}} = M\dot{q}\,,$$
$$L_{q\dot{q}} = 0\,, \quad L_{\dot{q}\dot{q}} = M\,, \quad L_{qq} = -V_{qq}\,.$$

The corresponding Euler equation is

$$\frac{d}{dt}(M\dot{q}) = -V_q\,,$$

which is just Newton's equation.

Now we want to discuss exactly when a solution q of this equation minimises the action function

$$W(q) = \int_I L(t,q(t),\dot{q}(t))dt$$

for a time interval $I = [a,b]$.

Legendre's necessary condition requires that $L_{\dot{q}\dot{q}} = M$ be positive definite, and the strengthened form of Legendre's condition requires in addtition that M be strictly positive definite, or in other words, $M \geq m\mathbb{1}_n$, $m > 0$. Since M defines the quadratic form of the kinetic energy T of the system,

$$T = \frac{1}{2}\langle \dot{q}, M\dot{q}\rangle\,,$$

these are quite natural requirements for the Lagrangian given above. It is a lot more difficult to satisfy Jacobi's conditions. The functions $A = A^{(q)}$ and $B = B^{(q)}$ associated with a solution q of Euler's equation are given by

$$A(t) = L_{qq} = -V_{qq}(q(t))$$

and

$$B(t) = L_{\dot{q}\dot{q}} = M$$

in this case. The Jacobi equation associated with A and B, i.e. the Euler equation of the functional $J(h)$ on $E_1^0(I)^n$ where

$$J(h) = \int_I \left\{\langle h, Ah\rangle + \langle \dot{h}, B\dot{h}\rangle\right\} dt = J_I \left\{\langle \dot{h} M \dot{h}\rangle - \langle h, V_{qq} h\rangle\right\} dt\,,$$

is accordingly,

$$-\frac{d}{dt}(M\dot{U}) - V_{qq}(q(t))U = 0\,.$$

If the matrix M is time-independent and strictly positive, then this equation can be written

$$\ddot{U} + M^{-1}V_{qq}(q(t))U = 0\,.$$

For a general potential, it is not easy to decide whether a nontrivial solution $U \in C^2(I\mathbb{R}^n)$ with $U(a) = 0$ and $U(t) \neq 0$, $\forall t \in (a,b]$, exists for this equation, i.e. whether no conjugate point with respect to the functions $A = M$ and $B = -V_{qq}$ exists in $(a,b]$. We shall thus discuss an example.

Let

$$V(q) = \frac{1}{2}\langle q, Kq\rangle\,, \quad K = K^*\,, \quad K \in \mathcal{M}_n\,.$$

It follows that $V_{qq}(q(t)) = K$, $\forall t \in I$, and one thus has the Jacobi equation

$$\ddot{U} + M^{-1}KU = 0\,.$$

Let the matrix $M^{-1}K \in \mathcal{M}_n$ be diagonalised by the matrix $G \in \mathcal{M}_n$, i.e.

$$GM^{-1}KG^{-1} = D = \begin{pmatrix} \lambda_1 & 0 & \cdots & 0 \\ 0 & \lambda & \cdots & 0 \\ 0 & 0 & \cdots & \lambda_n \end{pmatrix}$$

where $\lambda_j \in \mathbb{R}$. We then have for $h = GU$

$$\ddot{h} + Dh = 0$$

or $\ddot{h}_j + \lambda_j h_j = 0$, $j = 1, \ldots, n$. The solutions for the initial conditions $h_j(a) = 0$ are

$$h_j(t) = d_j \sin(\omega_j(t-a))\,, \quad \omega_j = +\sqrt{\lambda_j}\,, \quad \lambda_j > 0$$

or

$$h_j(t) = d_j \sinh(\omega_j(t-a))\,. \quad \omega_j = +\sqrt{-\lambda_j}\,, \quad \lambda_j < 0$$

where $d_j \in \mathbb{R}$.

Thus, if $\lambda_j < 0$ for $j = 1, \ldots, n$, no point conjugate to a exists in $(a,b]$ and the trajectory q minimises the action W. On the other hand, if $\lambda_j > 0$ for some $j \in \{1, 2, \ldots, n\}$, then we have to distinguish two cases.

If $(b-a)\max(\sqrt{\lambda_j}\,|j \in \{1, \ldots, n\}, \lambda_j > 0) < \pi$, then the corresponding solutions h_j have no further zeros in $(a,b]$. Then $(a,b]$ contains no point which is conjugate to a and in this case the action is a minimum.

However, if $(b-a)\max(\sqrt{\lambda_j}\,|j \in \{1, \ldots, n\}, \lambda_j > 0) > \pi$, then the corresponding solutions u_j have a least one zero in (a,b). It follows that solutions $h = (h_1, \ldots, h_n)$ exist with

$$h(a) = h(c) = 0\,, \quad a < c < b\,, \quad h \neq 0\,.$$

Therefore a point conjugate to a exists in (a,b). Thus, in this case the trajectory q does not minimise the action W.

Remark 5.2.5. Legendre's conditions, i.e.

$$L_{\dot{q}\dot{q}} \geq 0 \quad \text{or} \quad L_{\dot{q}\dot{q}} > 0$$

express a certain partial convexity of the Lagrangian $L(t, q(t), \dot{q}(t))$ in the variables $\dot{q}(t)$.

Remark 5.2.6. We have formulated all results only for a local minimum of the action functional W. Similar results also hold *mutatis mutandis* for a local maximum.

5.3 Symmetries and Conservation Laws in Classical Mechanics

5.3.1 Hamiltonian Formulation of Classical Mechanics

In the previous section we discussed in detail the variational approach to the Lagrangian formulation of classical mechanics. However, in physics the so-called Hamiltonian formulation of classical mechanics is also very important, since traditional formulations of quantum mechanics usually start from it. Thus we shall now describe briefly the variational approach to this formulation, which is just as important as the Lagrangian formulation. We start with the Lagrangian $L \in C^2(I, \mathbb{R}^{2n})$ of a system with $n \geq 1$ degrees of freedom. The *canonical momenta* of the system are then well defined by

$$p_k = p_k(t, q, \dot{q}) = \frac{\partial L}{\partial \dot{q}_k}(t, q, \dot{q}), \quad k = 1, \dots, n.$$

If we assume in addition that

$$\det L_{\dot{q}\dot{q}} = \det \left(\frac{\partial^2 L}{\partial \dot{q}_j \, \partial \dot{q}_k}(t, q, \dot{q}) \right) \neq 0,$$

then the equations $p_k = (\partial L / \partial \dot{q}_k)(t, q, \dot{q})$ can, with the aid of the implicit function theorem, be solved at least locally for $\dot{q}_j$, $j = 1, \dots, n$,

$$\dot{q}_j = \dot{q}_j(t, q, p).$$

In this case the *Hamiltionian* $H(t, q, p)$,

$$H(t, p, q) = \left\{ \sum_{k=1}^{n} p_k \dot{q}_k - L(t, q, \dot{q}) \right\} \Bigg|_{\dot{q} = \dot{q}(t, q, p)},$$

is obtained from the Lagrangian by means of a *Legendre transformation* (or *contact transformation*). The designation *Hamiltonian* is not quite legitimate, because Lagrange was the first to investigate this function – using the same notation H, moreover!

The Legendre transformation takes the Euler-Lagrange equations

$$\frac{\partial L}{\partial p_k} = \frac{d}{dt}\frac{\partial L}{\partial \dot{p}_k}$$

into the so-called *Hamiltonian equations of motion*

$$\dot{q}_k = \frac{\partial H}{\partial p_k}, \quad \dot{p}_k = -\frac{\partial H}{\partial p_k}, \quad -\frac{\partial L}{\partial t} = \frac{\partial H}{\partial t}, \quad k = 1, \ldots, n,$$

since, on the one hand, the total differential of H is

$$dH = \frac{\partial H}{\partial t}dt + \sum_{k=1}^{n}\left\{\frac{\partial H}{\partial q_k}dq_k + \frac{\partial H}{\partial p_k}dp_k\right\}$$

and, on the other hand, from our definition,

$$\begin{aligned} dH &= \sum_{k=1}^{n}\{\dot{q}_k dp_k + p_k d\dot{q}_k\} - \frac{\partial L}{\partial t}dt - \sum_{k=1}^{n}\left\{\frac{\partial L}{\partial q_k}dq_k + \frac{\partial L}{\partial \dot{q}_k}d\dot{q}_k\right\} \\ &= \sum_{k=1}^{n}\left\{\dot{q}_k dp_k - \frac{\partial L}{\partial q_k}dq_k\right\} - \frac{\partial L}{\partial t}dt, \end{aligned}$$

The Hamiltionian equations of motion follow easily, using

$$\frac{\partial L}{\partial p_k} = \frac{d}{dt}\frac{\partial L}{\partial \dot{q}_k} = \dot{q}_k.$$

Instead of n second-order Euler-Lagrange equations in configuration space we now have $2n$ first-order Hamilton equations in phase space.

In the variational approach to these equations we have to investigate the action as a function of trajectories in the *phase space* $\Gamma = \{(q,p) | q \in \mathbb{R}^n, p \in \mathbb{R}^n\}$. Let $I = [a, b]$ be a time interval and

$$\mathcal{E}^1(I) = \{(q,p) \in \mathcal{C}^1(I; \mathbb{R}^{2n}) | q(a) = q_{\mathrm{S}}, q(b) = q_{\mathrm{E}}\},$$

$$\mathcal{E}^1_0(I) = \{(q,p) \in \mathcal{C}^1(I; \mathbb{R}^{2n}) | q(a) = 0 = q(b)\}.$$

By virtue of $L\,dt = \sum_{k=1}^{n} p_k dq_k - H\,dt$, the action $W = \int_I L\,dt$ can also be written as a line integral in phase space ($(q,p) \in \mathcal{E}^1(I)$):

$$W(q,p) = \int_I \left\{\sum_{k=1}^{n} p_k(t)\dot{q}_k(t) - H(t, q(t), p(t))\right\} dt.$$

The Fréchet derivative of the function $W : \mathcal{E}^1(I) \to \mathbb{R}$ is the linear mapping $\mathcal{E}^1_0(I) \to \mathbb{R}$ defined by

$$DW(q,p)(h,g) = \int_I \sum_{k=1}^{n} \left\{ g_k(t) \left(\dot{q}_k(t) - \frac{\partial H}{\partial p_k} \right) + p_k(t)\dot{h}_k(t) - \frac{\partial H}{\partial q_k} h_k(t) \right\} dt$$

$$= \int_I \sum_{k=1}^{n} \left\{ g_k(t) \left(\dot{q}_k(t) - \frac{\partial H}{\partial p_k} \right) - h_k(t) \left(\dot{p}_k(t) + \frac{\partial H}{\partial q_k} \right) \right\} dt\,.$$

Accordingly, by our previous arguments, the action W being stationary for the trajectory $(q,p) \in \mathcal{E}^1(I)$ is equivalent to this trajectory being a solution of the Hamilton equations.

5.3.2 Coordinate Transformations and Integrals of Motion

We start with a short characterisation of the integrals of motion of a system in terms of the Hamiltonian. A function $f : \mathbb{R} \times \Gamma \to \mathbb{R}$ is called an *integral of motion* if f is constant on all trajectories in phase space. Hamilton's equations hold along a trajectory $(q,p) \in \mathcal{C}^1(\mathbb{R}, \mathbb{R}^{2n})$. Thus a $\mathcal{C}^1$-function $f(t,q,p)$ is an integral of motion if, and only if, $0 = (d/df)f(t,q(t),p(t))$ for every trajectory $t \to (q(t),p(t))$ in phase space. Now

$$\frac{d}{dt} f(t,q(t),p(t)) = \frac{\partial f}{\partial t} + \sum_{j=1}^{n} \left\{ \frac{\partial f}{\partial q_j} \dot{q}_j(t) + \frac{\partial f}{\partial p_j} \dot{q}_j(t) \right\}$$

$$= \frac{\partial f}{\partial t} + \sum_{j=1}^{n} \left\{ \frac{\partial f}{\partial q_j} \frac{\partial H}{\partial p_j} - \frac{\partial f}{\partial p_j} \frac{\partial H}{\partial q_j} \right\}$$

$$\equiv \frac{\partial f}{\partial t} + \{H, f\}\,,$$

if the *Poisson bracket* $\{f,g\}$ of two functions $f,g : \Gamma \to \mathbb{R}$ is defined by

$$\{f,g\} = \sum_{j=1}^{n} \left(\frac{\partial f}{\partial p_j} \frac{\partial g}{\partial q_j} - \frac{\partial f}{\partial q_j} \frac{\partial g}{\partial p_j} \right)\,.$$

It follows that a $\mathcal{C}^1$-function $f : \mathbb{R} \times \Gamma \to \mathbb{R}$ is an integral of motion if, and only if,

$$\frac{\partial f}{\partial t} + \{H, f\} = 0\,.$$

The variational formulation of the equations of motion as "points" of stationary action can be used if we want to investigate the invariance of the equations of motion under certain transformation and thus discover the integrals of motion. The fundamental result here is *E. Noether's Theorem* [5.1] which we shall now discuss.

First of all, let us recall that two Lagrangians L and L' which differ only in a summand of the form

$$\frac{d}{dt} G(t,q)$$

lead to action functions $W(q) = \int_I L\,dt$ and $W'(q) = \int L'\,dt$ which differ on $E^1(I)^n$ in an additive constant: from

$$L' = L + \frac{d}{dt}G$$

we have, for $q \in E^1(I)^n \equiv E^1(I; q_S, q_E)^n$,

$$W'(q) = W(q) + \int_I dt \frac{d}{dt} G(t, q(t)) = W(q) + G(b, q_E) - G(a, q_S).$$

As we learned earlier, the following relationship exists between the equations of motion

$$[L]_j(q)(t) = \left(\frac{\partial L}{\partial q_j} - \frac{d}{dt}\frac{\partial L}{\partial \dot{q}_j} \right)(t, q(t), \dot{q}(t)) = 0, \quad j = 1, \ldots, n$$

associated with a Lagrangian L, and the Fréchet derivative $DW(q)$ of the action function defined by L:

$$DW(q)(h) = \int_I \sum_{j=1}^{n} [L]_j(q)(t) h_j(t) dt,$$

for all $h \in E_0^1(I)^n$. Thus, the Lagrangians L and L' lead to the same equations of motion.

We begin our investigations of the invariance of Lagrangians under coordinate transformations by a few remarks about one-parameter groups of transformations in $\mathbb{R}^n$. Let $(h_s)_{s\in\mathbb{R}}$ be a one-parameter group of diffeomorphisms $h_s : \mathbb{R}^n \to \mathbb{R}^n$ of $\mathbb{R}^n$, where $h_0 = \mathrm{id}_{\mathbb{R}^n} = \mathbb{1}_n$. Let $(s, x) \to h_s(x)$ be a function in $\mathcal{C}^2(\mathbb{R} \times \mathbb{R}^n; \mathbb{R}^n)$. A mapping

$$H_s : E^1(I; q_S, q_E)^n \to E^1(I; h_s(q_S), h_s(q_E))^n$$

is induced by h_s by virtue of

$$q \to H_s(q) = h_s \circ q.$$

We shall now determine the derivative of this mapping. Let $q \in E^1(I; q_S, q_E)^n$ and $f \in E_0^1(I)^n$. The differentiability of h_s yields for all $t \in I$

$$\begin{aligned} H_s(q + f)(t) &= h_s(q(t) + f(t)) \\ &= H_s(q)(t) + \frac{\partial h_s}{\partial x}(q(t)) f(t) + o(f(t)). \end{aligned}$$

It follows that $DH_s(q)$ is the linear mapping $DH_s(q) : E_0^1(I)^n \to E_0^1(I)^n$ defined by

$$(DH_s(q)(f))(t) := \frac{\partial h_s}{\partial x}(q(t)) f(t).$$

Since h_s is a diffeomorphism, it has at all points $x \in \mathbb{R}^n$ an invertible Jacobi matrix

$$\frac{\partial h_s}{\partial x}(x) = \left(\frac{\partial h_s}{\partial x_k}(x), j, k = 1, \ldots, n \right).$$

It follows that $DH_s(q)$ is an invertible mapping $E_0^1(I)^n \to E_0^1(I)^n$ for all $q \in E^1(I)$.

Let $Q_s \in E^1(I; h_s(q_S), h_s(q_E))^n$ be defined by

$$Q_s(t) = H_s(q)(t) = h_s(q(t))$$

for $q \in E^1(I; q_S, q_E)^n$. It follows that

$$\dot{Q}_s(t) = \frac{d}{dt} Q_s(t) = DH_s(q)(\dot{q})(t),$$

if we use the notation $DH_s(q)$ for the natural extension of $DH_s(q)$ to $\mathcal{C}(I, \mathbb{R}^n)$.

Finally, let L be a Lagrangian with standard properties that is "essentially invariant" under $(h_s)_{s \in \mathbb{R}}$; in other words, let there be a $\mathcal{C}^2$-function

$$\mathbb{R} \times \mathbb{R} \times \mathbb{R}^n \ni (s, t, x) \to G_s(t, x) \in \mathbb{R}$$

such that for all $q \in E^1(I; q_S, q_E)^n$ and all $s \in \mathbb{R}$

$$L(t, Q_s(t), \dot{Q}(t)) = L(t, q(t), \dot{q}(t)) + \frac{d}{dt} G_s(t, Q_s(t)), \quad t \in I.$$

It then follows that

$$\begin{aligned} W(Q_s) = W(H_s(q)) &= \int_I L(t, Q_s(t), \dot{Q}_s(t)) dt \\ &= W(q) + G_s(b, Q_s(b)) - G_s(a, G_s(a)) \end{aligned}$$

and thus, for all $s \in \mathbb{R}$,

$$DW(q) = DW(H_s(q)) DH_s(q).$$

Since we saw above that $DH_s(q)$ is invertible, the invariance of the equations of motion under h_s, $s \in \mathbb{R}$, follows, viz. $DW(q) = 0$ if, and only if, $DW(H_s)(q)) = 0$ for all $s \in \mathbb{R}$, or, in more detail,

$$[L]_j(q) = 0, \quad j = 1, \ldots, n,$$

if, and only if,

$$[L]_j(Q_s) = 0, \quad j = 1, \ldots, n, \quad \text{for all } s \in \mathbb{R}.$$

We shall now show that the one-parameter group $(h_s)_{s \in \mathbb{R}}$ allows us, in the case above, to determine an integral of motion for the Lagrangian L. It follows from

$$L(t, q, \dot{q}) = L(t, Q_s, \dot{Q}_s) - \frac{d}{dt} G_s(t, Q_s)$$

that

$$0 = \frac{d}{ds} L(t, q, \dot{q})$$
$$= \frac{\partial L}{\partial q}(t, Q_s, \dot{Q}_s) \frac{dQ_s}{ds} + \frac{\partial L}{\partial \dot{q}}(t, Q_s, \dot{Q}_s) \frac{d}{ds} \dot{Q}_s - \frac{d}{ds} \frac{d}{dt} G_s(t, Q_s) .$$

Since G is a C^2-function, we can change the order of the derivatives in the last summand. Similarly, from the assumption for h_s,

$$\frac{d}{ds} \dot{Q}_s = \frac{d}{dt} \left(\frac{dQ_s}{ds} \right) .$$

Finally, let us consider the Euler-Lagrange equations for the trajectory q and thus, as we showed above, for the trajectories $Q_s = H_s(q)$. This gives

$$0 = \frac{\partial L}{\partial q}(t, Q_s(t), \dot{Q}_s(t)) \frac{d}{ds} Q_s(t) + \frac{\partial L}{\partial \dot{q}}(t, Q_s(t), \dot{Q}_s(t)) \frac{d}{dt} \frac{d}{ds} Q_s(t)$$
$$- \frac{d}{dt} \frac{d}{ds} G_s(t, Q_s(t))$$
$$= \frac{d}{dt} \left\{ \frac{\partial L}{\partial \dot{q}}(t, Q_s(t), \dot{Q}_s(t)) \frac{d}{ds} Q_s(t) - \frac{d}{ds} G_s(t, Q_s(t)) \right\} .$$

Thus,

$$\widehat{H}_L(t, q, \dot{q}) = \left\{ \frac{\partial L}{\partial \dot{q}}(t, q, \dot{q}) \frac{d}{ds} H_s(q) - \frac{d}{ds} G_s(t, H_s(q)) \right\} \bigg|_{s=0}$$

is an integral of motion, taking into account the fact that $H_0(q) = q$. We have thus proved the following theorem.

Theorem 5.3.1 (E. Noether). *Let $h_s : \mathbb{R}^n \to \mathbb{R}^n$, $s \in \mathbb{R}$, be a one-parameter group of diffeomorphisms of $\mathbb{R}^n$, where $h_0 = \mathrm{id}_{\mathbb{R}^n}$. Let $(s, x) \to h_s(x)$ be a function in $C^2(\mathbb{R} \times \mathbb{R}^n; \mathbb{R}^n)$. Let L be a Lagrangian for which we make the standard assumptions. Furthermore, let a C^2-function*

$$\mathbb{R} \times \mathbb{R} \times \mathbb{R}^n \ni (s, t, x) \to G_s(t, x) \in \mathbb{R}$$

exist such that for all $q \in E^1(I; q_S, q_E)^n$ and for all $s \in \mathbb{R}$,

$$L(t, Q_s(t), \dot{Q}_s(t)) = L(t, q(t), \dot{q}(t)) + \frac{d}{dt} G_s(t, Q_s(t)) , \quad t \in I ,$$

(let $Q_s = h_s \circ q$ and $\dot{Q}_s$ be as defined above). Then

$$\widehat{H}(t, q, \dot{q}) = \left\{ \sum_{j=1}^{n} \frac{\partial L}{\partial \dot{q}_j}(t, q, \dot{q}) \frac{d}{ds} h_{s,j}(q) - \frac{d}{ds} G_s(t, h_s(q)) \right\} \Bigg|_{s=0}$$

is an integral of motion for the Lagrangian L.

Remark 5.3.1. This theorem, which is due to E. Noether, has found many important applications particularly in physics [5.2]. For example, it can be used to show easily [5.3] that 10 integrals of motion exist for the class of Lagrangians of the form

$$L(t,q,\dot{q}) = \sum_{i=1}^{n} \frac{m_i}{2}\dot{q}_i^2 - \sum_{1\leq i<j\leq n} V(|q_i - q_j|)$$

where V is a smooth two-particle interaction. These are the 10 classical integrals of motion (conservation of energy, centre of gravity, total linear momentum, total angular momentum) which follow from the invariance of this Lagrangian under the ten-parameter Galilei group.

5.4 The Brachystochrone Problem

As we mentioned in the introduction the brachystochrone problem is one of the oldest problems in variational calculus. The first solution was given by Johann Bernoulli in 1696; we shall look at his solution briefly at the end of this section. There are other methods of solution put forward by, for example, Jacob Bernoulli, Leibniz and Newton.

The problem is to determine a curve C passing through two points A and B in a plane perpendicular to the earth's surface such that a point mass $m = 1$ moving along C under the influence of gravity travels from A to B in the shortest possible time.

For our variational treatment of the problem we choose a coordinate system in this plane, with the positive x-axis pointing in the direction of the gravitational force. Furthermore, we can take the coordinates of the points A and B in this system to be

$$A = (0,0)\,, \quad B = (b,b')\,, \quad b > 0\,, \quad b' \geq 0\,.$$

The motion of the point mass can be described by the laws of classical mechanics. As we have already seen, the time functional T which is to be minimised is the function $T(q)$ of the "trajectory $q \in E^1([0,b];0,b')$" given by

$$T(q) = \int_0^b \left(\frac{1+q'(x)^2}{2gx}\right)^{1/2} dx\,.$$

The "Lagrangian" for this problem is

$$L(x,q,q') = \left(\frac{1+q'(x)^2}{2gx}\right)^{1/2}\,.$$

It does not satisfy our standard requirements. However, since the singularity of L at $x = 0$ is integrable, it can be shown that the results obtained earlier can be extended to this case. We thus have

$$DT(q)(h) = \int_I \frac{q'(x)}{[2gx(1+q'(x)^2)]^{1/2}} h'(x)dx\,,$$

$$D^2T(a)(h,h) = \int_I \frac{1}{[2gx(1+q'(x)^2)]^{1/2}} \frac{h'(x)^2}{1+q'(x)^2} dx\,,$$

for all $h \in E_0^1(I)$ and all $q \in E^1(I;0,b')$, $I = [0,b]$. $D^2T(q)$ is thus strictly positive definite. Thus every stationary point q of T is a local minimising point (Theorem 3.3.1). A simple calculation shows that the function T is convex on $E^1(I;0,b')$. A local minimum is therefore exactly that absolute minimum of $T(q)$ which we wanted to determine (Theorem 1.1.3). It is thus sufficient to determine a stationary point of T. The condition

$$0 = DT(q)(h) = -\int_I \left[\frac{d}{dx}\left(\frac{q'(x)}{[2gx(1+q'(x)^2)]^{1/2}}\right)\right] h(x)dx$$

for all $h \in E_0^1(I)$ requires that

$$\frac{d}{dx}\left(\frac{q'(x)}{[2gx(1+q'(x)^2)]^{1/2}}\right) = 0 \quad \text{for all } x \in I$$

and thus that

$$\frac{q'(x)}{[2gx(1+q'(x)^2)]^{1/2}} = k = \text{const.} \quad \text{for all } x \in I\,.$$

It follows that

$$\frac{q'^2}{1+q'^2}(x) = 2gk^2x\,, \quad 0 \le x \le b\,,$$

and thus

$$2gk^2b = \frac{q'(b)^2}{1+q'(b)^2} \le 1\,,$$

so that $k^2 \le 1/2\,gb$. If we put $\mathcal{H} = 1/4\,gk^2$, then from

$$q'^2(1-2gk^2x) = 2gk^2x$$

we get

$$g'^2\left(1 - \frac{x}{2\mathcal{H}}\right) = \frac{x}{2\mathcal{H}}\,, \quad \text{where } \frac{b}{2\mathcal{H}} \le 1\,.$$

Since we are looking for solutions $q \in E^1(I;0,b)$ we can rule out the case $b/2\mathcal{H} = 1$. When $b/2\mathcal{H} < 1$, exactly one $t_0 \in (0,\pi)$ exists such that $b = \mathcal{H}(1-\cos t_0)$. The points $x \in [0,b]$ can then the parametrised:

$$x = x(t) = \mathcal{H}(1-\cos t)\,, \quad 0 \le t \le t_0\,.$$

This parametrisation also yields easily the solution of the Euler-Lagrange equation in parametric form. If we put $y(t) = q(x(t))$, then

$$\dot{y}(t) = q'(x(t))\dot{x}(t) = q'(x(t))\mathcal{H}\sin t\,.$$

It follows that

$$\dot{y}(t)^2 = q'(x(t))^2 \mathcal{H}^2 \sin^2 t = \frac{2\mathcal{H}}{1 - x(t)/2\mathcal{H}} \mathcal{H}^2 \sin^2 t$$
$$= \mathcal{H}^2 \frac{1 - \cos t}{1 + \cos t} \sin^2 t = \mathcal{H}^2 (1 - \cos t)^2$$

and thus

$$\dot{y}(t) = \pm \mathcal{H}(1 - \cos t) .$$

Using the initial condition $y(0) = 0$ we obtain the solution

$$y(t) = \pm \mathcal{H}(t - \sin t) .$$

We have thus obtained the solution to the Euler equation for the brachystochrone problem in parametric form:

$$x(t) = \mathcal{H}(1 - \cos t) ,$$
$$y(t) = \pm \mathcal{H}(t - \sin t) , \quad 0 \leq t \leq t_0 .$$

These are the equations of part of a cycloid. A *cycloid* is the curve in the plane which is obtained as the trajectory of a point on a circle (here of radius $\mathcal{H}$) rolling along the y-axis without slipping. This curve passes through the point $B = (b, b')$ if, and only if,

$$b = x(t_0) = \mathcal{H}(1 - \cos t_0) ,$$
$$b' = y(t_0) = \mathcal{H}(t_0 - \sin t_0) . \qquad (\times)$$

We can thus determine t_0 from the ratio $b' : b$.

$$\frac{b'}{b} = \frac{t_0 - \sin t_0}{1 - \cos t_0} , \quad 0 < t_0 < \pi . \qquad (\Delta)$$

$\mathcal{H}$ can then be found very easily.

The function

$$t \to f(t) = \frac{t - \sin t}{1 - \cos t}$$

is monotone increasing in $t \in (0, \pi)$ and its maximum value is $f(\pi) = \pi/2$. Thus there is exactly one solution of (Δ) for $b'/b < \pi/2$ and thus exactly one solution to the brachystochrone problem. This solution is given by the equations above as part of a cycloid through the points $A = (0, 0)$ and $B = (b, b')$. When $b'/b \geq \pi/2$ there is no solution to this problem.

In conclusion, we would like to look briefly at the solution to the brachystochrone problem suggested by Johann Bernoulli. We start with a property which is one of the first things we learn in mechanics, namely, that a point mass falling from an initial position A along an arbitrary curve C has at every point P a velocity which is proportional to $\sqrt{h}$, where h is the vertical distance between A and P. We now divide space into many thin horizontal layers of thickness d. Then we assume that the velocity changes from layer to layer in small steps. The velocity in the first layer is $c\sqrt{d}$, in the second layer it is $c\sqrt{2\,d}$ and in the nth layer it is $c\sqrt{nd}$. The trajectory is supposed to be a straight line within each layer. In order to minimise the total time for crossing all layers, *Snell's law of refraction* from geometrical optics is applied. This states that

$$\frac{\sin\alpha_1}{\sqrt{d}} = \frac{\sin\alpha_2}{\sqrt{2\,d}} = \frac{\sin\alpha_3}{\sqrt{3\,d}} = \ldots = \frac{\sin\alpha_n}{\sqrt{nd}} = \ldots$$

where α_i is the angle between the trajectory and the normal to the boundary between layer i and layer $i-1$. Now let us imagine, together with Johann Bernoulli, that the thickness d of the layers gets smaller and smaller. Taking this limit, the conditions stipulated by the law of diffraction still hold. It thus follows that the solution is a curve C with the property that, if α is the angle between the tangent and the vertical at an arbitrary point P of C and if h is the vertical distance between A and P, then $(\sin\alpha)/\sqrt{h}$ is constant for all points P on C. One can convince oneself very easily that this geometrical property characterises the cycloid. Johan Bernoulli's considerations are in no way a rigorous proof but they are very ingenious and illustrative.

Remark 5.4.1. The problem of the brachystochrone with Coulomb friction has been investigated by N. Ashby et al. [5.4]. In this case too, the solution of the Euler-Lagrange equations can be expressed with the aid of elementary functions.

5.5 Systems with Infinitely Many Degrees of Freedom: Field Theory

In many cases, the equations of motion of classical mechanics follow from a variational principle, namely Hamilton's principle, which is also known as the principle of (least) stationary action. The Lagrangian $L(q,\dot{q})$ for a system of $n \geq 1$ point masses, where $q = (q_1\,, \ldots, q_n)$, $q_i \in \mathbb{R}^d$, $i = 1\,, \ldots, n$, depends on the position q_i and the velocity $\dot{q}_i$ of each point mass. In this formulation of classical mechanics, the equations of motion are the Euler (-Lagrange) equations for the Lagrangian. These equations make evident the fact that the action functional

$$W(q) = \int_I L(q,\dot{q})dt\,, \quad I = [a,b]\,,$$

takes on a stationary value for a solution q of these equations.

In hydrodynamics, electrodynamics or in the theory of gravitation we are dealing with systems which possess an infinite number of degrees of freedom. The physical situation is then described by means of a *field* $\phi(x,t)$, $x \in \mathbb{R}^d$, $t \in \mathbb{R}$, or several fields. A *field theory* is a generalisation of classical mechanics in which the field variables $\phi(x,t)$ play the role of the dynamic variables $q_i(t)$, $i = 1, \ldots, n$. The discrete index i, $1 \le i \le n$, now becomes the continuous variable $x \in \mathbb{R}^d$ and $\sum_{i=1}^n$ is replaced by $\int_{\mathbb{R}^d} dx$. In making this analogy we would expect for a classical field theory that the action functional W should be given as an integral over space and time, or, in the language of differential forms, as the integral of the $(d+1)$-form $\mathcal{L}$ (the Lagrangian form, here identified with its density with respect to the volume form):

$$W = \int_{\mathbb{R}^{d+1}} \mathcal{L}\,.$$

One of the advantages of the Lagrangian formulation is that it yields a conservation law for every one-parameter group which leaves the Lagrangian density $\mathcal{L}$ invariant. This is E. Noether's theorem. If the Lagrangian density $\mathcal{L}$ at the point (t,x) depends only on the value of ϕ and a finite number of partial derivatives $D^\alpha\phi$ $|\alpha| \le p$, at that same point, then the Lagrangian density is said to be *local.* One also speaks of a *local, classical field theory.* From now on we shall assume that $\mathcal{L}$ depends only on ϕ and the first derivative $D\phi$. As we shall see, this implies that the field equations are partial differential equations of at most second order.

We saw earlier (Sect. 5.3) that in obtaining the equations of motion as Euler-Lagrange equations for $\mathcal{L}$, we had a great deal of freedom in choosing the Lagrangian $\mathcal{L}$ depending on the underlying space of potential trajectories. We shall have a corresponding freedom in classical local field theory as far as the underlying space of potential "field configurations" is concerned: the Lagrangian densities

$$\mathcal{L}(\cdot, \phi(\cdot), D\phi(\cdot))$$

and

$$\mathcal{L}'(\cdot, \phi(\cdot), D\phi(\cdot)) = \mathcal{L}(\cdot, \phi(\cdot), D\phi(\cdot)) + \operatorname{div} f(\cdot, \phi(\cdot))$$

will lead to the same field equations, which will be Euler equations (for $\mathcal{L}$ or $\mathcal{L}'$).

5.5.1 Hamilton's Principle in Local Field Theory

In this section we shall generalise Hamilton's principle from classical mechanics to classical, local, scalar field theory. Instead of a time interval $I = [a,b] \subset \mathbb{R}$ we now take a nonempty open subset $\Omega \subset \mathbb{R}^n$, $n > 1$, with compact closure $\overline{\Omega}$ and smooth boundary $\partial\Omega = \overline{\Omega} \setminus \Omega$. Our starting point is to take the "submanifold"

$$E^1(\Omega) = \{f \in \mathcal{C}^1(\Omega; \mathbb{R}) \cap \mathcal{C}^0(\overline{\Omega}; \mathbb{R}) \,|\, f \restriction_{\partial\Omega} = g\} = E^1(\Omega; g)$$

of the Banach space $\mathcal{C}^1(\Omega, \mathbb{R})$ as the space of all potential field configurations whose values on the boundary $\partial\Omega$ are given by a function $g : \partial\Omega \to \mathbb{R}$. As standard hypothesis for the Lagrangian densities, we shall assume

$$\mathcal{L} \in \mathcal{C}^2(\Omega \times \mathbb{R} \times \mathbb{R}^n; \mathbb{R}).$$

The corresponding fundamental action $W = W_{\mathcal{L}}$ is then well defined on $E^1(\Omega)$ as

$$W(\phi) = \int_\Omega \mathcal{L}(x, \phi(x), D\phi(x))dx\,, \quad \phi \in E^1(\Omega)\,. \tag{$*$}$$

If f_0 is some fixed element of $E^1(\Omega)$ and if $E_0^1(\Omega)$ denotes the subspace

$$E_0^1(\Omega) = \{f \in \mathcal{C}^1(\Omega; \mathbb{R}) \cap \mathcal{C}^0(\overline{\Omega}; \mathbb{R}) \mid f \restriction_{\partial\Omega} = 0\}$$

of $\mathcal{C}^1(\Omega; \mathbb{R})$, then the representation

$$E^1(\Omega) = f_0 + E_0^1(\Omega)$$

for $E^1(\Omega)$ results.

Thus, we can easily calculate the Fréchet derivative of $W : E^1(\Omega) \to \mathbb{R}$ in a fashion analogous to previous calculations. Let us take $\mathcal{L}$ to be a function of the variables x, ϕ and $p_j = \partial\phi/\partial x_j$, with $\mathcal{L}_\phi$ and $\mathcal{L}_{p_j}$ denoting the corresponding partial derivatives. The Fréchet derivative $DW(\phi)$ at a point $\phi \in E^1(\Omega)$ is then the continuous linear mapping from $E_0^1(\Omega)$ into $\mathbb{R}$ given by

$$DW(\phi)(h) = \int_\Omega \left\{\mathcal{L}_\phi h + \sum_{j=1}^n \mathcal{L}_{p_j} h_{x_j}\right\} dx = \int_\Omega (\mathcal{L}_\phi h + \nabla_x h \cdot \nabla_p \mathcal{L})\, dx\,.$$

Now, it is true that

$$\operatorname{div}_x(h\nabla_p\mathcal{L}) = \nabla_x h \cdot \nabla_p \mathcal{L} + h \operatorname{div}_x \nabla_p \mathcal{L}\,,$$

and therefore

$$\begin{aligned}\int_\Omega \nabla_x h \cdot \nabla_p \mathcal{L}\, dx &= \int_\Omega \operatorname{div}(h\nabla_p\mathcal{L})dx - \int_\Omega h \operatorname{div} \nabla_p \mathcal{L}\, dx \\ &= \int_{\partial\Omega} h\nabla_p\mathcal{L}\, d\sigma - \int_\Omega h \operatorname{div} \nabla_p\mathcal{L}\, dx\,.\end{aligned}$$

The condition $h \restriction_{\partial\Omega} = 0$ implies that

$$DW(\phi)(h) = \int_\Omega [\mathcal{L}_\phi - \operatorname{div} \nabla_p \mathcal{L}] h\, dx$$

for all $\phi \in E^1(\Omega)$ and for all $h \in E_0^1(\Omega)$.

The vanishing of the first Fréchet derivative of W for a function $\phi \in E^1(\Omega)$ is thus equivalent to this function satisfying the *Euler equation*

$$\mathcal{L}_\phi(x, \phi(x), \nabla\phi(x)) - \operatorname{div} \nabla_p \mathcal{L}(x, \phi(x), \nabla\phi(x)) = 0\,, \quad \forall x \in \Omega\,.$$

The second Fréchet derivative of $W(\phi)$ follows from Taylor's formula for functions of several variables, i.e.

$$\mathcal{L}(x, \phi + h, \nabla\phi + \nabla h) = \mathcal{L}(x, \phi, \nabla\phi) + \mathcal{L}_\phi h + \nabla_p \mathcal{L} \nabla h$$
$$+ \frac{1}{2!}\left[D_\phi^2 \mathcal{L}(h, h) + 2h D_\phi D_p \mathcal{L} \nabla h + (D_p^2 \mathcal{L})(\nabla h, \nabla h)\right] + o(h, \nabla h).$$

We thus have

$$D^2 W(\phi)(h, h) = \int_\Omega dx \left[(D_\phi^2 \mathcal{L})(h, h) + 2(D_\phi D_p \mathcal{L})(h, \nabla h) + (D_p^2 \mathcal{L})(\nabla h, \nabla h)\right]$$

where the argument of $D_\phi^2 \mathcal{L}$, $D_\phi D_p \mathcal{L}$ and $D_p^2 \mathcal{L}$ is $(x, \phi(x), \nabla\phi(x))$. Now we can again rearrange the second term:

$$\begin{aligned} 2\int_\Omega (D_\phi D_p \mathcal{L})(h, \nabla h) &= 2\sum_{j=1}^n \int_\Omega \mathcal{L}_{\phi p_j} h \partial_j h \, dx = \sum_{j=1}^n \int_\Omega dx \mathcal{L}_{\phi p_j} \partial_j (h^2) \\ &= \int_\Omega dx \sum_{j=1}^n \left[\partial_j (h^2 \mathcal{L}_{\phi p_j}) - h^2 \partial_j \mathcal{L}_{\phi p_j}\right] \\ &= \int_\Omega dx \operatorname{div}(h^2 D_p D_\phi \mathcal{L}) - \int_\Omega dx \, h^2 \operatorname{div} D_p D_\phi \mathcal{L} \\ &= -\int_\Omega dx \, h^2 \operatorname{div} D_p \mathcal{L}_\phi \quad \text{since } h \in E_0^1(\Omega)). \end{aligned}$$

This yields

$$D^2 W(\phi)(h, h) = \int_\Omega \left[(D_p^2 \mathcal{L})(\nabla h, \nabla h) + (\mathcal{L}_{\phi\phi} - \operatorname{div} D_p \mathcal{L}_\phi) h^2\right] dx,$$
$$D^2 W(\phi)(h, h) = \int_\Omega \{\langle \nabla h, \mathcal{L}_{pp} \nabla h\rangle + h(\mathcal{L}_{\phi\phi} - \operatorname{div} \nabla_p \mathcal{L}_p) h\} \, dx$$

for all $\phi \in E^1(\Omega)$ and all $h \in E_0^1(\Omega)$.

The second Fréchet derivative therefore has the same form as it does for systems with finitely many degrees of freedom, i.e.

$$J(h) = D^2 W(\phi)(h, h) = \int_\Omega \left\{A h^2 + \langle \nabla h, B \nabla h\rangle\right\} dx$$

where

$$A(x) = \frac{\partial^2 \mathcal{L}}{\partial \phi^2}(x, \phi(x), \nabla\phi(x)) - \sum_{j=1}^n \partial_j \frac{\partial^2}{\partial p_j \partial \phi} \mathcal{L}(x, \phi(x), \nabla\phi(x)),$$

$$B(x) = (B_{ij}(x), \quad 1 \le i \le n, \quad 1 \le j \le n,)$$
$$B_{ij}(x) = \frac{\partial^2 \mathcal{L}}{\partial p_i \partial p_j}(x, \phi(x), \nabla\phi(x)).$$

Using well-known arguments, for positive definite J on $E_0^1(\Omega)$, B has to be a positive definite matrix for all $x \in \Omega$. This is *Legendre's condition.*

Assuming appropriate differentiability properties of the functions involved, and taking into account

$$\langle \nabla h, B\nabla h\rangle = \operatorname{div}(hB\nabla h) - h \operatorname{div}(B\nabla h)$$

and then Gauß-Stokes' theorem, i.e.

$$\int_{\Omega} \operatorname{div}(hB\nabla h)dx = \int_{\partial\Omega} hB\nabla h \, d\sigma = 0\,,$$

we obtain the following representation for $J(h)$:

$$J(h) = \int_{\Omega} h[Ah - \operatorname{div}(B\nabla h)]dx\,.$$

The Euler equation associated with the quadratic functional $J(h)$ is therefore

$$Ah - \operatorname{div}(B\nabla h) = 0\,.$$

This is *Jacobi's equation* for the problem.

We shall assume Legendre's sufficient condition in our further investigations of the necessary and sufficient conditions for a local minimum of the functional

$$W(\phi) = \int_{\Omega} \mathcal{L}(x, \phi, \nabla\phi)dx\,.$$

This means that the matrix

$$B_{ij}(x) = \frac{\partial^2 \mathcal{L}}{\partial p_i\, \partial p_j}(x, \phi(x), \nabla\phi(x))$$

constructed for a solution ϕ of the Euler equation associated with $\mathcal{L}$ is strictly positive definite (for almost all $x \in \Omega$).

For every function $\omega \in C^1(\Omega, \mathbb{R}^n)$ and all $h \in E_0^1(\Omega)$, the identity

$$0 \equiv \int_{\partial\Omega} h^2\omega \, d\sigma = \int_{\Omega} \operatorname{div}(h^2\omega)dx = \int_{\Omega} (2\langle h\nabla h, \omega\rangle + h^2 \operatorname{div}\omega)dx\,,$$

holds. We can rearrange $J(h)$ using this identity, yielding

$$\begin{aligned} J(h) &= \int_{\Omega} \left(Ah^2 + \langle \nabla h, B\nabla h\rangle + 2\langle \nabla h, h\omega\rangle + h^2 \operatorname{div}\omega\right) dx \\ &= \int_{\Omega} \left[\|B^{1/2}\nabla h + B^{-1/2}h\omega\|^2 + (A + \operatorname{div}\omega - \langle \omega, B^{-1}\omega\rangle)h^2\right] dx\,. \end{aligned}$$

As soon as a solution $\phi_0 \in C^2(\Omega)$ of Jacobi's equation

$$A\phi_0 - \operatorname{div}(B\nabla\phi_0) = 0$$

which has no zeros in Ω exists, the equation

$$\omega = -B\phi_0^{-1}\nabla\phi_0$$

yields a solution for the partial differential equation

$$A + \operatorname{div}\omega - \langle \omega, B^{-1}\omega \rangle = 0$$

in $C^1(\Omega, \mathbb{R}^n)$. Indeed, for such a ϕ_0,

$$-A\phi_0 = -\operatorname{div}(B\nabla\phi_0) = \operatorname{div}(\phi_0\omega) = \langle \nabla\phi_0, \omega \rangle + \phi_0 \operatorname{div}\omega .$$

We thus have

$$A + \operatorname{div}\omega = -\frac{1}{\phi}\langle \nabla\phi_0, \omega \rangle = \langle \omega, B^{-1}\omega \rangle .$$

It then follows for all $h \in E_0^1(\Omega)$ that

$$J(h) = \int_\Omega \|B^{1/2}\nabla h + B^{-1/2}h\omega\|^2 dx \geq 0 .$$

Now if $J(h) = 0$ for one $h \in E_0^1(\Omega)$, then this function h satisfies the equations

$$\nabla h + B^{-1}h\omega = 0 ,$$

$$\nabla h - h\phi_0^{-1}\nabla\phi_0 = 0 ,$$

or

$$\nabla h - h\nabla \log\phi_0 = 0$$

in Ω. The solution of this equation is $h = c\phi_0$. If ϕ_0 is not an element of $E_0^1(\Omega)$, then $J(h)$ is strictly positive.

Remark 5.5.1. Our previous considerations have shown that under appropriate assumptions classical field equations have the form of the Euler-Lagrange equations of a suitable Lagrangian density $\mathcal{L}$, i.e., by virtue of Hamilton's principle, these equations express precisely the fact that the associated local action functional $W_{\mathcal{L},\Omega}$ (see page 118, equation $(*)$) is stationary. These considerations also showed exactly how to do this. *Globally*, that is for $\Omega = \mathbb{R}^n$, difficulties generally arise because $\mathcal{L}(x, \phi(x), D\phi(x))$ is not necessarily integrable over $\mathbb{R}^n$. In this case, we talk about the principle of stationary action in the following sense.

Let the field equations be given on $\mathbb{R}^n$ in the form of the Euler-Lagrange equations

$$\mathcal{L}_\phi(x, \phi(x), D\phi(x)) - \operatorname{div}\nabla_p\mathcal{L}(x, \phi(x), D\phi(x)) = 0$$

of a suitable Lagrangian density $\mathcal{L}$. These equations are then equivalent to the condition that *all local* action functionals

$$W_{\mathcal{L},\Omega}(\phi) = \int_\Omega \mathcal{L}(x, \phi(x), D\phi(x))dx ,$$

$\Omega \subset \mathbb{R}^n$ (Ω as above) be stationary in ϕ.

In special cases, one can obtain the field equations as conditions for the globally defined action functional

$$W_{\mathcal{L}}(\phi) = \int_{\mathbb{R}^n} \mathcal{L}(x, \phi(x), D\phi(x))dx$$

to take on stationary values. In this case, the choice of the space of possible field configurations ϕ depends sensitively on the Lagrangian density $\mathcal{L}$ (see Chap. 9).

5.5.2 Examples of Local Classical Field Theories

(a) *Nonlinear Elliptic Field Equations*

In general, it is not easy to decide when a function $\phi \in E^1(\Omega)$ minimises the functional

$$W(\phi) = \int_{\Omega} \mathcal{L}(x, \phi, \nabla\phi)dx \, .$$

For example, let

$$\mathcal{L}(x, \phi(x), \nabla\phi(x)) = \frac{1}{2}\langle\nabla\phi(x), M(x)\nabla\phi(x)\rangle + V(\phi)$$

where $V \in \mathcal{C}^2(\mathbb{R})$, and let $M \in \mathcal{C}^1(\Omega, \mathcal{M}_n(\mathbb{R}^n))$ be symmetric so that $M = M^*$. It follows that

$$L_{\phi} = V'(\phi) \, , \quad D_p\mathcal{L} = Mp \, ,$$

$$D_p^2\mathcal{L} = M \, , \quad \mathcal{L}_{\phi\phi} = V''(\phi) \, , \quad D_p\mathcal{L}_{\phi} = 0 \, .$$

Euler's equation, i.e. the field equation of the local field theory we are considering, is, in this case,

$$V'(\phi(x)) - \operatorname{div}(M(x)\nabla\phi(x)) = 0 \, .$$

Legendre's condition requires that

$$M(x) \geq 0 \, , \quad \forall x \in \Omega \, ,$$

and Jacobi's equation reads here

$$V''(\phi(x))h(x) - \operatorname{div}(M(x)h(x)) = 0 \, .$$

In many applications, the term $\frac{1}{2}\langle\nabla\phi, M\nabla\phi\rangle$ can be interpreted as the kinetic energy of the field ϕ. Then $M(x) > 0$, $\forall x \in \Omega$, seems to be a plausible requirement. But again, in general it is practically impossible to decide when Jacobi's equation for a solution ϕ of Euler's equation possesses a solution h which does not vanish in Ω, i.e. when conjugate surfaces do not exist.

Consider now a special case:

$$M(x) = \mathbb{1}_n \, , \quad \forall x \in \Omega \, ,$$

$$V(\phi) = -\frac{m^2}{2}\phi^2, \quad m \in \mathbb{R}.$$

Then Euler's equation is

$$m^2\phi + \Delta\phi = 0$$

and Jacobi's equation is

$$m^2 h + \Delta h = 0.$$

Our general results therefore assert in this case that if Jacobi's equation has a nontrivial solution which does not vanish in Ω, then the solution of Euler's equation minimises the functional

$$W(\phi) = \int_\Omega dx \left(\frac{1}{2}\|\nabla\phi\|^2 - \frac{m^2}{2}\phi^2\right).$$

(b) *Nonlinear Klein-Gordon Equation*

Let $\Omega \subset \mathbb{R}^4$. Consider the Lagrangian density

$$\mathcal{L}^0(\phi, \nabla\phi) = \frac{1}{2}\left[(\partial_\mu\phi)^2 - m^2\phi^2\right], \quad m \in \mathbb{R},$$

where

$$(\partial_\mu\phi)^2 = (\partial_t\phi)^2 - (\partial_{x_1}\phi)^2 - (\partial_{x_2}\phi)^2 - (\partial_{x_3}\phi)^2 = (\partial_t\phi)^2 - |\nabla_x\phi|^2.$$

This Lagrangian density gives the following Euler equation (the so-called *Klein-Gordon equation*):

$$(\Box + m^2)\phi(t,x) = 0$$

where

$$\Box = \partial_t^2 - \Delta.$$

If we want to have a source term $j(t,x)$ in the Klein-Gordon equation we must add a term $j\phi$ to the Lagrangian density. For $m = 0$ therefore, the wave equation results. If the scalar field ϕ interacts with itself (as it does, for example, in anharmonic vibrations of a crystal), then the simplest terms which can be added to $\mathcal{L}^0$ are proportional to ϕ^3 and ϕ^4. For example, for the Lagrangian density

$$\mathcal{L} = \mathcal{L}^0 + \frac{\lambda}{4}\phi^4 + j\phi$$

we have the Euler equation

$$(\Box + m^2)\phi(t,x) = \lambda\phi^3 + j.$$

In applications to physics the parameter m is interpreted as the mass of a particle.

In the case of the Klein-Gordon equation, i.e.

$$\partial_t^2 \phi = \ddot{\phi} = \Delta\phi - m^2\phi\,,$$

we can integrate the equation of motion immediately with the aid of a Fourier transform. Let

$$\widetilde{\phi}(t,p) = (2\pi)^{-3/2} \int_{\mathbb{R}^3} \phi(t,x) \mathrm{e}^{-ipx} dx\,.$$

Then

$$\ddot{\widetilde{\phi}} = -(m^2 + p^2)\widetilde{\phi}\,,$$

and for given initial values of $\phi(t,x)$ and $\partial_t\phi(t,x)$ at $t = 0$, the solution is

$$\widetilde{\phi}(t,p) = \cos\left(t\sqrt{m^2+p^2}\right)\widetilde{\phi}(0,p) + \frac{\sin\left(t\sqrt{m^2+p^2}\right)}{\sqrt{m^2+p^2}}\dot{\widetilde{\phi}}(0,p)\,,$$

$$\dot{\widetilde{\phi}}(t,p) = -\sqrt{m^2+p^2}\sin\left(t\sqrt{m^2+p^2}\right)\widetilde{\phi}(0,p) + \cos\left(t\sqrt{m^2+p^2}\right)\dot{\widetilde{\phi}}(0,p)\,.$$

Due to the factor $\sqrt{m^2+p^2}$ in the formula for $\dot{\widetilde{\phi}}$, these relations define a mapping

$$t \to (\phi(t,\cdot), \partial_t(\phi(t,\cdot))$$

from $\mathbb{R}_+ = [0,\infty)$ into $H^1(\mathbb{R}^3) \oplus L^2(\mathbb{R}^3)$, in other words, a "flow" in $H^1(\mathbb{R}^3) \oplus L^2(\mathbb{R}^3)$. For a definition of the Sobolev space $H^1(\mathbb{R}^3)$ see Chapter 9 or the Appendices.

5.6 Noether's Theorem in Classical Field Theory

We would like to use the variational formulation of the equation of motion to derive conservation laws from invariance properties of the Lagrangian density, just as we did in classical (Lagrangian) mechanics. Noether's theorem is concerned with just this. First of all, we want to explain that the role of the integrals of motion of classical mechanics is taken, in classical local field theory, by conserved charges.

If, say, $\phi = (\phi^1, \ldots, \phi^N) \in \mathcal{E}^1(\mathbb{R}^4; \mathbb{R}^N)$ are fields over space-time $\mathbb{R}^4 \cong \mathbb{R} \times \mathbb{R}^3$, then for every C^1-function $I : \mathbb{R}^4 \times \mathbb{R}^N \times \mathbb{R}^{4N} \to \mathbb{R}^4$, the C^1-function $I^\phi : \mathbb{R}^4 \to \mathbb{R}^4$,

$$I^\phi(x) = I(x, \phi(x), \nabla\phi(x))\,,$$

which is associated with ϕ, is called *a current of* ϕ. We interpret the points $x = (x_0, x_1, x_2, x_3)$ of $\mathbb{R}^4$ as time (x_0) and space $(\underline{x} = (x_1, x_2, x_3))$ coordinates, with corresponding notation for the components of $I^\phi = (I_0^\phi, \underline{I}^\phi)$.

Subject to well-known integrability conditions, the fact that the *"charge"*

$$Q^{I^\phi}(x_0) = \int_{\mathbb{R}^3} I_0^\phi(x_0, \underline{x}) d^3\underline{x}$$

is constant with respect to time is equivalent to the fact that the divergence of the current I^{ϕ} vanishes:

$$\sum_{j=0}^{3} \frac{\partial}{\partial x_j} I_j^{\phi} = 0 .$$

Thus, associated with every current I^{ϕ} satisfying $\operatorname{div} I^{\phi} = 0$, there is, in this sense, a conservation law, namely, the conservation of the "charge" ($x_0 \in \mathbb{R}$ arbitrary)

$$Q(I^{\phi}) \equiv Q^{I^{\phi}}(x_0) .$$

In a field theory, these conserved charges play the same role as the integrals of motion in classical mechanics. In this context we speak about a conservation law if a divergence-free current exists (with suitable integrability conditions).

Let us consider again, for a subset $\Omega \subset \mathbb{R}^n$ as above and a function $g : \partial\Omega \to \mathbb{R}^N$, the submanifold $E^1(\Omega, g) = \{\phi \in \mathcal{C}^1(\Omega; \mathbb{R}^N) \cap \mathcal{C}^0(\overline{\Omega}; \mathbb{R}^N) \mid \phi \restriction \partial\Omega = g\}$ of the Banach space $\mathcal{C}^1(\Omega; \mathbb{R}^N)$. As before, we associate a local action functional

$$W = W^{\mathcal{L}} : E^1(\Omega, g) \to \mathbb{R}, \quad W(\phi) = \int_{\Omega} \mathcal{L}(x, \phi(x), \nabla\phi(x)) d^n x$$

with a Lagrangian density $\mathcal{L} \in \mathcal{C}^2(\Omega \times \mathbb{R}^N \times \mathbb{R}^{nN}; \mathbb{R})$. The equation of motion for the fields ϕ, i.e.

$$[\mathcal{L}^{\phi}]_j(x) = \frac{\partial \mathcal{L}}{\partial y_j}(x, \phi(x), \nabla\phi(x)) - \operatorname{div} \frac{\partial \mathcal{L}}{\partial \underline{p}_j}(x, \phi(x), \nabla\phi(x)) = 0 , \quad j = 1, \ldots, N ,$$

in the variational formulation are just the Euler condition $DW(\phi) = 0$ for W to take on stationary values in $\phi \in E^1(\Omega, g)$. We shall use the notation

$$\underline{p} = (\underline{p}_1, \ldots, \underline{p}_N) \longleftrightarrow (\nabla_x \phi^1, \ldots, \nabla_x \phi^N)$$

and

$$\frac{\partial \mathcal{L}}{\partial y_j}(x, y, \underline{p}) \equiv \nabla_{y_j} \mathcal{L}(x, y, \underline{p}) .$$

and

$$\frac{\partial L}{\partial \underline{p}_j}(x, y, \underline{p}) \equiv \nabla_{\underline{p}_j} \mathcal{L}(x, y, \underline{p}) .$$

Now let an r-parameter Lie group of diffeomorphisms h_ε of $\mathbb{R}^n \times \mathbb{R}^N$ be given: let $\varepsilon = (\varepsilon_1, \ldots, \varepsilon_r)$ be the parameters of this Lie group. It is known that the finite transformations for a Lie group can be constructed from the infinitesimal transformations (in the connected component of the neutral element). It is thus sufficient if we study the effects of the infinitesimal transformations.

We shall restrict our considerations to those cases which are very important in physics. We thus assume that the diffeomorphisms h_ε have the form:

$$h_\varepsilon(x,y) = (\psi_\varepsilon(x), \varphi_\varepsilon(x,y)), \quad h_\varepsilon(x,y)|_{\varepsilon=0} = (x,y),$$

where ψ_ε is a diffeomorphism of $\mathbb{R}^n$. Put

$$\xi_\nu(x) := \frac{\partial \psi_\varepsilon}{\partial \varepsilon_\nu}(x)\Big|_{\varepsilon=0},$$
$$\Theta_\nu(x,y) := \frac{\partial \varphi_\varepsilon}{\partial \varepsilon_\nu}(x,y)\Big|_{\varepsilon=0},$$

which results in the following representation for $\varepsilon \to 0$:

$$\psi_\varepsilon(x) = x + \sum_{\nu=1}^{r} \xi_\nu(x)\varepsilon_\nu + o_1(\varepsilon; x),$$
$$\varphi_\varepsilon(x,y) = y + \sum_{\nu=1}^{r} \Theta_\nu(x,y)\varepsilon_\nu + o_2(\varepsilon; x, y).$$

It follows that

$$\frac{\partial \psi_\varepsilon}{\partial x}(x) = \mathbb{1}_n + \sum_{\nu=1}^{r} \frac{\partial \xi_\nu}{\partial x}(x)\varepsilon_\nu + \nabla o_1(\varepsilon; x)$$

and thus, after a simple calculation,

$$\det \frac{\partial \psi_\varepsilon}{\partial x}(x) = 1 + \operatorname{div}\left(\sum_{\nu=1}^{r} \xi_\nu(x)\varepsilon_\nu\right) + o_1'(\varepsilon; x).$$

Define for every transformation h_ε a mapping

$$H_\varepsilon : E^1(\Omega, g) \to E^1(\Omega_\varepsilon, g_\varepsilon)$$

by means of the following equations:

$$\begin{aligned}
\Omega_\varepsilon &= \psi_\varepsilon(\Omega), \\
g_\varepsilon : \partial\Omega_\varepsilon &= \psi_\varepsilon(\partial\Omega) \to \mathbb{R}^N, \\
g_\varepsilon(x_\varepsilon) : &= \varphi_\varepsilon(\psi_\varepsilon^{-1}(x_\varepsilon), g(\psi_\varepsilon^{-1}(x_\varepsilon))), \\
H_\varepsilon(\phi) : &= \phi_\varepsilon \in E^1(\Omega_\varepsilon, g_\varepsilon), \\
\phi_\varepsilon(x_\varepsilon) : &= \varphi_\varepsilon(\psi_\varepsilon^{-1}(x_\varepsilon), \phi(\psi_\varepsilon^{-1}(x_\varepsilon))).
\end{aligned}$$

It follows after an elementary calculation that

$$\begin{aligned}
\frac{\partial \phi_\varepsilon}{\partial x_\varepsilon}(x_\varepsilon) &= \frac{\partial \varphi_\varepsilon}{\partial x}(\cdot,\cdot)\frac{\partial \psi_\varepsilon^{-1}}{\partial x}(x_\varepsilon) + \frac{\partial \varphi_\varepsilon}{\partial y}(\cdot,\cdot)\frac{\partial \phi}{\partial x}(\psi_\varepsilon^{-1}(x_\varepsilon))\frac{\partial \psi_\varepsilon^{-1}}{\partial x}(x_\varepsilon) \\
&= \frac{\partial \phi}{\partial x}(x) + \sum_{\nu=1}^{r} \frac{\partial}{\partial x}\Theta_\nu(x, \phi(x))\varepsilon_\nu + \sum_{\nu=1}^{r} \frac{\partial \Theta_\nu(x, \phi(x))}{\partial y}\varepsilon_\nu \frac{\partial \phi}{\partial x}(x) \\
&\quad - \frac{\partial \phi}{\partial x}(x) \sum_{\nu=1}^{r} \frac{\partial \xi_\nu(x)}{\partial x}\varepsilon_\nu + o(\varepsilon; x).
\end{aligned}$$

Now we would like to investigate the consequences of the following invariance properties of the Lagrangian density $\mathcal{L}$ under the diffeomorphisms h_ε:

$$\mathcal{L}(\psi_\varepsilon(x), \phi_\varepsilon(\psi_\varepsilon(x)), \frac{\partial \phi_\varepsilon}{\partial x_\varepsilon}(\psi_\varepsilon(x)))\Big| \det \frac{\partial \psi_\varepsilon}{\partial x}(x)\Big|$$
$$= \mathcal{L}(x, \phi(x), \nabla\phi(x)) + \operatorname{div} f(\varepsilon, x, \phi(x)) \qquad (*)$$

where $f : \mathbb{R}^r \times \mathbb{R}^n \times \mathbb{R}^N \to \mathbb{R}^n$ is a $\mathcal{C}^2$ function such that $f(0;\cdot,\cdot) = 0$. In other words, $\mathcal{L}$ is invariant with respect to the diffeomorphism we are considering apart from the divergence of some vector field f.

We shall discuss the meaning of the invariance property $(*)$ later. Using the formulas above, we calculate the derivatives of the left-hand side of $(*)$

$$\frac{\partial}{\partial \varepsilon_\nu}\Big[\mathcal{L}\Big(\psi_\varepsilon(x), \phi_\varepsilon(\psi_\varepsilon(x)), \frac{\partial \phi_\varepsilon}{\partial x_\varepsilon}(\psi_\varepsilon(x))\Big)\Big| \det \frac{\partial \psi_\varepsilon}{\partial x}(x)\Big|\Big]\Big|_{\varepsilon=0}$$
$$= \mathcal{L}\Big(x, \phi(x), \frac{\partial \phi}{\partial x}(x)\Big)\frac{\partial}{\partial \varepsilon_\nu}\Big(\det \frac{\partial \psi_\varepsilon}{\partial x}(x)\Big)\Big|_{\varepsilon=0}$$
$$+ \frac{\partial}{\partial \varepsilon_\nu}\mathcal{L}\Big(\psi_\varepsilon(x), \phi_\varepsilon(\psi_\varepsilon(x)), \frac{\partial \phi_\varepsilon}{\partial x_\varepsilon}(\psi_\varepsilon(x))\Big)\Big|_{\varepsilon=0}$$
$$= \mathcal{L}^\phi(x)\operatorname{div}\xi_\varepsilon(x) + \frac{\partial \mathcal{L}}{\partial x}\Big(x, \phi(x), \frac{\partial \phi}{\partial x}(x)\Big) \circ \frac{\partial \psi_\varepsilon(x)}{\partial \varepsilon_\nu}\Big|_{\varepsilon=0}$$
$$+ \sum_{j=1}^N \frac{\partial \mathcal{L}}{\partial y_j}\Big(x, \phi(x), \frac{\partial \phi}{\partial x}(x)\Big)\frac{\partial}{\partial \varepsilon_\nu}\phi_\varepsilon^j(\psi_\varepsilon(x))|_{\varepsilon=0}$$
$$+ \sum_{j=1}^N \frac{\partial \mathcal{L}}{\partial \underline{p}_j}\Big(x, \phi(x), \frac{\partial \phi}{\partial x}(x)\Big)\frac{\partial}{\partial \varepsilon_\nu}\frac{\partial \phi_\varepsilon^j}{\partial x}(\psi_\varepsilon(x))|_{\varepsilon=0}$$
$$= \mathcal{L}^\phi(x)\operatorname{div}\xi_\varepsilon(x) + \frac{\partial \mathcal{L}^\phi}{\partial x}(x)\cdot\xi_\nu(x) + \sum_{j=1}^N \frac{\partial \mathcal{L}^\phi}{\partial y_j}(x)\Theta_\nu^j(x, \phi(x))$$
$$+ \sum_{j=1}^N \frac{\partial \mathcal{L}^\phi}{\partial \underline{p}_j}\Big\{\frac{\partial \Theta_\nu^j}{\partial x}(x, \phi(x)) + \frac{\partial \Theta_\nu^j}{\partial y}(x, \phi(x))\frac{\partial \phi}{\partial x}(x) - \frac{\partial \phi^j}{\partial x}(x)\frac{\partial \xi_\nu}{\partial x}(x)\Big\}$$
$$= \sum_{j=1}^N [\mathcal{L}^\phi]_j(x)\Theta_\nu^j(x, \phi(x))$$
$$+ \operatorname{div}\Big\{\mathcal{L}^\phi(x)\xi_\nu(x) + \sum_{j=1}^N \frac{\partial \mathcal{L}^\phi}{\partial \underline{p}_j}(x)\Theta_\nu^j(x, \phi(x)) - \sum_{j=1}^N \frac{\partial \mathcal{L}^\phi}{\partial \underline{p}_j}(x)\frac{\partial \phi^j}{\partial x}\xi_\nu(x)\Big\}$$
$$+ \sum_{j=1}^N \Big\{\operatorname{div}\mathcal{L}^\phi_{,\underline{p}_j}(x)\frac{\partial \phi^j}{\partial x}\xi_\nu(x) + \mathcal{L}^\phi_{,\underline{p}_j}(x)\xi_\nu(x)\operatorname{div}\frac{\partial \phi^j}{\partial x}(x)\Big\}.$$

Differentiating the right-hand side of (∗) gives

$$\operatorname{div} f_\nu(x,\phi(x))\,, \quad \text{where } f_\nu(x,\phi(x)) = \frac{\partial}{\partial \varepsilon_\nu} f(\varepsilon, x, \phi(x))|_{\varepsilon=0}\,.$$

Thus we have shown that

$$\sum_{j=1}^{N} [\mathcal{L}^\phi]_j(x)\Theta_\nu^j(x,\phi(x)) = \operatorname{div}\Big\{ \sum_{j=1}^{N} \frac{\partial \mathcal{L}^\phi}{\partial \underline{p}_j} \circ \frac{\partial \phi^j}{\partial x} \circ \xi_\nu(x) + f_\nu(x,\phi(x))$$
$$- \sum_{j=1}^{N} \frac{\partial \mathcal{L}^\phi}{\partial \underline{p}_j}(x)\Theta_\nu^j(x,\phi(x)) - \mathcal{L}^\phi(x)\xi_\nu(x)\Big\}$$
$$- \sum_{j=1}^{N} \Big\{ \operatorname{div} \mathcal{L}^\phi_{,\underline{p}_j}(x) \frac{\partial \phi^j}{\partial x} \xi_\nu(x) + \mathcal{L}^\phi_{,\underline{p}_j}(x)\xi_\nu(x) \operatorname{div} \frac{\partial \phi^j}{\partial x}(x) \Big\}$$

If we now put

$$I_\nu^\phi(x) := \sum_{j=1}^{N} \frac{\partial \mathcal{L}^\phi}{\partial \underline{p}_j}(x) \frac{\partial \phi^j}{\partial x}(x)\xi_\nu(x) + f_\nu(x,\phi(x))$$
$$- \mathcal{L}^\phi(x)\xi_\nu(x) - \sum_{j=1}^{N} \frac{\partial \mathcal{L}^\phi}{\partial \underline{p}_j}(x)\Theta_\nu^j(x,\phi(x)) \qquad (**)$$

for $\nu = 1,\ldots,r$, then the equation (∗) asserts that r linearly independent equations

$$\sum_{j=1}^{N} [\mathcal{L}^\phi]_j(x)[\Theta_\nu^j(x,\phi(x)) - \frac{\partial \phi^j}{\partial x}(x)\cdot \xi_\nu(x)] = \operatorname{div} I_\nu^\phi(x)\,, \quad \nu = 1,\ldots,r\,,$$

result, where Θ_ν^j is a matrix of maximal rank.

In this way, we obtain, for solutions ϕ of the field equations $[\mathcal{L}^\phi]_j(x) = 0$, $j = 1,\ldots,N$, r linearly independent conserved currents I_ν^ϕ, $\nu = 1,\ldots,r$, and thus, subject to the known integrability conditions, r independent conserved charges $Q(I_\nu^\phi)$, $\nu = 1,\ldots,r$.

We shall start from a variational form of the equations of motion in order to interpret the invariance properties (∗) of the Lagrangian density in terms of the transformation properties of the associated action functional.

We define an action functional on $E^1(\Omega_\varepsilon, g_\varepsilon)$ by

$$W_\varepsilon(\phi) = \int_{\Omega_\varepsilon} \mathcal{L}\Big(x_\varepsilon, \phi(x_\varepsilon), \frac{\partial \phi}{\partial x}(x_\varepsilon)\Big) d^n x_\varepsilon\,, \quad \phi \in E^1(\Omega_\varepsilon, g_\varepsilon)\,.$$

Then $W_\varepsilon \circ H_\varepsilon$ is an action functional on $E^1(\Omega, g)$. The invariance condition (∗) for the Lagrangian density $\mathcal{L}$ implies for $W_\varepsilon \circ H_\varepsilon$ that

$$W_\varepsilon \circ H_\varepsilon(\phi) = W(\phi) + \int_\Omega \operatorname{div} f(\varepsilon, x, \phi(x)) d^n x$$
$$= W(\phi) + \int_{\partial\Omega} f(\varepsilon, x, g(x))\, d\sigma\,, \quad \phi \in E^1(\Omega, g)\,,$$

and thus

$$DW(\phi) = D(W_\varepsilon \circ H_\varepsilon)(\phi)\,,$$

and this is, according to Hamilton's principle, just the transformation property of the equation of motion.

Then $D(W_\varepsilon \circ H_\varepsilon)$ can be determined by means of an elementary but tedious calculation. The result is a tranformation formula for the equations of motion:

$$\begin{aligned}[\mathcal{L}^\phi]_i &= \sum_{j=1}^N [\mathcal{L}^{\phi_\varepsilon}]_j(x) \frac{\partial \varphi_\varepsilon^j}{\partial y_i}(x, \phi(x)) + \sum_{j=1}^N \operatorname{div} \frac{\partial \mathcal{L}^{\phi_\varepsilon}}{\partial \underline{p}_i} \frac{\partial \varphi_\varepsilon^j}{\partial y_i}(x, \phi(x)) \\ &\quad - J_\varepsilon(x) \sum_{j=1}^N \sum_{\nu=1}^n \left\{ \frac{\partial \mathcal{L}^{\phi_\varepsilon}}{\partial p_j^\nu}(x) \sum_{\mu=1}^n \frac{\partial I_\varepsilon^{\mu\nu}}{\partial x_\mu}(x) + \sum_{\mu=1}^n \frac{\partial}{\partial x_\mu} \left(\frac{\partial \mathcal{L}^{\phi_\varepsilon}}{\partial p_j^\nu}(x) \right) I_\varepsilon^{\mu\nu}(x) \right\} \\ &\quad \times \frac{\partial \varphi_\varepsilon^j}{\partial y_i}(x, \phi(x)) - \sum_{\mu=1}^n \frac{\partial J_\varepsilon(x)}{\partial x_\mu} \sum_{j=1}^n \sum_{\nu=1}^n \frac{\partial \mathcal{L}^{\phi_\varepsilon}}{\partial p_j^\nu}(x) I_\varepsilon^{\mu\nu}(x) \frac{\partial \varphi_\varepsilon^j(x, \phi(x))}{\partial y_i}\,. \end{aligned} \tag{Δ}$$

where we have used the abbreviations

$$[\mathcal{L}^{\phi_\varepsilon}]_j(x) = \left(\frac{\partial \mathcal{L}}{\partial y_j} - \operatorname{div} \frac{\partial \mathcal{L}}{\partial \underline{p}_j} \right) \left(\psi_\varepsilon(x), \phi_\varepsilon(\psi_\varepsilon(x)), \frac{\partial \phi_\varepsilon}{\partial x}(\psi_\varepsilon(x)) \right),$$
$$J_\varepsilon(x) = \left| \det \frac{\partial \psi_\varepsilon}{\partial x}(x) \right|, \quad I_\varepsilon(x) = \frac{\partial \psi_\varepsilon^{-1}}{\partial x}(\psi_\varepsilon(x))\,.$$

When $J_\varepsilon(x) = 1$ and $I_\varepsilon^{\mu\nu}(x) = \alpha_\nu(\varepsilon)\delta_{\mu\nu}$ the equations of motion are invariant, since then

$$[\mathcal{L}^\phi]_i(x) = \sum_{j=1}^N [\mathcal{L}^{\phi_\varepsilon}]_j(x) \frac{\partial \phi_\varepsilon^j}{\partial y_i}(x, \phi(x))\,,$$

and the matrix $(\partial\phi/\partial y)(x,y)$ is invertible for all x, y. Thus, it follows that

$$[\mathcal{L}^\phi]_i(x) = 0\,, \quad i = 1, \ldots, N\,,$$

if, and only if, $[\mathcal{L}^{\phi_\varepsilon}]_j(x) = 0$, $j = 1, \ldots, N$.

Let us summarise the above.

Theorem 5.6.1 (E. Noether). *Let $\mathcal{L} \in C^2(\mathbb{R}^n, \mathbb{R}^N, \mathbb{R}^{nN})$ be a Lagrangian density which possesses the transformation property $(*)$ under an r-parameter Lie group of diffeomorphisms h_ε. Then there are r linearly independent combinations of the Euler expressions $[\mathcal{L}^\phi]_j$, $j = 1, \ldots, N$, every one of which is the divergence of a current I_ν^ϕ of ϕ:*

$$\sum_{j=1}^{N}[\mathcal{L}^{\phi}]_j(x)\left[\Theta_{\nu}^{j}(x,\phi(x)) - \frac{\partial \phi^j}{\partial x}(x)\cdot \xi_{\nu}(x)\right] = \operatorname{div} I_{\nu}^{\phi}(x)\,, \quad \nu = 1\,,\ldots,r\,.$$

$$I_{\nu}^{\phi}(x) = \sum_{j=1}^{N}\frac{\partial \mathcal{L}^{\phi}}{\partial \underline{p}_j}(x)\frac{\partial \phi^j}{\partial x}(x)\xi_{\nu}(x) + f_{\nu}(x,\phi(x)) - \mathcal{L}^{\phi}(x)\xi_{\nu}(x)$$
$$-\sum_{j=1}^{N}\frac{\partial \mathcal{L}^{\phi}}{\partial \underline{p}_j}(x)\Theta_{\nu}^{j}(x,\phi(x))\,.$$

The transformation of the equations of motion of the field ϕ under the diffeomorphisms h_{ε} is determined by the formula (Δ).

Remark 5.6.1. As we have emphasised several times, it is possible to obtain from every conserved current J a quantity $Q^J(x_0) = \int J_0(x_0,\underline{x})d^3x$ which is constant in time. This corollary of Noether's theorem is used in many physical applications. See [5.3, 5–8] for explicit applications of Noether's theorem to field theory.

5.7 The Principle of Symmetric Criticality

In many areas, including mathematics and physics, it has proved extremely useful to look for symmetries and to exploit them, if they exist, in problem solving. The success of this procedure is based on the "principle of symmetric criticality". In order to illustrate the idea of this principle, let us describe a typical situation where it is used (usually implicitly).

Suppose a classical field theory, on $\mathbb{R}^3$ let us say, is described in terms of a rotationally invariant action functional $W(\phi)$ of the type considered in Sect. 5.5 and one looks for rotationally symmetric critical points of W, e.g. one looks for field configurations ϕ which are invariant under the action of the rotation group SO(3) and which are critical points of W. We take as an "ansatz" that the field components are functions of the Euclidean norm $r = |x|$ of the points $x \in \mathbb{R}^3$ and then we compute the variation of W with respect to all fields that also satisfy this ansatz, for instance along the lines of Sect. 5.2. Setting the first variation of W, calculated in this way, equal to zero yields a system of *ordinary* differential equations for the field components, which can be solved explicitly in some cases. Now this system is used as a necessary and sufficient condition for a rotationally symmetric field configuration to be a critical point of W. Obviously this system represents a necessary condition for a field configuration to be a critical point of W under these circumstances. But the converse is a claim that needs proof. Actually, it can be proved that the above conditions are also sufficient; this then means that a critical rotationally symmetric field configuration is a rotationally symmetric critical field configuration for W. This last statement is just the statement of the principle of symmetric criticality, in the context of this example. A suggestive formulation of this principle now is that "any critical symmetric point is a symmetric critical point". Unfortunately, the principle is *not* valid in this general form, as we shall see later.

The principle has been used in many applications of the calculus of variations, in particular in theoretical physics, without being particularly noticed. An early, typical example of the implicit use of this principle can be found in Weyl's derivation of the Schwarzschild solution of the Einstein field equations [Ref. 5.9, p. 252 or Ref. 5.10, p. 165]. About fifteen years ago this principle was made reference to explicitely by Coleman [Ref. 5.11., App. 4]. However, it was not until 1979 that R. S. Palais gave a precise formulation of it and a proof of its validity in fairly general circumstances [5.12]. We discuss here a simplified version of his presentation.

The example given leads by abstraction to the following mathematical framework in which the above heuristic discussion will attain a precise meaning (the notions used here are made more precise later on).

Let M be a smooth manifold on which a group G acts by diffeomorphisms, and suppose $F : M \to \mathbb{R}$ is a G-invariant smooth function on M, e.g. $F(g \cdot p) = F(p)$, $\forall g \in G$, $\forall p \in M$. The set of *symmetric points* is, by definition,

$$\Sigma = \Sigma(G, M) = \{p \in M |\ g \cdot p = p,\ \forall g \in G\} . \tag{5.7.1}$$

The restriction of F to Σ is denoted by $F_\Sigma = F \restriction \Sigma$. The set of critical points of F (or F_Σ) is

$$\begin{aligned} K(F) &= \{p \in M |\ D_p F = 0\} \\ (\text{or} \quad K(F_\Sigma) &= \{p \in \Sigma |\ D_p F_\Sigma = 0\}) . \end{aligned} \tag{5.7.2}$$

The *principle of symmetric criticality* then claims that

$$K(F_\Sigma) \subseteq \Sigma \cap K(F) , \tag{P}$$

at least under appropriate hypotheses on the triple (M, G, F).

Naturally, in order to be able to speak about critical symmetric points, i.e. critical points of F_Σ, Σ has to be a differentiable manifold, and this is assured by ensuring that Σ is a smooth submanifold of M. Thus a first hypothesis on (M, G) has to ensure that the set Σ of symmetric points is a smooth submanifold of M.

If this is the case, the tangent space $T_p\Sigma$ of Σ at $p \in \Sigma$ is welldefined and the tangent space T_pM of M at $p \in \Sigma$ naturally splits into $T_p\Sigma$ and a part $T_p^\perp \Sigma$ transversal to $T_p\Sigma$:

$$T_pM = T_p\Sigma + T_p^\perp \Sigma . \tag{5.7.3}$$

Furthermore, it follows that the differential D_pF_Σ of F_Σ at a point $p \in \Sigma$ is given by

$$D_pF_\Sigma = D_pF \restriction T_p\Sigma . \tag{5.7.4}$$

Thus we see immediately:

(i) Any symmetric critical point of F, e.g. any point of $\Sigma \cap K(F)$, is a critical symmetric point;
(ii) In order for (P) to hold, the implication

$$D_pF \restriction T_p\Sigma = 0 \Rightarrow D_pF \restriction T_p^{\perp}\Sigma = 0 \tag{5.7.5}$$

has to be realised, e.g. if $D_pF(v)$ vanishes for all directions v parallel to Σ then it also vanishes for all directions v transverse to Σ.

Because of (i) we could also formulate (P) as

$$K(F_\Sigma) = \Sigma \cap K(F)\,. \tag{P$'$}$$

Before we derive some general results, let us discuss two classes of examples. Within the first very simple class there are examples for which (P) does not hold. The second class is a particular case of that discussed at the beginning of this section. It represents an illustration of a typical and quite effective application of (P).

Example 5.7.1. A Class of Counterexamples. Let the additive group $\mathbb{R}$ act on $M = \mathbb{R}^2$ by

$$\phi_t(x, y) = (x + y^k t, y)$$

for all $p = (x, y) \in M$ and all $t \in \mathbb{R}$ for some fixed $k \in \mathbb{N}$. Obviously this is a smooth action of $\mathbb{R}$ on $\mathbb{R}^2$ which is linear in the case $k = 1$. The set of symmetric points of $G = \{\phi_t \,|\, t \in \mathbb{R}\}$ is the subspace

$$\Sigma = \Sigma(G, M) = \{p = (x, 0) \,|\, x \in \mathbb{R}\}$$

and a smooth function $F : \mathbb{R}^2 \to \mathbb{R}$ is G-invariant if, and only if, there is a smooth function $f : \mathbb{R} \to \mathbb{R}$ such that

$$F(x, y) = f(y) \quad \forall (x, y) \in \mathbb{R}^2\,.$$

It follows that $F_\Sigma(x, 0) = f(0)$, and thus $D_pF_\Sigma = 0$ for all $p = (x, 0) \in \Sigma$, e.g. $K(F_\Sigma) = \Sigma$.

On the other hand we have

$$K(F) = \{p = (x, y) \in \mathbb{R}^2 \,|\, D_pF = 0\} = \{(x, y) \in \mathbb{R}^2 \,|\, f'(y) = 0\}$$

and therefore

$$\Sigma \cap K(F) = \begin{cases} \emptyset & \text{if } f'(0) \neq 0\,, \\ \Sigma & \text{if } f'(0) = 0\,. \end{cases}$$

The cases of functions f with $f'(0) \neq 0$ thus provide counterexamples to (P). Some less trivial counterexamples are discussed in [5.12].

Example 5.7.2. A Class of Examples for (P) *to Hold.* For $0 < r_1 < r_2 < \infty$ write $\mathcal{R} = \{x \in \mathbb{R}^n \mid r_1 \leq |x| \leq r_2\}$ where $|x|$ is the Euclidean norm of $x \in \mathbb{R}^n$. The boundaries of the set $\mathcal{R}$ are the spheres $S_1 = \{x \in \mathbb{R}^n \mid |x| = r_i\}$, $i = 1, 2$. The rotation group $G = \mathrm{SO}(n)$ acts linearly on the Banach space

$$\mathcal{C}^2(\mathcal{R}) = \{u : \mathcal{R} \to \mathbb{R} \mid \|u\| = \sup\{|D^\alpha u(x)| \; |x \in \mathcal{R}, |\alpha| \leq 2\} < \infty\}$$

by $(g \cdot u)(x) = u(g^{-1}x)$ for all $x \in \mathcal{R}$, $g \in G$ and $u \in \mathcal{C}^2(\mathcal{R})$.

The subset of functions in $\mathcal{C}^2(\mathcal{R})$ having prescribed constant values $C_1, C_2 \in \mathbb{R}$ on the boundary of $\mathcal{R}$ is a smooth (Banach) manifold, denoted by $M = M(C_i)$:

$$M = \{u \in \mathcal{C}^2(\mathcal{R}) \mid u \restriction S_i = C_i \, , \; i = 1, 2\} \, .$$

The tangent space of M at $u \in M$ is

$$T_u M = \{v \in \mathcal{C}^2(\mathcal{R}) \mid v \restriction S_i = 0 \, , \; i = 1, 2\} \, .$$

Now we want to give a convenient description of the set $\Sigma = \Sigma(G, M)$ of symmetric points of M under the above action of the rotation group G. To this end consider the Banach space $\mathcal{C}^2([r_1, r_2])$ of twice continuously differentiable real functions on the interval $[r_1, r_2]$ with norm

$$\|u_0\|_0 = \sup\{|u_0^{(j)}(r)| \; |r_1 \leq r \leq r_2 \, , \; j = 0, 1, 2\}$$

and the submanifold

$$\Sigma_0 = \{u_0 \in \mathcal{C}^2([r_1, r_2]) \mid u_0(r_i) = C_i \, , \; i = 1, 2\} \, .$$

The Banach space $\mathcal{C}^2([r_1, r_2])$ is mapped onto the subspace of G-invariant elements of $\mathcal{C}^2(\mathcal{R})$ by the map

$$\tau : \mathcal{C}^2([r_1, r_2]) \to \mathcal{C}^2(\mathcal{R}) \, , \quad (\tau u_0)(x) = u_0(|x|) \, , \quad x \in \mathcal{R} \, , \quad u_0 \in \mathcal{C}^2([r_1, r_2]) \, .$$

Thus we obviously have

$$\tau \Sigma_0 \subseteq \Sigma = \{u \in M \mid g \cdot u = u, \; \forall g \in G\} \, .$$

Conversely it can be shown easily that for every G-invariant $u \in \mathcal{C}^2(\mathcal{R})$ there exists a $u_0 \in \mathcal{C}^2([r_1, r_2])$ such that $u = \tau u_0$ [5.7, p. 5]. Define $u_0(r)$, $r \in [r_1, r_2]$, simply by $u_0(r) = u(re)$ where e is any fixed unit vector in $\mathbb{R}^n$. By invariance we then have $\tau u_0 = u$. It follows that

$$\Sigma = \tau \Sigma_0 \, .$$

The tangent space of Σ_0 in $u_0 \in \Sigma_0$ is

$$T_{u_0} \Sigma_0 = \{v_0 \in \mathcal{C}^2([r_1, r_2]) \mid v_0(r_1) = v_0(r_2) = 0\} \, .$$

Now it is not hard to show that Σ is a smooth submanifold of M and that the following relation for the tangent spaces holds. If $u = \tau u_0$, $u_0 \in \Sigma_0$, then

$$T_u\Sigma = \tau(T_{u_0}\Sigma_0) = \{v = \tau v_0 \mid v_0 \in \mathcal{C}^2([r_1, r_2]),\ v_0(r_1) = v_0(r_2) = 0\}.$$

As G-invariant function F on M we consider the Dirichlet functional

$$F(u) = \frac{1}{2}\int_{\mathcal{R}} |\nabla u(x)|^2\,dx\,, \quad u \in M\,.$$

Its differential in $u \in M$, $D_uF : T_uM \to \mathbb{R}$, is given by

$$D_uF(v) = \int_{\mathcal{R}} \underline{\nabla} u(x) \cdot \underline{\nabla} v(x)\,dx\,, v \in T_uM$$

and therefore a critical point $u \in M$ of F is characterised by

$$0 = \int_{\mathcal{R}} \underline{\nabla} u \cdot \underline{\nabla} v\,dx = \int_{\mathcal{R}} \underline{\nabla} \cdot (v\underline{\nabla} u)\,dx - \int_{\mathcal{R}} v\Delta u\,dx$$

for all $v \in T_uM$. By Gauß' theorem we have

$$\int_{\mathcal{R}} \underline{\nabla} \cdot (v\underline{\nabla} u)\,dx = \int_{\partial\mathcal{R}} v \cdot \underline{\nabla} u \cdot d\underline{o} = 0$$

as $v \restriction \partial\mathcal{R} = 0$ for all $v \in T_uM$. Thus

$$K(F) = \{u \in M \mid D_uF = 0\} = \{u \in M \mid \Delta u = 0\}$$

as $0 = \int_{\mathcal{R}} v(x)\Delta u(x)\,dx$ for all $v \in \mathcal{C}^2(\mathcal{R})$, $v \restriction \partial\mathcal{R} = 0$, implies $\Delta u(x) = 0$ for $x \in \mathcal{R}$.

A *harmonic function* u on $\mathcal{R}$ is by definition an element $u \in \mathcal{C}^2(\mathcal{R})$ such that $\Delta u = 0$. Therefore

$$K(F) = \{u \in H(\mathcal{R}) \mid u \restriction S_i = C_i\,,\ i = 1, 2\}$$

if $H(\mathcal{R})$ denotes the set of harmonic functions on $\mathcal{R}$. Now for $u \in \Sigma$ we have $u = \tau u_0$ with $u_0 \in \Sigma_0$ and thus

$$\underline{\nabla} u(x) = u_0'(|x|)\frac{1}{|x|}x\,.$$

This allows us to give a simplified expression for

$$F_\Sigma = F \restriction \Sigma$$

by using polar coordinates:

$$F_\Sigma(u) = \frac{1}{2}\int_{\mathcal{R}} |\underline{\nabla}(\tau u_0)(x)|^2\,dx = \frac{1}{2}\omega_n \int_{r_1}^{r_2} u_0'(r)^2 r^{n-1}\,dr = F_0(u_0)$$

where ω_n denotes the volume of the unit sphere in $\mathbb{R}^n$, e.g. as $\Sigma = \tau\Sigma_0$, it follows that $F_\Sigma \circ \tau = F_0 : \Sigma_0 \to \mathbb{R}$. A simple calculation implies for $u = \tau u_0$, $u_0 \in \Sigma_0$, and all $v = \tau v_0$, $v_0 \in T_{u_0}\Sigma_0$ that

$$(D_u F_\Sigma)(v) = (D_{u_0} F_0)(v_0) = w_n \int_{r_1}^{r_2} u_0'(r) v_0'(r) r^{n-1}\, dr\,.$$

Therefore,

$$K(F_\Sigma) = \{u \in \Sigma \mid D_u F_\Sigma = 0\} = \tau\{u_0 \in \Sigma_0 \mid D_{u_0} F_0 = 0\} = \tau K(F_0)$$

if $K(F_0)$ denotes the set of critical points of $F_0 : \Sigma_0 \to \mathbb{R}$. But $K(F_0)$ is easily determined:

$$\begin{aligned} u_0 \in K(F_0) \Leftrightarrow 0 &= \int_{r_1}^{r_2} u_0'(r) v_0'(r) r^{n-1}\, dr \\ &= -\int_{r_1}^{r_2} \left(\frac{d}{dr}\{r^{n-1} u_0'(r)\} \right) v_0(r)\, dr \end{aligned}$$

for all $v_0 \in \mathcal{C}^2([r_1, r_2])$, $v_0(r_1) = v_0(r_2) = 0$ i.e.,

$$u_0 \in K(F_0) \Leftrightarrow \frac{d}{dr}\{r^{n-1} u_0'(r)\} = 0\,,$$

and this ordinary differential equation is easily solved:

$$n = 2 : u_0(r) = a + b \log r\,,$$

$$n \geq 3 : u_0(r) = a + b r^{2-n}\,,$$

where the constants a and b are fixed by the boundary values of $u_0 : u_0(r_i) = C_i$, $i = 1, 2$.

Obviously we have in both cases

$$\Delta \tau u_0 = \sum_{j=1}^{n} \partial_j^2 \tau u_0 = 0\,,$$

i.e. the function τu_0 determined by the ordinary differential equation above and the boundary values is harmonic. Therefore,

$$K(F_\Sigma) = \tau K(F_0) \subseteq \Sigma \cap K(F)$$

and (P) holds in this case.

Suppose conversely that (P) holds (for instance by general arguments such as Corollary 7.5.3). Then we know

$$K(F_\Sigma) = \Sigma \cap K(F) = \Sigma \cap H(\mathcal{R})\,.$$

As we have seen above $K(F_\Sigma)$ is easily determined by solving a simple ordinary differential equation with prescribed boundary values. Therefore (P) determines all rotationally invariant harmonic functions in $\mathcal{R}$ in a simple way! (Let the boundary values C_i vary.)

We now present some general results. These results rely on simplifying assumptions on the pair (M, G). Suppose E is a real Banach space on which a group G is represented by continuous linear maps $T_g : E \to E$; i.e. $T : G \to \mathcal{L}(E, E)$ satisfies

$$T_{g_1} T_{g_2} = T_{g_1 g_2} \quad \text{for all } g_i \in G\,.$$

Then the set of symmetric points

$$\Sigma = \{x \in E \,|\, T_g x = x \ \forall g \in G\}$$

is a closed subspace of E. Now if $F \in \mathcal{C}^1(E, \mathbb{R})$ is a G-invariant function, the equation

$$F(T_g x) = F(x) \quad \forall g \in G \ \ \forall x \in E$$

implies by the chain rule and the linearity of the action of G that

$$(D_{T_g x} F) \circ T_g = D_x F \quad \forall g \in G \ \ \forall x \in E$$

and in particular, for $x \in \Sigma$,

$$D_x F \circ T_g = D_x F \quad \forall g \in G\,,$$

e.g. $D_x F$ is a G-invariant continuous linear functional on E:

$$D_x F \in \Sigma' = \{l \in \mathcal{L}(E, \mathbb{R}) = E' \,|\, l \cdot T_g = l, \ \forall g \in G\}\,.$$

As usual we denote by Σ^0 the annihilator of the subspace Σ in E':

$$\Sigma^0 = \{l \in E' \,|\, l \restriction \Sigma = 0\}\,.$$

Thus ge get

$$\begin{aligned} K(F_\Sigma) &= \{x \in \Sigma \,|\, D_x F_\Sigma = (D_x F) \restriction \Sigma = 0\} \\ &= \{x \in \Sigma \,|\, D_x F \in \Sigma' \cap \Sigma^0\} \end{aligned}$$

and a first simple but quite effective result follows.

Theorem 5.7.1. *Let a group G be represented by continuous linear maps on a Banach space E and use the notation introduced above. Then*

$$\Sigma' \cap \Sigma^0 = \{0\} \quad \text{implies (P)}$$

for all G-invariant functions $F \in \mathcal{C}^1(E, \mathbb{R})$.

Proof. $\Sigma \cap K(F) = \{x \in \Sigma \,|\, D_x F = 0\}$. By the above expression for $K(F_\Sigma)$ the statement follows. □

Corollary 5.7.2. *If $E = \mathcal{H}$ is a Hilbert space and $g \mapsto T_g$ a unitary representation of a group G on $\mathcal{H}$ then (P) holds for all G-invariant functions $F \in \mathcal{C}^1(\mathcal{H}, \mathbb{R})$.*

Proof. If $\langle \cdot\,,\cdot \rangle$ denotes the scalar product of the Hilbert space $\mathcal{H}$, an anti-isomorphism $\mathcal{J}$ of $\mathcal{H}$ onto its topological dual space $\mathcal{H}'$ is defined by $y \mapsto \mathcal{J}(y) = l_y \in \mathcal{H}'$:

$$l_y(x) := \langle y, x \rangle \quad \forall x \in \mathcal{H}\,.$$

Under this isomorphism $\mathcal{J}$ the closed subspace Σ of symmetric points is mapped onto the subspace Σ' of G-invariant functionals:

$$\Sigma' = \mathcal{J}(\Sigma) \subseteq \mathcal{H}'$$

as T_g is unitary, and the orthogonal complement

$$\Sigma^\perp = \{ y \in \mathcal{H} \mid \langle y, x \rangle = 0 \ \forall x \in \Sigma \}$$

of Σ is mapped onto the annihilator of Σ in $\mathcal{H}'$:

$$\Sigma^0 = \mathcal{J}(\Sigma^\perp)\,.$$

But in a Hilbert space we always have

$$\Sigma \cap \Sigma^\perp = \{0\}\,,$$

and thus $\Sigma' \cap \Sigma^0 = \{0\}$ follows and the corollary is proved by Theorem 5.7.1.

□

Corollary 5.7.3. *If a compact group G is represented on a Banach space E by continuous linear maps $T_g : E \to E$ and if this representation is strongly continuous (i.e. if $g \mapsto T_g x$ is a continuous map $G \to E$ for each $x \in E$) then* (P) *holds for all G-invariant functions $F \in C^1(E, \mathbb{R})$.*

Proof. Introduce the spaces Σ, Σ' and Σ^0 as above and suppose that there is an $l \in \Sigma' \cap \Sigma^0$, $l \neq 0$. Then there is an $x_0 \in E$ with $l(x_0) = 1$. The compact group G has a normalized Haar measure μ which is invariant [5.13]. By assumption, $g \mapsto T_g x_0$ is a continuous function on the compact space G; thus this function is integrable:

$$\int_G T_g x_0 \, d\mu(g) = \overline{x}_0 \in E\,.$$

Also, $l \in \Sigma' \subset E'$ implies

$$l(\overline{x}_0) = \int_G l(T_g x_0)\, d\mu(g) = \int_G l(x_0)\, d\mu(g) = 1\,,$$

and the invariance of μ implies for all $g' \in G$:

$$T_{g'} \overline{x}_0 = \int_G T_{g'} T_g x_0 \, d\mu(g) = \int_G T_{g' \circ g} x_0 \, d\mu(g) = \overline{x}_0\,,$$

i.e. $\overline{x}_0 \in \Sigma$. But for $l \in \Sigma^0$, $\Sigma \subseteq \operatorname{Ker} l$. This contradiction proves $\Sigma' \cap \Sigma^0 = \{0\}$ and again Theorem 5.7.1 proves this corollary. □

Remark 5.7.3. There is another approach to Corollaries 5.7.2 and 5.7.3 which is a little less direct but which shows their common basis. This basis is the following observation:

If the Banach space E and the group G satisfy the hypothesis of Theorem 5.7.1 and if every closed G-invariant hyperplane H in E, $0 \in H$, has a G-invariant complemented subspace, then

$$\Sigma' \cap \Sigma^0 = \{0\}$$

follows.

The proof of this relies on the fact that each such hyperplane H is the kernel of an element $l \in \Sigma'$, $H = \operatorname{Ker} l$, and conversely. The corollaries then follow by proving the hypotheses of the above observation.

In actual application one quite often has to deal with a G-invariant smooth function F on the smooth manifold M modelled on a Banach space E instead of a function on the Banach space itself. Without explaining all necessary details about Banach-manifolds [5.14] we indicate the basic ideas about how to reduce this more general case to the simpler situation discussed above.

The basic fact about a smooth manifold M modelled on a Banach space E which we have to use here is the following. For every $p \in M$ there is an open set U of M containing p and a diffeomorphism φ from U onto an open neighborhood $\varphi(U) \subset E$ such that $\varphi(p) = 0 \in E$. The pair (U, φ) is called a coordinate system or a chart at $p \in M$. It follows that the differential $D_p\varphi$ of φ at p is an isomorphism from the tangent space T_pM of M in p onto the Banach space E.

Suppose now that a (Lie) group G acts smoothly on M, i.e. for each $g \in G$

$$p \mapsto g \circ p$$

is a smooth map $M \to M$ of class C^1, let us say. Then the differential D_pg of this map in $p \in M$ is a continuous linear map of the tangent space T_pM of M in p into the tangent space $T_{g \circ p}M$ of M in $g \circ p$. The chain rule implies

$$D_p(g_1 g_2) = (D_{g_2 \circ p} g_1) \circ D_p g_2 \,.$$

Thus for a fixed symmetric point

$$p \in \Sigma = \{p' \in M \mid g \circ p' = p', \ \forall g \in G\}$$

we get that $g \mapsto D_pg$ is a representation of G by continuous linear maps $T_pM \to T_pM$ and thus

$$g \longmapsto D_p\varphi \circ D_pg \circ D_p\varphi^{-1}$$

is a representation of G by continuous linear maps $E \to E$.

The basic definition now is the following. The action of G on M is called *linearisable* at $p \in M$ if, and only if, there is a chart (U, φ) at p such that the action looks linear:

$$\bigwedge_{g \in G} \bigvee_{\tilde{g} \in L(E,E)} \tilde{g} \restriction \varphi(U) = \varphi \circ g \circ \varphi^{-1} .$$

As usual, $L(E, E)$ denotes the space of linear maps $E \to E$. A linear map on E is uniquely determined by its values on an open neighborhood of $0 \in E$. Thus it follows that

(i) there is at most one such $\tilde{g} \in L(E, E)$,
(ii) $\tilde{g}$ is continuous,
(iii) $g \longmapsto \tilde{g}$ is a representation of G by elements of $\mathcal{L}(E, E)$, the space of continuous linear maps $E \to E$.

Differentiation at $0 \in E$ implies

$$\begin{aligned}\tilde{g} &= D_0(\varphi \circ g \circ \varphi^{-1}) = D_{g \circ p}\varphi \circ D_p g \circ D_0 \varphi^{-1} \\ &= D_{g \circ p}\varphi \circ D_p g \circ (D_p \varphi)^{-1}\end{aligned}$$

and thus, for $p \in \Sigma$,

$$\tilde{g} = D_p \varphi \circ D_p g \circ (D_p \varphi)^{-1} =: T_g^p$$

and $g \mapsto T_g^p \in \mathcal{L}(E, E)$ is a representation of G. For $p \in \Sigma$ we also introduce the G-invariant closed subspace Σ_p of E by

$$\Sigma_p = \Sigma_p(G, E) = \{x \in E \,|\, T_g^p x = x, \ \forall g \in G\} .$$

For $p' \in U \cap \Sigma$ it follows that

$$T_g^p \varphi(p') = \varphi \circ g \circ \varphi^{-1}(\varphi(p')) = \varphi(p')$$

and conversely, for $x \in \Sigma_p \cap \varphi(U)$, we get for $p' = \varphi^{-1}(x)$:

$$g \circ p' = \varphi^{-1} \circ \varphi \circ g \circ \varphi^{-1}(x) = \varphi^{-1} \circ T_g^p x = \varphi^{-1}(x) = p' .$$

This calculation shows that

$$\varphi : U \cap \Sigma \to \varphi(U) \cap \Sigma_p .$$

Thus, in a suitable neighborhood U of a symmetric point $p \in \Sigma$, the symmetric points are mapped by φ onto the symmetric points in $\varphi(U) \subset E$ with respect to the action T_g^p of G on E.

As Σ_p is a closed subspace of E it follows that Σ is a smooth submanifold of M and that the tangent space of Σ in p is given by

$$T_p \Sigma = (D_p \varphi)^{-1} \Sigma_p .$$

Therefore, locally we are back to the situation where a group G acts on a Banach space E, i.e. in the situation of Theorem 5.7.1. We proceed accordingly and introduce

$$\Sigma'_p = \{l \in E' \mid l \circ T_g^p = l \ \forall g \in G\},$$
$$\Sigma_p^0 = \{l \in E' \mid \Sigma_p \subseteq \operatorname{Ker} l\}.$$

Suppose that a function $F \in C^1(M, \mathbb{R})$ is G-invariant. Then $F(g \circ p) = F(p)$ implies

$$D_{g \circ p} F \circ D_p g = D_p F \quad \forall g \in G.$$

Thus for $p \in \Sigma$ it follows that

$$D_p F \circ (D_p \varphi)^{-1} \circ T_g^p p = D_p F \circ (D_p \varphi)^{-1} \quad \forall g \in G,$$

i.e.

$$l_p = D_p F \circ (D_p \varphi)^{-1} \in \Sigma'_p.$$

As, again, $D_p F_\Sigma = D_p F \restriction T_p \Sigma$, the set of critical symmetric points is given by

$$K(F_\Sigma) = \{p \in \Sigma \mid l_p \in \Sigma'_p \cap \Sigma_p^0\}$$

and therefore the assumption

$$\Sigma'_p \cap \Sigma_p^0 = \{0\} \quad \forall p \in \Sigma$$

implies $K(F_\Sigma) \subseteq \Sigma \cap K(F)$, i.e. (P).

Theorem 5.7.4. *Suppose M is a smooth manifold modelled on a Banach space E with a C^1 action of a Lie group G on it.*

(a) *If the action of G on M is linearisable at every*

$$p \in \Sigma = \Sigma(G, M) = \{p \in M \mid g \circ p = p, \ \forall g \in G\},$$

then the set Σ of symmetric points is a smooth submanifold of M.

(b) *For $p \in \Sigma$ define Σ_p, Σ'_p, and Σ_p^0 as above. If $\Sigma'_p \cap \Sigma_p^0 = \{0\}$ holds for all $p \in \Sigma$ then (P) holds for all G-invariant functions $F \in C^1(M, \mathbb{R})$.*

We want to apply Theorem 5.7.4 to prove the corresponding results of Corollaries 5.7.2 and 5.7.3 for the case of manifolds. Thus, corresponding to Corollary 5.7.2, we assume M to be a Riemann manifold, i.e. M is supposed to be locally diffeomorphic to a Hilbert space $\mathcal{H}$ and on each tangent space $T_p M$ there is a scalar product $\langle \cdot, \cdot \rangle$ that varies smoothly with p. Furthermore, we assume that a Lie group G acts on M by isometries, i.e., for each $p \in M$, $D_p g$ is supposed to be an isometry, $T_p M \to T_{g \circ p} M$. Then for a symmetric point $p \in \Sigma$,

$$g \mapsto D_p g$$

is a unitary representation on the Hilbert space $T_p M$. In order to see that the action of G is linearisable at every symmetric point p, we use the crucial fact that in the present case of a Riemann manifold the exponential map $\psi \equiv \exp$ exists, which maps a neighborhood V of zero of the Hilbert space T_pM diffeomorphically onto a neighborhood $\psi(V)$ of p in M with $\psi(0) = p$. Thus $(\psi(V), \psi^{-1})$ can be regarded as a chart in p (geodesic normal coordinates at p). For $p \in \Sigma$ the crucial property of the exponential map

$$\psi \circ D_p g(v) = g \circ \psi(v) \quad \forall v \in V$$

implies

$$D_p g(v) = \psi^{-1} \circ g \circ \psi(v)$$

for $v \in V$, i.e. the action of G on M is indeed linearisable. Therefore Theorem 5.7.4 implies the following corollary.

Corollary 5.7.5. *Let M be a Riemann manifold and suppose that a group G acts on M by isometries. Then the following holds:*

(a) *The set $\Sigma = \{p \in M \mid g \circ p = p, \ \forall g \in G\}$ of symmetric points is a smooth submanifold of M;*
(b) (P) *holds for every G-invariant function $F \in C^1(M, \mathbb{R})$.*

Furthermore, corresponding to Corollary 5.7.3, Theorem 5.7.4 also implies Corollary 5.7.6.

Corollary 5.7.6. *If M is a smooth Banach manifold on which a compact Lie group acts smoothly, then*

(a) *the set Σ of symmetric points is a smooth submanifold of M,*
(b) (P) *holds for all G-invariant functions $F \in C^1(M, \mathbb{R})$.*

The proof of this consists in showing first that compactness of G and continuity of $G \times M \to M$, $(g, p) \mapsto g \circ p$, imply the existence of arbitrarily small G-invariant neighborhoods of symmetric points and then using such a neighborhood to construct a chart which linearises the action of G. Then Theorem 5.7.4 is applied.

6. The Variational Approach to Linear Boundary and Eigenvalue Problems

One of the first areas in which the basic variational concepts and methods are applied is the solution of large, important classes of linear boundary and eigenvalue problems. The results discussed in this chapter follow in part from the results on nonlinear boundary and eigenvalue problems which we shall discuss in Chap. 8; nonetheless we present proofs in this chapter which emphasise the simplicity and elementary character of the variational approach. We start with a variational proof of the spectral theorem for compact self-adjoint operators, followed by a general version of the projection theorem (for convex closed sets).

We need to prepare for tackling the problem of determining the spectrum of an elliptic second-order operator by first of all reformulating the problem so that variational concepts and methods (quadratic forms) are applicable. Then we must prove a few results specific to this problem. The solutions of particular elliptic boundary and eigenvalue problems are then fairly simple special cases of these results.

6.1 The Spectral Theorem for Compact Self-Adjoint Operators. Courant's Classical Minimax Principle. Projection Theorem

We shall show first of all how we can quite easily obtain a well-known result from the spectral theory of linear operators in Hilbert spaces (see [6.1], for example) with the aid of the general variational-calculus results we have already obtained. As a preparatory exercise, we suggest you prove the spectral theorem for Hermitian matrices $A \neq 0$ by determining the critical points of the function

$$f(x) = \langle x, Ax \rangle$$

subject to the constraint $\phi(x) = \langle x, x \rangle - 1 = 0$. The same idea also yields a proof of the spectral theorem for compact self-adjoint operators. Recall that a compact operator A in a separable Hilbert space $\mathcal{H}$ is characterised by the property that, for every weakly convergent sequence $(x_n)_{n \in \mathbb{N}}$ in $\mathcal{H}$, the sequence of images $(Ax_n)_{n\in\mathbb{N}}$ converges strongly. This definition simply states that the functions

$$x \to \|Ax\| \quad \text{and} \quad x \to \langle x, Ax\rangle$$

are weakly continuous on $\mathcal{H}$. This is immediately obvious for the first function. The weak continuity of the function $x \to \langle x, Ax\rangle$ can be proved as follows. If $x_0 = w - \lim_{n\to\infty} x_n$, then it follows that $Ax_0 = s - \lim_{n\to\infty} Ax_n$ and thus

$$\langle x_0, Ax_0\rangle - \langle x_n, Ax_n\rangle = \langle x_0 - x_n, Ax_0\rangle + \langle x_n, Ax_0 - Ax_n\rangle$$

converges towards zero because $(x_n)_{n\in\mathbb{N}}$ is strongly bounded.

Theorem 6.1.1. Spectral Theorem for Compact Self-Adjoint Operators (Hilbert-Schmidt, Riesz-Schauder). *Let $\mathcal{H}$ be a separable Hilbert space and $A \neq 0$ a compact self-adjoint operator on $\mathcal{H}$. Then A possesses an orthonormal system $\{e_j\}_{j\in\mathbb{N}}$ of eigenvectors with real eigenvalues $(\lambda_j)_{j\in\mathbb{N}}$, $Ae_j = \lambda_j e_j$, which has the following properties.*

(i) *The eigenvalues are arranged in descending order of magnitude: $|\lambda_1| \geq \cdots \geq |\lambda_j| \geq |\lambda_{j+1}| \geq \cdots$.*
(ii) *Either finitely many of the eigenvalue are nonzero, or all eigenvalues are nonzero, and $\lim_{j\to\infty} \lambda_j = 0$ (0 is the only clusterpoint of the λ_j, $j \in \mathbb{N}$).*
(iii) *The multiplicity of each nonzero eigenvalue is finite:*

$$\dim \operatorname{Ker}(A - \lambda_j I) < \infty \quad \textit{for all} \quad j \in \mathbb{N}\,.$$

(iv) *The orthonormal system of eigenvectors $\{e_j,\ j \in \mathbb{N}\}$ is complete if, and only if, A is injective.*

Proof. (a) We start off by determining the eigenvalue of greatest magnitude. The following holds:

$$\|A\| = \sup_{u\in\overline{B}_1(\mathcal{H})} \|Au\| = \sup_{u\in S_1(\mathcal{H})} |\langle u, Au\rangle| = \|A\|_{B(\mathcal{H})}\,,$$

where

$$\overline{B}_1(\mathcal{H}) = \{x \in \mathcal{H} \,|\, \|x\| \leq 1\} \quad \text{and} \quad S_1(\mathcal{H}) = \{x \in \mathcal{H} \,|\, \|x\| = 1\}\,.$$

The closed unit ball $\overline{B}_1(\mathcal{H})$ of a Hilbert space is weakly (sequentially) compact and $u \to \|Au\|$ and $u \to |\langle u, Au\rangle|$ are weakly continuous. Theorem 1.1.2 then gives the existence of a maximising point $e_1 \in \overline{B}_1(\mathcal{H})$, i.e.

$$\|A\| = \|Ae_1\| = \sup_{u\in\overline{B}_1(\mathcal{H})} \|Au\|\,.$$

$A \neq 0$ implies $e_1 \neq 0$, and so $\|e_1\| = 1$, since if $0 < \|e_1\| < 1$, then

$$\widehat{e}_1 = \frac{1}{\|e_1\|} e_1 \in S_1(\mathcal{H})$$

and it would follow that

$$\|Ae_1\| < \|A\hat{e}_1\| = \frac{1}{\|e_1\|}\|Ae_1\|,$$

which is a contradiction.

(b) The function $x \to Q(x) = \langle x, Ax\rangle$ is differentiable and has the Fréchet derivative

$$Q'(x)(h) = 2\langle Ax, h\rangle, \quad \forall h \in \mathcal{H}.$$

The following holds:

$$\sup_{x\in S_1(\mathcal{H})} Q(x) = \pm\|A\|.$$

The constraint $\phi(x) = \langle x, x\rangle - 1 = 0$ is defined by a mapping ϕ which is regular at all points of the level surface $\phi^{-1}(0) = S_1(\mathcal{H})$, since $\phi'(x)(h) = 2\langle x, h\rangle$ for $h \in \mathcal{H}$.

We established in (a) that $e_1 \in S_1(\mathcal{H})$ is an extremal point of the function Q. Theorem 4.2.3 (Ljusternik) then gives the existence of a $\lambda_1 \in \mathbb{R}$ such that

$$Q'(e_1) = \lambda_1\phi'(e_1), \quad \text{i.e.} \quad Ae_1 = \lambda e_1,$$

and $|\lambda_1| = \|A\|$.

(c) Let $\mathcal{H}_1 = \{e_1\}^\perp \subset \mathcal{H}$ be the subspace of $\mathcal{H}$ orthogonal to e_1. Since A is self-adjoint, it maps the subspace $\mathcal{H}_1$ into itself: for $x \in \mathcal{H}_1$,

$$\langle Ax, e_1\rangle = \langle x, Ae_1\rangle = \langle x, \lambda e_1\rangle = 0, \quad \text{i.e.} \quad Ax \in \{e_1\}^\perp.$$

Thus the restriction $A_1 = A \upharpoonright \mathcal{H}_1$ of A to the subspace $\mathcal{H}_1$ is a well defined operator on the Hilbert space $\mathcal{H}_1$. This operator is likewise self-adjoint and compact, and its norm is not larger than that of A:

$$\|A_1\|_{\mathcal{B}(\mathcal{H}_1)} \leq \|A\|_{\mathcal{B}(\mathcal{H})}.$$

If $A_1 \neq 0$, then our considerations from (a) and (b) can be applied to the compact self-adjoint operator A_1 in the Hilbert space $\mathcal{H}_1$. They give the existence of a normed eigenvector $e_2 \in \mathcal{H}_1$ of A_1 with eigenvalue λ_2, where

$$|\lambda_2| = \|A_1\|_{\mathcal{B}(\mathcal{H}_1)} \leq \|A\|_{\mathcal{B}(\mathcal{H})} = |\lambda_1| \quad \text{and} \quad \langle e_2, e_1\rangle = 0.$$

(d) The reduction step described in (c) can be repeated successively. It is necessary to differentiate between two cases, depending on whether the procedure stops after finitely many steps or not. However, no further variational arguments are needed to conclude the proof, and we thus refer the reader to the literature [6.1]. □

Theorem 6.1.1 establishes the existence and some of the properties of the eigenvalues of a compact operator. Furthermore, a method for calculating these eigenvalues is hinted at in the proof. Working out this method in detail leads to the *classical minimax principle of Courant-Weyl-Fischer-Poincaré.*

Theorem 6.1.2. Minimax Principle. *Let $\mathcal{H}$ be a real, separable Hilbert space and $A \geq 0$ a self-adjoint operator on $\mathcal{H}$ with spectrum $\sigma(A) = \{\lambda_m \mid m \in \mathbb{N}\}$, $\lambda_{m+1} \geq \lambda_m$. Let $\mathcal{E}_m$ denote the family of all m-dimensional subspaces E_m of $\mathcal{H}$. Then the eigenvalues $(\lambda_m)_{m\in\mathbb{N}}$ can be calculated as follows:*

$$\lambda_m = \min_{E_m \in \mathcal{E}_m} \max_{v \in E_m} \frac{\langle v, Av\rangle}{\langle v, v\rangle}.$$

Proof. We determine the value distribution of the Rayleigh quotient

$$R(v) = \frac{\langle v, Av\rangle}{\langle v, v\rangle}$$

by expanding the vectors $v \in \mathcal{H}$ in terms of the eigenvectors $(e_j)_{j\in\mathbb{N}}$ of A as determined in Theorem 6.1.1. For

$$v = \sum_{j=1}^{\infty} \alpha_i e_i\,, \quad \langle v, v\rangle = \sum_{i=1}^{\infty} \alpha_i^2 < \infty\,,$$

we get

$$\langle v, Av\rangle = \sum_{i,j=1}^{\infty} \langle \alpha_i e_i, A\alpha_j e_j\rangle = \sum_{j=1}^{\infty} \lambda_j \alpha_j^2$$

and so

$$R(v) = \frac{\sum_{i=1}^{\infty} \lambda_i \alpha_i^2}{\sum_{i=1}^{\infty} \alpha_i^2}$$

Now let $V_m = \text{lin}\{e_1, \ldots, e_m\}$ denote the subspace generated by the first m eigenvectors. It follows that

$$\max_{v\in V_m} R(v) = \max_{(\alpha_1, \ldots, \alpha_m)\in\mathbb{R}^m} \frac{\sum_{j=1}^{m} \lambda_j \alpha_j^2}{\sum_{j=1}^{m} \alpha_j^2} = \lambda_m = R(e_m)\,,$$

and thus it remains for us to show that $\max_{v\in E_m} R(v) \geq \lambda_m$ for every other subspace $E_m \in \mathcal{E}_m$. Let $E_m \neq V_m$ be such a subspace; then $E_m \cap V_m^{\perp} \neq \{0\}$ and it follows that

$$\begin{aligned}\max_{v\in E_m} R(v) &\geq \max_{v\in E_m\cap V_m^{\perp}} R(v)\\ &= \max_{v=\sum_{j\geq m+1} a_j e_j \in E_m} \frac{\sum_{j\geq m+1} \lambda_j \alpha_j^2}{\sum_{j\geq m+1} \alpha_j^2} \geq \lambda_{m+1} \geq \lambda_m\,,\end{aligned}$$

which completes the proof. □

Remark 6.1.1. Later on, when we discuss the Ljusternik-Schnirelman theory, we shall take the following simple reformulation of Theorem 6.1.2 as our starting point for a very important generalisation of the classical minimax principle.

Let $\mathcal{S}_{\{m\}}$ denote the set of unit spheres S_m in m-dimensional subspaces E_m of $\mathcal{H}$,

$$S_m = E_m \cap S_1(\mathcal{H}), \quad S_1(\mathcal{H}) = \{v \in \mathcal{H}, \|v\| = 1\}, \quad E_m \in \mathcal{E}_m .$$

Then according to Theorem 6.1.2,

$$\lambda_m = \min_{S_m \in \mathcal{S}_m} \max_{v \in S_m} \langle v, Av \rangle .$$

The following version of the projection theorem for closed convex sets of a Hilbert space represents a generalisation of the well-known projection theorem for subspaces and it shows that the essence of the proof is a "variational" argument. Although a direct proof is known (see, for example [6.2]), we want to present the variational proof here.

Theorem 6.1.3. Projection Theorem for Convex Closed Sets (Beppo Levi). *Let $\mathcal{H}$ be a real Hilbert space and $K \subset \mathcal{H}$ a nonempty convex closed subset. Then there is a function $p_K : \mathcal{H} \to K$ which has the following properties:*

(a) $\|x - p_K(x)\| = \inf_{z \in K} \|x - z\|$, $\forall x \in \mathcal{H}$.
(b) *$p_K(x)$ is characterised for $x \in \mathcal{H}$ by*

$$\langle x - p_K(x), z - p_K(x) \rangle \le 0, \quad \forall z \in K .$$

(c) *p_K is (Lipschitz) continuous:*

$$\|p_K(x_1) - p_K(x_2)\| \le \|x_1 - x_2\|, \quad \forall x_1, x_2 \in \mathcal{H} .$$

Proof. We intend to define the function p_K by property (a). For this we have to show that for arbitrary, but fixed, $x \in \mathcal{H}$, the function

$$\varphi_x : K \to \mathbb{R}, \quad \varphi_x(z) = \|x - z\| \quad \text{for} \quad z \in K$$

has exactly one minimising point in K, which we shall call $p_K(x)$. The first step is to show the existence of minimising points: to this end, let $x \in \mathcal{H}$ be arbitrary, but fixed. Since φ_x is obviously a convex continuous function on K, the set

$$M = \{x \in K \mid \varphi_x(z) \le \varphi_x(z_0)\}$$

is, for arbitrary, but fixed, $z_0 \in K$, a nonempty convex closed set. If K is unbounded, then $\varphi_x(z) \to +\infty$, $\|z\| \to \infty$, and it follows that M is also bounded and thus, by Theorem 1.2.2, weakly compact. By Lemma 1.2.4, Theorem 1.2.5 gives the existence of at least one minimising point $u_x \in K$:

$$\varphi(u_x) = \inf_{z \in K} \varphi_x(z) .$$

The uniqueness of u_x follows from the fact that Hilbert spaces are *strictly normed* (i.e. whenever $x, y \in \mathcal{H}$ satisfy $\|x+y\| = \|x\| + \|y\|$, there exists a $\lambda > 0$ such that $y = \lambda x$. Let $u'_x \in K$ be another minimising point; then

$$\varphi_x(u'_x) = \varphi_x(u_x) \leq \varphi_x\left(\frac{1}{2}(u_x + u'_x)\right) \leq \frac{1}{2}\varphi_x(u_x) + \frac{1}{2}(u'_x)$$

due to the convexity of φ_x. Thus

$$\varphi_x\left(\frac{1}{2}u_x + \frac{1}{2}u'_x\right) = \frac{1}{2}\varphi_x(u_x) + \frac{1}{2}\varphi_x(u'_x),$$

which means that

$$\left\|\frac{1}{2}(u_x - x) + \frac{1}{2}(u'_x - x)\right\| = \left\|\frac{1}{2}(u_x - x)\right\| + \left\|\frac{1}{2}(u'_x - x)\right\|$$

$$\frac{1}{2}(u_x - x) = \lambda\frac{1}{2}(u'_x - x)$$

for some $\lambda > 0$. $\|u'_x - x\| = \|u_x - x\|$ implies $\lambda = 1$, and thus $u'_x = u_x$. This shows that $p_K : \mathcal{H} \to K$ is well defined by

$$p_K(x) := \begin{cases} u_x\,, & \text{for } x \in \mathcal{H},\ x \notin K\,, \\ x\,, & \text{for } x \in K\,. \end{cases}$$

Remark 6.1.2. This proof works in every strictly normed reflexive Banach space and thus also yields the existence of best approximation element for these spaces.

In the next step we show that all minimising points of φ_x are characterised by the inequality (b). If u_x is a minimising point of φ_x in K, then it follows for all $z \in K$ and all $t \in [0,1]$ that

$$0 \leq \|x - u_x + t(u_x - z)\|^2 - \|x - u_x\|^2 = 2t\langle x - u_x, u_x - z\rangle + t^2\|u_x - z\|^2$$

and thus $\langle x - u_x, u_x - z\rangle \geq 0$, which is (b).

Conversely, if u_x is a point in K for which this inequality holds for all $z \in K$, then it follows that

$$\begin{aligned} \|x - u_x\|^2 - \|x - z\|^2 &= \langle x - u_x, z - u_x\rangle + \langle x - z, z - u_x\rangle \\ &\leq \langle x - z, z - u_x\rangle \\ &= \langle x - u_x, z - u_x\rangle + \langle u_x - z, z - u_x\rangle \\ &\leq 0\,, \end{aligned}$$

i.e. $\|x - u_x\| \leq \|x - z\|$ for all $z \in K$ and thus

$$\|x - u_x\| = \inf_{z \in K}\|x - z\|\,.$$

This proves (b).

If $x_1, x_2 \in \mathcal{H}$, then inequality (b) states that

$$\langle x_1 - p_K(x_1), p_K(x_2) - p_K(x_1) \rangle \leq 0$$

and

$$\langle x_2 - p_K(x_2), p_K(x_1) - p_K(x_2) \rangle \leq 0 .$$

Therefore

$$\langle x_1 - x_2 + p_K(x_2) - p_K(x_1), p_K(x_2) - p_K(x_1) \rangle \leq 0$$

and thus

$$\begin{aligned} \|p_K(x_2) - p_K(x_1)\|^2 &\leq \langle x_2 - x_1, p_K(x_2) - p_K(x_1) \rangle \\ &\leq \|x_2 - x_1\| \, \|p_K(x_2) - p_K(x_1)\| , \end{aligned}$$

which proves (c). □

6.2 Differential Operators and Forms

Generally speaking, the possibility of solving linear partial differential equations using the direct methods of the variational calculus is based on the simple idea of representing the particular differential operator as the Fréchet derivative of a quadratic form on a suitable function space; this form is then investigated with the aid of general theorems and, in particular, its critical points are determined. The boundary conditions for potential solutions are already taken into account through the choice of function space on which the quadratic form is being investigated. This simple formulation leads to our goal for both linear boundary value and eigenvalue problems. If the quadratic form is replaced by a general function, then this treatment can also be successfully applied to *nonlinear* boundary and eigenvalue problems (Chaps. 7 and 8). We shall demonstrate these methods explicitly for linear and quasi-linear second-order differential operators (which are the most important in applications). It should then become sufficiently clear what is to be done in the case of higher-order differential operators.

The approach is based upon suitable classes of function spaces and their embedding relations. Specifically, the function spaces used here are the so-called Sobolev spaces; the most important properties and embedding relations of these spaces are discussed in Appendix D. The theory of Sobolev spaces was developed essentially to be used in this approach. A newer, more complete exposition of this theory can be found in [6.3].

Obviously, not all (linear) differential operators can be treated following the approch described above: only those which have "divergence form" can be treated thus. The reason for this restriction will very soon become clear. A further restriction consists in the fact that, up to now, it has only been possible to give the most interesting results for differential operators on bounded sets. This restriction essentially stems from the fact that embedding relations between Sobolev spaces are used. We shall later discuss successful attempts of extending this approach to differential operators on $\mathbb{R}^n$.

To get down to details, let $\Omega \subset \mathbb{R}^n$ be an open nonempty bounded set and let A be a linear differential operator in divergence form on Ω, i.e. A acts as follows on functions $u : \Omega \to \mathbb{R}$:

$$Au = a_0 u - \sum_{j=1}^{n} \partial_j \Big(\sum_{j=1}^{n} a_{ji} \partial_i u \Big), \quad a_{ij} = a_{ji}. \tag{6.2.1}$$

The *eigenvalue problem* for the differential operator A then consists in finding numbers λ and functions $u_\lambda \neq 0$ on Ω such that

$$Au_\lambda = \lambda u_\lambda \tag{6.2.2}$$

Appropriate choice of the space in which we want to look for the eigenfunctions u_λ is naturally a decisive factor for the solubility of this problem. In deciding which space of potential solutions to choose, an important viewpoint results from establishing the behaviour of the potential solutions on the boundary $\Gamma = \partial\Omega$ of Ω, i.e. from the *boundary conditions* for the differential operator A.

Let the coefficients of the differential operator A be essentially bounded on Ω, that is $a_0, a_{ij} \in L^\infty(\Omega)$. Then A is certainly defined on $\mathcal{D}(\Omega) = \mathcal{C}_0^\infty(\Omega)$ and for $\varphi, \psi \in \mathcal{D}(\Omega)$, we have

$$\begin{aligned}\langle \varphi, A\psi \rangle_2 &= \langle \varphi, a_0 \psi \rangle_2 - \Big\langle \varphi, \sum_{j=1}^{n} \partial_j \Big(\sum_{i=1}^{n} a_{ji} \partial_i \psi \Big) \Big\rangle_2 \\ &= \langle \varphi, a_0 \psi \rangle_2 + \sum_{j=1}^{n} \Big\langle \partial_j \varphi, \sum_{i=1}^{n} a_{ji} \partial_i \psi \Big\rangle_2 - \int_\Omega d^n x \sum_{j=1}^{n} \partial_j \Big(\varphi \sum_{i=1}^{n} a_{ji} \partial_i \psi \Big)\end{aligned}$$

where $\langle \cdot\,, \cdot \rangle$ is the scalar product of $L^2(\Omega)$. If Ω has a boundary sufficiently smooth for Gauß' integral theorem to be applicable, we have

$$\int_\Omega d^n x \sum_{j=1}^{n} \partial_j \Big\{ \varphi \sum_{i=1}^{n} a_{ji} \partial_i \psi \Big\} = \int_{\partial\Omega} \sum_{j=1}^{n} d\sigma_j \varphi \sum_{i=1}^{n} a_{ij} \partial_i \psi = 0\,.$$

In this case, therefore

$$\langle \varphi, A\psi \rangle_2 = \langle \varphi, a_0 \psi \rangle_2 + \langle \nabla\varphi, a \nabla\psi \rangle_2 \tag{6.2.3}$$

for all $\varphi, \psi \in \mathcal{D}(\Omega)$. The right-hand side of this equation defines a quadratic form $Q = Q_A$ on $\mathcal{D}(\Omega)$. This quadratic form possesses many extensions. A natural upper bound for these extensions in $L^2(\Omega)$ has the domain

$$\{ u \in L^2(\Omega) \,|\, \partial_j u \in L^2(\Omega),\ j = 1, \ldots, n \} = H^1(\Omega)\,, \tag{6.2.4}$$

where the derivative is meant in the sense of distributions. $H^1(\Omega)$ is a Hilbert space with the scalar product $\langle u, v\rangle_2 + \langle \nabla u, \nabla v\rangle_2$.

We shall show in Appendix D that for $\Omega \subset \mathbb{R}^n$ with sufficiently smooth boundary

$$\begin{aligned} H_0^1(\Omega) &= \text{closure of } \mathcal{D}(\Omega) \text{ in } H^1(\Omega) \\ &= \{v \in H^1(\Omega) \mid v \upharpoonright \partial\Omega = 0\}\,. \end{aligned} \tag{6.2.5}$$

We thus expect from the derivation of (6.2.3) that

$$Au_\lambda = \lambda u_\lambda\,, \quad u_\lambda \in H_0^1(\Omega)\,, \tag{6.2.6 a}$$

if, and only if,

$$Q(v, u_\lambda) - \lambda(v, u_\lambda) = 0 \tag{6.2.6 b}$$

for all $v \in H_0^1(\Omega)$, that is if u_λ is a critical point of the function

$$u \to Q_\lambda(u) = Q(u, u) - \lambda\langle u, u\rangle_2 \tag{6.2.6 c}$$

on $H_0^1(\Omega)$.

Remark 6.2.1. Equation (6.2.6 a) expresses the fact that u_λ is an eigenfunction of the differential operator A in the Hilbert space $L^2(\Omega)$ with *Dirichlet boundary conditions* ($u_\lambda \upharpoonright \partial\Omega = 0$). We shall go into the formulation of other boundary conditions in more detail later on, when we come to discuss boundary value problems. Equation (6.2.6 b) is also referred to as *the variational form of the eigenvalue problem* (6.2.6 a).

The proof of the equivalence assertion (6.2.6 a–c) is simple. First of all

$$Q'_\lambda(u)(h) = 2Q_\lambda(h, u)\,, \quad u, h \in H_0^1(\Omega)\,,$$

is the Fréchet derivative of the function Q_λ on $H_0^1(\Omega)$ (all functions are real-valued). Thus u_λ is a critical point of Q_λ if, and only if, (6.2.6 b) holds, i.e. if

$$0 = \langle v, (a_0 - \lambda)u_\lambda\rangle_2 + \langle \nabla v, a\nabla u_\lambda\rangle_2 \tag{$*$}$$

for all $v \in H_0^1(\Omega)$, $u_\lambda \in H_0^1(\Omega)$ implies that $a\nabla u_\lambda \in L^2(\Omega)^{\times n}$ and it is thus differentiable in the sense of distributions. It follows that u_λ is a critical point of Q_λ if, and only if,

$$(a_0 - \lambda)u_\lambda - \nabla\cdot(a\nabla u_\lambda) = 0 \quad \text{in} \quad \mathcal{D}'(\Omega) \tag{$\times$}$$

since $\mathcal{D}(\Omega)$ is dense in $H_0^1(\Omega)$. $(a_0 - \lambda)u_\lambda \in L^2(\Omega)$ implies that $\underline{\nabla}\cdot(a\underline{\nabla}u_\lambda)$ also belongs to $L^2(\Omega)$. Therefore, ($\times$) also holds in $L^2(\Omega)$, i.e.

$$Au_\lambda = \lambda u_\lambda \quad \text{in} \quad L^2(\Omega)\,,$$

and u_λ is an eigenfunction of the differential operator A in $L^2(\Omega)$ with Dirichlet boundary conditions. The converse statement results in just the same way, taking into consideration in the derivation of (6.2.3) that $\overline{\mathcal{D}(\Omega)} = H_0^1(\Omega)$.

We have not been able to say very much above the solubility of the eigenvalue problem (6.2.6 a–c) in the general situations we have treated up to now. The so-called *ellipticity conditions* prove to be sufficiently general and, as far as the solubility is concerned, very effective hypotheses; we shall present them here in their simplest form: The differential operator A is (strongly) elliptic if positive numbers $0 < m \leq M < \infty$ exist such that for almost all $x \in \Omega \subset \mathbb{R}^n$ and all $\xi \in \mathbb{R}^n$

$$m \sum_{j=1}^{n} \xi_j^2 \leq \sum_{j,i=1}^{n} \xi_j a_{ji}(x) \xi_i \leq M \sum_{j=1}^{n} \xi_j^2 . \tag{6.2.7}$$

Moreover, let

$$0 \leq a_0(x) \leq K < \infty$$

hold for almost all $x \in \Omega$. If (6.2.7) holds, then the estimate

$$\begin{aligned} m \int_\Omega d^n x \sum_{j=1}^{n} (\partial_j u)^2(x) &\leq \int_\Omega d^n x \sum_{j,i=1}^{n} \partial_j u(x) a_{ji}(x) \partial_i u(x) \\ &\leq M \int_\Omega d^n x \sum_{j=1}^{n} (\partial_j u)^2(x) \end{aligned}$$

results, and thus

$$m\langle \nabla u, \nabla u \rangle_2 \leq Q(u,u) \leq M\langle \nabla u, \nabla u \rangle_2 + K\langle u, u \rangle_2 . \tag{6.2.8}$$

The second inequality in (6.2.8) implies that the quadratic form $u \to Q(u,u)$ is continuous on the Hilbert space $H_0^1(\Omega)$.

Let us consider the *Poincaré inequality* (Appendix D)

$$\|v\|_2 \leq \lambda \|\nabla v\|_2 , \quad \forall v \in H_0^1(\Omega) , \tag{6.2.9}$$

where $\lambda = \lambda(\Omega) < \infty$. Then the first inequality in (6.2.8) implies that the quadratic form Q on $H_0^1(\Omega)$ is also coercive:

$$Q(u,u) \geq \frac{m}{1+\lambda^2} (\|u\|_2^2 + \|\nabla u\|_2^2) . \tag{6.2.10}$$

Now $H_0^1(\Omega)$ is continuously and densely embedded in $L^2(\Omega)$. Moreover, Rellich's theorem for open bounded sets $\Omega \subset \mathbb{R}^n$ with smooth boundaries states that the identical embedding of $H_0^1(\Omega)$ into $L^2(\Omega)$ is compact. We are then able to recognise the eigenvalue problem (6.2.6 a) in the variational formulation (6.2.6 b) as a special case of the following general eigenvalue problem:

(E) Two real Hilbert spaces $\mathcal{H}_1$ and $\mathcal{H}_2$ with scalar products $\langle\cdot,\cdot\rangle_1$ and $\langle\cdot,\cdot\rangle_2$ are given. It is assumed that
(i) $\mathcal{H}_1$ is densely contained in $\mathcal{H}_2$;
(ii) the identical embedding $\mathrm{i} : \mathcal{H}_1 \to \mathcal{H}_2$ is compact.
We have to determine, for a continuous symmetric bilinear form Q on $\mathcal{H}_1$, all real numbers λ for which there is a $\mathrm{u}_\lambda \in \mathcal{H}_1$, $\mathrm{u}_\lambda \neq 0$, such that

$$Q(u, u_\lambda) = \lambda\langle u, u_\lambda\rangle_2$$

for all $\mathrm{u} \in \mathcal{H}_1$.

We shall solve the eigenvalue problem (6.2.1) for the differential operator A in $L^2(\Omega)$ by solving the *general eigenvalue problem* (E). This solution is based on several abstract results which will be formulated and proved in the next section.

6.3 The Theorem of Lax-Milgram and Some Generalisations

The original theorem of Lax-Milgram is a simple result for linear operators in Hilbert spaces. We shall use it to prove a result which, with the aid of the spectral theorem for compact operators, fairly easily yields the solution of the general eigenvalue problem.

Lemma 6.3.1. *Let $\mathcal{H}$ be a (complex) Hilbert space and $a : \mathcal{H} \times \mathcal{H} \to \mathbb{C}$ a continuous sesquilinear form on $\mathcal{H}$ with*

$$a(u,u) \geq c\|u\|^2, \quad \forall u \in \mathcal{H}, \quad c > 0.$$

Then there is exactly one bounded linear operator $\mathcal{A}$ on $\mathcal{H}$ such that

$$a(u,v) = (\mathcal{A}u, v), \quad \forall u, v \in \mathcal{H}, \tag{6.3.1}$$

where $(\cdot,\cdot)$ denotes the scalar product of $\mathcal{H}$. $\mathcal{A}$ is bijective and has a continuous inverse operators $\mathcal{A}^{-1}$.

Proof. By virtue of the theorem of Riesz-Fréchet, every continuous sesquilinear form a has the representation

$$a(u,v) = (\mathcal{A}u, v),$$

where $\mathcal{A} : \mathcal{H} \to \mathcal{H}$ is a uniquely determined bounded linear operator [6.1]. In the case of a coercive sesquilinear form this operator is injective, as is its adjoint:

$$c\|u\|^2 \leq a(u,u), \quad (\mathcal{A}u, u) \leq \|u\| \, \|\mathcal{A}u\|, \quad \text{and} \quad (u, \mathcal{A}^*u) \leq \|u\| \, \|\mathcal{A}^*u\|;$$

therefore

$$c\|u\| \le \|\mathcal{A}u\|, \|\mathcal{A}^*u\|. \tag{+}$$

The well-known relation

$$(\operatorname{Ran}\mathcal{A})^{\perp} = \operatorname{Ker}\mathcal{A}^*$$

yields

$$\overline{\operatorname{Ran}\mathcal{A}} = (\operatorname{Ran}\mathcal{A})^{\perp\perp} = \{0\}^{\perp} = \mathcal{H}.$$

It follows easily from $c\|u\| \le \|\mathcal{A}u\|$ that $\operatorname{Ran}\mathcal{A}$ is closed. Thus $\operatorname{Ran}\mathcal{A} = \mathcal{H}$ and $\mathcal{A} : \mathcal{H} \to \mathcal{H}$ is bijective. The estimate $c\|u\| \le \|\mathcal{A}u\|$ implies also that $\mathcal{A}^{-1}$ is bounded and thus continuous

$$\|\mathcal{A}^{-1}v\| \le \frac{1}{c}\|v\|, \quad \forall v \in \mathcal{H}.$$

□

Theorem 6.3.2. *Let $\mathcal{H}$ be a real Hilbert space and E a Hausdorff locally convex topological vector space over $\mathbb{R}$. Suppose the following:*

(i) *$\mathcal{H} \subset E$ is dense in E.*
(ii) *The identical embedding $i : \mathcal{H} \to E$ is continuous.*
(iii) *$a : \mathcal{H} \times \mathcal{H} \to \mathbb{R}$ is a coercive continuous bilinear form.*

Then for every continuous linear functional f on E there is exactly one $u_f \in \mathcal{H}$ such that

$$a(u_f, v) = f \circ i(v) \tag{6.3.2}$$

holds for all $v \in \mathcal{H}$.

Proof. (a) Let $f \in E'$ be given. Because of (ii), $f \circ i : \mathcal{H} \to \mathbb{R}$ is a continuous linear functional on $\mathcal{H}$ such that the theorem of Riesz-Fréchet guarantees the existence of exactly one $\xi = \xi_f \in \mathcal{H}$ with

$$f \circ i(v) = (\xi_f, v), \quad \forall v \in \mathcal{H}.$$

We have in this way a well-defined mapping $J : E' \to \mathcal{H}$, $J(f) = \xi_f$. It is evident that it is linear; but it is also injective since according to (i) $\mathcal{H}$ is dense in E.

(b) Define

$$D(A) = \{u \in \mathcal{H} \mid v \to a(u, v) \text{ is an } E\text{-continuous linear form on } \mathcal{H}\}.$$

The fact that $u \in D(A)$ therefore means that $\mathcal{H} \ni v \to a(u, v)$ is a linear form on $\mathcal{H}$ which is continuous with respect to the topology induced by E on $\mathcal{H}$. Let $u \in D(A)$; since $\mathcal{H}$ is dense in E, the linear form $a(u, v)$ has a unique extension to a continuous linear form Au on E: we have

$$a(u, v) = \langle Au, i(v) \rangle, \quad \forall v \in \mathcal{H}. \tag{6.3.3}$$

where $\langle \cdot, \cdot \rangle$ denotes the duality between E and the topological dual space E'. This defines a mapping $A : D(A) \to E'$. A is obviously linear. $Au = 0$ implies $a(u, v) = 0$, $\forall v \in \mathcal{H}$, and thus $(v = u)$ $u = 0$. Thus, A is injective.

(c) It remains for us to show that A is also surjective. Let $f \in E'$ be given. Put $u_f := \mathcal{A}^{-1} J(f)$ where $\mathcal{A} : \mathcal{H} \to \mathcal{H}$ denotes the bijective operator from Lemma 6.3.1. It follows that $u_f \in D(A)$ since because

$$a(u_f, v) = (\mathcal{A} u_f, v) = (J(f), v) = f \circ i(v), \quad \forall v \in \mathcal{H},$$

$v \to a(u_f, v)$ is an E-continuous linear form on $\mathcal{H}$. Since $\mathcal{H}$ is dense in E, it follows from

$$\langle A u_f, i(v) \rangle = a(u_f, v) = \langle f, i(v) \rangle, \quad \forall v \in \mathcal{H},$$

that

$$A u_f = f.$$

A is thus surjective. □

Theorem 6.3.2 is mostly applied in the more specialised version in which E is a Banach space or a Hilbert space. We therefore note the following.

Corollary 6.3.3. *Let $\mathcal{H}_1$ and $\mathcal{H}_2$ be two real Hilbert spaces with scalar products $\langle \cdot, \cdot \rangle_1$ and $\langle \cdot, \cdot \rangle_2$ where*

(i) *$\mathcal{H}_1$ is dense in $\mathcal{H}_2$;*
(ii) *the identical embedding of $\mathcal{H}_1$ in $\mathcal{H}_2$ is continuous.*

If $a : \mathcal{H}_1 \times \mathcal{H}_1 \to \mathbb{R}$ is now a coercive continuous bilinear form on $\mathcal{H}_1$, then for every $\xi \in \mathcal{H}_2$ there is exactly one $u_\xi \in \mathcal{H}_1$ such that

$$a(u_\xi, v) = \langle \xi, i(v) \rangle_2$$

for all $v \in \mathcal{H}_1$.

The analogy to Browder's results for more general, not necessarily quadratic, coercive functionals becomes clear in the more general version of this corollary in the form of Theorem 7.2.3 (cf. Chap. 7). We note only that the coercivity of the bilinear form $a : \mathcal{H} \times \mathcal{H} \to \mathbb{R}$ implies the *monotonicity* of the linear operators $\mathcal{A} : \mathcal{H} \to \mathcal{H}$ and $A : D(A) \subset \mathcal{H} \to E'$ defined by Lemma 6.3.1 and Theorem 6.3.2, respectively. For we have

$$(\mathcal{A}u - \mathcal{A}v, u - v) = a(u - v, u - v) \geq c\|u - v\|^2 \geq 0, \quad \forall u, v \in \mathcal{H},$$

and for all $u, v \in D(A)$

$$\langle A(u) - A(v), i(u - v) \rangle = a(u - v, u - v) \geq c\|u - v\|^2 \geq 0.$$

In the case of not necessarily quadratic functionals, a weakened coercivity condition and a monotonicity condition of the above type are used.

The complete solution of the general eigenvalue problem will now be described in the following theorem.

Theorem 6.3.4. *Let $\mathcal{H}_1$ and $\mathcal{H}_2$ be two real Hilbert spaces with scalar products $\langle\cdot,\cdot\rangle_1$ and $\langle\cdot,\cdot\rangle_2$, $\dim\mathcal{H}_1 = \infty$, and assume the following:*

(i) *$\mathcal{H}_1$ is dense in $\mathcal{H}_2$;*
(ii) *the identical embedding $i : \mathcal{H}_1 \to \mathcal{H}_2$ is (continuous and) compact.*

Then for every symmetric bilinear form Q on $\mathcal{H}_1$ which satisfies

(iii) *Q is continuous,*
(iv) *Q is coercive, i.e.*

$$Q(u,u) \geq c\|u\|^2\,, \quad c>0\,, \quad \forall u \in \mathcal{H}_1\,,$$

there is a monotone increasing sequence (λ_m) of eigenvalues,

$$0 < \lambda_1 \leq \lambda_2 \leq \lambda_m \underset{m\to\infty}{\longrightarrow} +\infty\,,$$

and an orthonormal basis $\{e_m\}_{m\in\mathbb{N}} \subset \mathcal{H}_1$ of $\mathcal{H}_2$ such that

$$Q(e_m, v) = \lambda_m\langle e_m, v\rangle_2\,, \quad \forall v \in \mathcal{H}_1\,, \quad \forall m \in \mathbb{N}\,.$$

$\{v_m\} = \{\lambda_m^{-1/2} e_m\}$ is an orthonormal basis of $\mathcal{H}_1$ with respect to the scalar product $\langle\cdot,\cdot\rangle = Q(\cdot,\cdot)$ on $\mathcal{H}_1$.

Proof. (a) By Lemma 6.3.1, $Q(u,v) = \langle u, Au\rangle_1$ where A is a self-adjoint operator:

$$A : \mathcal{H}_1 \to \mathcal{H}_1\,, \quad c \leq A\,, \quad A^{-1} \leq c^{-1}\,.$$

It follows that

$$c\|u\|_1^2 \leq Q(u,u) \leq \|A\|\;\|u\|_1^2 \quad \text{for all} \quad u \in \mathcal{H}_1\,.$$

Thus

$$(u,v) \to \langle u,v\rangle := Q(u,v)$$

defines a scalar product on $\mathcal{H}_1$ equivalent to $\langle\cdot,\cdot\rangle_1$.

(b) If we identify $\mathcal{H}_2$ with its dual space $\mathcal{H}_2'$, then Corollary 6.3.3 states that for every $\xi \in \mathcal{H}_2$ there is exactly one $u_\xi \in \mathcal{H}_1$ such that $Q(u_\xi, v) = \langle\xi, i(v)\rangle_2$ for all $v \in \mathcal{H}_1$. A mapping

$$B : \mathcal{H}_2 \to \mathcal{H}_1$$

is therefore well defined by

$$B\xi := u_\xi\,, \quad \xi \in \mathcal{H}_2\,.$$

The relations

$$\langle B\xi, v\rangle = Q(B\xi, v) = \langle\xi, i(v)\rangle_2$$

and

$$\langle B\xi, v\rangle_1 = \langle \xi, A^{-1}v\rangle_2, \quad \forall v \in \mathcal{H}_1,$$

imply that B is an injective continuous linear operator from $\mathcal{H}_2$ into $\mathcal{H}_1$.

(c) Since the identical embedding $i : \mathcal{H}_1 \to \mathcal{H}_2$ is compact,

$$C = B \circ i : \mathcal{H}_1 \to \mathcal{H}_1$$

is a continuous compact operator on $(\mathcal{H}_1, \langle \cdot\,, \cdot\rangle)$. C is also self-adjoint and positive since

$$\langle Cu, v\rangle = \langle i(u), i(v)\rangle_2 = \langle u, Cv\rangle.$$

Since B is injective, C is also injective since $\mathcal{H}_1$ is dense in $\mathcal{H}_2$. Thus, by Theorem 6.1.1, there is an orthonormal basis $\{v_m\}_{m\in\mathbb{N}}$ of $(\mathcal{H}_1, \langle \cdot\,, \cdot\rangle)$ and a sequence of numbers

$$(\mu_m)_{m\in\mathbb{N}} \text{ where } \mu_m \leq \mu_{m-1}, \quad \mu_m \underset{m\to\infty}{\longrightarrow} 0, \quad \text{and} \quad Cv_m = \mu_m v_m, \quad \forall m \in \mathbb{N}.$$

It follows that the sequence of numbers $\lambda_m = 1/\mu_m$ satisfies

$$\lambda_m \underset{m\to\infty}{\longrightarrow} +\infty \quad \text{and} \quad \lambda_m \leq \lambda_{m+1}, \quad \forall m \in \mathbb{N}.$$

For the vectors $e_m := \lambda_m^{1/2} v_m \in \mathcal{H}_1$ we have

$$Q(e_m, v) = \langle e_m, v\rangle = \lambda_m \langle Ce_m, v\rangle = \lambda_m \langle i(e_m), i(v)\rangle_2$$

and thus

$$\langle e_n, e_m\rangle_2 = \langle i(e_n), i(e_m)\rangle_2 = \langle v_n, v_m\rangle = \delta_{nm}.$$

$\{e_m\}_{m\in\mathbb{N}} \subset \mathcal{H}_1$ is therefore an orthonormal system in $\mathcal{H}_2$ consisting of "eigenvectors" of Q. Since $\{v_m\}_{m\in\mathbb{N}}$ is an orthonormal basis of $(\mathcal{H}_1, \langle \cdot\,, \cdot\rangle)$, it follows from $f \in \mathcal{H}_1 \cap \{e_n \,|\, n \in \mathbb{N}\}^\perp$ that

$$0 = \langle i(e_n), i(f)\rangle_2 = \lambda_n^{-1/2} \langle v_n, f\rangle, \quad \forall n \in \mathbb{N}$$

and so $f = 0$. Since $\mathcal{H}_1$ is dense in $\mathcal{H}_2$, the system $\{e_m\}$, $m \in \mathbb{N}$, is also complete in $\mathcal{H}_2$. □

6.4 The Spectrum of Elliptic Differential Operators in a Bounded Domain. Some Problems from Classical Potential Theory

The solution of the eigenvalue problem (6.2.2) for the differential operator (6.2.1) in the variational form (6.2.6 a–c) becomes rather simple if we use the abstract results of the last section. The solution of our original problem proves to be a special case of the solution in Theorem 6.3.4 of the general eigenvalue problem (E).

Theorem 6.4.1. *Let $\Omega \subset \mathbb{R}^n$ be a nonempty open bounded set with a sufficiently smooth boundary $\Gamma = \partial\Omega$. Then assuming the ellipticity condition (6.2.7) holds, the eigenvalue problem*

$$Au = a_0 u - \sum_{i,j=1}^{n} \partial_j(a_{ij}\partial_i u) = \lambda u\,, \quad u \restriction \Gamma = 0$$

has a monotone increasing sequence of eigenvalues λ_j:

$$0 < \lambda_1 \leq \lambda_2 \leq \cdots\,, \quad \lambda_j \underset{j\to\infty}{\longrightarrow} +\infty$$

and a sequence $\{e_j\}_{j\in\mathbb{N}} \subset H_0^1(\Omega)$ of eigenvectors e_j in $H_0^1(\Omega)$ which form an orthonormal basis of $L^2(\Omega)$. $\{\lambda_j^{1/2} e_j \,|\, j \in \mathbb{N}\}$ is an orthonormal basis of $H_0^1(\Omega)$ with respect to the scalar product $\langle\cdot\,,\cdot\rangle = Q(\cdot\,,\cdot)$. The eigenvalues can be calculated with the aid of the minimax principle:

$$\lambda_m = \min_{E_m \in \mathcal{E}_m} \max_{v \in E_m} \frac{Q(v,v)}{\|v\|_2^2}\,,$$

$$Q(u,v) = \langle u, a_0 v\rangle_2 + \sum_{i,j=1}^{n} \langle \partial_i u, a_{ij}\partial_j v\rangle_2\,, \tag{6.4.1}$$

where $\mathcal{E}_m$ is the family of all m-dimensional subspaces of $H_0^1(\Omega)$.

Proof. We show that the hypotheses of Theorem 6.3.4 are satisfied in the situation described above. According to Appendix D, $\mathcal{H}_1 = H_0^1(\Omega)$ is dense in $\mathcal{H}_2 = L^2(\Omega)$ and the identical embedding $i : \mathcal{H}_1 \to \mathcal{H}_2$ is compact. Theorem 6.3.4 guarantees that a continuous symmetric coercive bilinear form Q on $\mathcal{H}_1 = H_0^1(\Omega)$ is defined by (6.4.1). Finally, we show that the eigenvalues can be calculated by means of the minimax principle. This is obvious firstly for the eigenvalues μ_m of the compact self-adjoint operator $C \geq 0$ on $\mathcal{H}_1$, defined by $C = B \circ i$, $Q(Bu, v) = \langle u, i(v)\rangle_2$, $\forall u \in \mathcal{H}_2$ and $\forall v \in \mathcal{H}_1$, by virtue of the proof of Theorem 6.3.4. The eigenvalues λ_m of A that we require are characterised by (6.2.6 b) and can be calculated from $\lambda_m = 1/\mu_m$; the assertion follows. □

Remark 6.4.1. Theorem 6.4.1 determines, among other things, the spectrum of the Laplace operator ($a_0 = 0, a_{ij} = \delta_{ij}$) with Dirichlet boundary conditions on open bounded sets with sufficiently smooth boundaries.

In conclusion let us discuss the variational solution of a few simple linear boundary value problems which, in their simplest forms, have played an important role in classical potential theory. The abstract method of solution of these problems is indicated by Theorem 6.3.2 and Corollary 6.3.3. Let us consider a differential operator A in the form (6.2.1) over an open set $\Omega \subset \mathbb{R}^n$

with smooth boundary, subject to the assumptions of Sect. 6.2. The following problem has to be solved for a given function $f : \Omega \to \mathbb{R}$.

Determine a function $u : \overline{\Omega} \to \mathbb{R}$ such that

$$\text{(i)}\ Au = a_0 u - \sum_{j=1}^{n} \partial_j \Big(\sum_{i=1}^{n} a_{ji} \partial_i u \Big) = f \ \text{ in } \Omega \tag{6.4.2 a}$$

and

$$\text{(ii)}\ u \restriction \partial\Omega = 0. \tag{6.4.2 b}$$

A slight generalisation of our treatment in Sect. 6.2 results in the variational formulation and precise specification of this problem in the following form.

For given $T \in H_0^1(\Omega)'$, determine $u_T \in H_0^1(\Omega)$ such that

$$Q(u_T, v) = T(v) \quad \text{for all} \quad v \in H_0^1(\Omega)\,, \tag{6.4.2 a$'$}$$

where Q is defined by

$$Q(u,v) = \langle u, a_0 v\rangle_2 + \sum_{i,j=1}^{n} \langle \partial_i u, a_{ij} \partial_j v\rangle_2\,, \quad u, v \in H_0^1(\Omega)\,. \tag{6.4.2 b$'$}$$

The boundary condition $u \restriction \partial\Omega = 0$ is again satisfied here simply by the choice of the space of potential solutions (see Appendix D). Problem (6.4.2) always has a solution in the version (6.4.2$'$).

Theorem 6.4.2. *Let $\Omega \subset \mathbb{R}^n$ be an open bounded set with smooth boundary and let A be a linear differential operator in divergence form which satisfies the ellipticity condition (6.2.7). Then the following holds. For every continuous linear form T on $H_0^1(\Omega)$ there is exactly one $u_T \in H_0^1(\Omega)$ such that*

$$Q(u_T, v) = T(v)\,, \quad \forall v \in H_0^1(\Omega)\,.$$

The mapping $T \to u_T$ from $(H_0^1(\Omega))'$ into $H_0^1(\Omega)$ is linear, continuous and bijective. u_T is the minimising point of the functional

$$E_T(v) = \frac{1}{2} Q(v,v) - T(v)$$

on $H_0^1(\Omega)$, i.e.

$$E_T(u_T) = \inf\{E_T(v) \mid v \in H_0^1(\Omega)\}\,.$$

Proof. Due to the boundedness of Ω, the ellipticity condition (6.2.7) implies, as above, that the symmetric bilinear form Q on the Hilbert space $\mathcal{H} = H_0^1(\Omega)$ is well defined, continuous and (strictly) coercive. Applying Theorem 6.3.2 for $\mathcal{H} = E = H_0^1(\Omega)$, the first part of the assertion follows immediately. Following the method of proof of Theorem 6.3.2, we have $u_T = A^{-1} J T$, which shows the second part of the assertion, since J here is the canonical isomorphism of the dual space $\mathcal{H}'$ on $\mathcal{H}$. The identity

$$E_T(u_T + h) = E_T(u_T) + Q(u_T, h) - T(h) + \frac{1}{2}Q(h,h) = E_T(u_T) + \frac{1}{2}Q(h,h)$$

shows immediately that the critical point u_T is indeed a minimising point of E_T. □

Remark 6.4.2. This theorem can also be extended to unbounded domains if the coercivity of Q on $H_0^1(\Omega)$ can be established using stronger hypotheses for A, and without using the Poincaré inequality. It is also possible to do without assuming the boundary of Ω is "smooth", even if an interpretation of the boundary condition is then not so obvious.

When $n = 3$ and $Au = -\Delta u$, Theorem 6.4.2 asserts the unique solubility of the *classical Poisson problem.* It is well known [6.4] that the unique solubility of this problem is equivalent to the unique solubility of the *classical Dirichlet problem*: "determine a function $u : \overline{\Omega} \to \mathbb{R}$ such that $-\Delta u = 0$ in Ω and $u \restriction \partial\Omega = \sigma$ for a given 'surface charge' $\sigma : \partial\Omega \to \mathbb{R}$". Finally, the unique solubility of this problem is also equivalent to the *existence* of exactly *one Green's function* $= G_\Omega^\Delta : \overline{\Omega} \times \overline{\Omega} \to \mathbb{R}$ for the Laplace operator Δ and the domain Ω, that is for arbitrary but fixed $y \in \Omega$, the function G must satisfy the conditions

(i) $-\Delta_x G(x, y) = \delta(x - y)$
and
(ii) $G(x, y) = 0,\ \forall x \in \partial\Omega$.

Thus Theorem 6.4.2 (for more general elliptic differential operators as well) can be used to show the unique solubility of the corresponding Dirichlet problem and the existence of a Green's function.

Remark 6.4.3. If $n = 1$, this is a boundary value problem for ordinary second-order differential equations. J. Sturm and his friend J. Liouville were the first to systematically investigate such problems, in 1836; these problems are known nowadays as *Sturm-Liouville boundary value problems.* For more details we refer to [6.5, 6].

6.5 Variational Solution of Parabolic Differential Equations. The Heat Conduction Equation. The Stokes Equations

As we saw in the previous section, elliptic boundary and eigenvalue problems can be solved relatively easily using the methods of variational calculus. We shall show in this section that one can also use the methods to solve parabolic differential equations. To this end, we shall discuss two specific problems, the heat conduction equation and the Stokes equations of hydrodynamics, in more detail. These will illustrate and provide us with a reason for seeking a general framework for variational solutions of parabolic differential equations. A treatment which goes somewhat further than ours can be found in [6.7, 8].

Parabolic differential equations are mainly used in physics to describe the behaviour of systems over time. The heat conduction equation is a simple and not untypical example which, in convenient units, takes on the form

$$(\mathrm{W}) \begin{cases} \dfrac{\partial u}{\partial t}(t,x) - \Delta_x u(t,x) = f(t,x)\,, \quad (t,x) \in (0,T) \times \Omega \subset \mathbb{R}^4 & (6.5.1\,\mathrm{a}) \\ u(0,x) = u_0(x)\,, \quad x \in \Omega \subset \mathbb{R}^3\,, & (6.5.1\,\mathrm{b}) \\ u(t,x) = 0\,, \quad x \in \partial\Omega,\ t \in [0,T]\,, & (6.5.1\,\mathrm{c}) \end{cases}$$

for isotropic homogeneous media.

The problem (W) then consists in determining the time behaviour $t \to u(t,\cdot)$ of the temperature distribution in Ω, for a given heat source f and initial temperature distribution u_0, such that the temperature at the boundary $\partial\Omega$ of Ω stays constant (and equal to zero). Thus, as far as the time behaviour is concerned, this is an initial value problem according to (6.5.1 b), and for the spatial coordinates it is a boundary value problem according to (6.5.1 c) for the solution of a linear inhomogeneous partial differential equation.

We have already seen how to treat boundary value problems like (6.5.1 c): we satisfy (6.5.1 c) "automatically" if we search for solutions of (6.5.1 a) which are "functions" $[0,T] \ni t \to u_t$ with values in the Sobolev space $H_0^1(\Omega)$. (We obviously make the implicit assumption that the initial conditions also belong to the Sobolev space $H_0^1(\Omega)$.) We can express the differential equation (6.5.1 a), as before, by means of an equation of the form

$$\frac{d}{dt}\langle u_t, v\rangle_2 + Q(u_t, v) = \langle f_t, v\rangle_2\,, \quad \forall v \in H_0^1(\Omega)\,, \quad 0 \le t \le T\,, \tag{6.5.2}$$

if we put

$$\langle f_t, v\rangle_2 = \int_\Omega f(t,x) v(x) d^3 x\,,$$

$$Q(u_t, v) = \int_\Omega \nabla_x u(t,x) \cdot \nabla_x v(x) d^3 x = \langle \nabla u_t, \nabla v\rangle_2\,. \tag{6.5.2$'$}$$

Conversely, if $t \to u_t$ is a solution of (6.5.2) with $u_{t=0} = u_0$, then $(t,x) \to u_t(x)$ is also a weak solution of (6.5.1 a) with the correct initial and boundary conditions (6.5.1 b) and (5.5.1 c).

The strategy which leads to the solution of (6.5.2) can be generalised without too much effort. Let us first of all discuss a general framework for the treatment of parabolic differential equations. We shall obtain as a special case the solution of (6.5.2) and thus the weak solution of (6.5.1).

6.5.1 A General Framework for the Variational Solution of Parabolic Problems

The framework introduced here proves to be a natural generalisation of the ideas which lead to a solution of the Dirichlet boundary value problem. The following notation and facts serve as preparation for a formulation of this framework.

We denote by $\mathcal{C}([0,T],X)$, for a Hilbert space X with scalar product $\langle\cdot,\cdot\rangle_X$ and norm $\|\cdot\|_X$ and a positive number T, the space of all continuous functions on the interval $[0,T]$ with values in X. The space $\mathcal{C}([0,T],X)$ with norm

$$\|v\|_{\mathcal{C}([0,T],X)} = \sup_{0\le t\le T} \|v(t)\|_X$$

is a Banach space. Correspondingly, $L^2([0,T],X)$ denotes the space of all (equivalence classes of) Lebesgue-measurable square-integrable functions on $[0,T]$ with value in X, that is functions $[0,T]\to X$ for which

$$t \to \|v(t)\|_X$$

is Lebesgue-measurable and

$$\|v\|_{L^2([0,T],X)} = \left(\int_0^T \|v(t)\|_X^2 \, dt\right)^{1/2}$$

finite. It follows by polarisation for two such functions u and v that

$$t \to \langle u(t), v(t)\rangle_X$$

is also Lebesgue-measurable and integrable over $[0,T]$. Thus

$$\langle u, v\rangle_{L^2([0,T],X)} = \int_0^T \langle u(t), v(t)\rangle_X \, dt$$

defines a scalar product on $L^2([0,T],X)$ and makes this space a Hilbert space.

We now come to a formulation of a general framework for the variational solution of parabolic problems. We assume we have the following:

(a) Two real Hilbert spaces $\mathcal{H}_1$ and $\mathcal{H}_2$ with the following properties:
  (i) $\mathcal{H}_1$ is dense in $\mathcal{H}_2$;
  (ii) the identical embedding of $\mathcal{H}_1$ into $\mathcal{H}_2$ is continuous.

The scalar products and norms on $\mathcal{H}_1$, $i=1,2$, are denoted by $\langle\cdot,\cdot\rangle_i$ and $\|\cdot\|_i$.

(b) A continuous bilinear form Q on $\mathcal{H}_1$.
(c) An "initial condition" $u_0 \in \mathcal{H}_2$.
(d) An "external force" $f \in L^2([0,T],\mathcal{H}_2)$.

We are looking for all functions u which have the following properties.

(α) $u \in C([0,T],\mathcal{H}_2) \cap L^2([0,T],\mathcal{H}_1)$.
(β) $u(0) = u_0$.
(γ) For all $v \in \mathcal{H}_1$,

$$\frac{d}{dt}\langle u(t), v\rangle_2 - Q(u(t), v) = \langle f(t), v\rangle_2 \tag{6.5.3}$$

holds in the sense of distributions on $(0,T)$.

Remark 6.5.1. The restriction of solutions to those in $C([0,T],\mathcal{H}_2)$ allows us to formulate the initial condition (β). The assumptions that u should belong to $L^2([0,T],\mathcal{H}_1)$ and that $\mathcal{H}_1$ is continuously embedded in $\mathcal{H}_2$ imply that the equation (γ) makes sense in $\mathcal{D}'((0,T))$.

The heat conduction problem (6.5.1) fits into this framework in the reformulated form (6.5.2) and (6.5.2′). One need only put $\mathcal{H}_1 = H_0^1(\Omega)$ and $\mathcal{H}_2 = L^2(\Omega)$. The condition $f \in L^2([0,T],\mathcal{H}_2)$ should then be regarded as making our assumptions specific.

In the version above, with its weak assumptions, one is able to say very little about the solubility of the general parabolic problem given by the conditions (a)–(γ). However, if we make the same assumptions as in our treatment of general elliptic boundary value problems, then the following useful and quite general theorem results.

Theorem 6.5.1. *The general parabolic problem given by conditions* (a) *to* (γ), *subject to the following hypotheses, always possesses exactly one solution.*

(E) *The bilinear form Q is strictly coercive on $\mathcal{H}_1$, i.e. a positive number α exists such that $Q(v,v) \geq \alpha \|v\|_1^2$, $\forall v \in \mathcal{H}_1$.*
(K) *The identical embedding $i : \mathcal{H}_1 \to \mathcal{H}_2$ is compact.*
(S) *The bilinear form Q is symmetric.*

Proof. (a) According to Theorem 6.3.4, we know that under the above assumptions, there is an increasing sequence $\{\lambda_i\}_{i\in\mathbb{N}}$ of eigenvalues

$$0 < \lambda_1 \leq \lambda_2 \leq \cdots, \quad \lambda_i \underset{i\to\infty}{\longrightarrow} +\infty,$$

of Q and an orthonormal basis $\{w_i\}_{i\in\mathbb{N}}$ of $\mathcal{H}_2$ whose elements w_i are all in $\mathcal{H}_1$ such that

$$Q(v, w_i) = \lambda_i \langle v, w_i\rangle_2, \quad \forall v \in \mathcal{H}_1, \quad \forall i \in \mathbb{N}.$$

Moreover, $v_i = \lambda_i^{1/2} w_i$ is an orthonormal basis of $\mathcal{H}_1$ with respect to the scalar product $\langle\cdot,\cdot\rangle = Q(\cdot,\cdot)$.

(b) Instead of determining the one function we are looking for, $u \in L^2([0,T],\mathcal{H}_1) \cap C([0,T],\mathcal{H}_2)$ which satisfies (β) and (γ), let us first of all determine all components

$$u_i(t) = \langle w_i, u(t)\rangle_2 \,, \quad i = 1,2, \ \dots \,,$$

of this function with respect to the orthonormal basis $\{w_i\}$ of $\mathcal{H}_2$. According to (α)–(γ), these components have to satisfy the conditions

(α') $u_i \in C([0,T],\mathrm{I\!R})$, $\sum_{i=1}^{\infty} \int_0^T |u_i(t)|^2 \, dt < \infty$,
(β') $u_i(0) = u_i^0 = \langle w_i, u_0\rangle_2$,
(γ') $\frac{d}{dt}u_i(t) + \lambda_i u_i(t) = f_i(t) = \langle w_i, f(t)\rangle_2$,

where (γ') can be obtained from

$$Q(w_i, u(t)) = \lambda_i \langle w_i, u(t)\rangle_2 \,.$$

This problem is easily and uniquely solved by

$$\begin{aligned} u_i(t) &= \exp(-\lambda_i t)u_i^0 + \int_0^t f_i(\tau)\exp[-\lambda_i(t-\tau)]\, d\tau \\ &\equiv x_i(t) + y_i(t) \,, \quad 0 \le t \le T \,, \quad i = 1,2, \ \dots \,. \end{aligned}$$

$x_i(t) = \exp(-\lambda_i t)u_i^0$ is a C^∞-function and $y_i(t)$ is absolutely continuous; thus u_i is certainly a solution of the differential equation (γ') in the sense of L^1 functions. The initial conditions (β') and $u_i \in C([0,T],\mathrm{I\!R})$ are obvious. Furthermore, the following estimates hold for $0 \le t \le T$:

$$|x_i(t)| \le |u_i^0| \,, \quad \lambda_i \int_0^T |x_i(t)|^2 \, dt < \frac{1}{2}|u_i^0|^2 \,,$$

$$|y_i(t)| = \left| \int_0^t f_i(\tau)\exp[-\lambda_i(t-\tau)]\, d\tau \right| \le \left\{ \int_0^T |f_i(\tau)|^2 \, d\tau \frac{1-\exp(-2\lambda_i T)}{2\lambda_i} \right\}^{1/2} ,$$

$$\lambda_i \int_0^T |y_i(t)|^2 \, dt \le T \int_0^T |f_i(t)|^2 \, dt \,.$$

It follows that

$$\sum_{i=1}^{\infty} \lambda_i \int_0^T |x_i(t)|^2 \, dt \le \frac{1}{2} \sum_{i=1}^{\infty} |u_i^0|^2 = \frac{1}{2}\|u_0\|_2^2$$

and

$$\sum_{i=1}^{\infty} \lambda_i \int_0^T |y_i(t)|^2 \, dt \le T \sum_{i=1}^{\infty} \int_0^T |f_i(t)|^2 \, dt = T\|f\|^2_{L^2([0,T],\mathcal{H}_2)} \,,$$

and thus by virtue of (a) it is also certain that

$$\sum_{i=1}^{\infty} \int_0^T |u_i(t)|^2 \, dt < \infty ,$$

so that

$$t \to u(t) := \sum_{i=1}^{\infty} u_i(t) w_i \in L^2([0,T], \mathcal{H}_2)$$

follows. We then also have

$$t \to \sum_{i=1}^{\infty} \lambda_i u_i(t) w_i \in L^2([0,T], \mathcal{H}_2) .$$

(c) We have already shown in (b) that

$$u = \lim_{n \to \infty} S_n \quad \text{in} \quad L^2([0,T], \mathcal{H}_2)$$

where

$$S_n(\cdot) = \sum_{i=1}^{n} u_i(\cdot) w_i \in \mathcal{C}([0,T], \mathcal{H}_1) \subseteq \mathcal{C}([0,T], \mathcal{H}_2) .$$

The estimates given in (b) for $x_i(t)$ and $y_i(t)$ easily yield the fact that $(S_n)_{n \in \mathbb{N}}$ is a Cauchy sequence in $\mathcal{C}([0,T], \mathcal{H}_2)$. It follows that there is a uniquely determined element $u^2 \in \mathcal{C}([0,T], \mathcal{H}_2)$ where

$$u^2 = \lim_{n \to \infty} S_n \quad \text{in} \quad \mathcal{C}([0,T], \mathcal{H}_2) .$$

Now, for all $v \in \mathcal{C}([0,T], \mathcal{H}_2)$,

$$\|v\|_{L^2([0,T],\mathcal{H}_2)} \leq T^{1/2} \|v\|_{\mathcal{C}([0,T],\mathcal{H}_2)}$$

holds. It therefore follows for all $n \in \mathbb{N}$ that

$$\begin{aligned} \|u^2 - u\|_{L^2([0,T],\mathcal{H}_2)} &\leq \|u^2 - S_n\|_{L^2([0,T],\mathcal{H}_2)} + \|S_n - u\|_{L^2([0,T],\mathcal{H}_2)} \\ &\leq T^{1/2} \|u^2 - S_n\|_{\mathcal{C}([0,T],\mathcal{H}_2)} + \|S_n - u\|_{L^2([0,T],\mathcal{H}_2)} \end{aligned}$$

and thus that

$$u^2 = u \quad \text{in} \quad L^2([0,T], \mathcal{H}_2) ,$$

which means that

$$u \in \mathcal{C}([0,T], \mathcal{H}_2) .$$

(d) We shall show in a similar fashion that $u \in L^2([0,T], \mathcal{H}_1)$ also holds. We shall use the coerciveness of Q and the fact that $(w_i/\lambda_i^{1/2})_{i \in \mathbb{N}}$ is an orthonormal basis of $\mathcal{H}_1$ with respect to $Q(\cdot,\cdot)$ in order to show that the sequence $(S_n)_{n \in \mathbb{N}}$ in $L^2([0,T], \mathcal{H}_1)$ is a Cauchy sequence. Thus it follows for $n > m$ that

$$\begin{aligned}
\|S_n - S_m\|_{L^2([0,T],\mathcal{H}_1)} &= \int_0^T \left\| \sum_{i=m+1}^{n} u_i(t) w_i \right\|_{\mathcal{H}_1}^2 dt \\
&\le \int_0^T \frac{1}{\alpha} Q\Big(\sum_{i=m+1}^{n} u_i(t) w_i, \sum_{i=m+1}^{n} u_i(t) w_i \Big) dt \\
&= \frac{1}{\alpha} \sum_{i=m+1}^{n} \lambda_i \int_0^T |u_i(t)|^2 \, dt \\
&\le \frac{2}{\alpha} \sum_{i=m+1}^{n} \lambda_i \int_0^T \{ |x_i(t)|^2 + |y_i(t)|^2 \} \, dt \,.
\end{aligned}$$

The estimates proved in (b) therefore show that $(S_n)_{n \in \mathbb{N}}$ is also a Cauchy sequence in $L^2([0,T],\mathcal{H}_1)$. There is thus a uniquely determined element $u^1 \in L^2([0,T],\mathcal{H}_1)$ satisfying

$$u^1 = \lim_{n \to \infty} S_n \quad \text{in} \quad L^2([0,T],\mathcal{H}_1)\,.$$

Since $\mathcal{H}_1$ is continuously embedded in $\mathcal{H}_2$, a constant $c > 0$ exists such that $\|v\|_{\mathcal{H}_2} \le c \|v\|_{\mathcal{H}_1}$, $\forall v \in \mathcal{H}_1$. It follows that

$$\|v\|_{L^2([0,T],\mathcal{H}_2)} \le c \|v\|_{L^2([0,T],\mathcal{H}_1)}$$

for all $v \in L^2([0,T],\mathcal{H}_1)$ and thus, for all $n \in \mathbb{N}$,

$$\begin{aligned}
\|u^1 - u\|_{L^2([0,T],\mathcal{H}_2)} &< \|u^1 - S_n\|_{L^2([0,T],\mathcal{H}_2)} + \|S_n - u\|_{L^2([0,T],\mathcal{H}_2)} \\
&\le c \|u^1 - S_n\|_{L^2([0,T],\mathcal{H}_1)} + \|S_n - u\|_{L^2([0,T],\mathcal{H}_2)}
\end{aligned}$$

and consequently $u^1 = u$ in $L^2([0,T],\mathcal{H}_1)$, which means u also belongs to $L^2([0,T],\mathcal{H}_1)$.

This yields, with (c),

$$u \in \mathcal{C}([0,T],\mathcal{H}_2) \cap L^2([0,T],\mathcal{H}_1) \subset L^2([0,T],\mathcal{H}_2)\,.$$

(e) Since $y_i(0) = 0$ and $x_i(0) = u_i^0$, $\forall i \in \mathbb{N}$, it follows immediately that $u(t) = \sum_{i=1}^{\infty} u_i(t) w_i$ satisfies the initial condition. It thus remains for us to show the validity of the differential equation (γ). The u_i are solutions of equation (6.8′) in $\mathcal{D}'((0,T))$, which means that $\forall \varphi \in \mathcal{D}((0,T))$,

$$\int_0^T [-\varphi'(t) u_i(t) + \lambda_i u_i(t) \varphi(t)] \, dt = \int_0^T \varphi(t) f_i(t) \, dt \,.$$

For arbitrary $v \in \mathcal{H}_1$ and for all $N \in \mathbb{N}$, it follows that

$$\int_0^T \varphi(t) \sum_{i=1}^N \langle v, w_i\rangle_2 f_i(t)\, dt$$
$$= \int_0^T \Big\{ -\varphi'(t) \sum_{i=1}^N \langle v, w_i\rangle_2 u_i(t) + \sum_{i=1}^N \lambda_i \langle v, w_i\rangle_2 u_i(t)\varphi(t)\Big\}\, dt$$
$$= \int_0^T \Big\{ -\varphi'(t) \sum_{i=1}^N \langle v, w_i\rangle_2 u_i(t) + Q\Big(v, \sum_{i=1}^N u_i(t) w_i\Big)\varphi(t)\Big\}\, dt\,.$$

Since $\{w_i\}_{i\in\mathbb{N}}$ is an orthonormal basis of $\mathcal{H}_2$, $\sum_{i=1}^N \langle v, w_i\rangle_2 f_i(t)$ converges in $L^2([0,T])$ to $\langle v, f(t)\rangle_2$. Since $\sum_{i=1}^N u_i(\cdot) w_i$ converges to $u(\cdot)$ in $L^2([0,T],\mathcal{H}_1)$ and $L^2([0,T],\mathcal{H}_2)$, then $\sum_{i=1}^N \langle v, w_i\rangle_2 u_i(\cdot)$ converges to $\langle v, u(\cdot)\rangle_2$ in $L^2([0,T])$ and $Q(v, \sum_{i=1}^N u_i(\cdot) w_i)$ converges to $Q(v, u(\cdot))$ in $L^2([0,T])$, since Q is continuous on $\mathcal{H}_1$. We can thus take the limit $N \to \infty$ in the above equation and we obtain

$$\int_0^T [-\varphi'(t)\langle v, u(t)\rangle_2 + Q(v, u(t))] = \int_0^T \varphi(t)\langle v, f(t)\rangle_2\, dt\,,$$

which is the differential equation (γ).

The uniqueness of u follows from the uniqueness of the components u_i with respect to the orthonormal basis $\{w_i\}_{i\in\mathbb{N}}$ of $\mathcal{H}_2$. This completes the proof of Theorem 6.5.1. □

6.5.2 The Heat Conduction Equation

The heat conduction equation (6.5.1) can now be solved quite simply in its variational form (6.5.2) and (6.5.2′).

Theorem 6.5.2. *Let $\Omega \subset \mathbb{R}^n$ be an open, bounded set with a smooth boundary. Then the heat conduction problem* (6.5.1) *has exactly one solution*

$$u = u_{f,u_0} \in \mathcal{C}([0,T], L^2(\Omega)) \cap L^2([0,T], H_0^1(\Omega))$$

for all initial conditions $u_0 \in H_0^1(\Omega)$ and for all "heat sources" $f \in L^2([0,T], L^2(\Omega))$. In the spectral representation of the Laplace operator in Ω with Dirichlet boundary conditions, this solution is given by

$$u(t) = \sum_{i=1}^\infty \{\exp(-\lambda_i t)\langle w_i, u_0\rangle_{L^2(\Omega)}$$
$$+ \int_0^t \exp[-\lambda_i(t-\tau)]\langle w_i, f(t)\rangle_{L^2(\Omega)}\, d\tau\} w_i\,,$$

where $(-\lambda_i, w_i)_{i\in\mathbb{N}}$ are the eigenvalues and eigenfunctions of the Laplace operator Δ^{D} with Dirichlet boundary conditions.

Proof. We establish in Appendix D the fact that $\mathcal{H}_1 = H_0^1(\Omega)$ is, subject to certain assumptions, compact and densely embedded in $\mathcal{H}_2 = L^2(\Omega)$. $Q(u,v) = \langle \nabla u, \nabla u\rangle_{L^2(\Omega)}$ is obviously a continuous symmetric bilinear form on $H_0^1(\Omega)$, since we are considering real-valued functions, and by virtue of our previous considerations (Sects. 4.2 and 4.4). Q is also coercive on $\mathcal{H}_1$. Thus, Theorem 6.5.1 is applicable. In this theorem, λ_i and w_i are eigenvalues and eigenfunctions of the bilinear form Q:

$$Q(v, w_i) = \lambda_i \langle v, w_i\rangle_{\mathcal{H}_2}\,, \quad \forall v \in H_0^1(\Omega)\,;$$

here this means that

$$\langle \nabla v, \nabla w_i\rangle_{L^2(\Omega)} = \lambda_1 \langle v, w_i\rangle_{L^2(\Omega)}\,, \quad \forall v \in H_0^1(\Omega)\,,$$

where $w_i \in H_0^1(\Omega)$. Theorem 6.5.2 follows. □

Remark 6.5.2. Formally speaking, equations (6.5.1 a) and (6.5.1 b) can be integrated using the function

$$u(t,x) = \exp(t\Delta^{\mathrm{D}})u_0(x) + \int_0^t \exp[(t-\tau)\Delta^{\mathrm{D}}]f(\tau,x)\,d\tau\,,$$

where Δ^{D} is the Laplace operator on Ω with Dirichlet boundary conditions. A more exact investigation of the semigroup of operators in $L^2(\Omega)$, $(\exp(t\Delta^{\mathrm{D}}))_{t\geq 0}$, which is generated by Δ^{D} shows that some meaning can be attached to the formal solution above. However, this "operator solution" of (6.5.1 a) and (6.5.1 b) will not necessarily satisfy the boundary condition (6.5.1 c) for arbitrary $f \in L^2([0,T]), L^2(\Omega))$. Substituting the spectral representation for $\exp(t\Delta^{\mathrm{D}})$, also proved in Theorem 6.5.2,

$$\exp(t\Delta^{\mathrm{D}}) = \sum_{i=1}^{\infty} \exp(-\lambda_i t)|w_i\rangle\langle w_i|\,, \quad w_i \in H_0^1(\Omega)\,,$$

$|w_i\rangle\langle w_i|$ being the orthogonal projector onto the one dimensional subspace spanned by w_i, the operator solution yields the solution given in Theorem 6.5.2.

6.5.3 The Stokes Equations in Hydrodynamics

We now want to briefly discuss another application of the general results of Sect. 6.5.1, namely the solubility of the so-called Stokes equations. A bounded domain $\Omega \subset \mathbb{R}^n$ with a "smooth" boundary $\Gamma = \partial\Omega$, an n-tuple $u_0 = (u_{0_1}, \ldots, u_{0_n})$ of functions $u_{0_j} : \Omega \to \mathbb{R}$, an n-tuple $(f_1, \ldots, f_n)$ of functions

$$f_j : \Omega_T \to \mathbb{R}\,, \quad \Omega_T = (0,T) \times \Omega\,,$$

and a positive constant μ are given. We are looking for an n-tuple $u = (u_1, \ldots, u_n)$ of functions $u_j : \Omega_T \to \mathbb{R}$ and a function $p : \Omega_T \to \mathbb{R}$ which are solutions of the system of equations (S):

$$\text{(S)} \qquad \begin{aligned} &\partial_t u - \mu \Delta u + \operatorname{grad} p = f && \text{in} \quad \Omega_T\,, \\ &\operatorname{div} u = 0 && \text{in} \quad \Omega_T\,, \\ &u = 0 && \text{on} \quad [0,T] \times \Gamma\,, \\ &u(0,\cdot) = u_0(\cdot) && \text{in} \quad \Omega\,. \end{aligned}$$

A solution of this system of equations is interpreted as the velocity and pressure field (u,p) of an incompressible viscous fluid with kinematic viscosity μ, in a domain Ω during the time interval $[0,T]$, and on which an external force with density f is acting. It is assumed here that neither the velocity u nor the initial velocity u_0 is too large; the motion of the fluid is then sufficiently slow so that the system of *Stokes equations* (S) can actually be regarded as a linear approximation of the *Navier-Stokes equations*

$$\partial_t u - \mu \Delta u + \sum_{i=1}^{n} u_i \partial_i u + \operatorname{grad} p = f \quad \text{in} \;\; \Omega_T$$

with the same constraint $\operatorname{div} u = 0$ in Ω_T and the same boundary and initial conditions. The Stokes and Navier-Stokes equations are treated in detail in [6.8].

So that we shall be able to discuss the problem of the solubility of the Stokes equations within the general framework given above, we define first of all suitable Hilbert spaces $\mathcal{H}_1$ and $\mathcal{H}_2$ and a continuous symmetric bilinear form Q and $\mathcal{H}_1$, and then we check that the assumptions of Theorem 6.5.1 are satisfied.

An obvious choice for $\mathcal{H}_1$ seems to be

$$\mathcal{H}_1 = \{v \in H_0^1(\Omega)^n \mid \operatorname{div} v = 0\}\,.$$

If we choose $\mathcal{H}_1$ thus, the boundary and divergence conditions for the Stokes equations will be automatically taken into account in a weak sense. The Hilbert space $\mathcal{H}_2$ is essentially fixed by virtue of the fact that the hypotheses of Theorem 6.5.1 must be satisfied: $\mathcal{H}_2$ must be chosen to be

$$\mathcal{H}_2 = \text{closure of } \mathcal{H}_1 \text{ in } L^2(\Omega)^n\,.$$

Then $\mathcal{H}_1$ is continuous and densely embedded in $\mathcal{H}_2$.

We need a more exact characterisation of $\mathcal{H}_2$ for further analysis, and we take this from the following (see [6.8]).

Lemma 6.5.3. *If $\Omega \subset \mathbb{R}^n$ is a bounded domain with a sufficiently smooth boundary $\Gamma = \partial\Omega$, then*

$$\mathcal{H}_2 = \{u \in L^2(\Omega)^n \mid \operatorname{div} u = 0, \underline{n} \cdot u \restriction \Gamma = 0\}$$

where $\underline{n}$ is the unit normal vector on Γ. If E denotes the orthogonal projector from $L^2(\Omega)^n$ onto the subspace $\mathcal{H}_2$, it follows that

$$\mathcal{H}_2 = EL^2(\Omega)^n \quad \textit{and} \quad \mathcal{H}_1 = EH_0^1(\Omega)^n .$$

$H_0^1(\Omega)$ is compactly embedded in $L^2(\Omega)$ for bounded domains Ω (Rellich's theorem); it follows that $H_0^1(\Omega)^n$ is also compactly embedded in $L^2(\Omega)^n$.

This representation of $\mathcal{H}_1$ and $\mathcal{H}_2$ shows easily that, since the product of a compact operator and a bounded operator is also compact, $\mathcal{H}_1$ is also compactly embedded in $\mathcal{H}_2$. A suitable bilinear form Q on $\mathcal{H}_1$ is

$$Q(u,v) = \mu\langle \nabla u, \nabla v\rangle_{L^2(\Omega)^n}$$

which is obviously symmetric and continuous on $\mathcal{H}_1$. By virtue of earlier considerations, Q is also coercive on $\mathcal{H}_1$. Thus Theorem 6.5.1 can be applied and we obtain the following.

Theorem 6.5.4. *For all $f \in L^2([0,T], L^2(\Omega)^n)$ and for all $u_0 \in \mathcal{H}_2$, there is exactly one*

$$u = u_{f,u_0} \in \mathcal{C}([0,T],\mathcal{H}_2) \cap L^2([0,T],\mathcal{H}_1)$$

such that for all $v \in \mathcal{H}_1$,

$$\partial_t\langle v, u(t)\rangle_{\mathcal{H}_2} + \mu\langle \nabla v, \nabla u(t)\rangle_{\mathcal{H}_2} = \langle v, f(t)\rangle_{L^2(\Omega)} \quad \textit{in} \quad \mathcal{D}'((0,T))$$

and $u(0) = u_0$.

Proof. $Ev = v$ holds for $v \in \mathcal{H}_1$ and thus so does

$$\langle v, f(t)\rangle_{L^2(\Omega)^n} = \langle v, Ef(t)\rangle_{L^2(\Omega)^n} = \langle v, g(t)\rangle_{\mathcal{H}_2}$$

where $g = Ef \in L^2([0,T],\mathcal{H}_2)$; we are then indeed dealing with the situation of Theorem 6.5.1.

It remains for us to clarify in which sense Theorem 6.5.4 describes the solution of the Stokes equations (S). Since $\mathcal{D}(\Omega)^n$ is not contained in $\mathcal{H}_1$, a distributional interpretation of the solution does not immediately follow canonically. According to Theorem 6.5.4,

$$S = \partial_t u - \mu\Delta_x u - f$$

is an n-tuple of distributions on $\Omega = (0,T) \times \Omega$. It follows from Theorem 6.5.4 for all $v \in \mathcal{D}(\Omega)^n$, where $\operatorname{div} v = 0$, i.e. $v \in \mathcal{H}_1$, and for all $\psi \in \mathcal{D}((0,T))$ that

$$\begin{aligned} S(\psi \otimes v) &= \int_0^T \psi(t)\{\partial_t\langle v, u(t)\rangle_{L^2(\Omega)^n} + \mu\langle \nabla v, \nabla u\rangle_{L^2(\Omega)^n} \\ &\qquad - \langle v, f(t)\rangle_{L^2(\Omega)^n}\}\, dt \\ &= 0 . \end{aligned}$$

Making use of a fundamental result due to G. de Rham [6.8], it follows from

$$S(\psi \otimes v) = 0\,, \quad \forall v \in \mathcal{D}(\Omega)^n\,, \quad \operatorname{div} v = 0\,, \quad \forall \psi \in \mathcal{D}((0,T))\,,$$

that a distribution $p \in \mathcal{D}'(\Omega_T)$ exists such that

$$S = -\operatorname{grad}_x p\,.$$

Thus, the solution u of Theorem 6.5.4 is given by

$$\partial_t u - \mu \Delta_x u + \operatorname{grad}_x p = f \quad \text{in } \mathcal{D}'(\Omega_T)^n\,,$$

i.e. this theorem yields the existence and uniqueness of a solution $u \in L^2([0,T],\mathcal{H}_1) \cap \mathcal{C}([0,T],\mathcal{H}_2)$ which satisfies the Stokes differential equation and the condition that there should be no sources, in the sense of distributions on $(0,T) \times \Omega$. The fact that the boundary condition $u = 0$ on $[0,T] \times \Omega$ is satisfied in the weak sense is contained in the statement $u \in L^2([0,T],\mathcal{H}_1) \subset L^2([0,T],H_0^1(\Omega)^n)$. It follows from $u \in \mathcal{C}([0,T],\mathcal{H}_2)$ that the initial condition $u(0) = u_0$ is satisfied in $L^2(\Omega)^n$ in the sense of $\lim_{t\to 0} u(t) = u_0$. □

Remark 6.5.3. If one makes additional assumptions about the regularity of Ω, u_0 and f, it is possible to obtain corresponding assertions about the regularity for the solution (u,p) [6.8].

7. Nonlinear Elliptic Boundary Value Problems and Monotonic Operators

7.1 Forms and Operators – Boundary Value Problems

In the previous chapter we saw how to solve linear elliptic boundary value problems using variational methods. In this chapter we shall discuss some of *F. E. Browder's* generalisations to nonlinear, or, more specifically, quasi-linear elliptic boundary value problems [7.1]. Our attempts at solving boundary value problems for quasi-linear differential equations in "divergence form",

$$A(u)(x) \equiv A_0(x, u(x), \nabla u(x)) - \sum_{j=1}^{n} \partial_j A_j(x, u(x), \nabla u(x)) = f(x) \quad (7.1.1)$$

using the methods of variational calculus start off in a very similar fashion to our attempts at solving linear differential equations. Here too, we start by considering a suitable *"generalised Dirichlet form"* a associated with A, on a distribution space which takes into account the boundary conditions for the potential solutions of (7.1.1). We shall limit our investigations to second-order differential equations (see [7.1] for a natural generalisation to differential equations of order $2m$, $m > 1$) and thus we consider the following *generalised Dirichlet form* on the Sobolev space $W^{1,p}(G)$, where $G \subset \mathbb{R}^n$ is open:

$$a(u, v) = \langle A_0(\cdot, y(u)), v \rangle_2 + \sum_{j=1}^{n} \langle A_j(\cdot, y(u)), \partial_j v \rangle_2 \quad (7.1.1')$$

where

$$y(u) = (y_0(u), y_1(u), \ldots, y_n(u)) \equiv (u, \partial_1 u, \ldots, \partial_n u)$$

and

$$\langle f, g \rangle_2 = \int_G f(x) g(x) d^n x ,$$

assuming all values to be in $\mathbb{R}$.

A first group of hypotheses for the coefficient functions $A_j : G \times \mathbb{R}^{n+1} \to \mathbb{R}$ will establish the following property of a:

$$|a(u, v)| \leq h(\|u\|_{1,p}) \|v\|_{1,p} , \quad u, v \in W^{1,p}(G) \quad (7.1.2)$$

$$\|u\|_{1,p} = \left(\sum_{\|\alpha\| \leq 1} \int_G |D^\alpha u(x)|^p \, dx \right)^{1/p}$$

where the function $h : \mathbb{R}_+ \to \mathbb{R}_+$ is bounded on bounded sets. If (7.1.2) is satisfied, then for arbitrary, but fixed, $u \in W^{1,p}(G)$, $v \to a(u,v)$ is a continuous linear form $T(u)$ on $X \equiv W^{1,p}(G)$ for which

$$\langle T(u), v \rangle = a(u,v) \,, \quad \forall u, v \in X \,, \tag{7.1.3}$$

holds, where $\langle \cdot \,, \cdot \rangle$ denotes the duality of X and X'. Thus (7.1.3) defines a mapping T from X into X', assuming that (7.1.2) holds.

We shall introduce boundary conditions in the following, initially abstract, way. We choose a closed subspace V of $W^{1,p}(G)$ which satisfies

$$W_0^{1,p}(G) \subseteq V \subseteq W^{1,p}(G) \,. \tag{7.1.4}$$

We can then also take a to be a generalised Dirichlet form on V which, due to (7.1.2), defines a mapping $T : V \to V'$ by means of the duality (7.1.3). (V' is the topological dual space of the Banach space V with norm $\| \cdot \|'$.)

We can now formulate the following problem with these conditions.

Variational Boundary Value Problem. For given V, a and $f \in V'$, determine all $u \in V$ which satisfy

$$a(u,v) = \langle f, v \rangle \,, \quad \forall v \in V \,. \tag{7.1.5}$$

Equation (7.1.5) and the requirement that u be in V not only mean that u must satisfy (at least in a weak sense, since $\mathcal{D}(G) \subseteq V$) the differential equation

$$A(u)(x) \equiv A_0(x, y(u)(x)) - \sum_{j=1}^{n} \partial_j A_j(x, y(u)(x)) = f(x) \,, \tag{7.1.6}$$

but also that u satisfy certain boundary conditions. The origin of these boundary conditions is:

(i) The choice of V; this boundary condition can be expressed as "$u \in V$". It becomes effective particularly when V "is very much smaller than $W^{1,p}(G)$". This is the only boundary condition for $V = W_0^{1,p}(G)$, i.e. for the Dirichlet problem.

(ii) On the other hand, if V "is considerably larger than $W_0^{1,p}(G)$", then (7.1.5) gives the so-called *natural boundary conditions* for u, since the equations $A(u) = f$ and $a(u,v) = \langle f, v \rangle$, $\forall v \in V$, assert that the boundary terms occuring in partial integration of the equation must vanish. These natural boundary conditions can be regarded as being nonlinear analogues of Neumann's boundary conditions for nonlinear differential operators in divergence form,

$$v(x)\underline{n}(x) \cdot \underline{A}(x, y(u)(x)) = v(x) \sum_{j=1}^{n} n_j(x) A_j(x, y(u)(x)) = 0$$

for all $x \in \partial G$ and all $v \in V$. Here $\underline{n}(x)$ denotes the outer unit normal to ∂G at the point x. Obviously, we shall not be able to express these natural boundary conditions in the explicit form given above without making some assumptions about the smoothness of the boundary of G. The variational formulation then seems to be a suitable generalisation of this case.

Further hypotheses on the coefficient functions A_j imply additional properties of the mapping $T : V \to V'$: the *monotonicity*

$$\langle T(v) - T(u), v - u \rangle \geq 0 , \quad \forall v, u \in V , \tag{7.1.7}$$

and the *coerciveness*

$$\langle T(u), u \rangle \geq C(\|u\|)\|u\| , \quad \forall u \in V , \tag{7.1.8}$$

where the function $C : [0, \infty) \to \mathbb{R}$ has the property $C(s) \to +\infty$ for $s \to +\infty$.

The surjectivity of the mapping T then follows by means of a result due to Browder (and Minty), and hence the solubility of the variational boundary value problem.

These abstract results will be proved in the following section together with a few generalisations. Finally, in Sect. 7.3, the abstract properties of the monotonicity and coerciveness given above will be transformed into concrete hypotheses for the coefficient functions A_j. To this end, we shall first of all recall the properties of Niemytski operators. The remaining part of the proof of the solubility of the variational boundary value problem then consists more or less of concrete estimates, using well-known inequalities.

Remark 7.1.1. A complete discussion of the boundary conditions with specific examples can be found in [7.2–4].

7.2 Surjectivity of Coercive Monotonic Operators. Theorems of Browder and Minty

Let E be a real Banach space and T a mapping from E into the topological dual space E' of E. We emphasise that the "operator" T is not assumed to be linear. However, in the sense of the definitions (7.1.7) and (7.1.8) we assume T be monotone and coercive. The underlying Banach space E will always be separable in our applications and we shall therefore prove results about the surjectivity of coercive monotone operators for the separable case only. Generalisations to nonseparable Banach spaces have been shown (by Browder and Carrol).

The basic theorem on the surjectivity of continuous coercive monotone operators was discovered independently by *F. E. Browder* and *G. Minty* in 1963. This theorem is the nonlinear analogue of the Lax-Milgram theorem.

Since we consider this chapter to be simply an illustration of the effectiveness of variational methods, we shall be content just to discuss the simplest versions of Browder's results [7.5]. First of all we consider the following two lemmas.

Lemma 7.2.1. *Let E be a Banach space and $T : E \to E'$ a monotone mapping.*

(a) *Let T be continuous. Then for a given $f \in E'$, every solution of the equation $T(u) = f$ is characterised by*

$$\langle T(v) - f, v - u\rangle \geq 0\,, \quad \forall v \in E\,.$$

The solution set $T^{-1}(f)$ for the equation $T(u) = f$ is closed and convex.

(b) *Let T be coercive, i.e. a function $C : [0, \infty) \to \mathbb{R}$ exists, where $C(r) \to +\infty$ for $r \to +\infty$, such that*

$$\langle T(u), u\rangle \geq C(\|u\|)\|u\|\,, \quad \forall u \in E\,.$$

Then the solution set $T^{-1}(f)$ is bounded for every $f \in E'$:

$$T^{-1}(f) \subseteq \{u \in E|\ \|u\| \leq \widehat{C}(\|f\|')\}\,,$$

where

$$\widehat{C}(s) := \sup\{r \,|\, C(r) \leq s\}\,.$$

Proof. (a) The monotonicity of T implies that the inequality above must hold for a solution u. Conversely, if this inequality holds, then for arbitrary, but fixed, $z \in E$ we can put $v = u + tz$, $t > 0$. It follows that $\langle T(u+tz) - f, tz\rangle \geq 0$ and thus $\langle T(u + tz) - f, z\rangle \geq 0$. Since T is continuous, the limit $t \searrow 0$ gives $\langle T(u) - f, z\rangle \geq 0$. Since z is an arbitrary point in E it follows that $T(u) - f = 0$. Thus $T^{-1}(f)$ can be written as

$$T^{-1}(f) = \bigcap_{v \in E} \{u \in E \,|\, \langle T(v) - f, v - u\rangle \geq 0\}\,.$$

$T^{-1}(f)$ is thus the intersection of closed half-spaces

$$\{u \in E \,|\, \langle T(v) - f, v - u\rangle \geq 0\}$$

and therefore a closed convex set.

(b) For $u \in T^{-1}(f)$,

$$\|u\|C(\|u\|) \leq \langle T(u), u\rangle = \langle f, u\rangle \leq \|f\|'\|u\|\,.$$

It follows that

$$C(\|u\|) \leq \|f\|' \quad \text{or} \quad \|u\| \leq \widehat{C}(\|f\|')\,. \qquad \square$$

Roughly speaking, the surjectivity of continuous coercive monotone operators is implied by the combination of the following two facts:

(i) continuous coercive operators on finite-dimensional Banach spaces are surjective;
(ii) a generalised Galerkin approximation converges as a result of the monotonicity of the operators.

This indicates the central role played by the following lemma.

Lemma 7.2.2. *Let $T : F \to F'$ be a continuous coercive mapping on the finite-dimensional Banach space F. Then T is surjective: $T(F) = F'$.*

Proof. (a) To show that $T(F) = F'$ it is sufficient to show that $0 \in T(F)$, since $T(u) = f$ if, and only if, $T_f(u) = T(u) - f = 0$. T_f is continuous and coercive if, and only if, T has these properties.

Note, furthermore, that the hypotheses of this lemma remain true if we pass to an equivalent Banach space (possibly with another function $C : [0, \infty) \to \mathbb{R}$ from the same class). Since we have assumed that $\dim F < \infty$, we can also assume that F is a Hilbert space; F and F' can thus be identified.

(b) For sufficiently large $R > 0$ we know $C(R) > 0$. For such a number R we thus have, for $s(u) = u - T(u)$

$$\langle s(u), u \rangle < \|u\|^2 , \quad \forall u \in F , \quad \|u\| = R .$$

Let $B_R = \{u \in F \mid \|u\| \leq R\}$. The *radial retraction* r from F to B_R is given by

$$r(v) = \begin{cases} v , & \text{for } v \in B_R , \\ \frac{R}{\|v\|} v , & \text{for } v \notin B_R . \end{cases}$$

It follows that $r(B_R^c) \subseteq \partial B_R$ and, furthermore, that

$$f(u) = r \circ s(u)$$

is a continuous mapping from B_R into B_R. Brouwer's fixed point theorem states that f has a fixed point u_0 in B_R. If $\|u\|_0 < R$, then the definition of r and the fact that $u_0 = f(u_0)$ show that

$$f(u_0) = s(u_0) = u_0 ,$$

and thus

$$T(u_0) = 0 .$$

The case when $\|u_0\| = R$ can be excluded, since it follows first of all from $\|u_0\| = R$ and $u_0 = f(u_0)$ that

$$s(u_0) \in B_R^c \cup \partial B_R ,$$

and thus

$$\rho = \|s(u_0)\| \geq R .$$

Then,

$$f(u_0) = r(s(u_0)) = \frac{R}{\rho}s(u_0),$$

which, however, gives a contradiction:

$$R^2 = \|u_0\|^2 = \langle f(u_0), u_0\rangle = \left\langle \frac{R}{\rho}s(u_0), u_0 \right\rangle = \frac{R}{\rho}\langle s(u_0), u_0\rangle < R^2 \qquad \square$$

The following theorem, due to Browder and Minty, can now be proved quite easily. First of all, we introduce a suitable generalised Galerkin approximation. Then, Lemma 7.2.2 can be used for the approximating mappings. Finally, we can pass to the limit, using the monotonicity property and the characterisation of the solution set $T^{-1}(f)$ by Lema 7.2.1.

Theorem 7.2.3. (Browder-Minty). *Let E be a separable, reflexive Banach space with dual space E' and let $T : E \to E'$ be a continuous, coercive, monotone mapping. Then T is surjective: $T(E) = E'$.*

For fixed $f \in E'$, $T^{-1}(f)$ is a bounded, closed, convex subset of E.

Proof. To show that $T(E) = E'$ it is sufficient, as in Lemma 7.2.2, to show that $0 \in T(E)$.

1st step: Definition of a suitable Galerkin approximation. Since E is separable, there is an increasing sequence $(E_n)_{n\in\mathbb{N}}$ of finite-dimensional subspaces E_n of E whose union is dense in E. Let $\phi_n : E_n \to E$ denote the identical embedding and let $\phi_n' : E' \to E_n'$ denote the adjoint projection. For given $T : E \to E'$, put

$$T_n = \phi_n' \circ T \circ \phi_n , \quad n \in \mathbb{N} . \tag{7.2.1}$$

It follows that all T_n have the properties of T. This is obvious as far as the continuity is concerned. The monotonicity of the T_n on E_n can be shown as follows. For arbitrary $u, v \in E_n$ we have

$$\begin{aligned}\langle T_n(u) - T_n(v), u - v\rangle_{E_n} &= \langle \phi_n'(T \circ \phi_n(u) - T \circ \phi_n(v)), u - v\rangle_{E_n} \\ &= \langle T(\phi_n(u)) - T(\phi_n(v)), \phi_n(u) - \phi_n(v)\rangle \\ &\geq 0 .\end{aligned}$$

The "uniform" coerciveness of the T_n can be shown similarly. For all $u \in E_n$, the coerciveness of T on E implies

$$\begin{aligned}\langle T_n(u), u\rangle_{E_n} &= \langle T(\phi_n)), \phi_n(u)\rangle \geq C(\|\phi_n(u)\|)\|\phi_n(u)\| \\ &= C(\|u\|)\|u\| ,\end{aligned}$$

where $\langle , \rangle_{E_n}$ denotes the duality of E_n and E_n'.

2nd step: Proof of the surjectivity of the T_n. Since $T_n : E_n \to E'_n$ is a continuous coercive mapping of the finite-dimensional Banach space E_n, Lemmas 7.2.2 and 7.2.1 give the existence of a $u_n \in E_n$ such that

$$T_n(u_n) = 0 \quad \text{and} \quad \|u_n\| \leq M = \widehat{C}(0) < \infty\,.$$

3rd step: Passage to the limit. Since E is a reflexive Banach space we see that the bounded sequence of the solutions u_n, has a weakly convergent subsequence $(u_{n_j})_{j\in\mathbb{N}}$,

$$u_{n_j} \underset{j\to+\infty}{\overset{w}{\longrightarrow}} u_0\,.$$

Now we show that $T(u_0) = 0$.

Let $v \in E_m$, $m \in \mathbb{N}$, be arbitrary, but fixed. We have $E_m \subseteq E_n$ for all $n \geq m$ and thus, by virtue of the monotonicity of T_n on E_n,

$$0 \leq \langle T_n(v) - T_n(u_n), v - u_n\rangle_{E_n} = \langle T_n(v), v - u_n\rangle_{E_n} = \langle T(v), v - u_n\rangle\,.$$

Passing to the limit gives

$$0 \leq \langle T(v), v - u_0\rangle$$

and thus

$$0 \leq \langle T(v), v - u_0\rangle\,, \quad \forall v \in \bigcup_{m\in\mathbb{N}} E_m\,.$$

Since T is continuous and $\bigcup_{m\in\mathbb{N}} E_m$ is dense in E, it follows that

$$0 \leq \langle T(v), v - u_0\rangle\,, \quad \forall v \in E\,,$$

and thus, by Lemma 7.2.1 (a), $T(u_0) = 0$. This shows that $0 \in T(E)$, and thus $T(E) = E'$. □

The monotonicity of T was not used above until the third step of the proof, to conclude from $u_0 = w - \lim_{j\to\infty} u_{n_j}$ that $T(u_0) = \lim_{j\to\infty} T(u_{n_j})$. From this it should be clear that every other property of T which allows to reach this conclusion also implies, using coerciveness, the surjectivity of T. A property of T like this which has proved itself useful in applications is given by the *Smale condition*:

$$\text{(S)}\quad \begin{cases} \text{For every sequence } (u_n)_{n\in\mathbb{N}} \subset E, \text{ the conditions} \\ \qquad u = w - \lim\limits_{n\to\infty} u_n \quad \text{and} \quad \langle T(u_n) - T(u), u_n - u\rangle \underset{n\to\infty}{\longrightarrow} 0 \\ \text{imply } u = s - \lim\limits_{n\to\infty} u_n\,. \end{cases}$$

The condition (S) can thus be considered as a generalised monotonicity condition.

The following theorem shows that the surjectivity of coercive mappings is still valid if, in the Browder-Minty theorem, the monotonicity condition is replaced by the Smale condition with T bounded. The boundedness of T means that T maps bounded sets onto bounded sets.

Theorem 7.2.4. *Let E be a separable reflexive Banach space and let $T : E \to E'$ be a continuous coercive mapping which is bounded and which satisfies the Smale condition. Then T is surjective:*

$$T(E) = E' .$$

Proof. Define the Galerkin approximation $(E_n, T_n)_{n \in \mathbb{N}}$ as in the proof of Theorem 7.2.3. Now let $f \in E'$ be given. By Lemmas 7.2.2 and 7.2.1 there is a sequence $(u_n)_{n \in \mathbb{N}}$ which has the following properties:

$$u_n \in E_n , \quad T_n(u_n) = \phi'_n(f) , \quad \|u_n\| \leq M < \infty , \quad \forall n \in \mathbb{N} .$$

Since E is reflexive, a weakly convergent subsequence $(u_{n_j})_{j \in \mathbb{N}}$ exists with the limit u_0. Since T is bounded, $\{T(u_n) \mid n \in \mathbb{N}\}$ is also bounded.

Now let $m \in \mathbb{N}$ and $v \in E_m$; $E_m \subseteq E_n$ for all $n \geq m$, and thus

$$\langle T(u_n), v \rangle = \langle T_n(u_n), v \rangle_{E_n} = \langle \phi'_n(f), v \rangle_{E_n} = \langle f, v \rangle .$$

Therefore

$$\lim_{n \to \infty} \langle T(u_n), v \rangle = \langle f, v \rangle .$$

Since $(T(u_n))_{n \in \mathbb{N}}$ is bounded in E' and $\bigcup_m E_m$ is dense in E, it follows that

$$f = w - \lim_{n \to \infty} T(u_n)$$

and thus also that

$$\begin{aligned} &\lim_{j \to \infty} \langle T(u_{n_j}) - T(u_0), u_{n_j} - u_0 \rangle \\ &\quad = \lim_{j \to \infty} \langle T(u_{n_j}), u_{n_j} \rangle - \lim_{j \to \infty} \langle T(u_0), u_{n_j} - u_0 \rangle - \lim_{j \to \infty} \langle T(u_{n_j}), u_0 \rangle \\ &\quad = \lim_{j \to \infty} \langle f, u_{n_j} \rangle - \langle f, u_0 \rangle = 0 . \end{aligned}$$

The Smale condition implies that $u_0 = s - \lim_{j \to \infty} u_{n_j}$, and by virtue of the continuity of T we also have $T(u_0) = s - \lim_{j \to \infty} T(u_{n_j})$. We have already shown that $f = w - \lim_{j \to \infty} T(u_{n_j})$ and so it follows that $f = T(u_0)$. Thus, T is surjective. □

7.3 Nonlinear Elliptic Boundary Value Problems. A Variational Solution

The general strategy for the solution of the variational boundary value problem has been indicated in Sects. 7.1 and 7.2 by Browder's abstract theorems, Theorems 7.2.3 and 7.2.4. We must now satisfy the hypotheses of these general theorems in terms of specific conditions on the coefficient functions $A_0 , \ldots , A_n$

of the generalised Dirichlet form. In a first step we ensure, under the most general hypotheses possible, that a generalised Dirichlet form a is well defined on $W^{1,p}(G)$ according to (7.1.1′) and that this then has the continuity property (7.1.2). The crux of the proof that the hypotheses formulated in (H_1) for A_j are indeed sufficient for this purpose can be found in [Ref. 7.6, paragraphs 18–20]. For convenience we have summarised the relevant results below.

For a given measurable subset $B \subset \mathbb{R}^m$ we shall refer to a function $g : B \times \mathbb{R}^n \to \mathbb{R}$ as a C *function (Carathéodory function)* if the following conditions hold:

(i) $y \mapsto g(x,y)$ is a continuous function on $\mathbb{R}^n$ for almost all $x \in B$.
(ii) $x \mapsto g(x,y)$ is a measurable function on B for all $y \in \mathbb{R}^n$.

If g is a C function on $B \times \mathbb{R}^n$ and if $v_1, \ldots, v_n$ are measurable functions on B which are finite almost everywhere, then $\widehat{g}v$, given by

$$\widehat{g}v(x) = g(x, v_1(x), \ldots, v_n(x)) \tag{7.3.1}$$

for almost all $x \in B$, is also a measurable function of B. If, moreover, B has a finite Lebesgue measure, then $v \mapsto \widehat{g}v$ is also continuous in measure.

If $B \subset \mathbb{R}^m$ is a Lebesgue-measurable set and if $g = (g_1, \ldots, g_N)$ are C functions on $B \times \mathbb{R}^n$ which define via (7.3.1) mappings $\widehat{g} = (\widehat{g}_1, \ldots, \widehat{g}_N)$ from $L^{p,n}(B) = L^p(B) \times \cdots \times L^p(B)$ (n times) into $L^{p_j}(B)$, $j = 1, \ldots, N$, then the mappings $\widehat{g}_j : L^{p,n}(B) \to L^{p_j}(B)$ are continuous and bounded. $\widehat{g}$ is thus a continuous bounded mapping from $L^{p,n}(B)$ into $L^{p_1}(B) \times \cdots \times L^{p_N}(B)$, $1 < p < \infty$, $1 \le p_j < \infty$.

The operator $\widehat{g}$ defined in this way is called the *Niemytski operator*, associated with g.

Remark 7.3.1. The statement above asserts in particular that the mappings $\widehat{g}_j$ associated with the C functions g_j are already continuous and bounded (as mappings from $L^{p,n}(B)$ into $L^{p_1}(B)$) if they are defined only on the whole of $L^{p,n}(B)$ and take values in $L^{p_j}(B)$. The following criteria express when this is the case:

(a) $\widehat{g}_j$ maps $L^{p,n}(B)$ into $L^{p_j}(B)$, $1 \le p_j < \infty$, if, and only if, a function $a_j \in L^{p_j}(B)$ and a constant $b \ge 0$ exist such that

$$|g_j(x,y)| \le a_j(x) + b\sum_{i=1}^{n} |y_i|^{p/p_j} \tag{7.3.2}$$

for almost all $x \in B$ and all $y = \mathbb{R}^n$.

(b) $\widehat{g}_j$ maps $L^{p,n}(B)$ into $L^\infty(B)$ if, and only if, a constant $c \ge 0$ exists such that

$$|g_j(x,y)| \le c \tag{7.3.2′}$$

for all $y \in \mathbb{R}^n$ and almost all $x \in B$.

The hypotheses (H_1) below for the coefficient functions $A_0, A_1, \dots, A_n$ are thus seen to be quite natural.

(H_1) (i) Let $A_0, A_1, \dots, A_n$ be C functions on $G \times \mathbb{R}^{n+1}$.
(ii) The following estimates for $j = 0, 1, \dots, n$ hold almost everywhere in $x \in G$ and for all $y \in \mathbb{R}^{n+1}$:

$$|A_j(x,y)| \leq \alpha_j(x) + \beta_j(x)|y_0|^{p_j} + \sum_{i=1}^{n} g_{ji}(x)|y_i|^{q_{ji}} + \sum_{i=1}^{n} f_{ji}(x)|y_0|^{r_{ji}}|y_i|^{s_{ji}} \tag{7.3.3}$$

subject to the following constraints: the exponents p_j, q_{ji}, r_{ji} and s_{ji} are all nonnegative and satisfy

$$p_j, q_{ji} \leq \frac{p}{p'} = p - 1, \quad r_{ji} + s_{ji} \leq p - 1,$$

and the nonnegative functions α_j, β_j, f_{ji} and g_{ji} satisfy

$$\alpha_{ji} \in L^{p'}(G), \quad p' = \frac{p}{p-1},$$

$$\beta_j \in L^{p/(p-1-p_j)}(G), \quad g_{ji} \in L^{p/(p-1-q_{ji})}(G),$$
$$f_{ji} \in L^{p/(p-1-s_{ji}-r_{ji})}(G).$$

Vainberg's results, which we mentioned above, now make it fairly easy to prove the next theorem.

Theorem 7.3.1. *If the functions $A_0, A_1, \dots, A_n$ satisfy the hypotheses (H_1), then (7.1.1′) gives a well-defined generalised Dirichlet form a on $W^{1,p}(G)$ which has the continuity property (7.1.2). The mapping thus defined by (7.1.3), $T : W^{1,p}(G) \to (W^{1,p}(G))'$, is continuous and bounded.*

Proof. (a) Since $u \to y(u) = (u, \partial_1 u, \dots, \partial_n u)$ is a continuous linear mapping from $W^{1,p}(G)$ into $L^{p,n+1}(G)$, we first of all show that the $\widehat{A}_j$ are well-defined mappings from $L^{p,n+1}(G)$ into $L^{p'}(G)$. Vainberg's results then ensure the continuity of these mappings. It follows that

$$u \to \widehat{A}_j u(u), \quad \widehat{A}_j y(u)(x) = A_j(x, y(u)(x)),$$

is a well-defined continuous mapping from $W^{1,p}(G)$ into $L^{p'}(G)$.

(b) For $g = (g_0, g_1, \dots, g_n) \in L^{p,n+1}(G)$, the inequality (7.3.3) gives the estimate

$$\|\widehat{A}_j g\|_{p'} \leq \|\alpha_j\|_{p'} + \|\beta_j |g_0|^{p_j}\|_{p'} + \sum_{i=1}^{n} \{\|g_{ji}|g_j|^{q_{ji}}\|_{p'} + \sum_{i=1}^{n} \|f_{ji}|g_0|^{r_{ji}}|g_i|^{s_{ji}}\|_{p'}\}.$$

The rest of the proof consists in repeated application of Hölder's inequality, taking into account the restrictions on the exponents formulated in (H_1).

Finally, we get the following estimate which shows that $\widehat{A}_j$ is a mapping from the whole of $L^{p,n+1}(G)$ into $L^{p'}(G)$:

$$\begin{aligned}\|\widehat{A}_j g\|_{p'} &\leq \|\alpha_j\|_{p'} + \|\beta_j\|_{p/(p-1-p_j)}\|g_0\|_p^{p_j} \\ &\quad + \sum_{i=1}^{n}\{\|g_{ji}\|_{p/(p-1-q_{ji})}\|g_j\|_p^{q_{ji}} + \|f_{ij}\|_{p/(p-1-r_{ji}-s_{ji})}\|g_0\|_p^{r_{ji}}\|g_j\|_p^{s_{ji}}\} \\ &\leq h_j(\|g\|_{L^{p,n+1}}) \end{aligned} \tag{7.3.4}$$

where the h_j are functions which are bounded on bounded sets.

(c) The estimate (7.3.4) proves that $u \mapsto \widehat{A}_j y(u)$ is well defined and a continuous mapping from $W^{1,p}(G)$ into $L^{p'}(G)$. It is now easy to show for arbitrary $u, v \in W^{1,p}(G)$ that

$$\begin{aligned}|a(u,v)| &= \left|\langle \widehat{A}_0 y(u), v\rangle_2 + \sum_{j=1}^{n}\langle \widehat{A}_j y(u), \partial_j v\rangle_2\right| \\ &\leq \left\{\sum_{j=0}^{n}\|\widehat{A}_j y(u)\|_{p'}^{p'}\right\}^{1/p'}\left\{\|v\|_p^p + \sum_{j=1}^{n}\|\partial_j v\|_p^p\right\}^{1/p}.\end{aligned}$$

Thus, with

$$h(s)^p := \sum_{j=0}^{n} h_j(s)^p\,,$$

and taking

$$\|y(u)\|_{L^{p,n+1}} = \|u\|_{1,p} \tag{7.3.5}$$

into account, we get the estimate

$$|a(u,v)| \leq h(\|u\|_{1,p})\|v\|_{1,p}\,. \tag{7.3.6}$$

(d) For arbitrary $u, u_0 \in W^{1,p}(G)$, we have, by (7.1.3),

$$\begin{aligned}\|T(u) - T(u_0)\|_{1,p}' &= \sup_{\substack{v\in W^{1,p}(G)\\ \|v\|_{1,p}\leq 1}} |a(u,v) - a(u_0,v)| \\ &\leq \sup_{\|v\|_{1,p}\leq 1}\Big\{|\langle \widehat{A}_0 y(u) - \widehat{A}_0 y(u_0), v\rangle_2| \\ &\quad + \sum_{j=1}^{n}|\langle \widehat{A}_j y(u) - \widehat{A}_j y(u_0), \partial_j v\rangle_2|\Big\} \\ &\leq \left\{\sum_{j=0}^{n}\|\widehat{A}_j y(u) - \widehat{A}_j y(u_0)\|_p^{p'}\right\}^{1/p'}.\end{aligned}$$

Since $u \mapsto \widehat{A}_j y(u)$ is continuous, the continuity of the mapping T also follows, and the proof of Theorem 7.3.1 is thus complete. It follows from (7.3.6) that

$$\|T(u)\|'_{1,p} \leq h(\|u\|_{1,p}) .$$

where the function h is bounded on bounded sets. □

To prove the solubility of the variational boundary value problem with the aid of Theorem 7.2.3, we need additional hypotheses which imply the monotonicity and coerciveness of the mapping T. The following hypotheses provide a simple basis for this.

(H_2) *Monotonicity.* For almost all $x \in G$ and all $y, y' \in \mathbb{R}^{n+1}$ we have

$$\sum_{j=0}^{n} [A_j(x, y) - A_j(x, y')][y_j - y'_j] \geq 0 . \tag{7.3.7}$$

(H_3) *Coerciveness.* Suppose there are a constant $\alpha > 0$ and nonnegative functions $g_j \in L^{p/(p-r_j)}$, $0 \leq r_j < p$, such that for almost all $x \in G$ and all $y \in \mathbb{R}^{n+1}$ we have

$$\sum_{j=0}^{n} A_j(x, y) y_j \geq \alpha \sum_{j=0}^{n} |y_j|^p - \sum_{j=0}^{n} g_j(x) |y_j|^{r_j} . \tag{7.3.8}$$

Remark 7.3.2. The coerciveness condition (H_3) can, in this general situation, be regarded as a suitable generalisation of the ellipticity condition assumed in the linear problem. Later on, we shall discuss a weakening of the coerciveness condition which we shall show to be sufficient in situations in which the Sobolev embedding theorem can be used. This weakening will consist in omitting the summands $A_0(\cdot) y_0$ and $|y_0|^p$ in the first two sums in the inequality (7.3.8), just as we did in the linear problem. The coerciveness hypotheses (H'_3) is, in this version, a considerable weakening of the ellipticity condition used in linear problems (cf. Chap. 6).

We now come to the proof of the solubility of the variational boundary value problem.

Theorem 7.3.2. *For every open, nonempty set $G \subset \mathbb{R}^n$, the variational boundary value problem* (7.1.5),

$$A(u) = f \quad \text{with respect to } V ,$$

has a solution for every closed subspace V satisfying

$$W_0^{1,p}(G) \subseteq V \subseteq W^{1,p}(G) ,$$

and for every $f \in V'$ if the coefficient functions $A_0, A_1, \dots, A_n$ satisfy the hypotheses (H_1), (H_2) *and* (H_3). *For fixed $f \in V'$, the family of all solutions is a closed, convex, bounded subset of V.*

Proof. (a) Since V is a subspace of $W^{1,p}(G)$ it has the topology induced by $W^{1,p}(G)$. Thus we can substitute V for $W^{1,p}(G)$ in Theorem 7.3.1. It follows that (7.1.1′) defines a generalised Dirichlet form on V which satisfies the estimate (7.1.2) for $u, v \in V$.

In addition,

$$\langle T(u), v\rangle := a(u,v)\,, \quad \forall v \in V \tag{7.3.9}$$

defines a continuous mapping from V into V'.

(b) The monotonicity of T follows simply from Hypothesis (H$_2$). If $u, v \in V$, then it follows form (7.3.7) and (7.1.1′) that

$$\begin{aligned}\langle T(u) - T(v), u - v\rangle &= a(u, u - v) - a(v, u - v)\\ &= \int_G d^n x \sum_{j=0}^{n} \{A_j(x, y(u)(x)) - A_j(x, y(v)(x))\}\\ &\quad \times \{y_j(u)(x) - y_j(v)(x)\}\\ &\geq 0\,.\end{aligned}$$

(c) The coerciveness of the mapping T follows from (H$_3$) applying Hölder's inequality. For arbitrary $u \in V$, we have

$$\begin{aligned}\langle T(u), u\rangle = a(u,u) &= \int_G d^n x \sum_{j=0}^{n} A_j(x, y(u)(x)) y_j(u)(x)\\ &\geq \int_G d^n x \left\{\alpha \sum_{j=0}^{n} |y_j(u)(x)|^p - \sum_{j=0}^{n} g_j(x)|y_j(u)(x)|^{r_j}\right\}\\ &= \alpha\|u\|_{1,p}^p - \sum_{j=0}^{n} \|g_j|y_j(u)|^{r_j}\|_1\\ &\geq \alpha\|u\|_{1,p}^p - \sum_{j=0}^{n} \|g_j\|_{p/(p-r_j)}\|y_j(u)\|_p^{r_j}\,.\end{aligned}$$

Since $\|y_j(u)\|_p \leq \|u\|_{1,p}$ and $0 \leq r_j < p$, this inequality implies, by virtue of the fact that $p - 1 > 0$, that

$$\langle T(u), u\rangle \geq C(\|u\|_{1,p})\|u\|_{1,p}$$

with a function $C : [0,\infty) \to \mathbb{R}$ which has the property that $C(s) \to +\infty$ for $s \to +\infty$.

(d) Since V is a reflexive Banach space, part (a)–(c) of the proof show that T satisfies the hypotheses of Theorem 7.2.3. Finally, applying this theorem proves Theorem 7.3.2. □

Remark 7.3.3. We would like to draw particular attention to the fact that, in Theorem 7.3.2, no special assumptions were used about G. It would, for example, be possible for G to be unbounded and not have a smooth boundary (meaning that Sobolev's embedding theorems could not be used).

If we assume that G is an open, bounded subset of $\mathbb{R}^n$, then for arbitrary $r, q \geq 1$ we have

$$\|f\|_r \leq |G|^{(1-1/q)/r}\|f\|_{rq} \quad \text{where} \quad |G| = \int_G d^n x\,,$$

which means that $L^{rq}(G)$ is continuously embedded in $L^r(G)$. If, in addition, we assume that G also has a smooth boundary, then we may use Sobolev's embedding theorems, i.e. for

$$0 \leq \frac{1}{p} - \frac{1}{n} \leq \frac{1}{r}\,,$$

$W^{1,p}(G)$ is continuously embedded in $L^r(G)$, and when

$$\frac{1}{p} - \frac{1}{n} < \frac{1}{r} < \infty\,,$$

this embedding is compact and the Sobolev inequality $\|f\|_r \leq K(|G|, p, r)$ $\|\nabla f\|_p$ holds. We can, of course, use this additional information to weaken Hypothesis (H_1) somewhat, one possibility being given by Browder [7.7]. However, it is much more important to take the opportunity to weaken Hypotheses (H_2) and (H_3) considerably and still be able to guarantee the solubility of the variational boundary value problem with the aid of Theorem 7.2.4.

We shall discuss the following version in some detail, since, firstly, it uses the direct generalisation (H_3') of the ellipticity condition assumed for linear problems and also because the proof of this version solves a problem explicitly which is typical in this variational calculus, i.e. it derives strong convergence of a sequence from its weak convergence and suitable additional hypotheses. In order to be able to conclude from hypothesis (H_3') that the mapping T is coercive, we need a strengthening of Hypothesis (H_1). We thus make the following assumptions.

(H_1'): Hypothesis (H_1) with the restrictions

$$0 \leq r_{0i}, s_{0i}\,, \quad r_{0i} + s_{0i} < p - 1\,,$$

$$0 \leq p_j, q_{0,i} < p - 1\,.$$

(H_2'): For all almost $x \in G$ and all $y_0 \in \mathbb{R}$ we have

$$\sum_{j=1}^{n} \{A_j(x, y_0, \underline{y}) - A_j(x, y_0, \underline{y}')\}\{y_j - y_j'\} > 0$$

for $\underline{y} \neq \underline{y}'$.

(H_3'): A constant $\alpha > 0$ and a nonnegative function

$$g_0 \in L^{p/(p-r_0)}(G)\,, \quad 0 \leq r_0 < p\,,$$

exist such that for almost all $x \in G$ and all $y \in \mathbb{R}^{n+1}$ we have

$$\sum_{j=1}^{n} A_j(x,y)y_j \geq \alpha \sum_{j=1}^{n} |y_j|^p - g_0(x)|y_0|^{r_0} .$$

We can thus prove the following theorem.

Theorem 7.3.3. *For every open nonempty set $G \subset \mathbb{R}^n$ which is bounded and which has a "smooth boundary", the variational boundary value problem (7.1.5), $A(u) = f$ with respect to V, has a solution for every closed subspace V where $W_0^{1,p}(G) \subseteq V \subseteq W^{1,p}(G)$ and for every $f \in V'$ if the coefficient functions satisfy the hypotheses* (H_1'), (H_2') *and* (H_3').

Proof. (a) Hypothesis (H_1') implies, as above, that (7.1.1′) and (7.1.3) define a continuous mapping T from V into V' which is bounded on bounded subsets of V. This time, however, it is rather more difficult to prove that T is coercive. (H_3') gives

$$\begin{aligned}
\sum_{j=1}^{n} \langle \widehat{A}_j y(u), y_j(u) \rangle_2 &= \int_G d^n x \sum_{j=1}^{n} A_j(x, y(u)(x)) y_j(u)(x) \\
&\geq \int_G d^n x \left\{ \alpha \sum_{j=1}^{n} |y_j(u)(x)|^p - g_0(x)|u(x)|^{r_0} \right\} \\
&= \alpha \|\nabla u\|_p^p - \|g_0 |u|^{r_0}\|_1 \geq \alpha \|\nabla u\|_p^p - \|g_0\|_{p/(p-r_0)} \|u\|_p^{r_0} .
\end{aligned}$$

It follows that

$$\begin{aligned}
\langle T(u), u \rangle &= \langle \widehat{A}_0 y(u), u \rangle_2 + \sum_{j=1}^{n} \langle \widehat{A}_j y(u), y_j(u) \rangle_2 \\
&\geq \alpha \|\nabla u\|_p^p - \|g_0\|_{p/(p-r_0)} \|u\|_p^{r_0} - \|\widehat{A}_0 y(u)\|_{p'} \|u\|_p .
\end{aligned}$$

Estimate (7.3.4) shows that under Hypothesis (H_1') we have

$$\|\widehat{A}_0 y(u)\|_{p'} \leq b(\|u\|_{1,p}) \|u\|_{1,p}^s$$

where b is a bounded function $b : \mathbb{R}_+ \to \mathbb{R}_+$ and $0 \leq s < p - 1$. The Sobolev inequality states that $\|\nabla f\|_p$ and $\|f\|_{1,p}$ are equivalent norms, i.e.

$$\|\nabla f\|_p \leq \|f\|_{1,p} \leq C \|\nabla f\|_p$$

with a suitable $C \in \mathbb{R}_+$. It follows that

$$\begin{aligned}
\frac{\langle T(u), u \rangle}{\|\nabla u\|_p} &\geq \alpha \|\nabla u\|_p^{p-1} - \|g_0\|_{p/(p-r_0)} \frac{\|u\|_p^{r_0}}{\|\nabla u\|_p} - b(\|u\|_{1,p}) \frac{\|u\|_{1,p}^{s+1}}{\|\nabla u\|_p} \\
&\geq \alpha \|\nabla u\|_p^{p-1} - C_2 \|\nabla u\|_p^{r_0 - 1} - b(\|u\|_{1,p}) C^{s+1} \|\nabla u\|_p^s \\
&\to +\infty \quad \text{for} \quad \|\nabla u\|_p \to \infty ,
\end{aligned}$$

since $p-1 > r_0 - 1$ and $p-1 > s$. It also follows that

$$\lim_{\|u\|_{1,p}\to\infty} \frac{\langle T(u), u\rangle}{\|u\|_{1,p}} = +\infty\,,$$

and thus that T is coercive.

(b) Since V is a separable reflexive Banach space, Theorem 7.3.3 follows from Theorem 7.2.4 if we can prove that T also satisfies the condition (S).

Let $\{u_i\}_{i\in\mathbb{N}} \subset V$ be a weakly convergent sequence with the limit u, for which

$$\langle T(u_i) - T(u), u_i - u\rangle \underset{i\to\infty}{\longrightarrow} 0\,. \tag{7.3.10}$$

For $1 \le p < n$, $W^{1,p}(G)$ is compactly embedded in $L^p(G)$ (for $p = n$ in $L^r(G)$, $1/r > 1/p - 1/n = 0$) and so V is also compactly embedded in $L^p(G)$. The weak convergence in V implies that $\{u_i \,|\, i \in \mathbb{N}\}$ is strongly bounded in V. Then $\{u_i \,|\, i \in \mathbb{N}\}$ is relatively compact in $L^p(G)$, i.e. a subsequence $\{u_{i_k}\}_{k\in\mathbb{N}}$ and a set $N \subset G$ of measure zero exist such that

$$\left.\begin{array}{l}\text{(i) } u_{i_k} \underset{k\to\infty}{\longrightarrow} u \text{ strongly in } L^p(G)\,,\\ \text{(ii) } u_{i_k}(x) \underset{k\to\infty}{\longrightarrow} u(x) \text{ pointwise in } G\setminus N\,.\end{array}\right\} \tag{$*$}$$

We shall simplify the notation by again denoting this subsequence by $\{u_i\}_{i\in\mathbb{N}}$. Its image under the mapping $y : W^{1,p}(G) \to L^{p,n+1}(G)$ has the following properties:

$$\begin{array}{rl}\text{(i)} & \|y(u_i)\|_{L^{p,n+1}(G)} \le M < \infty\,,\ \forall i \in \mathbb{N}\,,\\ \text{(ii)} & \{y_0(u_i)\}_{i\in\mathbb{N}} \text{ satisfies } (*)\,,\\ \text{(iii)} & \{y_j(u_i)\}_{i\in\mathbb{N}} \text{ converges weakly in } L^p(G) \text{ to } y_j(u),\ j = 1,\dots,n\,.\end{array} \tag{7.3.11}$$

We still have to show that the sequences $\{y_1(u_i), \dots, y_n(u_i)\}_{i\in\mathbb{N}}$ converge strongly in $L^p(G)$, and to this end it is helpful to recall the following facts.

(α) A weakly convergent sequence in a normed space converges strongly if, and only if, it contains a strongly convergent subsequence.

(β) Since G has a finite Lebesgue measure, Vitali's convergence theorem [Ref. 7.8, Theorem III.6.15] states that $\{y_j(u_i)\}_{i\in\mathbb{N}}$ converges strongly in $L^p(G)$ if the following conditions are satisfied:

(i) $\{y_j(u_i)\}_{i\in\mathbb{N}}$ converges pointwise almost everywhere.

(ii) The sequence $\{y_j(u_i)\}_{i\in\mathbb{N}}$ is uniformly absolutely continuous, i.e. for every $\varepsilon > 0$ there is a $\delta_\varepsilon > 0$ such that for all measurable subsets $A \subset G$ with measure $|A| < \delta_\varepsilon$ and all $i \in \mathbb{N}$,

$$\int_A |y_j(u_i)(x)|^p\, d^n x < \varepsilon\,.$$

The strong convergence of the sequence $\{y_j(u_i)\}_{i\in\mathbb{N}}$ in $L^p(G)$ follows if we can prove the latter two conditions for a suitable subsequence.

(c) Our first step is to reduce the condition (7.3.10). Since $\{\widehat{A}_0 y(u_i)\}_{i\in\mathbb{N}}$ is bounded in $L^{p'}(G)$ (inequality (7.3.4)), the strong convergence of $\{y_0(u_i)\}_{i\in\mathbb{N}}$ in $L^p(G)$ gives

$$\lim_{i\to\infty} \langle \widehat{A}_0 y(u_i) - \widehat{A}_0 y(u), u_i - u\rangle_2 = 0\,.$$

It thus follows that

$$\lim_{i\to\infty} \sum_{j=1}^{n} \langle \widehat{A}_j y(u_i) - \widehat{A}_j y(u), y_j(u_i) - y_j(u)\rangle_2 = 0\,.$$

Since

$$\widehat{A}_j : L^{p,n+1}(G) \to L^{p'}(G)$$

is continuous, we also have

$$\lim_{i\to\infty} A_j(\cdot, y_0(u_i), \underline{y}(u)) = A_j(\cdot, y_0(u), y_0(u))$$

in $L^{p'}(G)$, so that (7.3.10) finally reduces to the following assertion of convergence:

$$\lim_{i\to\infty} \int_G d^n x \sum_{j=1}^{n} \{A_j(x, u_i(x), \underline{y}(u_i)) - A_j(x, u_i(x), \underline{y}(u))\} \times \{y_j(u_i) - y_j(u)\} = 0\,. \tag{7.3.12}$$

By Hypothesis (H_2'), the integrand $I_i(x)$ in (7.3.12) is nonnegative, $I_i(x) \geq 0$, so that (7.3.12) expresses exactly the strong convergence of $\{I_i\}_{i\in\mathbb{N}}$ in $L^1(G)$:

$$I_i \underset{i\to\infty}{\longrightarrow} 0 \text{ in } L^1(G)\,, \quad I_i(x) \geq 0\,. \tag{7.3.13}$$

It follows that a subsequence $\{I_{i_k}\}_{k\in\mathbb{N}}$ and a set $N' \subset G$ of measure zero exist such that

$$I_{i_k}(x) \underset{k\to\infty}{\longrightarrow} 0\,, \quad \forall x \in G \setminus N'\,. \tag{7.3.13$'$}$$

(d) The next step is to show that Condition (i) of (β) above holds for a suitable subsequence of $\{y_j(u_i)\}_{i\in\mathbb{N}}$. A simple rearrangement gives

$$\begin{aligned}\sum_{j=1}^{n} A_j(x, y(u_i)) y_j(u_i) = &\sum_{j=1}^{n} A_j(x, y(u_i)) y_j(u)\\ &+ \sum_{j=1}^{n} A_j(x, y_0(u_i), \underline{y}(u))[y_j(u_i) - y_j(u)]\\ &+ I_i(x)\,.\end{aligned} \tag{7.3.14}$$

Thus, by virtue of (H_2'), we can make the following estimate:

$$\sum_{j=1}^{n} |y_j(u_i)|^p \le \frac{1}{\alpha} \sum_{j=1}^{n} A_j(x, y(u_i)) y_j(u_i) + \frac{1}{\alpha} g_0(x) |y_0(u_i)|^{r_0}$$
$$= \frac{1}{\alpha} g_0(x) |y_0(u_i)|^{r_0} + \frac{1}{\alpha} \sum_{j=1}^{n} A_j(x, y(u_i)) y_j(u)$$
$$+ \frac{1}{\alpha} I_i(x) + \frac{1}{\alpha} \sum_{j=1}^{n} A_j(x, y_0(u_i), \underline{y}(u))[y_j(u_i) - y_j(u)] . \quad (7.3.15)$$

Estimate (7.3.3) for the A_j can also be represented for $|y_i| \le 1$, in the following way. We have, for suitable $0 \le \sigma, \rho, \sigma + \rho \le p - 1$,

$$|A_j(x, y)| \le \alpha_j(x) + \beta_j(x) |y_0|^{p-1} + \gamma_j(x) |\underline{y}|^{p-1} + \delta_j(x) |y_0|^{\rho} |\underline{y}|^{\sigma} ,$$

where the functions $\alpha_j, \ldots, \delta_j$ are all finite outside a common set $N_0 \subset G$ of measure zero. If we substitute this estimate in (7.3.15), then it follows for $x \in G \setminus N_0$ that

$$\sum_{j=1}^{n} |y_j(u_i)(x)|^p \le \frac{1}{\alpha} g_0(x) |y_0(u_i)(x)|^{r_0} + \frac{1}{\alpha} I_i(x)$$
$$+ F(x, y_0(u_i)(x), \underline{y}(u)(x)) |\underline{y}(u_i)(x)|^{p-1} \quad (7.3.16)$$

where F is a function which is bounded on bounded sets in $G \times \mathbb{R}^{n+1}$. However, since I_{ik} and $y_0(u_{ik})$ converge at all points $x \in G \setminus N_0 \cup N \cup N'$, estimate (7.3.16) shows that the sequence $\{\underline{y}(u_{ik})\}_{k \in \mathbb{N}}$ is bounded at all these points. For an arbitrary, but fixed, point $x \in G \setminus N_0 \cup N \cup N'$, let $\underline{y}(u_{ikl}(x))$ be a subsequence which converges to a point $\underline{y} \in \mathbb{R}^n$.

Since we also have $u_{ik}(x) \to u(x)$, it follows from (7.3.13′) and the fact that the A_j are continuous in $y \in \mathbb{R}^{n+1}$ that

$$\sum_{j=1}^{n} [A_j(x, u(x), \underline{y}) - A_j(x, u(x), \underline{y}(u)(x))][y_j - y_j(u)(x)] = 0 .$$

If we had $\underline{y} \ne \underline{y}(u)(x)$, then there would be a contradiction to (H_2'). We therefore have

$$\lim_{l \to \infty} \underline{y}(u_{ikl}(x)) = \underline{y}(u)(x) .$$

This shows that every convergent subsequence $\{\underline{y}(u_{ikl})(x)\}_{l \in \mathbb{N}}$ of the bounded subsequence $\underline{y}(u_{ik})(x)$ has the same limit, i.e. $\underline{y}(u)(x)$. Then $\{\underline{y}(u_{ik})(x)\}_{k \in \mathbb{N}}$ itself converges to $\underline{y}(u)(x)$. Since we chose $x \in G \setminus N_0 \cup N \cup N'$ to be arbitrary, then the pointwise converges in $G \setminus N_0 \cup N \cup N'$ follows, and thus so does the condition (i) in (β) above for $\{\underline{y}(u_{ik})\}_{k \in \mathbb{N}}$.

(e) In the final step we shall show the uniform absolute continuity of the sequences $\{\underline{y}(u_{ik})\}_{k \in \mathbb{N}}$ in $L^p(G)$. We integrate (7.3.15) over an arbitrary measurable subset $H \subset G$ and then estimate termwise:

$$\int_H d^n x \sum_{j=1}^n |y_j(u_i)(x)|^p \le \frac{1}{\alpha}\int_H d^n x I_i(x) + \frac{1}{\alpha}\int_H d^n x g_0(x)|u_i(x)|^{r_0}$$
$$+\frac{1}{\alpha}\int_H d^n x \sum_{j=1}^n A_j(x, y(u_i))y_j(u)(x)$$
$$+\frac{1}{\alpha}\int_H \sum_{j=1}^n A_j(x, u_i, \underline{y}(u))[y_j(u_i) - y_j(u)]d^n x\,. \tag{7.3.17}$$

If $\varepsilon > 0$ is now given, then an $i_\varepsilon \in \mathbb{N}$ exists such that for all $i \ge i_\varepsilon$ we have

$$\frac{1}{\alpha}\int_G d^n x I_i(x) \le \frac{\varepsilon}{4}\,,$$

since $0 \le I_i$ converges to zero in $L^1(G)$. This estimate then holds for all measurable subsets $H \subset G$. Let us therefore choose a $\delta_0 > 0$ such that for all measurable $H \subset G$, where $|H| \le \delta_0$, we have

$$\frac{1}{\alpha}\int_H d^n x I_i(x) \le \frac{\varepsilon}{4}\,, \quad i = 1, \ldots, i_\varepsilon - 1\,,$$

which implies

$$\frac{1}{\alpha}\int_H d^n x I_i(x) \le \frac{\varepsilon}{4}$$

for all $i \in \mathbb{N}$ and all such $H \subset G$. By virtue of

$$\int_H d^n x g_0(x)|u_i(x)|^{r_0} \le \|g_0\|_{L^\sigma(H)}\|u_i\|_p^{r_0} \le M_1\|g_0\|_{L^\sigma(H)}\,,$$

where

$$\sigma = \frac{p}{p - r_0} \quad \text{and} \quad g_0 \in L^\sigma(G)\,,$$

a $\delta_1 > 0$ exists such that for measurable subsets $H \subset G$, with $|H| \le \delta_1$, it follows that

$$\frac{1}{\alpha}\int_H d^n x g_0(x)|u_i(x)|^{r_0} \le \frac{\varepsilon}{4}\,, \quad \forall i \in \mathbb{N}\,.$$

The estimate of the third term in (7.3.17) follows in a similar fashion:

$$\left|\int_H d^n x \sum_{j=1}^n A_j(x, y(u_i))y_j(u)\right| \le h(\|u_i\|_{1,p})\|u\|_{W^{1,p}(H)}\,,$$

and the boundedness of $\{h(\|u_i\|_{1,p})\}_{i\in\mathbb{N}}$ shows that a $\delta_2 > 0$ exists such that for measurable $H \subset G$, $|H| \le \delta_2$, it follows that

$$\frac{1}{\alpha}\left|\int_H d^n x \sum_{j=1}^n A_j(x, y(u_i))y_j(u)\right| \le \frac{\varepsilon}{4}\,, \quad \forall i \in \mathbb{N}\,.$$

Since an estimate for the last term in (7.3.17) can also be done similarly, i.e.

$$\left| \frac{1}{\alpha} \int_H d^n x \sum_{j=1}^{n} A_j(x, u_i, \underline{y}(u))[y_j(u_i) - y_j(u)] \right|$$
$$\leq \frac{1}{\alpha} \sum_{j=1}^{n} \|A_j(\cdot, u_i, \underline{y}(u))\|_{L^{p'}(H)} \{\|y_j(u_i)\|_p + \|y_j(u)\|_p\}$$
$$\leq M_3 \sum_{j=1}^{n} \|A_j(\cdot, u_i, \underline{y}(u)\|_{L^{p'}(H)} ,$$

the previous explicit estimates of $\|A_j(\cdots)\|_{p'}$ and the convergence of $\{u_i\}_{i \in \mathbb{N}}$ in $L^p(G)$ show that a $\delta_3 > 0$ exists such that this term is also less than or equal to $\varepsilon/4$ for all measurable subsets $H \subset G$, $|H| < \delta_3$, and all $i \in \mathbb{N}$. If we choose $\delta = \min\{\delta_0, \delta_1, \delta_2, \delta_3\}$, then by (7.3.17) it does indeed follow for all $i \in \mathbb{N}$ and all measurable $H \subset G$, $|H| \leq \delta$, that

$$\int_H d^n x \sum_{j=1}^{n} |y_j(u_i)(x)|^p \leq \varepsilon ,$$

which is Condition (ii) of (β) for $\{y(u_i)\}_{i \in \mathbb{N}}$, and thus also for the subsequence $\{\underline{y}(u_{ik})\}_{k \in \mathbb{N}}$.

(f) It follows that the remarks (α) and (β) above imply that $\{y_j(u_i)\}_{i \in \mathbb{N}}$ converges strongly in $L^p(G)$ to $y_j(u)$ and thus that the sequence $\{u_i\}_{i \in \mathbb{N}}$ converges strongly in $W^{1,p}(G)$ to u. The mapping T thus has the property (S) and Theorem 7.2.4 completes the proof. □

Remark 7.3.4. Theorems 7.3.2 and 7.3.3 solve a large class of boundary value problems for the quasi-linear differential equation (7.1.1) in the sense of distributions of G. However, these theorems say nothing about the important problem of the regularity of these weak solutions, namely whether the solutions u in V, $W_0^{1,p}(G) \subseteq V \subseteq W^{1,p}(G)$, are also solutions in the classical sense.

Example 7.3.5. Let $G \subset \mathbb{R}^n$ be an open, nonempty, bounded subset with a smooth boundary, and let V be a closed subspace of $W^{1,2}(G) = H^1(G)$, where $H_0^1(G) \subseteq V$. Then, with the aid of Theorem 7.3.3, it is very easy to solve the variational boundary value problem for $n \geq 2$,

$$-\Delta u = f ,$$

for arbitrary $f \in V'$ with boundary conditions defined via V. We only have to put $p = 2$ in Theorem 7.3.3. All hypotheses (H_i'), $i = 1, 2, 3$, are then trivially satisfied and the solubility follows.

Obviously, solution by means of the linear theory of Chap. 6 is even simpler. This is, however, not the case if a simple nonlinear modification is made to this equation, say for $n = 2$. Let the Carathéodory functions $\rho_{ij}, i, j = 1, 2$, be given on $G \times \mathbb{R}$ with the following properties:

(i) There is an $\alpha > 0$ such that for all $(x, y_0) \in G \times \mathbb{R}$ and all $z \in \mathbb{R}^2$ we have

$$\langle z, \rho(x, y_0)z\rangle = \sum_{i,j=1}^{2} z_i \rho_{ij}(x, y_0) z_j \geq \alpha \sum_{j=1}^{2} z_j^2 .$$

(ii) Functions $f_{ij} \in L^\infty(G)$ exist such that for all $y_0 \in \mathbb{R}$ and almost all $x \in G$ we have

$$|\rho_{ij}(x, y_0)| \leq f_{ij}(x) .$$

Now we want to show, with the aid of Theorem 7.3.3, the solubility of the boundary value problem

$$-\sum_{j=1}^{2} \partial_j A_j(x, u(x), \nabla u(x)) = f(x)$$

with boundary conditions defined by V and for given $f \in V'$ for the case when

$$A_j(x, y_0, y) = \sum_{i=1}^{2} \rho_{ji}(x, y_0) y_i .$$

To this end, we show that the hypotheses (H_i'), $i = 1, 2, 3$ are satisfied. Condition (i) above easily implies the monotonicity hypothesis (H_2') and the coerciveness hypothesis (H_3') for $p = 2 = n$. It follows form (ii) that (H_1') is also satisfied for $p = 2$. Theorem 7.3.3 thus guarantees the solubility of this boundary value problem, which reduces to the linear boundary value problem $-\Delta u = f$ when $\rho_{ij}(x, y_0) = \delta_{ij}$.

8. Nonlinear Elliptic Eigenvalue Problems

8.1 Introduction

As a further application of the direct methods of the calculus of variations let us discuss a special class of nonlinear eigenvalue problems. As far as the technical framework is concerned, we proceed here as in our treatment of nonlinear boundary value problems in Chap. 7, which means, that if we are seeking solutions in a region $G \subset \mathbb{R}^n$, we work in an appropriate Sobolev space $W^{m,p}(G) = E$. The starting point of this method of solution is the following simple application of Theorem 4.2.3 on Lagrange multipliers. If f and h are two C^1 functions on E with the derivatives $Df = f'$ and $Dh = h'$, then we can solve the nonlinear eigenvalue equation

$$f'(u) = \lambda h'(u), \quad u \in E, \quad \lambda \in \mathbb{R}, \tag{8.1.1}$$

in a simple way by determining the critical points of the function h on suitable level surfaces $f^{-1}(c)$ of f or, conversely, by determining the critical points of f on sutiable level surfaces $h^{-1}(c)$ of h. The eigenvalue λ appears thereby as a Lagrange multiplier.

To locate the critical points precisely, it is now necessary to impose certain restrictions for the level surfaces and the functions whose critical points on the level surfaces are to be determined. This introduces a certain asymmetry in the roles of f and h, which is also manifest in the hypotheses for f and h.

For our particular application in this direction, we start with an elliptic differential operator A of second order in divergence form and an operator B of zero order. This means that A and B can be expressed as follows.

For a function $u : G \to \mathbb{R}$ we put

$$y(u) = (y_0(u), y_1(u), \ldots, y_n(u)) = (u, \partial_1 u, \ldots, \partial_n u). \tag{8.1.2}$$

For the functions $F_j : G \times \mathbb{R}^{n+1} \to \mathbb{R}$ we have, therefore,

$$\widehat{F}_j(y(u))(x) = F_j(x, y(u)(x)), \quad \forall x \in G, \tag{8.1.3 a}$$

and similarly, for a function $H_0 : G \times \mathbb{R}$, we have

$$\widehat{H}_0(u)(x) = H_0(x, u(x)), \quad \forall x \in G. \tag{8.1.3 b}$$

Now let

$$A(u) = \widehat{F}_0(y(u)) - \sum_{j=1}^{n} \partial_j F_j(y(u)) \tag{8.1.4 a}$$

and

$$B(u) = \widehat{H}_0(u). \tag{8.1.4 b}$$

Here we assume that the functions F_j are the partial derivatives of a function $F : G \times \mathbb{R}^{n+1} \to \mathbb{R}$, i.e.

$$F_j(x,y) = \frac{\partial F}{\partial y_j}(x,y) \tag{8.1.5 a}$$

and, similarly,

$$H_0(x,y_0) = \frac{\partial H}{\partial y_0}(x,y_0) \tag{8.1.5 b}$$

where H is a function $H : G \times \mathbb{R} \to \mathbb{R}$.

Now it is necessary to state precisely what we understand by a solution of the quasi-linear elliptic eigenvalue equation

$$A(u) = \lambda B(u) \tag{8.1.6}$$

on G. This is given in the following definition.

Definition 8.1.1. Consider an open set $G \subset \mathbb{R}^n$ and two quasi-linear partial differential operators A and B on G, as given by the (8.1.2–8.1.5) in divergence form. Furthermore, let V be a closed subspace of the Sobolev space $E = W^{1,p}(G)$. Then u is said to be a *solution of the eigenvalue problem* (8.1.6) *for the variational boundary conditions defined by* V if there exists a real number $\lambda = \lambda(u)$ such that

$$a(u,v) = \lambda b(u,v), \quad \forall v \in V. \tag{8.1.7}$$

Here $(\cdot,\cdot)$ is the *generalised Dirichlet form for the operator* A *on* V and similarly b is the generalised Dirichlet form for the operator B on V; in other words, for all $u, v \in V$ we have

$$a(u,v) = \sum_{j=0}^{n} \langle \widehat{F}_j(y(u)), y_j(v)\rangle_2 = \sum_{j=0}^{n} \int_G d^n x F_j(x, y(u)(x)) y_j(v)(x), \tag{8.1.8 a}$$

$$b(u,v) = \langle \widehat{H}_0(u), v\rangle_2 = \int_G d^n x H_0(x, u(x)) v(x). \tag{8.1.8 b}$$

The connection with the eigenvalue problem given by (8.1.1) can be established if we can show under suitable hypotheses that

$$a(u,v) = \langle f'(u), v\rangle, \tag{8.1.9 a}$$
$$b(u,v) = \langle h'(u), v\rangle \quad \forall u, v \in V, \tag{8.1.9 b}$$

where we set

$$f(u) = \int_G d^n x F(x, (y)(u)(x)) \tag{8.1.10 a}$$

and

$$h(u) = \int_G d^n x H(x, u(x)) \,. \tag{8.1.10 b}$$

In (8.1.9), $\langle \cdot , \cdot \rangle$ denotes the duality between V' and V where V' is the topological dual of V.

Remark 8.1.1. There is no difficulty in principle in extending the theoretical framework to eigenvalue problems of the order $2m$, $m > 1$, and this extension has been explicitly carried out by Browder [8.1, 2]. We shall, however, restrict ourselves to the simpler case of $m = 1$, because

(i) its treatment is much more transparent and brings out clearly the general strategy;
(ii) in most of the applications, particularly in physics, this is the case needed.

An overview will make the orientation easier.

Section 8.2 deals with the determination of a solution to the eigenvalue problem (8.1.6). In analogy to linear eigenvalue theory, this can be regarded as equivalent to the problem of determining the ground state. The approach takes two steps. First we develop an abstract theory for the boundary value problem. Then we state the hypotheses for the coefficient functions for the differential operators A and B in the form (8.1.4), from which we deduce the hypotheses for the abstract theory. We shall then give explicit examples to illustrate the different hypotheses. In this way the solution for the ground state of (8.1.6) will be obtained by determining the minimum of h on the level surface $f^{-1}(c)$ of f (or the maximum of $h_c = h \restriction f^{-1}(c)$).

The determination of eigenfunctions other than the solution of the ground state is already indicated by linear eigenvalue theory. The classical minimax principle of Poincaré-Fischer-Weyl-Courant et al. gives the most convenient machinery known for the derivation and investigation of eigenvalues of higher order in the linear case, by utilising variational methods. For example, the minimax principle enables one to find the eigenvalues and the beginning of the essential spectrum of a self-adjoint operator A bounded from below in a Hilbert space $\mathcal{H}$ [Ref. 8.3, Theorem XIII.1].

An essentially topological version of the minimax principle which covers in particular nonlinear problems was first proposed by Ljusternik [8.4] and then further developed by him and others. This version of the minimax principle shows the relation between the theory of critical points and the Ljusternik-Schnirelman theory of "category" of subsets of manifolds on which the critical points are sought [8.5, 6].

Very important steps for the further development of Ljusternik-Schnirelman theory, were the extension to smooth Hilbert manifolds by *J. T. Schwartz* and the extensions to smooth Banach manifolds (Finsler manifolds) by *R. S. Palais* [8.7]. *F. E. Browder* [8.1, 2] proposed a further extension of the LS-theory on Finsler manifolds including investigations of concrete eigenvalue problems. We refer the reader to Palais [8.7], Browder [8.1] and Rabinowitz [8.8] for a detailed list of sources on the development of LS-theory.

If we want to realise the level surfaces $f^{-1}(c)$, $c \in \mathbb{R}$, of a C^1 function $f : W^{1,p}(G) \to \mathbb{R}$ as Finsler manifolds and then apply the fairly demanding LS-theory for Finsler manifold, then we must assume $p \geq 2$ as shown in [8.2]. This means that additional conditions are needed for the coefficients F and F_j. We therefore prefer *F. E. Browder's* method of Galerkin approximation [8.9], which

(i) only requires the LS-theory for C^1 manifolds and hence is much simpler;
(ii) allows finite-dimensional approximations for an infinite-dimensional eigenvalue problem, a property which is very useful not only from a practical point of view;
(iii) is also capable of treating the cases of C^1 functions f on $W^{1,p}(G)$ where $1 < p < \infty$.

The Ljusternik-Schnirelman theory for compact manifold in $\mathbb{R}^n$ is discussed in Sect. 8.3.

In Sect. 8.4 we begin with the investigation of special level surfaces $f^{-1}(c)$, for which, finally, a "Galerkin approximation" is developed. This approximation then enables us to implement the results of the LS-theory for compact manifolds in $\mathbb{R}^n$ to suitable level surfaces in a separable Banach space. In this way we obtain an abstract existence theorem which gives the lower bound for the number of solutions of equation (8.1.6). Finally, a class of concrete quasi-linear elliptic eigenvalue problems is treated. A few explicit examples are given to illustrate this quite general result.

8.2 Determination of the Ground State in Nonlinear Elliptic Eigenvalue Problems

8.2.1 Abstract Versions of Some Existence Theorems

We present here two abstract existence theorems for the eigenvalue problem (8.1.1). Roughly speaking, the two theorems differ in the fact that the two functions f and h exchange their roles in the variational approach (maximising or minimising, respectively, under constraints). In the second result, the solution is characterised by a minimality property and can therefore be directly interpreted as the *"ground state solution"*. It is characteristic for the proof of both theorems that one again applies a generalised Galerkin or, more precisely, a Rayleigh-Ritz approximation.

Theorem 8.2.1. *Let E be a real, separable, reflexive Banach space whose dual space is denoted by E'. Let the functions $f, h \in C^1(E, \mathbb{R})$ have the following properties:*

(a) *Hypothesis for the constraint f. For $c \in \mathbb{R}$ let the following be true:*
   (a_1) $M_c(f) = f^{-1}(c) = \{u \in E \mid f(u) = c\}$ *is bounded.*
   (a_2) $\langle f'(u), u\rangle \neq 0$, $\forall u \in M_c(f)$.
   (a_3) $f' : E \to E'$ *is bounded.*
   (a_4) f' *satisfies Condition (S) (Sect. 7.2).*

(b) *Hypothesis for the function h which is to be maximised on $M_c(f)$.*
   (b_1) $h' : E \to E'$ *is compact.*
   (b_2) h *is weakly continuous on bounded sets of E.*

(c) *Hypothesis of "relative boundedness" of h' and f'. For every sequence $\{u_n\}_{n\in\mathbb{N}} \subset M_c(f)$ there exists a $c_0 > 0$ such that*

$$|\langle h'(u_n), u_n\rangle| \geq c_0 |\langle f'(u_n), u_n\rangle| , \quad \forall n \in \mathbb{N}$$

is true.

It follows then that there exists an $u \in M_c(f)$ and a $\lambda \in \mathbb{R}$ satisfying

(i) $h(u) = \max_{v\in M_c(f)} h(v)$,
(ii) $f'(u) = \lambda h'(u)$.

Proof. Step 1. Rayleigh-Ritz Approximation. Since E is separable there exists a total sequence $\{x_n\}_{n\in\mathbb{N}}$ in E which contains a total subsequence in $M_c(f)$. Then we have

$$E_n = \mathrm{lin}\{x_1, \ldots, x_n\},$$

a sequence of finite-dimensional subspaces of E, with

$$E_n \subseteq E_{n+1} \quad \text{and} \quad E = \overline{\bigcup_{n\in\mathbb{N}} E_n}\,. \tag{8.2.1}$$

Now, setting

$$M_{c,n} = M_c(f) \cap E_n\,, \quad n \in \mathbb{N}\,, \tag{8.2.2}$$

we get similarly

$$M_{c,n} \subseteq M_{c,n+1} \quad \text{and} \quad M_c(f) = \overline{\bigcup_{n\in\mathbb{N}} M_{c,n}}\,. \tag{8.2.3}$$

Here the relation

$$\overline{\bigcup_{n\in\mathbb{N}} M_{c,n}} \subseteq M_c(f)$$

is evident, and the assumption

$$M_c(f) \setminus \overline{\bigcup_{n\in\mathbb{N}} M_{c,n}} \neq \emptyset$$

can be easily shown to lead to contradiction.

Let $\varphi_n : E_n \to E$ denote the identical embedding and $\varphi'_n : E' \to E'_n$ the dual projection of φ_n. Then, setting

$$f_n = f \circ \varphi_n = f \restriction E_n\,, \quad n \in \mathbb{N} \tag{8.2.4 a}$$

it follows that

$$f'_n = \varphi'_n \circ f' \circ \varphi_n \tag{8.2.4 b}$$

because $f \in \mathcal{C}^1(E, \mathbb{R})$ means

$$f(x+y) - f(x) = \langle f'(x), y\rangle + o(y)\,, \quad \forall x, y \in E\,.$$

In particular, for $x = \varphi_n(u)$, $y = \varphi_n(h)$, $u, h \in E_n$,

$$\begin{aligned} f_n(u+h) - f_n(u) &= f(\varphi_n(u) + \varphi_n(h)) - f(\varphi_n(u)) \\ &= \langle f'(\varphi_n(u)), \varphi_n(h)\rangle + o(\varphi_n(h)) \\ &= \langle \varphi'_n \circ f' \circ \varphi_n(u), h\rangle + o(\varphi_n(h)) \end{aligned}$$

and hence (8.2.4 b) holds.

As a consequence of (a_2), $M_c(f)$ is a $\mathcal{C}^1$ manifold in E of codimension 1.

According to (8.2.2) and (8.2.4 a), we have $M_{c,n} = M_c(f_n)$, so that by (8.2.4) and (a_2), $M_{c,n}$ is again a $\mathcal{C}^1$ submanifold of codimension 1 in E_n since

$$\langle f'_n(u), u\rangle \neq 0\,, \quad \forall u \in M_{c,n}\,. \tag{8.2.5}$$

Setting also $h_n = h \restriction E_n$, we get

$$h'_n = \varphi'_n \circ h' \circ \varphi_n\,. \tag{8.2.6}$$

Step 2. Solution of the Finite-Dimensional Eigenvalue Problem. According to (a_1), $M_c(f)$ is bounded; therefore, all $M_{c,n}$ in the finite-dimensional Banach space E_n are bounded. Further, as they are all closed, it follows that all $M_{c,n}$ are compact $\mathcal{C}^1$ submanifolds in E_n.

Weierstrass' theorem gives therefore that for all $n \in \mathbb{N}$ there exists

$$u_n \in M_{c,n} \quad \text{such that } h_n(u_n) = \sup_{u\in M_{c,n}} h(u)\,. \tag{8.2.7}$$

According to (8.1.5) the theorem of Lagrange multipliers is applicable. It follows therefore that for all $n \in \mathbb{N}$ there exists $\lambda_n \in \mathbb{R}$ such that

$$h'_n(u_n) = \lambda_n \varphi'_n(u_n) \tag{8.2.8}$$

is true.

Step 3. The Limit. This part of the proof is the real core of the theorem. While in the two foregoing steps we have utilised only the easier part of the hypothesis, we shall see now that the hypotheses of Theorem 8.2.1 are strong enough in their entirety to control the limiting process.

(α) From (8.1.8) we get

$$\langle h_n'(u_n), u_n\rangle = \lambda_n \langle f_n'(u_n), u_n\rangle, \quad \forall n \in \mathbb{N}.$$

The hypothesis (c) gives us that $(\xi_n = 1/\lambda_n)_{n\in\mathbb{N}}$ is bounded. Since the sequence $\{u_n\}_{n\in\mathbb{N}}$ is bounded in norm and E is reflexive, there exist subsequences $\{u_{n_j}\}_{n\in\mathbb{N}}$ and $\{\xi_{n_j}\}_{j\in\mathbb{N}}$ so that

$$u_{n_j} \xrightarrow[j\to\infty]{w} u \text{ in } E \quad \text{and} \quad \xi_{n_j} \xrightarrow[j\to\infty]{} \xi \in \mathbb{R}. \tag{8.2.9}$$

The hypothesis (b_2) gives

$$h(u) = \lim_{j\to\infty} h(u_{n_j}).$$

According to (8.2.6) and (8.2.7), we have for all $j \in \mathbb{N}$

$$h(u_{n_j}) = h_{n_j}(u_{n_j}) = \sup_{v\in M_{c,n_j}} h(v) \le \sup_{v\in M_{c,n_j+1}} h(v) = h(u_{n_j+1}),$$

and therefore

$$h(u) = \sup_{j\in\mathbb{N}} \sup_{v\in M_{c,n_j}} h(v) = \sup_{v\in M_c(f)} h(v). \tag{8.2.10}$$

This shows that the function h on the level surface $M_c(f)$ attains its supremum at a point $u \in E$. This point can be obtained as the weak limit of the eigenfunction $\{u_{n_j}\}_{j\in\mathbb{N}}$ of the finite-dimensional approximation of the eigenvalue problem.

(β) It remains for us to show that this point u really lies on the level surface. As soon as we can prove $u \in M_c(f)$, it follows easily that u is also an eigenfunction of the pair f', h'; because from (8.2.10) we then have

$$h(u) = \max_{v\in M_c(f)} h(v),$$

so that by assumption (a_2) the theorem of Lagrange multipliers, guarantees the existence of a number $\rho \in \mathbb{R}$ such that $\rho f'(u) = h'(u)$.

We show that $u \in M_c(f)$ by proving that the sequence u_{n_j}, $j \in \mathbb{N}$ converges not only weakly but also strongly towards u. Condition (S), which f' satisfies according to (a_1), suffices for this.

(γ) Verification of the hypothesis of Condition (S) for the sequence $\{u_{n_j}\}_{j\in\mathbb{N}}$. Let $m \in \mathbb{N}$ and $v \in E_n$ be arbitrary, but fixed. The hypothesis (b_1) guarantees the strong convergence of the sequence $\{h'(u_{n_j})\}_{j\in\mathbb{N}}$ in E', say towards $w \in E'$. For $n_j \ge m$ it follows from (8.2.4 b), (8.2.6) and (8.2.18) that

$$\begin{aligned}\langle f'(u_{n_j}), u_{n_j} - v\rangle &= \langle \varphi'_{n_j} \circ f' \circ \varphi_{n_j}(u_{n_j}), u_{n_j} - v\rangle \\ &= \langle \xi_{n_j}\varphi'_{n_j} \circ h' \circ \varphi_{n_j}(u_{n_j}), u_{n_j} - v\rangle \\ &= \xi_{n_j}\langle h'(u_{n_j}), u_{n_j} - v\rangle ,\end{aligned}$$

and therefore by (8.1.9) and the strong convergence $h'(u_{n_j}) \underset{j\to\infty}{\longrightarrow} w$,

$$\begin{aligned}\lim_{j\to\infty}\langle f'(u_{n_j}), u_{n_j} - v\rangle &= \lim_{j\to\infty}\{\xi_{n_j}\langle h'(u_{n_j}), u_{n_j} - v\rangle\} \\ &= \xi\langle w, u - v\rangle .\end{aligned}$$

Thus, for all $v \in \bigcup_{n\in\mathbb{N}} E_n$,

$$\lim_{j\to\infty}\langle f'(u_{n_j}), u_{n_j} - v\rangle = \xi\langle w, u - v\rangle \tag{8.2.11}$$

We now show that (8.2.11) is valid for all $v \in E$. We have by (a_3) that $C_1 = \sup_{j\in\mathbb{N}} \|f'(u_{n_j})\|'$ is finite. Now let $v \in E$ and $\varepsilon > 0$ be given. According to (8.2.1) there exists a $v_0 \in \bigcup_{n\in\mathbb{N}} E_n$ such that

$$\|v - v_0\| \le \varepsilon' := \frac{1}{2(C_1 + |\xi|\ \|w\|')}\varepsilon .$$

By (8.2.11) there exists a $j_0 = j_0(v_0, \varepsilon')$ corresponding to v_0 such that for all $j \ge j_0$

$$|\langle f'(u_{n_j}), u_{n_j} - v_0\rangle - \xi\langle w, u - v_0\rangle| \le \frac{\varepsilon}{2} .$$

Hence it follows for all $j \ge j_0$ that

$$\begin{aligned}&|\langle f'(u_{n_j}), u_{n_j} - v\rangle - \xi\langle w, u - v\rangle| \\ &\quad \le |\langle f'(u_{n_j}), u_{n_j} - v_0\rangle - \xi\langle w, u - v_0\rangle| + |\langle f'(u_{n_j}) - \xi w, v_0 - v\rangle| \\ &\quad \le \frac{\varepsilon}{2} + \{\|f'(u_{n_j})\|' + |\xi|\ \|w\|'\}\|v_0 - v\| \le \frac{\varepsilon}{2} + \frac{\varepsilon}{2} = \varepsilon .\end{aligned}$$

Therefore (8.2.11) is valid for all $v \in E$. For $v = u$, in particular, we get

$$\lim_{j\to\infty}\langle f'(u_{n_j}), u_{n_j} - u\rangle = 0$$

and thus by (8.2.9)

$$\lim_{j\to\infty}\langle f'(u_{n_j}) - f'(u), u_{n_j} - u\rangle = 0 .$$

This means that the weakly convergent sequence $\{u_{n_j}\}_{j\in\mathbb{N}} \subset E$ satisfies the hypothesis of Condition (S) for the function f'.

This condition therefore implies the strong convergence of this sequence

$$u = s - \lim_{j\to\infty} u_{n_j} .$$

(δ) Since f is continuous we have

$$f(u) = \lim_{j\to\infty} f(u_{n_j}) = c ,$$

which means that $u \in M_c(f)$. It follows in addition that

$$\langle h'(u), u\rangle = \lim_{j\to\infty} \langle h'(u_{n_j}), u_{n_j}\rangle = \langle w, u\rangle\,,$$
$$\langle f'(u), u\rangle = \lim_{j\to\infty} \langle f'(u_{n_j}), u_{n_j}\rangle\,.$$

Hypothesis (c) with $c_0 = c_0(\{u_{n_j}\}_{j\in\mathbb{N}})$ and (a_2) allow us to conclude that

$$\frac{|\langle h'(u), u\rangle|}{|\langle f'(u), u\rangle|} = \lim_{j\to\infty} \frac{|\langle h'(u_{n_j}), u_{n_j}\rangle|}{|\langle f'(u_{n_j}), u_{n_j}\rangle|} \geq c_0 > 0\,,$$

and therefore that

$$\rho = \frac{\langle h'(u), u\rangle}{\langle f'(u), u\rangle} \neq 0\,.$$

Thus we obtain the assertion (ii) where $\lambda = 1/\rho$. Finally, it follows that $\rho = \xi = \lim_{j\to\infty} \xi_n$ and hence that $\lambda = \lim_{j\to\infty} \lambda_{n_j}$. □

The following theorem proves the existence of a solution of a nonlinear eigenvalue problem of the form $f'(u) = \lambda h'(u)$ by determing the "ground state" of f on the level surface of h in accordance with variational theory. The hypothesis and the methods of proof are therefore similar to those of Theorem 8.2.1.

Theorem 8.2.2. *Let E be a real, separable, reflexive Banach space whose dual space is E'. Let the functions $f, h \in C^1(E, \mathbb{R})$ have the following properties:*

(a) *Hypothesis for "constraint". For $c \in \mathbb{R}$ let the following hold:*
   (a_1) $\langle h'(u), u\rangle > 0$, $\forall u \in M_c(h) = \{u \in E \mid h(u) = c\}$.
   (a_2) $\forall R > 0$, $\exists c(R) > 0 : \langle h'(u), u\rangle \geq c(R)$, $\forall u \in M_c(h) \cap \overline{B}_R(0)$.
   (a_3) *$h' : E \to E'$ is compact.*
   (a_4) *h is weakly continuous on bounded sets.*

(b) *Hypothesis for the function to be minimised:*
   (b_1) *$f(u) \to \infty$ for $\|u\| \to +\infty$.*
   (b_2) *$f' : E \to E'$ is bounded.*
   (b_3) *$f' : E \to E'$ satisfies the condition* (S).

Then it follows that there exists a $u \in M_c(h)$ and a $\lambda \in \mathbb{R}$ such that

(i) $$f(u) = \min_{v\in M_c(h)} f(v)\,,$$

(ii) $$f'(u) = \lambda h'(u)\,.$$

Proof. Step 1. Rayleigh-Ritz Approximation. Let us define the finite-dimensional subspaces E_n as in Step 1 for the proof of Theorem 8.2.1. Then it follows similarly that

$$M_c(h) = \overline{\bigcup_{n\in\mathbb{N}} M_{c,n}}\,, \quad M_{c,n} = M_c(h) \cap E_n \subseteq M_{c,n+1}\,. \tag{8.2.12}$$

$M_c(h)$ is a C^1 submanifold of codimension 1 in E and so is $M_{c,n}$ in E_n for sufficiently large $n \in \mathbb{N}$. Further, (8.2.4) and (8.2.6) are valid.

Step 2. Solution of the Finite-Dimensional Eigenvalue Problem. The formula

$$f(u) - f(v) = \int_0^1 d\tau \langle f'(v + \tau(u-v)), u - v\rangle$$

and the hypothesis (b_2) imply that f is bounded on all bounded sets of E. By (b_1) there exists an $R_1 > 0$ such that

$$\inf\{f(u)\,|\,u \in M_c(h)\} = \inf\{f(u)\,|\,u \in M_c(h) \cap \overline{B_{R_1}(0)}\} \equiv m > -\infty\,. \tag{8.2.13}$$

For all $n \in \mathbb{N}$ it follows that

$$m_{n-1} \geq m_n := \inf\{f(u)\,|\,u \in M_{c,n}\} \geq m\,. \tag{8.2.14}$$

Let $v_1 \in M_{c,1}$. According to (b_1) there exists an $R > 0$ such that $\|v\| \geq R$ implies $f(v) \geq f(v_1)$. It follows then for all $n \in \mathbb{N}$ that

$$m_n = \inf\{f_n(u)\,|\,u \in M_{c,n}\} = \inf\{f_n(u)\,|\,u \in M_{c,n} \cap \overline{B_R(0)}\} \tag{8.2.15}$$

But again, $M_{c,n} \cap B_R(0)$ is a compact subset of E_n. Therefore it follows in the usual way that for all $n \in \mathbb{N}$ there exists a $u_n \in M_{c,n} \cap B_R(0)$ and a $\lambda_n \in \mathbb{R}$ satisfying

$$f_n(u_n) = \min_{v\in M_{c,n}} f_n(v)\,, \tag{8.2.16 a}$$

$$f_n'(u_n) = \lambda_n h_n'(u_n)\,. \tag{8.2.16 b}$$

Step 3. The Limit. By (8.2.4 c) and (8.2.6) we have

$$f(u_n) = f_n(u_n) = \min_{v\in M_{c,n}} f(v) \geq \min_{v\in M_{c,n+1}} f(v) = f(u_{n+1})\,.$$

Since

$$M_c(h) = \overline{\bigcup_{n\in\mathbb{N}} M_{c,n}}$$

holds, it follows that $f(u_n) \searrow_{n\to\infty} m$. For $R > 0$ fixed as in (8.2.15) choose $c(R) > 0$ by (a_2). It follows then that

$$\langle h_n'(u_n), u_n\rangle \geq c(R)\,, \quad \forall n \in \mathbb{N}\,,$$

and therefore by (8.2.16 b) and (b_2) that

$$|\lambda_n| = \left| \frac{\langle f_n'(u_n), u_n \rangle}{\langle h_n'(u_n), u_n \rangle} \right| \leq \frac{R}{c(R)} \sup_{v \in B_R(0)} \|f'(v)\|' < \infty$$

for all $n \in \mathbb{N}$.

In the usual manner we then obtain the subsequences $\{u_{n_j}\}_{j \in \mathbb{N}}$ and $\{\lambda_{n_j}\}_{j \in \mathbb{N}}$ with

$$u_{n_j} \xrightarrow[j \to \infty]{w} u \in E \quad \text{and} \quad \lambda_{n_j} \xrightarrow[j \to \infty]{} \lambda \in \mathbb{R}. \tag{8.2.17}$$

The strong convergence of $h'(u_{n_j})$ follows from (a_3), namely,

$$w = \lim_{j \in \mathbb{N}} h'(u_{n_j}) \in E'.$$

(a_4) implies $h(u_{n_j}) \xrightarrow[j \to \infty]{} h(u)$ and thus $u \in M_c(h)$. As in (γ) in the proof of Theorem 8.2.1, the strong convergence of the sequence $\{u_{n_j}\}_{n \in \mathbb{N}}$ follows from (b_3). We therefore obtain

$$f(u) = \lim_{j \to \infty} f(u_{n_j}) = m = \inf_{v \in M_c(h)} f(v).$$

Thus $u \in M_c(h)$ is a minimising point of f on $M_c(h)$. It follows further that

$$w = \lim_{j \to \infty} h'(u_{n_j}) = h'(u),$$

$$\lim_{j \to \infty} f'(u_{n_j}) = f'(u).$$

For $v \in \bigcup_{n \in \mathbb{N}} E_n$, say $v \in E_n$, we have from (8.2.4), (8.2.6) and (8.2.16 b)

$$\begin{aligned} \langle f'(u), v \rangle &= \lim_{j \to \infty} \langle f'(u_{n_j}), v \rangle = \lim_{j \to \infty} \langle f'(\varphi_{n_j}(u_{n_j})), \varphi_{n_j}(v) \rangle \\ &= \lim_{j \to \infty} \langle f_{n_j}'(u_{n_j}), v \rangle = \lim_{j \to \infty} \langle \lambda_{n_j} h'(u_{n_j}), v \rangle \\ &= \lim_{j \to \infty} \lambda_{n_j} \langle h'(u_{n_j}), v \rangle = \lambda \langle h'(u), v \rangle . \end{aligned}$$

Since $\bigcup_{n \in \mathbb{N}} E_n$ is dense in E_n, it follows that $f'(u) = \lambda h'(u)$, which is the assertion (ii). □

Addendum to Theorem 8.2.1 *and Theorem* 8.2.2. The eigenfunction u and the eigenvalue λ whose existence is assured by Theorems 5.2.1 and 5.2.2, can be obtained as strong limits of subsequences $(u_{n_j})_{j \in \mathbb{N}}$ and $(\lambda_{n_j})_{j \in \mathbb{N}}$ of the sequences of eigenfunctions $(u_n)_{n \in \mathbb{N}}$ and the eigenvalues $(\lambda_{n_j})_{n \in \mathbb{N}}$ of the approximating finite-dimensional eigenvalue problem

$$f_n'(u_n) = \lambda_n h_n'(u_n), \quad u_n \in E_n, \quad \lambda_n \in \mathbb{R}, \quad n \in \mathbb{N},$$

so that

$$u = s - \lim_{j \to \infty} u_{n_j}, \quad \lambda = \lim_{j \to \infty} \lambda_{n_j}.$$

Proof. u and λ are defined in this manner. □

Before applying these existence theorems to concrete nonlinear eigenvalue problems, let us briefly discuss the hypothesis of relative boundedness of f' and h', i.e. (c) in Theorem 8.2.1. This hypothesis has an important role in the proof: it guarantees the boundedness of the sequence of eigenvalues for the approximating finite-dimensional problems, and consequently contributes in a decisive way to the proof of the hypothesis of Condition (S).

Naturally, the hypothesis of relative boundedness of f' and h' can be realised in different ways with the other hypotheses.

In Theorem 14 in [8.2], Theorem 8.2.1 (c) has been replaced by the following hypothesis (c′):

(c′) *There exists a* $u_0 \in M_c(f)$ *and an* $r > 0$ *such that for all* $u \in M_c(f)$ *with* $h(u) \geq h(u_0)$, *the inequality* $\langle h'(u), u\rangle \geq r$ *holds.*

Hypothesis (c′) gives the desired result for the sequence of eigenfunctions $\{u_n\}_{n\in\mathbb{N}}$ of the approximating finite-dimensional eigenvalue problems as can be easily verified, since we may assume, without restriction, that u_0 in (c′) belongs to E_1; consequently, it follows for all $n \in \mathbb{N}$ that

$$h_n(u_n) = h(u_n) \geq h(u_0)$$

and therefore

$$\langle h_n'(u_n), u_n\rangle = \langle h'(u_n), u_n\rangle \geq r > 0 \quad \text{for all } n \in \mathbb{N}\,.$$

Since $M_c(f)$ is bounded and f' is bounded on bounded sets, it follows for the sequence $u_n \in M_{c,n} \subset M_c(f)$ that

$$|\lambda_n| = \left|\frac{\langle f'(u_n), u_n\rangle}{\langle h'(u_n), u_n\rangle}\right| \leq \frac{1}{r} \sup_{n\in\mathbb{N}} \|f'(u_n)\|'\|u_n\| < \infty\,.$$

Similarly, it should be clear that Theorem 8.2.2 remains valid when we replace (a_2) in that theorem by (c) or (c′) for the relative boundedness of f' and h'.

Another version of the hypothesis of relative boundedness of f' and h' will be used in Theorem 8.4.5, (B_1). Theorem 8.2.1 remains valid with this hypothesis instead of (c).

8.2.2 Determining the Ground State Solution for Nonlinear Elliptic Eigenvalue Problems

If the hypotheses of Theorems 8.2.1 and 8.2.2 for the coefficient functions of nonlinear elliptic eigenvalue problems are used, one obtains the existence and a method for calculation of the solutions.

The starting point of this problem is again the quasi-linear differential operators $A(u)$ and $B(u)$, determined by the coefficient functions $F_0, F_1, \dots, F_n$ and H_0, respectively.

Theorem 8.2.3. *Let $G \subset \mathbb{R}^n$ be an open, bounded subset with smooth boundary. Let the $n+1$ C functions $F_0, F_1, \ldots F_n : G \times \mathbb{R}^{n+1} \to \mathbb{R}$ be the partial derivatives $\partial F/\partial y_j$ of the C function $F : G \times \mathbb{R}^{n+1} \to \mathbb{R}$.*

Let the functions F and F_j satisfy Hypotheses (D_1), (D_2), (K) *and* (M) *of Sect.* 8.4 *for a $p \in (1, n)$. Further, let $H : G \times \mathbb{R} \to \mathbb{R}$ be a C function whose derivative*

$$H_0(x,y) = \frac{\partial H(x,y)}{\partial y_0}$$

is again a C function. The functions H and H_0 satisfy Hypotheses (D_3) *and* (D_4) *of Sect.* 8.4 *for the same $p \in (1,n)$. Finally, let V be a closed subspace of the Sobolev space $W^{1,p}(G)$ with $W_0^{1,p}(G) \subseteq V$. Now,*

$$f(u) = \int_G F(x, y(u)(x))d^n x \quad \text{and} \quad h(u) = \int_G H(x, u(x))d^n x\,, \tag{8.2.18}$$

are continuously Fréchet-differentiable functions on V and possess the following derivatives $(u, v \in V)$:

$$\langle f'(u), v\rangle = \int_G d^n x \sum_{j=0}^{n} F_j(x, y(u)(x)) y_j(v)(x) \tag{8.2.19 a}$$

and

$$\langle h'(u), v\rangle = \int_G d^n x H_0(x, u(x)) v(x)\,. \tag{8.2.19 b}$$

The following two existence results are then valid for the eigenvalue problem $A(u) = \lambda B(u)$:

(A) *For $c \in \mathbb{R}$ let*

$$\langle f'(u), u\rangle \neq 0\,, \quad \forall u \in M_c(f) = f^{-1}(c)\,.$$

Further, let a $c_0 > 0$ and a $u_1 \in M_c(f)$ exist such that for all $u \in M_c(f)$ where $h(u) \geq h(u_0)$ one has always

$$\langle h'(u), u\rangle > c_0\,.$$

Then the function h attains its maximum on the level surface $M_c(f)$ at a point $u \in M_c(f)$ which solves the eigenvalue problem $A(u) = \lambda B(u)$, $\lambda \in \mathbb{R}$, under the boundary conditions defined through V which means

$$\langle f'(u), v\rangle = \lambda \langle h'(u), v\rangle\,, \quad \forall v \in V\,.$$

(B) *For a $c \in \mathbb{R}$ let the following be true for the level surface $M_c(h) = h^{-1}(c)$:*

(i) $\langle h'(u), u\rangle > 0$, $\forall u \in M_c(h)$,
(ii) $\forall R > 0$, $\exists c(R) > 0 : \langle h'(u), u\rangle \geq c(R)$, $\forall u \in M_c(h) \cap \overline{B_R(0)}$.

Under these conditions the function f attains its minimum on $M_c(h)$ at a point $u \in M_c(h)$. This point is a solution of the eigenvalue problem $A(u) = \lambda B(u)$, $\lambda \in \mathbb{R}$, with the variational boundary condition defined by V.

Proof. (a) The hypotheses for the coefficient functions F, F_j, H and H_0 enable us to apply Lemmas 8.4.6 and 8.4.6′. From Lemma 8.4.6 we find that $f : V \to \mathbb{R}$ is bounded on bounded sets and is continuously Fréchet-differentiable with derivative (8.2.19 a). $f' : V \to V'$ is continuous, coercive and bounded on bounded sets. Condition (S) is satisfied by f'.

For h, Lemma 8.4.6′ gives $h : V \to \mathbb{R}$ is well defined, bounded on bounded sets and continuously Fréchet-differentiable with derivative (8.2.19 b). $h' : V \to V'$ is completely continuous, bounded on bounded sets and weakly continuous.

It follows from Theorem 8.4.2 that $f(u) \to \infty$ for $\|u\| \to \infty$ that $M_c(f)$ is sphere-like for sufficiently large c.

(β) The coerciveness of f' implies $\langle f'(u), u\rangle > 0$ for all $u \in M_c(f)$, if c is sufficiently large. The rest of the hypothesis in (A) is exactly the hypothesis of relative boundedness of f' and h' in the fom (c'). We have proved all the hypotheses of Theorem 8.2.1 in (α) and therefore the assertion (A) follows.

(γ) The additional hypothesis in (B) together with the properties of f, f', h and h' proved in (α) enable us to apply Theorem 8.2.2 so that the assertion follows. □

Remark 8.2.1. The statements formulated in the addendum to Theorems 8.2.1 and 8.2.2 are obviously also valid in the context Theorem 8.2.3.

8.3 Ljusternik-Schnirelman Theory for Compact Manifolds

8.3.1 The Topological Basis of the Generalised Minimax Principle

Let X be a topological space and $K \subset X$ a closed subset. Further, let $h : X \to \mathbb{R}$ be a continuous real function on X and $\mathcal{F}$ a family of subsets of X on which h is bounded. For such a family we set

$$m(h, \mathcal{F}) = \inf_{A\in\mathcal{F}} \sup_{x\in A} h(x) = \inf_{A\in\mathcal{F}} \sup h(A) \,. \tag{8.3.1}$$

Then $m(h, \mathcal{F})$ is called the *minimax value of h with respect to $\mathcal{F}$*. Our aim is to isolate effective conditions for which the minimax value is a value of the function, and in fact the value of h at a point in K. This is achieved in the following way. Let ψ be a *deformation of X*, which means that

$$(x, t) \to \psi_t(x)$$

is a continuous mapping $X \times \mathbb{R}_+ \to X$ with $\psi_0 = \mathrm{id}_x$, i.e.

$$\psi(x) = x\,, \quad \forall x \in X\,.$$

Let ψ effectively decrease the values of h in $X \setminus K$, i.e. let ψ satisfy Hypothesis (A), following.

For all $-\infty < a < b < +\infty$ with $h^{-1}([a,b]) \cap K = \emptyset$, let there be a $t_0 > 0$ such that the following is true:

$$\psi_{t_0}(h^b) \subseteq h^a := h^{-1}([-\infty, a])\,. \tag{A}$$

Under this assumption we have the following theorem.

Theorem 8.3.1 (Generalised Minimax Principle). *Let $\{X, K, h, F, \psi_t\}$ be as defined above; let $\mathcal{F}$ be ψ_t-invariant, i.e. $A \in \mathcal{F}$, $t > 0 \Rightarrow \psi_t(A) \in \mathcal{F}$; let $h(K) \subset \mathbb{R}$ be closed and $m(h, \mathcal{F})$ finite. Then there exists an $x_0 \in K$, so that*

$$m(h, \mathcal{F}) = h(x_0)\,. \tag{8.3.2}$$

Proof. The very simple proof of this theorem will be given indirectly. Assume $m(h, \mathcal{F}) \in \mathbb{R} \setminus h(K)$; then $a, b \in \mathbb{R}$ exist such that

(i) $a < m(h, \mathcal{F}) < b$,
(ii) $K \cap h^{-1}([a,b]) = \emptyset$.

According to (i) there exists an $A \in \mathcal{F}$, such that $m(h, \mathcal{F}) < \sup h(A) \le b$, and thus $A \subseteq h^b$. According to (ii) there exists a $t > 0$ such that $\psi_t(h^b) \subseteq h^a$ holds. It follows then that $\psi_t(A) \subseteq h^a$, a contradiction, since $\psi_t(A) \in \mathcal{F}$ implies

$$m(h, \mathcal{F}) \le \sup h(\psi_t(A)) \le \sup h(h^a) = a\,.$$

Therefore it follows that $m(h, \mathcal{F}) \in h(K)$, which means exactly (8.3.2). □

Remark 8.3.1. (a) The assumption that $h(K)$ is closed will be realised in applications by showing that

$$K \cap h^{-1}([a,b]) \text{ is compact in } X \text{ for all } -\infty < a < b < \infty\,.$$

(b) The proof of Theorem 8.3.1 shows that Hypothesis (A) on the decrease of the values of the function h in $X \setminus K$ by ψ can be weakened to the hypothesis (A′), below.

For all $-\infty < a < b < \infty$ with $h^{-1}([a,b]) \cap K = \emptyset$ and all $B \in \mathcal{F}$ let there be a $t_0 = t_0(B, a, b) > 0$ such that

$$\psi_{t_0}(B \cap h^b) \subseteq h^a \tag{A′}$$

Example 8.3.2. (a) For $\mathcal{F} = \{X\}$ one has $m(h, \mathcal{F}) = \sup_{x \in X} h(x)$.

(b) For $\mathcal{F} = \{\{x\} \mid x \in X\}$ one has $m(h, \mathcal{F}) = \inf_{x \in X} h(x)$.

(c) Let S^k be an k-dimensional unit sphere and ξ a homotopy class of maps $S^k \to X$. For such a class ξ we set

$$\mathcal{F}(\xi) = \{A \subset X \mid A = g(S^k) \text{ for a } g \in \xi\}.$$

Then $\mathcal{F}(\xi)$ is invariant for all deformations in X.

An especially important application of Theorem 8.3.1 is the determination of critical points. Let X be a $\mathcal{C}^1$ manifold and $h \in \mathcal{C}^1(X, \mathbb{R})$ a $\mathcal{C}^1$ function on X. Then the set

$$K = K(h) = \{x \in X \mid h'(x) = 0\}$$

of critical points of h is closed in X. Now if the pair $(\mathcal{F}, \psi)$ satisfies the hypotheses of Theorem 8.3.1, then the minimax value of h with respect to $\mathcal{F}$ is a critical point, and h possesses at least one critical point. It is clear, then, that in order to determine as many critical points as possible we construct

(i) deformations ψ of X with the property (A) with respect to $K = K(h)$;
(ii) ψ-invariant families $\mathcal{F}$ of subsets of X which give a measure for the number of ciritical points.

The next two sections deal with this construction.

8.3.2 The Deformation Theorem

The deformation theorem shows how to construct deformations with the property (A). The main technique here is that of "gradient" (or gradient-like) flows on X, i.e. the "method of steepest (or sufficiently steep) descent". We shall discuss this technique here.

In order to keep the framework introductory and elementary, we shall prove the deformation theorem only for the simple case of *a compact* $\mathcal{C}^1$ *manifold* X. The more difficult extension of this theorem to the case of a "Finsler manifold" X can be found in [8.7] and [8.1]. Thanks to the very effective reduction technique ("Galerkin approximation") of *Browder* [8.1] our elementary version of the deformation theorem is sufficient for treating the most far-reaching applications in nonlinear eigenvalue theory.

Let X be a compact $\mathcal{C}^1$ manifold and $h : X \to \mathbb{R}$ a $\mathcal{C}^1$ function on X. Then the gradient ∇h of h is the vector field on X defined by $x \to \nabla h(x)$ where

$$\langle \nabla h(x), v_x \rangle_x = D_x h(v_x), \quad \forall v_x \in T_x(X).$$

Here $T_x(X)$ is the tangent space of X at the point x and $\langle \cdot\,, \cdot \rangle$ the scalar product on $T_x(X)$. The Schwarz inequality shows that h decreases fastest at the point $x \in X$ in the direction $-\nabla h(x)$:

$$D_x h(-\nabla h(x)) = -\langle \nabla h(x), \nabla h(x) \rangle_x .$$

Let $t \to \phi_t$ be the one-parameter group of diffeomorphisms of X generated by the vector field $-\nabla h$, which means ϕ_t is the flow of the vector field $-\nabla h$ and as such is characterised by the following properties [Ref. 8.10, paragraph 8; 8.11]

$$\frac{d}{dt}\phi_t(x) = -\nabla h(\phi_t(x)), \quad \phi_0(x) = x, \quad \forall (x,t) \in X \times \mathbb{R}. \tag{8.3.3}$$

Consequently,

$$\frac{d}{dt}h(\phi_t(x)) = D_{\phi_t(x)}h\Big(\frac{d}{dt}\phi_t(x)\Big) = -\|\nabla h(\phi_t(x))\|^2_{\phi_t(x)} \tag{8.3.4}$$

and we have the following alternatives:

(a) $\nabla h(x) = 0$, i.e. x is a critical point of h and therefore $\phi_t(x) = x$, $\forall t \in \mathbb{R}$.
(b) $\nabla h(x) \neq 0$, and then

$$\begin{aligned} &\text{(i)} \qquad h(\phi_t)(x)) \leq h(x), \quad \forall t > 0. \\ &\text{(ii)} \qquad h(\phi_t(x)) \geq h(x), \quad \forall t < 0. \end{aligned} \tag{8.3.5}$$

In particular, (8.3.5 a) contains the following statement:

The *critical points* $K(h)$ of $h \in C^1(X, \mathbb{R})$ are exactly the *fixed points of the flow* ϕ *associated with* $-\nabla h$, $\phi_t : X \to X$ for $t \neq 0$.

Our aim is to show now that the flow ϕ_t can play the role of ψ_t in Theorem 8.3.1 and thus possesses the property (A). The following lemma is a preparation for the proof.

Lemma 8.3.2. *Let $x_0 \in X$ be a regular point of h (i.e. $\nabla h(x_0) \neq 0$), and let $c = h(x_0)$ be the corresponding regular value. Then there exists an $\varepsilon > 0$ and an open neighborhood $U = U(x_0)$ of x_0 in X, so that $\phi_1(U) \subseteq h^{c-\varepsilon}$ holds.*

Proof. According to 8.3.4 we have

$$\frac{d}{dt}(h(\phi_t(x_0)) = -\|\nabla h(\phi_t(x_0))\|^2_{\phi_t(x_0)} \leq 0 \quad \text{and} \quad \frac{d}{dt}h(\phi_t(x_0))|_{t=0} < 0.$$

Thus

$$t \to h(\phi_t(x_0)), \quad t \geq 0,$$

decreases strictly monotonically in a neighborhood of $t = 0$, i.e.

$$h(\phi_1(x_0)) < h(x_0) = c,$$

or $h(\phi_1(x_0)) < c - 2\varepsilon$ for a suitable $\varepsilon > 0$. The continuity or $x \to h(\phi_1(x))$ now guarantees the existence of an open neighborhood $U = U(x_0)$ of x_0 in X such that $h(\phi_1(x)) < c - \varepsilon$, $\forall x \in U$, i.e. $\phi_1(U) \subseteq h^{c-\varepsilon}$. □

Thus, the central point of the so-called deformation theorem, namely, the effective decrease of the values of the function in the neighborhood of a regular point, has already been proved by applying the method of trajectories of steepest descent.

Theorem 8.3.3 (Deformation Theorem). *Let X be a compact C^1 manifold and $h : X \to \mathbb{R}$ a C^1 function. Let ϕ be the flow defined on X by $-\nabla h$. Then there exists for given $c \in \mathbb{R}$ and a given open neighborhood U of $K_c = h^{-1}(x) \cap K(h)$ in X an $\varepsilon > 0$ so that one has*

$$\phi_1(h^{c+\varepsilon} \setminus U) \subseteq h^{c-\varepsilon}. \tag{8.3.6}$$

In particular, for every regular value c of h there is an $\varepsilon > 0$ for which

$$\phi_1(h^{c+\varepsilon}) \subseteq h^{c-\varepsilon} \tag{8.3.6'}$$

holds.

Proof. Every point of $Y = h^{-1}(c) \setminus U$ is a regular point of h, so that Lemma 8.3.2 guarantees the existence of an open neighborhood U_x of x in X and the existence of an $\varepsilon_x > 0$ with the property $\phi_1(U_x) \subseteq h^{c-\varepsilon_x}$. Now Y is closed in X and is therefore compact, and $\bigcup_{x\in Y} U_x$ covers Y. Thus there exist points $x_1, \ldots, x_n \in Y$, such that $Y \subset \bigcup_{j=1} U_{x_j}$. Setting $\varepsilon_0 = \min\{\varepsilon_{x_1}, \ldots, \varepsilon_{x_n}\}$, we get

$$\bigcup_{j=1}^{N} \phi_1(U_{x_j}) \subseteq h^{c-\varepsilon_0}.$$

Now $W = U \cup \bigcup_{j=1}^{N} U_{x_j}$ is an open covering of $h^{-1}(c)$. The continuity of h ensures the existence of an $\varepsilon \in (0, \varepsilon_0)$ with $h^{-1}([c-\varepsilon, c+\varepsilon]) \subseteq W$. Observing that $h^{c+\varepsilon} \subseteq h^{c-\varepsilon} \bigcup h^{-1}([c-\varepsilon, c+\varepsilon])$ we get

$$h^{c+\varepsilon} \setminus U \subseteq h^{c-\varepsilon} \cup W \setminus U \subseteq h^{c-\varepsilon} \cup \bigcup_{j=1}^{N} U_{x_j}$$

and therefore

$$\phi_1(h^{c+\varepsilon} \setminus U) \subseteq \phi_1(h^{c-\varepsilon}) \cup \bigcup_{j=1}^{N} \phi_1(U_{x_j}) \subseteq h^{c-\varepsilon_0} \cup h^{c-\varepsilon},$$

since according to (8.3.5) one has $\phi_1(h^{c-\varepsilon}) \subseteq h^{c-\varepsilon}$. □

Under suitable assumptions the deformation theorem can also be proved for noncompact manifolds, and even for certain infinite-dimensional manifolds, the so-called Finsler manifolds [8.1, 7 and references therein]. There the main problem is that the gradient field $-\nabla h$ does not necessarily possess a global flow on noncompact manifolds X. Fortunately, we only need a realisation of a deformation of X with the property (A), and not necessarily trajectories of steepest descent such as those associated with $-\nabla h$, but only those with "sufficiently steep" descent. This enables us to change the gradient field $-\nabla h$ to a vector field having a global flow ψ on X and satisfying the condition (A) (pseudo-gradient field in [8.7], quasi-gradient field in [8.1]).

For the case when X is not a compact manifold one needs some further restriction for h, including *condition* (C) *of Palais and Smale* [8.12]. This states:

(PS) If the sequence of function values $\{h(x_n)\}_{n\in\mathbb{N}}$ is bounded for a sequence $\{x_n\}_{n\in\mathbb{N}}$ and also

$$\|\nabla h(x_n)\|_{x_n} \underset{n\to\infty}{\longrightarrow} 0\,,$$

then there exists a convergent subsequence $\{x_{n(j)}\}_{j\in\mathbb{N}}$.

As a consequence of the continuity of ∇h, the limiting point x of the subsequence $\{x_{n(j)}\}_{j\in\mathbb{N}}$ is a critical point of h.

With reference to Remark 8.3.1, we therefore state in this context:

$$\text{If } h \text{ satisfies the PS-condition, then } h(K) \text{ is closed.} \tag{8.3.7}$$

This is because the PS-condition implies, as can easily be seen, that for any $a < b$, $K(h) \cap h^{-1}([a,b])$ is compact. The PS-condition can therefore be regarded as a generalised compactness criterion.

8.3.3 The Ljusternik-Schnirelman Category and the Genus of a Set

In this section we shall study different deformation-invariant classes of subsets of a C^1 manifold X in connection with the minimax principle. These different classes of sets will be characterised by a quantity which does not decrease under arbitrary (odd) deformations. This finally enables us to derive lower bounds on the number of critical points by applying Theorem 8.3.1.

Originally, and in much important later work, the concept of *Ljusternik-Schnirelman category*, $\operatorname{cat}(A;X)$, of a set A in a topological space X was introduced. We define $\operatorname{cat}(\phi, X) = 0$ and $\operatorname{cat}(A;X) = 1$ if A is closed and contractible in X to a point. In other words, $\operatorname{cat}(A;X) = 1$ if, and only if, the embedding $i : A \to X$ is homotopic to a constant mapping. For an arbitrary set $A \subset X$ which can be covered by finitely many closed sets which are contractible in X to a point, we define $\operatorname{cat}(A;X)$ as the minimal number of such sets required for the covering. If a subset does not have any such covering, then we put $\operatorname{cat}(A;X) = +\infty$.

The most important properties of this function on the power set $\mathcal{P}(X)$,

$$\operatorname{cat}(\cdot\,;X) : \mathcal{P}(X) \to \mathbb{N} \cup \{+\infty\}\,,$$

are its monotony, subadditivity and its behaviour under deformations:

(a) If $G : X \times [0,1] \to X$ is a homotopy of id_X, then

$$\mathrm{cat}(g_1(A); X) \geq \mathrm{cat}(A; X), \quad \forall A \in \mathcal{P}(X),$$

where $g_1(x) = G(x,1)$.

(b) If g is a homeomorphism from X onto X, then

$$\mathrm{cat}(g(A); X) = \mathrm{cat}(A; X), \quad \forall A \in \mathcal{P}(X).$$

Now defining for $k = 1, 2, \ldots,$

$$\mathcal{F}_k^c(X) := \{A \subset X \mid \mathrm{cat}(A; X) \geq k\},$$

we obtain deformation-invariant families of subsets of X which can be used in Theorem 8.3.1. This is done in the work of Ljusternik, Schnirelman, Palais and Browder, and leads to results such as: an even function $h \in \mathcal{C}^1(X; \mathbb{R})$ on X possesses at least $\mathrm{cat}(X; X)$ critical points.

The determination of the category of a subset A of X is, in general, a rather complicated matter which is based on nontrivial results of cohomology and homotopy theory. We therefore prefer an elementary approach based on the concept of "genus" of a set and at the end of this section comment briefly on the relation between genus and category of a set.

For a real Banach space E, let $\Sigma(E)$ denote the family of all closed subsets $A \subset E \setminus \{0\}$ which have reflection symmetry ($x \in A \Rightarrow -x \in A$). If A and B are reflection-symmetric sets in the same (different) Banach space (s), then $\mathcal{C}_u(A, B)$ denotes the collection of all odd continuous mappings of A into B. Obviously, one can have also $\mathcal{C}_u(A, B) = \emptyset$.

Definition 8.3.4. Let E denote a real Banach space and $\Sigma(E)$ the family of all closed subsets of $E \setminus \{0\}$ which are symmetric under reflection. Define a function γ on $\Sigma(E)$ as follows:

(i) $\gamma(\emptyset) = 0$,
(ii) $\gamma(A) = +\infty$, if $\mathcal{C}_u(A, \mathbb{R}^n \setminus \{0\}) = \emptyset$, $\forall n \in \mathbb{N}$,
(iii) $\gamma(A) = \inf\{n \in \mathbb{N} \mid \mathcal{C}_u(A, \mathbb{R}^n \setminus \{0\}) \neq \emptyset\}$.

Then $\gamma(A)$ is called the *genus* of the set $A \in \Sigma(E)$.

Thus the genus $\gamma(A)$ of a set $A \in \Sigma(E)$ is the smallest integer $n \geq 1$ such that there exists a continuous odd mapping of A in $\mathbb{R}^n$ which does not have any zeros. The concept of the genus of a set was introduced by *Krasnoselskij* [8.13].

Example 8.3.3. (a) Let $A \in \Sigma(E)$ be finite: a simple explicit construction shows that $\mathcal{C}_u(A, \mathbb{R} \setminus \{0\}) \neq \phi$; thus, $\gamma(A) = 1$. Let $A = \{\pm x_1, \ldots, \pm x_n\}$. We set $f(\pm x_i) = \pm 1$; $f : A \to \mathbb{R} \setminus \{0\}$ is continuous and odd.

(b) For $x \in E \setminus \{0\}$, let $B_r(x) = \{y \in E \mid \|y - x\| < r\}$; if $0 < r < \|x\|$, then $B_r(x) \subset E \setminus \{0\}$ holds. Setting $C_x = B_r(x) \bigcup B_r(-x)$, it follows that $\overline{C}_x \in$

$\Sigma(E)$. Again, a simple explicit construction shows that $C_u(\overline{C}_x, \mathbb{R}^1 \setminus \{0\}) \neq \emptyset$ and hence $\gamma(\overline{C}_x) = 1$.

(c) Let $A \in \Sigma(E)$ where $A \cap (-A) = \emptyset$ is assumed. Then the genus of $A \bigcup (-A)$ is exactly 1. For the proof it suffices to define a nonzero constant function on A and to set

$$\phi(x) = \begin{cases} \alpha\,, & \text{for all } x \in A\,, \\ -\alpha\,, & \text{for all } x \in -A,\ \alpha \neq 0\,. \end{cases}$$

Then ϕ is an odd function from $A \bigcup (-A)$ into $\mathbb{R} \setminus \{0\}$ which is also continuous due to $A \cap (-A) = \emptyset$. Thus $\gamma(A \bigcup (-A)) = 1$.

The following lemma gives the most important properties of the genus of sets. One can find further information in [8.14].

Lemma 8.3.5. *The following properties hold for $A, B \in \Sigma(E)$:*

(i) $C_u(A, B) \neq \emptyset \Rightarrow \gamma(A) \leq \gamma(B)$,

(ii) $A \subset B \Rightarrow \gamma(A) \leq \gamma(B)$,

(iii) $C_u(A, B) \neq \emptyset$, $C_u(B, A) \neq \emptyset \Rightarrow \gamma(A) = \gamma(B)$.

In particular, one also has $\gamma(A) = \gamma(B)$ if there exists an odd homeomorphism of A onto B.

(iv) $\gamma(A \cup B) \leq \gamma(A) + \gamma(B)$.

(v) $\gamma(B) < \infty$, $\gamma(\overline{A \setminus B}) \geq \gamma(A) - \gamma(B)$.

(vi) *For compact $A \in \Sigma(E)$, $\gamma(A) < \infty$ and $\exists \delta > 0$ such that $\gamma(A_\delta) = \gamma(A)$ ($A_\delta = \{y \in E \mid \operatorname{dist}(y, A) \leq \delta\}$).*

(vii) *Let $V \subset E$ be a finite-dimensional subspace and p the linear continuous projection onto V. Then from $A \in \Sigma(E)$ and $\gamma(A) > k = \dim V$, it follows that $A \cap (\mathbb{1} - p)E \equiv A \cap V^\perp \neq \emptyset$.*

Proof. (i)–(iii) are trivial and (v) follows immediately from (iii) and (iv).

For (iv), when $\gamma(A) = +\infty$ or $\gamma(B) = +\infty$ the statement is trivial. Therefore, consider $\gamma(A) = n < +\infty$, $\gamma(B) = m < +\infty$. Consequently, there exists

$$\varphi_A \in C_u(A, \mathbb{R}^n \setminus \{0\}) \quad \text{and} \quad \varphi_B \in C_u(B, \mathbb{R}^m \setminus \{0\})\,.$$

Now set

$$\psi_A^0(x) = \frac{\varphi_A(x)}{1 + |\varphi_A(x)|}\,.$$

It follows that ψ_A^0 is a bounded function in $C_u(A, \mathbb{R}^n \setminus \{0\})$. The extension theorem of Tietze-Urysohn [8.15] gives an extension ψ_A of ψ_A^0 to a continuous function on all of E. Denoting by $\widehat{\psi}_A$ the odd part,

$$\widehat{\psi}_A(x) = \frac{1}{2}\{\psi_A(x) - \psi_A(-x)\}\,,$$

we have

$$\widehat{\psi}_A \in \mathcal{C}_u(E, \mathbb{R}^n) \quad \text{and} \quad \psi_A(x) = \psi_A^0(x) \quad \text{for } x \in A\,.$$

Similarly, let $\widehat{\psi}_B \in \mathcal{C}_u(E, \mathbb{R}^m)$ be constructed. Setting

$$\varphi := (\widehat{\psi}_A \restriction A \bigcup B, \widehat{\psi}_B \restriction A \bigcup B)\,,$$

we have

$$\varphi \in \mathcal{C}_u(A \bigcup B, \mathbb{R}^n \times \mathbb{R}^m)\,,$$

and φ does not have any zeros. For, if $x \in A \bigcup B$, say $x \in A$; then

$$\varphi(x) = (\psi_A^0(x), \widehat{\psi}_B(x))$$

and $\psi_A^0(x) \in \mathbb{R}^n \setminus \{0\}$ gives the assertion. Therefore, $\mathcal{C}_u(A \bigcup B, \mathbb{R}^{n+m} \setminus \{0\}) \neq \emptyset$, and hence $\gamma(A \bigcup B) \leq n + m$.

For (vi), for $x \in A$ let us choose $0 < r_x < \|x\|$. The sets $C_x = B_{r_x}(x) \bigcup B_{r_x}(-x)$, $x \in A$, are an open covering of $A : A \subset \bigcup_{x \in A} C_x$. Thus $C_{x_1}, \ldots, C_{x_n}$ cover the compact set A. It follows that $A \subset \bigcup_{j=1}^{n} \overline{C_{x_j}}$. From (ii) and Example 8.3.3 (b) we obtain

$$\gamma(A) \leq \sum_{j=1}^{n} \gamma(\overline{C}_{x_j}) = n < \infty\,.$$

Let, for example $\gamma(A) = m \leq n$. Then there exists a $\varphi_A \in \mathcal{C}_u(A, \mathbb{R}^m \setminus \{0\})$ which can be extended, in analogy to (iv), to a continuous odd mapping $\widehat{\varphi} \in \mathcal{C}_u(E, \mathbb{R}^m)$. $\widehat{\varphi} \restriction A = \varphi_A$ implies as a consequence of the continuity of $\widehat{\varphi}$ and the compactness of A the existence of a $\delta > 0$, such that $\widehat{\varphi} \restriction A_\delta \neq 0$, since $\varphi_A \restriction A \neq 0$. Thus $\mathcal{C}_u(A_\delta, \mathbb{R}^m \setminus \{0\}) \neq \emptyset$ and therefore $\gamma(A_\delta) \leq m$. With the help of (ii) one then obtains

$$\gamma(A_\delta) = m = \gamma(A)\,.$$

For (vii), as shown for example in [8.16, Chap. V] there exists a topological complement $V^\perp$ of V and is itself the image of E under a linear continuous mapping $p : V = p(E)$. If one had $A \cap (\mathbb{1} - p)E = \emptyset$, i.e. $A \subset V$, then one would have $p \in \mathcal{C}_u(A, V \setminus \{0\}) \neq \emptyset$, and therefore $\gamma(A) \leq k$ by referring to (iii). □

For our applications it is necessary to know the genus of special sets. We consider this next.

Theorem 8.3.6 (Genus of Spheres in Real Banach Spaces). *Let E be a real Banach space and $S_1(E) = \{x \in E \,|\, \|x\| = 1\}$ the unit sphere in E. Then one has*

$$\dim E = \gamma(S_1(E))\,. \tag{8.3.8}$$

Proof. (a) First we prove the theorem for finite-dimensional real Banach spaces E_n. From Lemma 8.3.5 (ii), we can assume $E_n = \mathbb{R}^n$. For $m \geq n$, $\mathcal{C}_u(S_1(\mathbb{R}^n), \mathbb{R}^m \setminus \{0\})$ is not empty, because the mapping

$$(x_1, \ldots, x_n) \mapsto (x_1, \ldots, x_n, 0, \ldots, 0) \in \mathbb{R}^m, \quad m \geq n,$$

is odd and continuous and maps $S_1(\mathbb{R}^n)$ in $\mathbb{R}^m \setminus \{0\}$. A version of the Borsuk-Ulam theorem [8.17] shows that $\mathcal{C}_u(S_1(\mathbb{R}^n), \mathbb{R}^m \setminus \{0\})$ is empty for $m < n$. This theorem states that if Ω is a bounded, open (reflection) symmetric neighborhood of 0 in $\mathbb{R}^n$ and $\phi \in \mathcal{C}_u(\partial\Omega, \mathbb{R}^m)$ with $m < n$, then there exists an $x \in \partial\Omega$ such that $\phi(x) = 0$.

(b) Now let $\dim E = +\infty$. Then E contains subspaces E_n of dimension $n = 1, 2, \ldots$. Since $S_1(E) \supseteq S_1(E) \cap E_n = S_1(E_n)$ we obtain by Lemma 8.3.5 (ii) and (a)

$$\gamma(S_1(E)) \geq \gamma(S_1(E_n)) = n, \quad \forall n \in \mathbb{N},$$

which means that

$$\gamma(S_1(E)) = \infty = \dim E. \qquad \square$$

With the help of the genus of sets, the following families of subsets of a real Banach space can be distinguished:

$$\mathcal{F}_k = \mathcal{F}_k(E) = \{A \in \Sigma(E) \,|\, \gamma(A) \geq k\}, \quad k = 1, 2, \ldots.$$

One sees easily that

$$\mathcal{F}_{k+1} \subseteq \mathcal{F}_k, \quad \forall k.$$

Lemma 8.3.5 (i) states that each of these families $\mathcal{F}_k$ is invariant under odd deformations. Moreover, Theorem 8.3.6 shows that when $\dim E = +\infty$, all $\mathcal{F}_k$, $k \in \mathbb{N}$, are not empty.

Now let X be a sphere-like $\mathcal{C}^1$ submanifold of the real Banach space E, i.e. there exists an odd $\mathcal{C}^1$ diffeomorphism $d : X \to S_1(E)$ of X on the unit space of E. Similarly,

$$\mathcal{F}_k(X) = \{A \subset X \,|\, A \in \Sigma(E), \gamma(A) \geq k\}, \quad k = 1, 2 \ldots.$$

By Lemma 8.3.5 (ii) it follows that

$$\gamma(X) = \gamma(S_1(E)) = \dim E.$$

Therefore,

$$\mathcal{F}_k(X) \neq \emptyset, \quad k = 1, 2, \ldots, \dim E, \tag{8.3.9}$$

and, when $\dim E < \infty$,

$$\mathcal{F}_k(X) = \emptyset \tag{8.3.9'}$$

for $k > \dim E$.

8.3.4 Minimax Characterisation of Critical Values of Ljusternik-Schnirelman

In Sects. 8.3.2 and 8.3.3, we have established the central elements of the minimax characterisation of critical values according to Theorem 8.3.1. Now it will be relatively easy to obtain lower bounds for the number of critical points of an even $\mathcal{C}^1$ function h on a compact $\mathcal{C}^1$ manifold X. This is achieved by applying Theorems 8.3.1 and 8.3.3, and using the property of the genus of sets according to Lemma 8.3.5. To do this effectively, we must restrict ourselves to sphere-like manifolds. Now a (compact) $\mathcal{C}^1$ manifold X in the Banach space E is said to be *sphere-like* if it is $\mathcal{C}^1$ diffeomorphic to the unit space $S_1(E)$ in E. A more precise characterisation will be given in the next section.

Theorem 8.3.7. *Let X_n be a sphere-like $\mathcal{C}^1$ submanifold of an n-dimensional real Banach space E_n and $h \in \mathcal{C}^1(X_n, \mathbb{R})$ an even function on X_n, i.e. $h(-x) = h(x)$, $\forall x \in X_n$. In particular, X_n will be assumed to be reflection-invariant. Now, for $k = 1, 2, \ldots, n = \dim E_n$, set*

$$m_k(h) = m(h, \mathcal{F}_k(X_n)) = \inf_{A \in \mathcal{F}_k(X_n)} \sup h(A) . \tag{8.3.10}$$

Then the following statements are true:

(a) $$\inf_{x \in X_n} h(x) = m_1(h) \leq m_2(h) \leq \ldots \leq m_n(h) = \sup_{x \in X_n} h(x) . \tag{8.3.11}$$

(b) *h possesses at least n different pairs $\{x_k, -x_k\}$ of critical points x_k on X_n, $k = 1, 2, \ldots, n$, and the corresponding critical value $h(x_k)$ are precisely the minimax values of h with respect to $\mathcal{F}_k(X_n)$. This means that for $k = 1, , \ldots, n$ there are points $x_k \in K(h)$ such that*

$$h(x_k) = m_k(h) .$$

(c) *If for certain $l, k \in \{1, \ldots, n-1\}$ with $k + l - 1 \leq n$ the following equalities hold:*

$$m_n(h) = m_{k+1}(h) = \ldots = m_{k+l-1}(h) \equiv c ,$$

then the set of critical points $K_c = K(h) \cap h^{-1}(c)$ on the c-level of h has at least the genus l:

$$\gamma(K_c) \geq l .$$

K_c is thus infinite for $l \geq 2$.

Proof. (a) The inequalities between the $m_k(h)$, $m_{k+1}(h)$ are obvious because of the relation $\mathcal{F}_{k+1}(X_n) \subseteq \mathcal{F}_k(X_n)$. The smallest sets in $\mathcal{F}_1$ are the pairs $\{x, -x\}$, $x \in X_n$. Since h is even, it follows that

$$m_1(h) = \inf_{A \in \mathcal{F}_1} \sup_{x \in A} h(x) = \inf_{x \in X_n} h(x) .$$

The equation $m_n(h) = \sup_{x \in X_n} h(X)$ follows from the fact that X_n is the only set in $\mathcal{F}_n(X_n)$. Once again, by (8.3.9) X_n belongs to $\mathcal{F}_n(X_n)$. Now if $A \subset X_n$, $A \in \Sigma(E_n)$, then we can assume without any restriction that

$$A = g^{-1}(B)\,, \quad B \subset S_1(\mathbb{R}^n)\,, \quad (0\,, \ldots, 0, \pm 1) \notin B$$

for an odd $\mathcal{C}^1$ diffeomorphism $g : X_n \to S_1(\mathbb{R}^n)$.

The projection $p : \mathbb{R}^n \to \mathbb{R}^{n-1}$, $p(x_1\,, \ldots, x_n) = (x_1\,, \ldots, x_{n-1})$, after restriction to B, belongs to $\mathcal{C}_u(B, \mathbb{R}^{n-1} \setminus \{0\})$. Therefore $p \circ g \in \mathcal{C}_u(A, \mathbb{R}^{n-1} \setminus \{0\})$, and hence $\gamma(A) \leq n-1$.

(b) The set $K = K(h)$ of critical points of h is closed in X_n and thus compact. The set $h(K)$ of the critical values of h is then compact in $\mathbb{R}$ and therefore also closed. Since h is even, ∇h is odd. Thus K is (reflection) symmetric, and the flow ϕ associated with the vector field $-\nabla h$ is odd ($\phi_t(-x) = -\phi_t(x)$, $\forall t \in \mathbb{R}, \forall x \in X_n$). The deformation theorem (Theorem 8.3.3) states therefore that the flow ϕ satisfies all the hypotheses of Theorem 8.3.1, since all the families $\mathcal{F}_k(X_n)$ are ϕ-invariant and all the $m_k(h)$ are finite according to (a). Therefore, it follows from Theorem 8.3.1 that all minimax values of h with respect to $\mathcal{F}_k(X_n)$ are critical values. This implies that $x_k \in K(h)$ exists such that

$$h(x_k) = h(-x_k) = m_k(h)\,, \quad k = 1, \ldots, n\,,$$

where we have used (8.3.9).

(c) Now $K_c = K(h) \cap h^{-1}(c)$ is compact, as it is a closed subset of X_n, and therefore by Lemma 8.3.5 (vi), a $\delta > 0$ exists so that $\gamma(K_{c,\delta}) = \gamma(K_c) < \infty$. The interior $K^0_{c,\delta}$ of $K_{c,\delta} = \{x \in X_n \mid \operatorname{dist}(x, K_c) < \delta\}$ is an open neighborhood of K_c in X_n. By Theorem 8.3.3 there exists therefore an $\varepsilon > 0$ such that

$$\phi_1(h^{c+\varepsilon} \setminus K^0_{c,\delta}) \subseteq h^{c-\varepsilon}\,.$$

The characterisation of the minimax values which can be easily verified,

$$m_k(h) = \inf\{a \in \mathbb{R} \mid h^a \in \mathcal{F}_k(X_n)\}\,, \tag{8.3.12}$$

gives

$$c = m_{k+l-1}(h) \Rightarrow \gamma(h^{c+\varepsilon} \geq k + l - 1\,,$$

$$c = m_k(h) \Rightarrow \gamma(h^{c-\varepsilon}) \leq k - 1\,.$$

Now observing that $\phi_1(-x) = -\phi_1(x)$ and applying Lemma 8.3.5, we get

$$\begin{aligned} \gamma(h^{c-\varepsilon} &\geq \gamma(\phi_1(h^{c+\varepsilon} \setminus K^0_{c,\delta})) \geq \gamma(h^{c+\varepsilon} \setminus K^0_{c,\delta}) \\ &= \gamma(\overline{h^{c+\varepsilon} \setminus K_{c,\delta}}) \geq \gamma(h^{c+\varepsilon}) - \gamma(K_{c,\delta})\,. \end{aligned}$$

Therefore we have

$$\gamma(K_c) = \gamma(K_{c,\delta}) \geq \gamma(h^{c+\varepsilon}) - \gamma(h^{c-\varepsilon}) \geq k + l - 1 - (k-1) = l\,,$$

which proves Theorem 8.3.7 completely. For $l \geq 2$, K_c contains infinitely many critical points, because a finite set has the genus 1. □

We close this section with a remark on the connection between genus and category of a set. This connection has been investigated in [8.18]. One finds the following.

If E denotes a real Banach space and $A \in \Sigma(A)$ is compact, then

$$\gamma(A) = \mathrm{cat}(\pi(A), P(E)) \tag{8.3.13}$$

where $P(E)$ is the projective space over E with the quotient topology with respect to the canonical projection $\pi : E \to P(E)$, $\pi(x) = \{x, -x\}$, $\forall x \in E \setminus \{0\}$.

8.4 The Existence of Infinitely Many Solutions of Nonlinear Elliptic Eigenvalue Problems

8.4.1 Sphere-Like Constraints

A $\mathcal{C}^1$ manifold X in a real Banach space is said to be *sphere-like* if there exists a $\mathcal{C}^1$ diffeomorphism g of X onto the unit sphere $S_1(E)$ of E. Naturally this presumes that $S_1(E)$ is itself a $\mathcal{C}^1$ manifold. For this to be the case, we also need the differentiability properties of the norm $q(\cdot) = \|\cdot\|$ of the Banach space E. We therefore assume

$$q \in \mathcal{C}^1(E \setminus \{0\}) .$$

Then it follows (Chap. 2) that the Fréchet derivative q' of q is a continuous mapping of $E \setminus \{0\}$ into the dual space E' of E, where

$$\|q'(x)\|_{E'} = 1 , \quad q'(\lambda x) = \lambda q'(x) , \quad \lambda > 0 ,$$

and

$$\langle q'(x), x \rangle = \|x\| , \quad \forall x \in E \setminus \{0\} .$$

The tangent space $T_y(S_1(E))$ of $S_1(E)$ at y can be identified with the kernel of $q'(y) : E \to \mathbb{R}$,

$$T_y(S_1(E)) = \mathrm{Ker}\, q'(y) , \quad \forall y \in S_1(E) = \{x \in E \,|\, q(x) = 1\} .$$

Now we are in a position to answer the question "when is the level surface $M_c = M_c(f) = \{x \in E \,|\, f(x) = c\}$ of a real function $f : E \to \mathbb{R}$ a sphere-like manifold?" Furthermore, looked at from this point of view the hypothesis of the following theorem appears natural.

Theorem 8.4.1 (Sphere-Like Constraints). *Let E be a real Banach space with a norm $q(\cdot) = \|\cdot\| \in C^1(E \setminus \{0\})$, $f : E \to \mathbb{R}$ a C^1 function on E. Then the level surface*

$$M_c = M_c(f) = \{x \in E \mid f(x) = c\}$$

of f is sphere-like, if

(i) $\langle f'(x), x\rangle \neq 0$, $\forall x \in M_c$ *and*
(ii) *every ray from the origin intersects M_c at exactly one point.*

The C^1 diffeomorphism r_c which maps M_c on $S_1(E)$ is given by the restriction of the mapping

$$r : E \setminus \{0\} \to S_1(E), \quad r(x) = \frac{1}{\|x\|}x,$$

to M_c and satisfies

$$\|D_x r_c\| \leq \frac{2}{\|x\|}, \qquad \forall x \in M_c.$$

Proof. (a) First let us investigate the mapping r. A simple calculation shows that r is Fréchet-differentiable on $E \setminus \{0\}$, and the derivative of r at the point $x \in E \setminus \{0\}$ is given by the following continuous linear mapping $E \to E$:

$$r'_x : E \to E, \qquad r'_x(h) = \frac{1}{\|x\|}h - \frac{\langle q'(x), h\rangle}{\|x\|^2}x, \tag{8.4.1}$$

from which one derives easily

$$\|r'_x\| \leq \frac{2}{\|x\|}, \qquad \forall x \in E \setminus \{0\}.$$

We obtain from $\langle q'(x), x\rangle = \|x\|$

$$r'_x : E \to \operatorname{Ker} q'(x) \quad (= T_x(S_1(E)) \text{ if } \|x\| = 1) \tag{8.4.1'}$$

for all $x \in E \setminus \{0\}$.

(b) According to Corollary 4.2.2 and by (i), $M_c(f)$ is a C^1 manifold in E, which does not contain the origin of E and whose tangent space at a point $x \in M_c(f)$ can be identified with $\operatorname{Ker} f'(x)$:

$$T_x(M_c(f)) = \operatorname{Ker} f'(x), \quad \forall x \in M_c(f).$$

Again by (ii), $r_c := r \restriction M_c(f)$ is a bijective mapping $M_c(f) \to S_1(E)$.

We shall now show that the derivative of this mapping is an isomorphism of the corresponding tangent space:

$$D_x r_c : T_x(M_c(f)) \to T_{r_c(x)}(S_1(E)).$$

To this end, it is sufficient to show that r'_x as given by (8.4.1) is an isomorphism $\operatorname{Ker} f(x) \to \operatorname{Ker} q'(r_c(x))$.

From (8.4.1′) we know that r'_x maps into $\operatorname{Ker} q'(x)$. Since, however, $q'(x) = q'(\lambda x)$, $\forall \lambda > 0$ and $\forall x \in E \setminus \{0\}$, it follows that r'_x maps linearly into $\operatorname{Ker} q'(r_c(x))$. This mapping is also continuous. For

$$v \in \operatorname{Ker} q'(r_c(x)) = \operatorname{Ker} q'(x)\,, \quad \forall x \in M_c\,,$$

set

$$u = q(x)v - q(x)\frac{\langle f'(x), v\rangle}{\langle f'(x), x\rangle}x\,.$$

The mapping $v \to u$ is again linear and continuous. Furthermore, it follows that $u \in \operatorname{Ker} f'(x)$, since

$$\langle f'(x), u\rangle = q(x)\langle f'(x), v\rangle - q(x)\frac{\langle f'(x), v\rangle}{\langle f'(x), x\rangle}\langle f'(x), x\rangle = 0\,,$$

and $r'_x(u) = v$ as can be verified by a simple calculation.

Thus $r'_x : \operatorname{Ker} f'(x) \to \operatorname{Ker} q'(r_c(x))$ is surjective; one can also easily see that r'_x is injective, since

$$r'_x(h) = 0 \Leftrightarrow h = \frac{\langle q'(x), h\rangle}{\|x\|}x\,.$$

Since x does not belong to $\operatorname{Ker} q'(x)$, this equality holds only for $h = 0$. Consequently,

$$D_x r_c : T_x(M_c(f)) \to T_{r_c(x)}(S_1(E))$$

is an isomorphism and therefore $r_c : M_c(f) \to S_1(E)$ a C^1 diffeomorphism [8.10.11]. Thus Theorem 8.4.1 is proved. □

For a function $f \in C^1(E, \mathbb{R})$ we say that $f(x) = c$, $x \in E$, defines a *sphere-like constraint* if the level surface $M_c(f) = f^{-1}(c)$ is sphere-like. Obviously, one would like to know more concrete conditions on f than those given in the hypotheses of Theorem 8.4.1 which would guarantee that the level surfaces $M_c(f)$ of f are sphere-like for certain values of c. The following theorem provides such conditions in an abstract form. These conditions have the advantage that for concrete eigenvalue problems for elliptic differential operators, which we will discuss later on, they can be formulated as very simple concrete requirements on the coefficient functions of these elliptic differential operators.

Theorem 8.4.2. *Let f be a C^1 function on the real Banach space E, whose norm q belongs to $C^1(E \setminus \{0\})$, with the following properties:*

(i) *f is bounded on bounded sets of E.*
(ii) *$f' : E \to E'$ is bounded, which means that f' maps the bounded subsets of E into the bounded subsets of E'.*
(iii) *$f' : E \to E'$ is coercive, which means*

$$\langle f'(x), x\rangle \underset{\|x\|\to\infty}{\longrightarrow} +\infty\,.$$

Then there exists a $c_0 > 0$ such that the level surfaces $M_c(f)$ for all $c \geq c_0$ are sphere-like.

Proof. The coercivity of f' implies that there exists $R > 0$ and $c_R > 0$ such that

$$\langle f'(x), x\rangle \geq c_R\,, \quad \forall x \in E\,, \quad \|x\| \geq R$$

holds. For $r \geq 1$ and $v \in E$, $\|v\| = R$ we then have

$$f(rv) - f(v) = \int_1^r ds \frac{d}{ds} f(sv) = \int_1^r ds \langle f'(sv), v\rangle = \int_1^r \frac{ds}{s} \langle f'(sv), sv\rangle$$
$$\geq \int_1^r \frac{ds}{s} c_R = c_R \log r\,.$$

Now according to (i), $m = \sup_{\|v\|=R} |f(v)|$ is finite; therefore $f(rv) \geq -m + c_R \log r$ and hence

$$f(x) \underset{\|x\|\to\infty}{\longrightarrow} +\infty \tag{$*$}$$

In particular, $M_c(f)$ is bounded in E for arbitrary $c \in \mathbb{R}$. The property $(*)$ also shows that there exists $R_0 < R_1$, $R < R_0$ and $c_0 < c_1$, such that for all $x \in E$, $R_0 \leq \|x\| \leq R_1$, one has

$$c_0 \leq f(x) \leq c_1\,.$$

For arbitrary, but fixed, $y \in S_1(E)$ consider the ray $x_t = ty$, $t \geq 0$. For $c \in [c_0, c_1]$ there exists a minimal $t_1 \geq R_0$ such that this ray intersects $M_c(f)$. As previously, it follows for $t > t_1 \geq R$ that

$$f(ty) - f(t_1 y) \geq c_R \log \frac{t}{t_1} > 0\,.$$

Thus this ray intersects $M_c(f)$ at exactly one point. □

8.4.2 Galerkin Approximation for Nonlinear Eigenvalue Problems in Separable Banach Spaces

In this section we wish to acquaint ourselves with an approximation method for nonlinear eigenvalue problems in separable Banach spaces, which will enable us to transfer our results on Ljusternik-Schnirelman theory for compact manifolds to the case of sphere-like level surfaces in infinite-dimensional separable Banach spaces. This method was developed by Browder [8.2]. It is a version of a Galerkin (or, precisely, a Rayleigh-Ritz) approximation and utilises the Ljusternik-Schnirelman theory to convert the given eigenvalue problem in an infinite-dimensional separable Banach space to a sequence of corresponding problems in finite-dimensional subspaces which can be then solved. Subsequently it is shown that the sequence of eigenvalues converges to the eigenvalues and the sequence of eigenfunctions to the eigenfunctions of the initial problem, just as was shown in the proof of the existence of the ground state in Theorems 8.2.1 and 8.2.2.

Let E be an infinite-dimensional, separable, real Banach space whose norm $\|\cdot\|$ in $\setminus\{0\}$ is continuously differentiable once. Further, let the level surfaces $M_c = M_c(f)$ of the function $f \in C^1(E)$ be sphere-like (f satisfies, Theorem 8.4.1 (i) and (ii)). Then there exists a sequence $\{e_j\}_{j\in\mathbb{N}}$ of linearly independent elements in E which is total and contains a total subsequence in $M_c(f) = M_c$. For $n \in \mathbb{N}$ we then set

$$E_n = \text{lin}\{e_1, \ldots, e_n\}. \tag{8.4.2}$$

Then $\{E_n\}_{n\in\mathbb{N}}$ is a sequence of finite-dimensional subspaces of E with the following properties:

$$E_n \subseteq E_{n+1}, \quad \forall n \in \mathbb{N} \quad \text{and} \quad \overline{\bigcup_{n\in\mathbb{N}} E_n} = E. \tag{8.4.2'}$$

Furthermore, set

$$M_{c,n} := M_c(f) \cap E_n \tag{8.4.3 a}$$

and

$$f_n := f \upharpoonright E_n. \tag{8.4.3 b}$$

Then it follows that

$$M_{c,n} = M_c(f_n) \tag{8.4.4}$$

is a sphere-like manifold in E_n. This is because Theorem 8.4.1 (i) and (ii) are also valid for f_n. Moreover,

$$\overline{\bigcup_{n\in\mathbb{N}} M_{c,n}} = M_c(f). \tag{8.4.5}$$

Obviously, one has

$$\overline{\bigcup_{n\in\mathbb{N}} M_{c,n}} \subseteq M_c(f).$$

Now the assumption

$$M_c(f) \setminus \overline{\bigcup_{n\in\mathbb{N}} M_{c,n}} \neq \emptyset$$

can be shown to lead easily to a contradiction. If f is even, $M_c(f)$ and all $M_{c,n}$ are reflection-invariant.

The following lemma provides the topological basis for our approximation. The decisive point of this lemma is that it gives us a tool for controlling the genus of compact sets for this approximation.

Lemma 8.4.3. *If $K \subset M_c(f)$ is a reflection-invariant compact set and $\delta > 0$ is given, then there exists an $n = n(K, \delta) \in \mathbb{N}$ and a reflection-invariant compact subset $K_n \subseteq M_{c,n}$ such that*

(i) $C_u(K, K_n) \neq \emptyset$,
(ii) $\text{dist}_{M_c}(K, K_n) \leq \delta$.

Proof. (a) Since we know that M_c and $M_{c,n}$ are sphere-like, we can apply Theorem 8.4.1 and obtain odd C^1 mappings (in the notation of the theorem)

$$g = r_c^{-1} \circ r : E \setminus \{0\} \to M_c \,, \quad g_n : E_n \setminus \{0\} \to M_{c,n} \,,$$

where

$$g_n = g \restriction E_n \setminus \{0\} \,.$$

As a consequence of $g \restriction M_c = \mathrm{id}_{M_c}$ there exists an $\varepsilon > 0$, $\varepsilon < \delta/2$, such that

$$\|g(x) - x\| \le \delta/2 \,, \quad \forall x \in K_\varepsilon = \{x \in E \mid \mathrm{dist}(x, K) \le \varepsilon\} \subset E \setminus \{0\} \,.$$

The compactness and the reflection-invariance of K imply the existence of a covering of the form

$$K \subseteq \bigcup_{j=1}^{2m} B_\varepsilon(x_j) \cap M_c \quad \text{with } x_j \in K \,, \quad 0 \notin B(x_j) \,, \ \forall j \,, \tag{8.4.6}$$

and

$$x_{j+m} = -x_j \quad \text{for } j = 1, \ldots, m \,.$$

The points $\{x_1 \ldots, x_{2m}\}$ lie in a finite-dimensional subspace E_n of E.

(b) Let us now choose a continuous function $\beta : \mathbb{R}_+ \to \mathbb{R}_+$ with support in $[0, \varepsilon]$ and $\beta(t) > 0$ for $0 \le t < \varepsilon$ and set, for $i = 1, \ldots, 2m$,

$$\beta_i(x) = \beta(\|x - x_i\|) \,.$$

Then $\beta_i : E \to \mathbb{R}_+$ is continuous, has its support in $\overline{B_\varepsilon(x_i)}$ and is positive in $B_\varepsilon(x_i)$. According to (8.4.6), $\sum_{i=1}^{2m} \beta_i(x)$ is thus always positive on K. Consequently, the $\alpha_i : K \to \mathbb{R}_+$ defined by

$$\alpha_i(x) = \beta_i(x) \Big/ \sum_{j=1}^{2m} \beta_j(x)$$

are continuous functions on K with

$$0 \le \alpha_i(x) \le 1 \,, \quad \sum_{i=1}^{2m} \alpha_i(x) = 1 \,, \quad \forall x \in K \,,$$

and

$$\mathrm{supp}\, \alpha_i = \mathrm{supp}\, \beta_i \quad \text{and} \quad \alpha_{i+m}(x) = \alpha_i(-x) \,, \quad i = 1, \ldots, m \,,$$

due to the symmetry $x_{i+m} = -x_i$. Therefore, an odd continuous function $\chi : K \to E_n$ is defined by

$$\chi(x) = \sum_{i=1}^{2m} \alpha_i(x) x_i = \sum_{i=1}^{m} \{\alpha_i(x) - \alpha_i(-x)\} x_i \,.$$

By construction it follows for $x \in K$ that

$$\|\chi(x) - x\| = \left\| \sum_{i=1}^{2m} \alpha_i(x) x_i - \sum_{i=1}^{2m} \alpha_i(x) x \right\|$$
$$\leq \frac{1}{\sum_{j=1}^{2m} \beta(\|x - x_j\|)} \sum_{i=1}^{2m} \beta(\|x - x_i\|) \|x - x_i\| \leq \varepsilon$$

and hence $\chi(K) \subseteq K_\varepsilon \cap E_n$.

The function $g \circ \chi : K \to M_c$ is continuous and odd. Since χ maps into E_n and g leaves E_n invariant, $g \circ \chi$ maps the set K into E_n. Thus $K_n := g \circ \chi(K)$ is a reflection-invariant compact set in $M_{c,n}$.

Statement (i) now follows:

$$g \circ \chi \in \mathcal{C}_u(K, K_n).$$

As a consequence of

$$\|g \circ \chi(x) - x\| \leq \|g(\chi(x)) - \chi(x)\| + \|\chi(x) - x\| \leq \delta/2 + \delta/2 = \delta$$

for all $x \in K$, by choosing $\varepsilon < \delta/2$ it also follows that

$$\operatorname{dist}_{M_c(f)}(g \circ \chi(K), K) \leq \sup_{x \in K} \|g \circ \chi(x) - x\| \leq \delta,$$

which means (ii). □

In connection with the Galerkin approximation, the following concepts are useful:

$$\mathcal{F}_k^{\text{cpt}}(M_c(f)) = \{K \subset M_c(f) \mid K \in \Sigma(E), K \text{ compact}, \gamma(K) \geq k\},$$
$$\gamma_{\text{cpt}}(M_c(f)) = \sup\{\gamma(K) \mid K \in \Sigma(E), K \subseteq M_c(f) \text{ compact}\}$$

and for an even C^1 function $h : E \to \mathbb{R}$,

$$m_{k,\text{cpt}}(h) := \inf_{K \in \mathcal{F}_k^{\text{cpt}}(M_c)} \sup h(K). \tag{8.4.7}$$

This is because of the following theorem.

Theorem 8.4.4. *Let E be an infinite-dimensional, separable, real Banach space with a norm $\|\cdot\| \in C^1(E \setminus \{0\})$. Let the even C^1 function $f : E \to \mathbb{R}$ satisfy Theorem* 8.4.1 *(i) and (ii) for the bounded level surface $M_c \equiv M_c(f)$. Then the finite-dimensional subspace E_n and the submanifold $M_{c,n}$ defined by* (8.4.2) *and* (8.4.3) *satisfy*

(a) (i) $\qquad \gamma(M_{c,n}) < \infty, \quad \forall n \in \mathbb{N},$

(ii) $\qquad \gamma(M_{c,n}) \underset{n \to \infty}{\nearrow} \gamma_{\text{cpt}}(M_c).$

(b) *If h is an even C^1 function $E \to \mathbb{R}$, then the minimax values*

$$m_k(h_n) = \inf_{A \in \mathcal{F}_k(M_{c,n})} \sup h_n(A)$$

for the functions $h_n = h \restriction E_n$, $1 \leq k \leq n$, $n \in \mathbb{N}$, satisfy

$$m_k(h_n) \underset{n \to \infty}{\searrow} m_{k,\text{cpt}}(h) .$$

Proof. (a) The $M_{c,n}$ are closed and bounded in the finite-dimensional subspace E_n and hence are compact. Therefore, $\gamma(M_{c,n}) < \infty$, $\forall n \in \mathbb{N}$. The inclusions $M_{c,n} \subseteq M_{c,n+1} \subseteq M_c$ and the definition of $\gamma_{\text{cpt}}(M_c)$ imply immediately the inequality

$$\gamma(M_{c,n}) \leq \gamma(M_{c,n+1}) \leq \gamma_{\text{cpt}}(M_c) .$$

Now let $K \subset M_c$, $K \in \Sigma(E)$ be compact, and $\gamma(K) = k_0 \in \mathbb{N}$. According to Lemma 8.4.3, there exists an $n \in \mathbb{N}$ and a reflection-invariant compact set $K_n \subseteq M_n$ with $\mathcal{C}_u(K, K_n) \neq \emptyset$. Now Lemma 8.3.5 (ii) on the genus of sets implies $\gamma(K_n) \geq \gamma(K) = k_0$ and hence the limit property (a) (ii).

(b) From

$$\mathcal{F}_k(M_{c,n}) \subseteq \mathcal{F}_k(M_{c,n+1}) \subseteq \mathcal{F}_k^{\text{cpt}}(M_c)$$

and

$$m_k(h_n) = \inf_{A \in \mathcal{F}_k(M_{c,n})} \sup h(A)$$

for $1 \leq k \leq n$, $n \in \mathbb{N}$, the relation

$$m_k(h_n) \geq m_k(h_{n+1}) \geq m_{k,\text{cpt}}(h) , \quad 1 \leq k \leq n , \ n \in \mathbb{N} ,$$

is obvious. Thus, it remains to be shown that

$$\forall \varepsilon > 0 , \quad \exists n_\varepsilon \in \mathbb{N} : m_k(h_{n_\varepsilon}) \leq m_{k,\text{cpt}}(h) + \varepsilon .$$

But by the definition of $m_{k,\text{cpt}}(h)$ there exists a $K_\varepsilon \in \mathcal{F}_k^{\text{cpt}}(M_c)$ with

$$\operatorname{supp} h(K_\varepsilon) \leq m_{k,\text{cpt}}(h) + \varepsilon/2 .$$

Due to the continuity of h and the compactness of K_ε there exists a $\delta = \delta_\varepsilon > 0$ such that

$$\forall x \in K_\varepsilon , \quad \forall x' \in \overline{B_\delta(x)} : |h(x) - h(x')| \leq \varepsilon/2 .$$

Now applying Lemma 8.4.3 to (K_ε, δ) we obtain an $n_\varepsilon \in \mathbb{N}$ and a reflection-invariant compact subset $K_{n_\varepsilon} \subseteq M_{c,n_\varepsilon}$ such that

(i) $\mathcal{C}_u(K_\varepsilon, K_{n_\varepsilon}) \neq 0$,

(ii) $\operatorname{dist}_{M_c}(K_\varepsilon, K_{n_\varepsilon}) \leq \delta$.

It follows from (i) that

$$\gamma(K_{n_\varepsilon}) \geq \gamma(K_\varepsilon) \geq k\,, \quad \text{and} \quad K_{n_\varepsilon} \in \mathcal{F}_k(M_{c,n_\varepsilon})\,.$$

Because of (ii), for every $x \in K_{n_\varepsilon}$ there exists an $x' = x'(x) \in K_\varepsilon$ such that $x' \in \overline{B_\delta(x)}$ holds. By applying (8.4.6) it follows then that

$$\begin{aligned}\sup_{x\in K_{n_\varepsilon}} h_{n_\varepsilon}(x) &= \sup_{x\in K_{n_\varepsilon}} \{h(x'(x)) + h(x) - h(x'(x))\} \\ &\leq \sup_{y\in K_\varepsilon} h(y) + \sup_{x\in K_{n_\varepsilon}} |h(x) - h(x'(x))| \leq \sup h(K_\varepsilon) + \varepsilon/2\end{aligned}$$

and therefore

$$m_k(h_{n_\varepsilon}) \leq m_{k,\mathrm{cpt}}(h) + \varepsilon\,,$$

which was to be shown. □

8.4.3 The Existence of Infinitely Many Critical Points as Solutions of Abstract Eigenvalue Problems in Separable Banach Spaces

To assure the existence of infinitely many solutions of nonlinear eigenvalue problems, we naturally need hypotheses which go beyond the assumptions in Theorem 8.2.2 which imply the existence of a ground state. The following long list of assumptions appears natural if you refer to the discussion of sphere-like constraints and the theorem on the existence of a ground state.

Theorem 8.4.5. *Let E be an infinite-dimensional, separable, reflexive real Banach space with a norm $\|\cdot\| \in C^1(E\setminus\{0\})$. Let f and h be even C^1 functions $E \to \mathbb{R}$ with the following properties:*

(A) *Hypothesis for the constraint f:*

(A_1) *Let $M_c = M_c(f) = \{x \in E \,|\, f(x) = c\}$ be sphere-like, which means*
(α) *$\langle f'(x), x\rangle \neq 0$, $\forall x \in M_c$;*
(β) *every ray from the origin in E intersects M_c in exactly one point.*

(A_2) *M_c is bounded.*

(A_3) *$f' : E \to E'$ is bounded.*

(A_4) *$f' : E \to E'$ satisfies the condition* (S).

(B) *Hypothesis for the function h to be minimised on M_c:*

(B_1) *For every subset $A \subseteq M_c$ on which h is bounded, there exists a $c_A > 0$ so that one has*

$$c_A|\langle f'(x), x\rangle| \leq \langle h'(x), x\rangle|\,, \quad \forall x \in A\,.$$

(B_2) *h' is compact on all subsets $A \subset M_c$ on which h is bounded.*

Then the eigenvalue problem $f'(x) = \lambda h'(x)$ possesses at least the following solutions:

(α) *For all $k \in \mathbb{N}$ with $m_{k,\text{cpt}}(h) < \infty$ there exists $x_k \in M_c(f)$ and $\lambda_k \in \mathbb{R}$ with*

(i) $m_{k,\text{cpt}}(h) = h(x_k)$,

(ii) $f'(x_k) = \lambda_k h'(x_k)$.

(β) *The above pair of eigenfunctions and eigenvalues (u_k, λ_k) can be obtained as strong limits of a sequence of solutions of corresponding finite-dimensional eigenvalue problems. If we define E_n and $M_{c,n}$ via (8.4.2) and (8.4.3) and set $f_n = f \restriction E_n$, $h_n = h \restriction E_n$, then there exist, for all $n \in \mathbb{N}$ and all $1 \le k \le n$, points $x_{k,n} \in M_{c,n}$ and numbers $\lambda_{k,n} \in \mathbb{R}$ such that*

(i) $h_n(x_{k,n}) = m_k(h_n) = \inf_{A \in \mathcal{F}_k(M_{c,n})} \sup h_n(A)$,

(ii) $f_n'(x_{k,n}) = \lambda_{k,n} h_n'(x_{k,n})$.

For all $k \in \mathbb{N}$ with $m_{k,\text{cpt}}(h) < \infty$ the following holds for some subsequences:

$$x_k = \lim_{j\to\infty} x_{k,n_j}\,, \qquad \lambda_k = \lim_{j\to\infty} \lambda_{k,n_j} \quad \text{and} \quad h_n(x_{k,n_j}) \underset{j\to\infty}{\searrow} h'(x_k)\,.$$

Proof. As a first step we define the approximating sequence by corresponding finite-dimensional eigenvalue problems. This can be done as in Sect. 8.4.2 by means of (8.4.2) and (8.4.3) and $h_n := h \restriction E_n$. Denoting by $\psi_n : E_n \to E$ the identical embedding of E_n in E and by $\psi_n' : E' \to E_n'$ its dual projection, a simple calculation shows

$$f_n' = \psi_n' \circ f' \circ \psi_n \quad \text{and} \quad h_n' = \psi_n' \circ h' \circ \psi_n\,. \tag{8.4.8}$$

It follows from this, in particular, that $M_{c,n}$ is sphere-like and hence a compact C^1 manifold in the finite-dimensional Banach space E_n which is reflection-invariant. Further, h_n is an even C^1 function on $M_{c,n}$.

As the second step we can now apply the Ljusternik-Schnirelman theory in the form of Theorem 8.3.7 and obtain for all $n \in \mathbb{N}$ and $1 \le k \le n$

$$\begin{aligned} &\text{(i)} && x_{k,n} \in M_{c,n} \quad \text{with } h_n(x_{k,n}) = m_k(h_n)\,, \\ &\text{(ii)} && h_n(x_{1,n}) \le h_n(x_{2,n}) \le \ldots \le h_n(x_{n,n})\,. \end{aligned} \tag{8.4.9 a}$$

The theorem on Lagrange multipliers gives the existence of numbers $\lambda_{k,n}$ such that

$$f_n'(x_{k,n}) = \lambda_{k,n} h_n'(x_{k,n})\,. \tag{8.4.9 b}$$

In the third, and most difficult, step, we have to establish the convergence of $(x_{k,n}, \lambda_{k,n})_{n\in\mathbb{N}}$.

By hypothesis (A_2) the sequence $\{x_{k,n}\}_{n\in\mathbb{N}} \subseteq M$ is bounded. The reflexivity of E guarantees the existence of a weakly convergent subsequence

$$\{x_{k,n_j}\}_{j\in\mathbb{N}} \quad \text{with } x_{k,n_j} \xrightarrow[j\to+\infty]{w} x_k \in E\,.$$

To simplify the notation, we shall set $n_j = j$. For $y \in E_n$ and $j \geq n$ it follows from $E_n \subseteq E_j$ and (8.4.8) and (8.4.9 b) that

$$\langle f_j'(x_{k,j}), y\rangle = \langle \psi_j' \circ f' \circ \psi_j(x_{k,j}), y\rangle = \langle f'(x_{k,j}), y\rangle$$

and similarly

$$\langle h_j'(x_{k,j}), y\rangle = \langle h'(x_{k,j}), y\rangle\,.$$

Therefore one obtains

$$\langle f'(x_{k,j}), y\rangle = \lambda_{k,j}\langle h'(x_{k,j}), y\rangle\,, \quad \forall y \in E_n\,,\ \forall j \geq n\,. \tag{0}$$

By Theorem 8.4.4 and (8.4.9 a) we have

$$m_k(h_j) = h_j(x_{k,j}) = h(x_{k,j}) \underset{j\to\infty}{\searrow} m_{k,\mathrm{cpt}}(h)\,.$$

Thus, for all $k \in \mathbb{N}$ for which $m_{k,\mathrm{cpt}}(h) < \infty$ holds, $\{x_{k,j}\}_{j\in\mathbb{N}}$ is a subset of M_c on which h is bounded. The hypothesis (B_1) implies the existence of a $c_k \in (0,\infty)$ such that for all $j \in \mathbb{N}$,

$$|\lambda_{k,j}| = \left|\frac{\langle f'(x_{k,j}), x_{k,j}\rangle}{\langle h'(x_{k,j}), x_{k,j}\rangle}\right| \leq c_k < \infty\,.$$

Now from (B_2) the existence of subsequences $\{\lambda_{k,j(i)}\}_{j\in\mathbb{N}}$ and $\{x_{k,j(i)}\}_{i\in\mathbb{N}}$ with

$$\lambda_{k,j(i)} \to \lambda_k \text{ in } \mathbb{R} \quad \text{and} \quad h'(x_{k,j(i)}) \underset{j\to\infty}{\longrightarrow} y_k' \text{ in } E'$$

follows. From (0) one verifies easily for arbitrary $y \in \bigcup_{n\in\mathbb{N}} E_n$ that

$$\lim_{i\to\infty}\langle f'(x_{k,j(i)}), y\rangle = \lim_{i\to\infty}\langle \lambda_{k,j(i)} h'(x_{k,j(i)}), y\rangle = \lambda_k\langle y_k', y\rangle\,. \tag{$\times$}$$

By hypothesis (A_3) the sequence $\{f'(x_{k,j(i)})\}_{i\in\mathbb{N}}$ is bounded in E', say

$$\|f'(x_{,j(i)})\|' \leq a_k < \infty \quad \forall i \in \mathbb{N}\,.$$

For arbitrary $y \in E$ and $\varepsilon > 0$, there exists a $y_\varepsilon \in \bigcup_{n\in\mathbb{N}} E_n$ such that

$$\|y - y_\varepsilon\| < \frac{\varepsilon}{3a_k} \wedge \frac{\varepsilon}{3|\lambda_k|\ \|y_k'\|'}\,.$$

By ($\times$) there exists an $i_\varepsilon \in \mathbb{N}$ such that

$$|\langle f'(x_{k,j(i)}) - \lambda_k y_k', y_\varepsilon\rangle| \leq \varepsilon/3\,, \quad \forall i \geq i_\varepsilon$$

holds. For $i \geq i_\varepsilon$ it follows then that

$$\begin{aligned}
&|\langle f'(x_{k,j(i)}), y\rangle - \lambda_k\langle y_k', y\rangle| \\
&\quad \leq |\langle f'(x_{k,j(i)}) - \lambda_k y_k', y_\varepsilon\rangle| + |\langle f'(x_{k,j(i)}), y - y_\varepsilon\rangle| + |\langle \lambda_k y_k', y - y_\varepsilon\rangle| \\
&\quad < \frac{\varepsilon}{3} + \frac{\varepsilon}{3} + \frac{\varepsilon}{3} = \varepsilon\,;
\end{aligned}$$

and thus

$$f'(x_{k,j(i)}) \xrightarrow[i\to\infty]{w} \lambda_k y'_k \quad \text{in } E' . \tag{+}$$

As previously, it follows that the hypotheses of Condition (S) are satisfied by f': now it follows immediately from (0) that

$$\begin{aligned}\langle f'(x_{k,j(i)}), x_{k,j(i)}\rangle &= \langle \lambda_{k,j(i)} h'(x_{k,j(i)}), x_{k,j(i)}\rangle \\ &= \langle \lambda_k y'_k, x_{k,j(i)}\rangle + \langle \lambda_{k,j(i)} h'(x_{k,j(i)}) - \lambda_k y'_k, x_{k,j(i)}\rangle \\ &\xrightarrow[i\to\infty]{} \lambda_k \langle y'_k, x_k\rangle + 0 .\end{aligned}$$

This is because $\{x_{k,j(i)}\}_{i\in\mathbb{N}}$ converges weakly to x_k and is strongly bounded, and $\lambda_{k,j(i)} h'(k_{k,j(i)})$ converges strongly in E' to $\lambda_k y'_k$. It thus follows directly that

$$\lim_{i\to\infty} \langle f'(x_{k,j(i)}) - f'(x_k), x_{k,j(i)} - x_k\rangle = 0 .$$

The hypothesis (A_4) implies the strong convergence of the sequence $\{x_{k,j(i)}\}_{i\in\mathbb{N}}$ to an $x'_k \in E$. Since the sequence $\{x_{k,j}\}_{j\in\mathbb{N}}$ converges already weakly to $x_k \in E$, the strong convergence of $\{x_{k,j}\}_{j\in\mathbb{N}}$ to $x_k \in E$ follows. The continuity of h' gives

$$h'(x_k) = s - \lim_{j\to\infty} h'(x_{k,j}) = y'_k .$$

The continuity of f implies

$$f(x_k) = \lim_{j\to\infty} f(x_{k,j}) = \lim_j c = c .$$

Therefore one has $x_k \in M_c(f)$.

Noting the continuity of f' we obtain

$$f'(x_k) = \lim_{j\to\infty} f'(x_{k,j}) = \lambda_k y'_k = \lambda_k h'(x_k) .$$

Therefore, for all $k \in \mathbb{N}$ with $m_{k,\text{cpt}} < \infty$, (x_k, λ_k) is in fact a solution of the eigenvalue problem $f'(x) = \lambda h'(x)$. Finally, $h(x_k)$ is equal to the minimax value of $m_{k,\text{cpt}}(h)$:

$$h(x_k) = \lim_{j\to\infty} h(x_{k,j}) = \lim_{j\to\infty} m_k(h_j) = m_{k,\text{cpt}}(h) .$$

This concludes the proof of the statements (α) and (β). □

8.4.4 The Existence of Infinitely Many Solutions of Nonlinear Eigenvalue Problems

The abstract results of the previous section are illustrated here by proving, as an application, the existence of infinitely many solutions for a sufficiently large class of nonlinear eigenvalue problems. For this purpose we start off just as in the treatment of nonlinear boundary value problems, namely, by collecting a long list of concrete hypotheses for the coefficient functions of the eigenvalue problem in a central lemma; these hypotheses fit in with Theorem 8.4.5. A simple consideration shows that the appearance of this list is very natural within the framework of our approach.

Hypotheses. Let $G \subseteq \mathbb{R}^n$ be an open, bounded subset with smooth boundary and let

$$F_j : G \times \mathbb{R}^{n+1} \to \mathbb{R}, \quad j = 0, 1, \ldots, n$$

be $n+1$ Carathéodory functions. Let these C-functions F_j be partial derivatives of a C-function

$$F : G \times \mathbb{R}^{n+1} \to \mathbb{R}, \quad F_j(x,y) = \frac{\partial F}{\partial y_j}(x,y), \quad j = 0, 1, \ldots, n.$$

Let F be even with respect to y : $F(x,-y) = F(x,y)$, $\forall (x,y) \in G \times \mathbb{R}^{n+1}$. Let there be a number p, $1 < p < n$, such that F and F_j satisfy the following bounds.

(D$_1$) There exist an $a \in L^1(G)$, $a \geq 0$, and numbers $p_j \in \mathbb{R}$, $b > 0$ with

$$1 \leq p_j \leq p, \quad j = 1, \ldots, n, \qquad \frac{1}{p} - \frac{1}{n} \leq \frac{1}{p_0}, \quad p_0 \geq 1, \tag{8.4.10 a}$$

such that for all $(x,y) \in G \times \mathbb{R}^{n+1}$ one has

$$|F(x,y)| \leq a(x) + b \sum_{j=0}^{n} |y_j|^{p_j}. \tag{8.4.10 b}$$

(D$_2$) For $j = 0, 1, \ldots, n$ there exist functions a_j with $0 \leq a_j \in L^{q_j'}(G)$ and numbers b_j, q_j and p_{ji}, $i = 0, 1, \ldots, n$ satisfying

$$\frac{1}{p} - \frac{1}{n} \leq \frac{1}{q_0}, \quad 1 \leq q_j \leq p, \quad b_j \geq 0, \tag{8.4.11 a}$$

and

$$\frac{1}{p} - \frac{1}{n} \leq \frac{1}{p_{j0}}, \quad 1 \leq p_{ji} \leq p, \quad i = 1, \ldots, n, \tag{8.4.11 b}$$

such that for all $(x,y) \in G \times \mathbb{R}^{n+1}$ one has

$$|F_j(x,y)| \leq a_j(x) + b_j \sum_{i=0}^{n} |y_i|^{p_{ji}/q_j'}. \tag{8.4.11 c}$$

Here q and q' are dual exponents, i.e. $\frac{1}{q'} + \frac{1}{q} = 1$.

(K) *Coerciveness (Ellipticity).* There exist an $\alpha > 0$ and an $r_0 \in [0,p]$ such that for all $(x,y) \in G \times \mathbb{R}^{n+1}$ with some $0 \leq g_0 \in L^{p/(p-r_0)}(G)$ the following bounds hold:

$$\sum_{j=1}^{n} F_j(x,y)y_j \geq \alpha \sum_{j=1}^{n} |y_j|^p - g_0(x)|y_0|^{r_0} \tag{8.4.12}$$

(M) *Monotony.* For all $(x,y_0) \in G \times \mathbb{R}^n$ and all $\underline{y}, \underline{y'} \in \mathbb{R}^n$, $\underline{y} \neq \underline{y'}$, the following is true:

$$\sum_{j=1}^{n} \{F_j(x,y_0,\underline{y}) - F_j(x,y_0,y')\}(y_j - y_j') > 0\,. \tag{8.4.13}$$

Lemma 8.4.6. *Let $G \subseteq \mathbb{R}^n$ be an open, bounded subset with smooth boundary. Let the $n+1$ Carathéodory functions $F_j : G \times \mathbb{R}^{n\,n+1} \to \mathbb{R}$, $j = 0,1,\ldots,n$, be the partial derivatives of a Carathéodory function $F : G \times \mathbb{R}^{n+1} \to \mathbb{R}$. Let there exist a p, $1 < p < n$, such that the functions F and F_j satisfy the hypotheses D_1, D_2, K and M. Then it follows that:*

(1) *A function $f : E \to \mathbb{R}$ can be defined on the real Banach space $E = W^{1,p}(G)$ by*

$$u \to f(u) = \int_G d^n x F(x, u(x), \partial_1 u(x), \ldots, \partial_n u(x)) \tag{8.4.14}$$

so that f is continuous and bounded on bounded sets of E. Furthermore, one has

$$|f(u)| \leq \|a\|_1 + b' \sum_{j=0}^{n} \|u\|_{1,p}^{p_j} \equiv B(\|u\|_{1,p})\,. \tag{8.4.14'}$$

(2) *f is Fréchet-differentiable on E and possesses the derivative*

$$\langle f'(u), v\rangle = \int_G d^n x \Big\{ \sum_{j=1}^{n} F_j(x, u(x), \partial_1 u(x), \ldots, \partial_n u(x))\partial_j v(x) + F_0(x, u(x), \partial_1 u(x), \ldots, \partial_n u(x)) v(x) \Big\}, \quad \forall u, v \in E\,. \tag{8.4.15}$$

(3) *The Fréchet derivative of f, $f' : E \to E'$, has the following properties:*

(i) *f' is continuous,*
(ii) *f' is coercive,*
(iii) *f' is bounded on bounded subsets of E,*
(iv) *f' satisfies the condition (S).*

Proof. (a) The method of proof is similar to the corresponding part on nonlinear boundary value problems. The most important elements of the proof are the properties of the embedding mappings:

$$y : W^{1,p}(G) \to \bigoplus_{j=0}^{n} L^{p_j}(G)\,, \tag{8.4.16}$$

$$u \to y(u) = (y_0(u), y_1(u), \dots, y_n(u))\,, \quad y_0(u) = u\,, \quad y_j(u) = \partial_j u\,,$$

for suitable exponents $(p_0, p_1, \dots, p_n)$, and also the properties of the Niemytski operators:

$$\widehat{g} : \bigoplus_{n=0}^{n} L^{p_j}(G) \to L^q(G) \tag{8.4.17}$$

($g : G \times \mathbb{R}^{n+1} \to \mathbb{R}$ being a C function) and

$$\widehat{g}(v_0, v_1, \dots, v_n)(x) := g(x, v_0), v_1(x), \dots, v_n(x))\,,$$

as explained in [8.19].

The properties of the embeddings are determined by the Sobolev embedding theorems. With the restriction (8.4.1 a) for the exponents, these embeddings are continuous:

$$\|y_j(u)\|_{p_j} = \|\partial_j u\|_{p_j} \le c_j \|u\|_{1,p}\,, \quad j = 0, 1, \dots, n\,, \tag{8.4.18 a}$$

with a constant

$$c_j = c_j(|G|, p, p_j)\,.$$

For the case

$$\frac{1}{p} - \frac{1}{n} < \frac{1}{p_0}\,, \quad p_0 \ge 1\,,$$

the embedding

$$y_0 : W^{1,p}(G) \to L^{p_0}(G) \tag{8.4.18 b}$$

is compact.

As we mentioned previously, the operator $\widehat{g}$ is well defined and continuous if and only if there exist a function $0 \le \alpha \le L^q(G)$ and a constant $b \ge 0$ such that

$$|g(x,y)| \le \alpha(x) + b \sum_{j=0}^{n} |y_j|^{p_j/q}\,, \quad \forall (x,y) \in G \times \mathbb{R}^{n+1}\,. \tag{8.4.19}$$

(b) The bound (8.4.10 b) thus states by (8.2.1), that

$$\widehat{F} : \bigoplus_{j=0}^{n} L^{p_j}(G) \to L^1(G)$$

is well defined and continuous, and that the following holds:

$$\|\widehat{F}v\|_1 = \int_G d^n x |F(x, v_0(x), v_1(x), \dots, v_n(x))| \leq \|a\|_1 + b\sum_{j=0}^{n} \|v_j\|_{p_j}^{p_j} .$$

From (8.4.16) and (8.4.18) one gets for $v = y(u)$, $u \in E = W^{1,p}(G)$:

$$\|\widehat{F}y(u)\|_1 \leq \|a\|_1 + b\sum_{j=0}^{n} \|y_j(u)\|_{p_j}^{p_j} \leq \|a\|_1 + b\sum_{j=0}^{n} \|u\|_{1,p}^{p_j} .$$

Thus

$$u \to \int_G d^n x \widehat{F}y(u)(x) = f(u)$$

is well defined on E and satisfies the bound (8.4.14). f is a composition of the continuous mappings $\widehat{F}$ and y and is thus itself continuous, which proves (1).

(c) From (8.4.11 a) and (8.4.11 b) we obtain the continuous linear embeddings

$$y : E \to \bigoplus_{i=0}^{n} L^{p_{ji}}(G), \quad j = 0, 1, \dots, n,$$

and

$$y : E \to \bigoplus_{i=0}^{n} L^{q_i}(G).$$

The hypothesis (8.4.11. c) together with the bound (8.4.19) imply that the Niemytski operator $\widehat{F}_j$ maps from $\bigoplus_{i=0}^{n} L^{p_{ji}}(G)$ into $L^{q_j'}(G)$ continuously. Therefore it follows that

$$u \to \widehat{F}_j y(u) \text{ maps } W^{1,p}(G) \text{ continuously into } L^{q_j'}(G) \tag{8.4.20}$$

for $j = 0, 1, \dots, n$.

For $v_j \in L^{q_j}(G)$ we have, according to (8.4.11 c),

$$|\langle \widehat{F}_j y(u), v_j \rangle_2| = \left| \int_G d^n x F_j(x, y(u)(x)) v_j(x) \right|$$
$$\leq \left\{ \|a_j\|_{q_j'} + b_j \sum_{i=0}^{n} \|y_i(u)\|_{p_{ji}}^{p_{ji}/q_j'} \right\} \|v_j\|_{q_j} ,$$

and thus

$$\|\widehat{F}_j y(u)\|_{q_j'} \leq \|a_j\|_{q_j'} + b_j \sum_{i=0}^{n} c_i \|u\|_{1,p}^{p_{ji}/q_j'} \equiv B_j(\|u\|_{1,p}) . \tag{8.4.21}$$

Consequently,

$$(u, v) \to a(u, v) := \sum_{j=0}^{n} \langle \widehat{F}_j y(u), y_j(v) \rangle_2 \tag{8.4.22}$$

is a well-defined function $E \times E \to \mathbb{R}$ with

$$|a(u,v)| \leq \sum_{j=0}^{n} \|\widehat{F}_j y(u)\|_{q_j'} \|y_j(v)\|_{q_j} \leq \sum_{j=0}^{n} c_j \|\widehat{F}_j y(u)\|_{q_j'} \|v\|_{1,p}$$
$$\leq \left(\sum_{j=0}^{n} c_j B_j(\|u\|_{1,p})\right) \|v\|_{1,p} \,. \tag{8.4.22'}$$

Since $v \to a(u,v)$ is also linear, for a fixed $u \in E$, $a(u,v)$ defines a continuous linear form on E and hence a function $T : E \to E'$ such that

$$a(u,v) = \langle T(u), v \rangle \,, \quad \forall u,v \in E \,. \tag{8.4.23}$$

From (8.4.22′) it follows that

$$\|T(u)\|' = \sup_{\substack{v \in E \\ \|v\|_{1,p}=1}} |\langle T(u), v\rangle| \leq B(\|u\|_{1,p}) \,.$$

Therefore T is bounded on bounded subsets of E.

The continuity of T follows easily from (8.4.20):

$$\|T(u_1) - T(u_2)\|' = \sup_{\substack{v \in E \\ \|v\|_{1,p}=1}} \left| \sum_{j=0}^{n} \langle \widehat{F}_j y(u_1) - \widehat{F}_j y(u_2), v \rangle_2 \right|$$
$$\leq \sum_{j=0}^{n} \|\widehat{F}_j y(u_1) - \widehat{F}_j y(u_2\|_{q_j'} \,. \tag{8.4.24}$$

Since the hypothesis (K) is identical to the hypothesis (H_3') and the hypothesis (M) to the hypothesis (H_2') for the coefficient functions of the boundary value problem, we can use the results of the proof of Theorem 7.3.3 and obtain

(i) T is coercive;
(ii) T satisfies Condition (S).

(d) The Fréchet-differentiability of f and the properties of the Fréchet derivative f' therefore follow from the fact that f possesses a Gâteaux differential $\delta_u f$ at all points $u \in E$, for which $\delta_u f = T(u)$ holds.

Let $u, v \in E$ and $t \in \mathbb{R}$. With $F_j = \partial F / \partial y_j$ it follows for almost all $x \in G$ that

$$F(x, y(u+tv)(x)) - F(x, y(u)(x)) = \int_0^1 d\tau \frac{d}{d\tau} F(x, y(u+\tau tv)(x))$$
$$= \int_0^1 d\tau \sum_{j=0}^{n} F_j(x, y(u+\tau tv)(x)) t y_j(v)(x)$$

and therefore, by using the theorem of Fubini, for $t \neq 0$

$$\begin{aligned}\frac{1}{t}\{f(u+tv)-f(u)\} &= \int_G d^n x \frac{1}{t}\{F(x,y(u+tv)(x)) - F(x,y(u)(x))\} \\ &= \int_G d^n x \int_0^1 d\tau \sum_{j=0}^n \widehat{F}_j(y(u+\tau tv))(x) y_j(v)(x) \\ &= \int_0^1 d\tau \sum_{j=0}^n \langle \widehat{F}_j(y(u+\tau tv)), y_j(v)\rangle_2 \\ &= \int_0^1 d\tau \langle T(u+\tau tv), v\rangle \,.\end{aligned}$$

Since $T : E \to E'$ is continuous and bounded, one obtains easily

$$\lim_{|t|\to 0} \frac{1}{t}\{f(u+tv)-f(u)\} = \langle T(u), v\rangle \,.$$

Furthermore, the Gâteaux derivative of f at the point $u \in E$ in the direction $v \in E$ is a linear and continuous function of v. Thus f possesses a Gâteaux differential $\delta_u f$ at all points $u \in E$ for which one has $\delta_u f = T(u)$. This means that $\delta_u f$ is a continuous function of $u \in E$, and hence $f'(u) = \delta_u f = T(u)$ follows. The properties of T are therefore the same as the properties of the Fréchet derivative f' of f. □

Lemma 8.4.6 specifies the assumptions on the constraint $f(u) = c$, $u \in E$. Thus, in order to solve the eigenvalue problem $f'(u) = \lambda h'(u)$ with the help of the Theorem on Lagrange multipliers we need the corresponding information about the function h. We proceed in a fashion similar to the investigation of the "constraint" f.

Let H be a Carathéodory function on $G \times \mathbb{R}$. Let the derivative of H with respect to y_0, i.e.

$$H_0(x,y_0) = \frac{\partial H}{\partial y_0}(x,y_0)$$

again be a Carathéodory function. The following properties of the growth of H and H_0 will be assumed:

(D_3) There exist an α, $0 \le \alpha \in L^1(G)$, and an $r \ge 1$ with

$$\frac{1}{p} - \frac{1}{n} \le \frac{1}{r} \,,$$

such that the following holds:

$$|H(x,y_0)| \le \alpha(x) + \beta |y_0|^r \,, \quad \beta \ge 0 \,, \quad (x,y_0) \in G \times \mathbb{R} \,. \tag{8.4.25}$$

(D_4) There exist an $r_0 \ge 1$ with

$$\frac{1}{p} - \frac{1}{n} \le \frac{1}{r_0} \,,$$

an $s \geq 1$ with

$$\frac{1}{p} - \frac{1}{n} < \frac{1}{s} \tag{8.4.26 a}$$

and α_0 and β_0, $0 \leq \alpha \in L^{r_0'}(G)$, $\beta_0 \geq 0$, such that fo all $(x, y_0) \in G$ one has

$$|H_0(x, y_0)| \leq \alpha_0(x) + \beta_0 |y_0|^{s/r_0'} . \tag{8.4.26 b}$$

As in Lemma 8.4.6 we then obtain the following statements.

Lemma (8.4.6′). *Let $G \subseteq \mathbb{R}^n$ be an open bounded subset with smooth boundary. Let the Carathéodory function $H : G \times \mathbb{R} \to \mathbb{R}$ possess a partial derivative*

$$H_0(x, y) = \frac{\partial H}{\partial y_0}(x, y_0)$$

which is again a Carathéodory function. Let H and H_0 satisfy (D_3) and (D_4). It then follows that

(1) *On the real Banach space $E = W^{1,p}(G)$ a function $h : E \to \mathbb{R}$ is well defined by*

$$u \to h(u) = \int_G d^n x H(x, u(x)) \tag{8.4.27}$$

h is continuous and bounded on bounded sets of E. One has

$$|h(u)| \leq \|\alpha\|_1 + \beta' \|u\|_{1,p}^r . \tag{8.4.27′}$$

(2) *h is Fréchet-differentiable on E and possesses the derivative*

$$\langle h'(u), v \rangle = \int_G d^n x H_0(x, u(x)) v(x) , \quad \forall u, v \in E . \tag{8.4.28}$$

(3) *The Fréchet derivative of h, namely $h' : E \to E'$, has the following properties: h' is*
   (i) *continuous,*
   (ii) *bounded on bounded sets,*
   (iii) *completely continuous.*

(4) *h is weakly continuous on bounded subsets of E.*

Proof. (a) It is obvious that the hypotheses (D_3) and (D_4) for H and H_0 will lead to the same results as the hypotheses (D_1) and (D_2) for the functions F and F_j in Lemma 8.4.6. Thus the arguments in the proof of that lemma establish the statements (1) and (2) and the properties (i) and (ii) from statement (3). The complete continuity of h' follows from the hypothesis (D_4) (8.4.26 a), which is sharper than (D_2), via Sobolev's embedding theorem or as shown below.

By (8.4.26 a) $\widehat{H}_0 : L^s(G) \to L^{r_0'}(G)$ is continuous. By (8.4.26 a) the identical embedding of $W^{1,p}(G)$ in $L^s(G)$ is completely continuous, i.e. if a sequence $(u^\alpha)_\alpha$ converges weakly in E to a $u \in E$ then $\{\widehat{H}_0(u^\alpha)\}_\alpha$ converges strongly to $\{\widehat{H}_0(u)\}$ in $L^{r_0'}(G)$. It follows, that $\{h'(u^\alpha)\}_\alpha$ converges strongly in E' to $h'(u)$, because

$$\begin{aligned}\|h'(u) - h'(u^\alpha)\|' &= \sup_{\substack{v\in E\\ \|v\|_{1,p}=1}} |\langle h'(u) - h'(u^\alpha), v\rangle| \\ &= \sup_{v\in S_1(E)} |\langle \widehat{H}_0(u) - \widehat{H}_0(u^\alpha), v\rangle_2| \\ &\le c_0\|\widehat{H}_0(u) - \widehat{H}_0(u^\alpha)\|\|)_{r_0'} .\end{aligned}$$

(b) Here we shall show only the weak sequential continuity. Let $\{u^\alpha\}_\alpha$ be a sequence converging weakly to a $u \in E$. Then $\{u^\alpha, u\}$ is strongly bounded. The complete continuity of h' gives the strong convergence of $\{h'(u^\alpha)\}_\alpha$ in E'. Thus we have

$$\begin{aligned}\lim_\alpha\{h(u^\alpha) - h(u)\} &= \lim_\alpha \int_0^1 d\tau \langle h'(u + \tau(u^\alpha - u)), u^\alpha - u\rangle \\ &= \int_0^1 d\tau \lim_\alpha \langle h'(u + \tau(u^\alpha - u)), u^\alpha - u\rangle = 0\,,\end{aligned}$$

for

$$\begin{aligned}&\lim_\alpha \langle h'(u + \tau(u^\alpha - u)), u^\alpha - u\rangle \\ &\quad = \lim_\alpha \{\langle h'(u), u^\alpha - u\rangle + \langle h'(u + \tau(u^\alpha - u)) - h'(u), u^\alpha - u\rangle\} = 0\,,\end{aligned}$$

since

$$u^\alpha \xrightarrow{w} u \quad \text{and} \quad h'(u + \tau(u^\alpha - u)) \xrightarrow{s} h'(u)\,. \qquad \square$$

Remark 8.4.1. If we assume in (D_3) that

$$\frac{1}{p} - \frac{1}{n} < \frac{1}{r}\,;$$

then $W^{1,p}(G)$ is compactly embedded in $L^r(G)$ and $\widehat{H}$ is a continuous mapping of $L^r(G)$ into $L^1(G)$. Then the weak (sequential) continuity of h follows very easily.

Now it is simple to prove the existence of infinitely many solutions for a certain class of nonlinear eigenvalue problems. Lemma 8.4.6 (8.4.6′) and Theorem 8.4.2 translate the concrete requirements for the coefficient functions into the hypotheses of the basic existence theorem, Theorem 8.4.5.

Theorem 8.4.7. *Let $G \subset \mathbb{R}^n$ be an open bounded subset with smooth boundary. Let the $n+1$ C-functions $F_j : G \times \mathbb{R}^{n+1} \to \mathbb{R}$, $j = 0, 1, \ldots, n$, be* *the partial derivatives* $\partial F/\partial y_j$ *of the* C*-functions* $F : G \times \mathbb{R}^{n+1} \to \mathbb{R}$ *with* $F(x,-y) = F(x,y)$. *The functions* F *and* F_j *satisfy the hypotheses* (D_1), (D_2), (K) *and* (M) *for a* $p \in (1,n)$. *Furthermore, let* $H : G \times \mathbb{R} \to \mathbb{R}$ *be a* C*-function with* $H(x,-y_0) = H(x,y_0)$. *Let the derivative* $H_0(x,y) = (\partial H/\partial y_0)(x,y_0)$ *of* H *again be a* C*-function and the functions* H *and* H_0 *satisfy the hypotheses* (D_3) *and* (D_4) *for the same* $p \in (1,n)$. *Then,*

(a) *On all closed subspaces* $V \subseteq W^{1,p}(G)$ *with* $\mathcal{D}(G) \subseteq V$

$$u \to f(u) = \int_G d^n x F(x, y(u)(x))$$

and

$$u \to h(u) = \int_G d^n x H(x, y(x))$$

are real, well-defined and continuous functions which are bounded on bounded sets. The following is true:

$$f(u) \to \infty \quad \text{for} \quad \|u\| \to \infty \quad (\|\cdot\| = \|\cdot\|_{1,p} \restriction V).$$

(b) f *and* h *are continuously Fréchet-differentiable on* V *with the derivatives*

$$\langle f'(u), v\rangle = \int_G d^n x \sum_{j=0}^{n} F_j(x, y(u)(x)) y_j(v)(x)$$

and

$$\langle h'(u), v\rangle = \int_G d^n x H_0(x, u(x)) v(x)\,, \quad \forall u, v, \in V\,.$$

(c) *If the following hypothesis* (B) *is satisfied for a sufficiently large* $c > 0$,

$$(\mathrm{B}) \begin{cases} \text{for every subset } A \subseteq M_c(f) = f^{-1}(c), \text{ on which } h \text{ is bounded,} \\ \text{there exists a } d_A > 0 \text{ such that} \\ \qquad d_A |\langle f'(u), u\rangle| \leq \langle h'(u), u\rangle|\,, \quad \forall u \in A \end{cases}$$

then there exist for all $k \in \mathbb{N}$ *eigenfunctions* $u_k \in M_c(f)$ *and eigenvalues* $\lambda_k \in \mathbb{R}$ *of the nonlinear eigenvalue equation*

$$f'(u) = \lambda h'(u)\,,$$

Proof. (a) First, $E = W^{1,p}(G)$ is an infinite-dimensional, separable, reflexive real Banach space whose norm is continuously differentiable in $E \setminus \{0\}$ due to $p > 1$. These properties are inherited by the closed subspace V with the induced norm.

Lemma 8.4.6 and 8.4.6′ justify directly the statements (a) and (b) up to the property $f(u) \to \infty$ for $\|u\| \to \infty$. The remaining statements of Lemma 8.4.6 together with Theorem 8.4.2 imply the statement $f(u) \to \infty$ for $\|u\| \to \infty$ and the rest of Hypothesis (A) in Theorem 8.4.5. Hypothesis (B_2) of Theorem 8.4.5 is satisfied due to Lemma 8.4.6′.

The hypotheses $F(x,-y) = F(x,y)$ and $H(x,-y_0) = H(x,y_0)$ finally imply the fact that f and h are even functions on V.

(b) According to Lemma 8.4.6 and Theorem 8.4.2, the level surface $M_c(f) = f^{-1}(c)$ of f is bounded and sphere-like for sufficiently large c. By (1), h is bounded on $M_c(f)$. Therefore, all the minimax values $m_{k,\mathrm{cpt}}(h)$ of h on $M_c(f)$, $k \in \mathbb{N}$, are finite. The statements (α) and (β) of Theorem 8.4.5 therefore imply our claim. □

The hypothesis of relative boundedness of f' and h' as used in Theorem 8.4.7 is not always easy to verify. We therefore recall here once again the remarks made at the end of the proof of Theorem 8.2.2, i.e. it follows that Theorem 8.4.7 remains valid if we replace the hypothesis (c) of Theorem 8.4.2 by the following hypothesis (c′) of relative boundedness of f' and h'.

(c′) There exist a $u_0 \in M_c(f)$ and an $r_0 > 0$ such that for all $u \in M_c(f)$ with $h(u) \geq h(u_0)$, $\langle h'(u), u\rangle \geq r_0$ is true.

Finally, we discuss a simple class of examples for Theorem 8.4.7. On the basis of the general results of Theorem 8.4.7 there are a few immediate generalisations of these examples which we shall, however, leave here.

Example 8.4.2. Let G be an open bounded subset of $\mathbb{R}^n$, $n \geq 3$, with smooth boundary. We wish to demonstrate the solvability of the eigenvalue equation

$$\Delta u + \alpha(x)u + \lambda\beta(x)u|u|^{\sigma-1} = 0$$

with variational boundary conditions under certain restrictions for the coefficient functions. We shall assume

$$\alpha, \beta \in L^\infty(G)\,, \quad 0 \leq \beta\,, \quad \beta \neq 0\,, \quad \alpha \leq 0\,, \quad 1 < \sigma < \frac{n+2}{n-2}\,.$$

Then

$$F(x, y_0, \underline{y}) = \frac{1}{2}\sum_{j=1}^{n} y_j^2 - \frac{1}{2}\alpha(x)y_0^2\,, \quad x \in G\,, \quad y = (y_0, \underline{y}) \in \mathbb{R}^{n+1}\,,$$

is a Carathéodory function whose derivatives

$$F_0(x, y_0, \underline{y}) = \frac{\partial F}{\partial y_0}(x, y_0, \underline{y}) = -\alpha(x)y_0\,,$$

$$F_j(x, y_0, \underline{y}) = \frac{\partial F}{\partial y_j}(x, y_0, \underline{y}) = y_j\,, \quad j = 1, \ldots, n\,,$$

are again Carathéodory functions. Since

$$\sum_{j=1}^{n} y_j F_j = \sum_{j=1}^{n} y_j^2\,,$$

the coercivity condition (K) "enforces" that this eigenvalue problem be treated in the Sobolev space $W^{1,2}(G) = H^1(G)$. Since $n \geq 3$ the exponent $p = 2$ also satisfies $1 < p < n$.

In this example the condition of monotonicity is trivially satisfied:

$$\sum_{j=1}^{n} \{F_j(x, y_0, \underline{y}) - F_j(x, y_0, \underline{y}')\}(y_j - y_j') = \sum_{j=1}^{n} (y_j - y_j')^2 .$$

A simple calculation shows that the hypotheses (D_1) and (D_2) are also satisfied. It follows that

$$f(u) = \int_G d^n x F(x, u(x), \nabla u(x)) = \frac{1}{2}\langle \nabla u, \nabla u\rangle_2 - \frac{1}{2}\langle u, \alpha u\rangle_2$$

is well defined on $E = H^1(G)$ and is continuous, bounded and Fréchet-differentiable with the derivative

$$\langle f'(u), v\rangle = \sum_{j=1}^{n} \langle \partial_j u, \partial_j v\rangle_2 - \langle u, \alpha v\rangle_2 , \quad \forall v \in E .$$

$f' : E \to E'$ is continuous, bounded and satisfies the condition (S). Similarly,

$$H(x, y_0) = \frac{\beta(x)}{\sigma + 1}(y_0^2)^{(\sigma+1)/2} , \quad x \in G , \quad y_0 \in \mathbb{R} ,$$

is a Carathéodory function, whose derivative

$$H_0(x, y_0) = \frac{\partial H}{\partial y_0}(x, y_0) = \beta(x) y_0 (y_0^2)^{(\sigma-1)/2}$$

is again a Carathéodory function. Once again we can easily verify that because of

$$1 < \sigma < \frac{n+2}{n-2} ,$$

the hypotheses (D_3) and (D_4) are satisfied. Consequently,

$$h(u) = \int_G d^n x H(x, u(x)) = \frac{1}{\sigma + 1}\int_G d^n x \beta(x)(u^2(x))^{(\sigma+1)/2}$$

is well defined on $E = H^1(G)$, continuous, bounded and Fréchet-differentiable with the derivative

$$\langle h'(u), v\rangle = \int d^n x \beta(x)(u^2(x))^{(\sigma-1)/2} u(x) v(x) , \quad \forall v \in E .$$

$h' : E \to E'$ is continuous, bounded, completely continuous and weakly continuous on bounded sets.

Finally, we prove the hypothesis of relative boundedness in the form (c'). We first note that

$$\langle f'(u), u\rangle = \langle \nabla u, \nabla u\rangle_2 - \langle u, \alpha u\rangle_2 = 2f(u)\,,$$
$$\langle h'(u), v\rangle = \int_G d^n x \beta(x)(u^2(x))^{(\sigma+1)/2} = (\sigma+1)h(u)\,.$$

Because of $\beta(x) \geq 0$ and $\beta \neq 0$ there exists a $u_1 \in H_0^1(G)$ with

$$h(u_1) > 0\,.$$

By Theorem 8.4.2, the properties of f verified above imply that there exists a positive number c_0 such that level surface $M_c(f)$ for $c \geq c_0$ are sphere-like. Now setting $u_r = ru_1$, $r > 0$, we get $f(u_r) = r^2 f(u_1)$. From $\alpha \leq 0$ and $u_1 \neq 0$, one has $f(u_1) > 0$. Therefore, $c_r = f(u_r)$ satisfies the condition $c_r \geq c_0$ for all $r \geq R_0$.

Now choosing a fixed $r \geq r_0$ and setting

$$r_0 = (\sigma+1)h(u_r) \quad (= (\sigma+1)r^{\sigma+1}h(u_1) > 0)\,,$$

we obtain for all $u \in M_{c_r}(f)$ with $h(u) \geq h(u_r)$ the estimate

$$\langle h'(u), u\rangle = (\sigma+1)h(u) \geq (\sigma+1)h(u_r) = r_0 > 0\,.$$

This shows condition (c') for the sphere-like level surfaces $M_{c_r}(f)$. Since f and h are even functions on E, we can apply Theorem 8.4.7 and obtain a sequence $\{u_j\}_{j\in\mathbb{N}}$ of eigenfunctions $u_j \in M_{c_r}(f)$ and a sequence $\{\lambda_j\}_{j\in\mathbb{N}}$ of eigenvalues $\lambda_j \in \mathbb{R}$:

$$f'(u_j) = \lambda_j h'(u_j)\,, \quad \forall j \in \mathbb{N}\,.$$

Here we may assume arbitrary variational boundary conditions which are induced by a subspace V of $H^1(G)$ with $H_0^1(G) \subseteq V$. Moreover, the above conditions hold for every fixed $r \geq R_0$. We therefore also obtain a sequence of eigenfunctions u_j, $j \in \mathbb{N}$, on all level surfaces $M_{c_r}(f)$, $r \geq R_0$.

Remark 8.4.3. The above example is naturally very special and simple with respect to the general formulation in Theorem 8.4.7. One therefore expects methods which are simpler and more direct to lead also to the solution of this eigenvalue problem. This is indeed the case, as shown as a special case in [Ref. 8.20, Theorem 6.3.9].

9. Semilinear Elliptic Differential Equations. Some Recent Results on Global Solutions

9.1 Introduction

All the results on concrete boundary and eigenvalue problems presented in Chaps. 6–8 use in an essential way at least one of the following two assumptions:

1. The domain $G \subset \mathbb{R}^d$, over which the problem is considered, is bounded.
2. The exponents which are used to bound the nonlinearities are strictly smaller than the *"critical Sobolev exponent"* p^* defined by

$$\frac{1}{p^*} = \frac{1}{p} - \frac{1}{d} \quad \text{or} \quad p^* = \frac{dp}{d-p}$$

   when one looks for solutions in $W^{1,p}(G)$, $1 \leq p < d$.

We recall that the reason for this restriction is due to the fact that in the proofs of these results compact Sobolev-embeddings had to be used. In this chapter we consider problems where at least one of these two assumptions is not satisfied and discuss some of their recent solutions. It is clear that the arguments used in Chaps. 6–8 will not suffice to conclude in these cases.

More concretely, we consider the following type of equation:

$$\begin{gathered} -\Delta u(x) = g(x, u(x)), \\ u : \Omega \to \mathbb{R}^n, \quad \Omega \subseteq \mathbb{R}^d, \quad d \geq 2, \quad n \geq 1, \end{gathered} \tag{9.1.1}$$

where in most cases the domain Ω equals $\mathbb{R}^d$. Under appropriate assumptions on the nonlinearity g we will show the existence of (at least) one solution.

Equations of the type (9.1.1) occur in many branches of science. They are used, for instance, in mathematical physics as global classical vector field equations. Stationary solutions of nonlinear Schrödinger and nonlinear Klein-Gordon equations are governed by such an equation. And special solutions of the Yang-Mills equation can be obtained by solving an equation of the form (9.1.1). Furthermore, stationary solutions of reaction-diffusion equations

$$\partial_t u - \Delta u = \widehat{g}(u)$$

have to satisfy an equation of the type (9.1.1). Note that reaction-diffusion equations themthelves have a huge area of application in biology, chemistry, physics, and reaction kinetics. For more details and further references see [9.1–7].

Such equations also occur in mathematics, for instance in analysis in the problem of determining the best constant S in Sobolev's inequality:

$$\|\varphi\|_{2^*} \leq S\|\nabla\varphi\|_2 , \quad \varphi \in \mathcal{D}(\mathbb{R}^d), \quad 2^* = \frac{2d}{d-2}, \quad d \geq 3 . \tag{9.1.2}$$

We present a solution to this problem in Sect. 9.4 where we show that it amounts to solving the equation

$$-\Delta u = au^{2^*-1} \tag{9.1.3}$$

on $\mathbb{R}^d$ for some constant $a > 0$ and $d \geq 3$. Here we have to solve a variational problem over the whole Euclidean space, $\Omega = \mathbb{R}^d$ in (9.1.1), involving the critical Sobolev exponent (see Proposition 9.7.4 for an explanation of why this is a "limit case" in the variational approach).

Finally, (9.1.1) is of importance in differential geometry. A solution of the "*Yamabe problem* for noncompact Riemannian manifold" (see, for instance [9.8]) is obtained by solving an equation of type (9.1.1). For some cases the solution is given by the results of Sect. 9.8.1.

Up to now, the only methods which have been available for solving equation (9.1.1) in some generality have been variational methods (critical point theory, minimisation). Clearly these methods are based on the assumption that (9.1.1) is of "potential" type. Therefore, we shall always assume that the nonlinearity g has a potential G, i.e., a function

$$G : \Omega \times \mathbb{R}^n \to \mathbb{R}$$

such that for almost all $x \in \Omega$ and all $y \neq 0$

$$g(x,y) = \operatorname{grad}_y G(x,y) \quad \text{and} \quad G(x,0) = 0 . \tag{9.1.4}$$

Also we always assume that g and G are C-functions on $\Omega \times \mathbb{R}^n$.

As we have already seen in the preceding sections, the choice of the space E of a priori solutions of (9.1.1) is of basic importance for a successful treatment. From the context in which these equations are used, the following conditions on a solution of (9.1.1) are natural. A solution u of (9.1.1) should

(a) have finite *kinetic energy*

$$K(u) = \frac{1}{2}\|\nabla u\|_2^2 = \frac{1}{2}\int_\Omega \sum_{j=1}^n \sum_{i=1}^d \left|\frac{\partial u_j}{\partial x_i}(x)\right|^2 dx , \tag{9.1.5}$$

(b) vanish on the boundary $\partial\Omega$ of Ω

$$u \restriction \partial\Omega = 0 .$$

Furthermore, in order to have the possibility of interpreting (9.1.1) in a distributional sense, the space $\mathcal{D}(\Omega; \mathbb{R}^n)$ of the Schwartz test functions with compact support should be densely and continuously embedded into the space E of a priori solutions and it should be assured by assumptions on the non-linearity g that $\widehat{g}(u)$ is locally integrable for any $u \in E$.

Thus, we are led in a natural way to define the space of a priori solutions E to be the completion

$$E^{1,2}(\Omega; \mathbb{R}^n) \tag{9.1.6}$$

of $\mathcal{D}(\Omega; \mathbb{R}^n)$ with respect to the *"energy norm"*

$$\|u\|_E = (2K(u))^{1/2} = \|\nabla u\|_2 \,. \tag{9.1.7}$$

If the dimension d of the underlying space, $\Omega \subseteq \mathbb{R}^d$, is not smaller than 3, $d \geq 3$, then by Sobolev's inequality (9.1.2) this completion can be realised as a subspace of the Lebesgue space

$$L^{2^*}(\Omega; \mathbb{R}^n)\,,$$

and hence forth we shall always consider $E^{1,2}(\Omega; \mathbb{R}^n)$ as a subspace of this Lebesgue space. As we will see later, this space of a priori solutions is indeed very useful for many problems on $\Omega \subset \mathbb{R}^d$, $d \geq 3$.

If, however, the dimension equals 2, i.e. $d = 2$, then Sobolev's inequality breaks down and as a consequence of this the completion $E^{1,2}(\mathbb{R}^2; \mathbb{R}^n)$ is not even a space of distributions, as has been shown by J. L. Lions [9.9]. Hence it is not very useful for our purposes, and we have to make another choice for the space of a priori solutions in two-dimensional problems. This is the main important reason why the case of two-dimensional vector field equations of type (9.1.1) is discussed separately in Sect. 9.9.

Section 9.2 provides technical preparation for the following sections. First, of all, some basic properties of the spaces $E^{1,2}(\Omega; \mathbb{R}^n)$, $\Omega \subseteq \mathbb{R}^d$, $d \geq 3$, and their two-dimensional counterparts are proved. Then some continuity properties for Niemytski operators $\widehat{G}$ on these spaces which are associated with the potential G will follow. This allows us to have the necessary control over the *"potential energy"* V defined on the subset $D(V) = \{u \in E \,|\, \widehat{G}(u) \in L^1(\Omega)\}$ by

$$V(u) = \int_\Omega \widehat{G}(u)(x)\,dx\,, \quad \widehat{G}(u)(x) = G(x, u(x))\,. \tag{9.1.8}$$

Section 9.2.2 recalls the definition and basic properties of the spherically symmetric rearrangement of functions. In Sect. 9.2.4, a simple but typical one-dimensional variational problem is solved in order to give a very transparent example of how the combination of the techniques presented thus far implies the solution. A particular case of this result will be used in Sect. 9.4.

Before we continue our brief description of the contents of Sects. 9.3–10, we want to make some comments on the development of the ideas which finally led to a solution of equations of type (9.1.1) in substantial generality. In order to do this conveniently some terminology has to be introduced.

If $n = 1$ in (9.1.1) we have a *scalar field equation*, while for $n > 1$ we call it a *vector field equation*. Whenever the nonlinearity g in (9.1.1) does not depend on the "space variable" $x \in \Omega$ but only on $y \in \mathbb{R}^n$, we speak about the *translation-invariant case*. Otherwise, we speak about the *non-translation-invariant case*. For translation-invariant equations there is another, at least historically quite important, distinction between the *positive mass case*, i.e. there is an $m > 0$ such that ($G_0^+ = \max\{0, G_0\}$)

$$G(y) = -\frac{m}{2}y^2 + G_0(y) \tag{9.1.9}$$

with

$$|y|^{-2}G_0^+(y) \to 0 \quad \text{for } |y| \to 0$$

and the *mass zero case*, i.e. (9.1.1) holds with $m = 0$.

In Chap. 8, for a bounded region Ω, we considered nonlinear elliptic eigenvalue problems of the form

$$f'(u) = \lambda h'(u)\,, \quad u \in E\,, \quad \lambda \in \mathbb{R}\,, \tag{9.1.10}$$

where f and h are two Fréchet-differentiable functions on some Banach space E of functions on Ω, by determining critical points of f on suitable level surfaces of h, i.e. in the simplest case by solving a *constrained minimisation problem* of the form

$$I_c = \inf\{f(u) \,|\, u \in E, h(u) = c\}\,.$$

For the case of *global field equations*, i.e. if in (9.1.1) Ω equals $\mathbb{R}^d$, $d \geq 2$, the arguments given in Chap. 8 do not suffice. For translation-invariant scalar field equations, *G. Talenti* [9.10] in 1976 and *W. Strauss* [9.11] in 1977 seem to have been the first to realise how to use spherically symmetric rearrangement of functions in such a global problem in order to solve a constrained minimisation problem for the functions K and V according to definitions (9.1.5) and (9.1.8):

$$I = \inf\{K(u) \,|\, u \in D(V), V(u) = 1\}\,. \tag{9.1.11}$$

By assumption on the potential it is assured that the domain $D(V)$ of V equals the space $E = H^1(\mathbb{R}^d)$ of a priori solutions and that the functions K and V are Fréchet-differentiable on E. Then a solution of the corresponding eigenvalue problem (9.1.10),

$$K'(u) = \lambda V'(u)\,, \tag{9.1.12}$$

follows just as in Chap. 8 with the help of the Lagrange multiplier theorem (Theorem 4.2.3). In this situation a solution of (9.1.1) is obtained by showing that this eigenvalue problem has the eigenvalue $\lambda = 1$.

For translation-invariant global field equations the associated functions K and V are always covariant under the following change of scale of the arguments of the function u:

$$K(u_\sigma) = \sigma^{2-d}K(u)\,, \quad V(u_\sigma) = \sigma^{-d}V(u)\,, \quad u_\sigma(\cdot) = u(\sigma\cdot) \tag{9.1.13}$$

This was first used for scalar field equations in [9.10] and in [9.12] by *V. Glaser* et al. to show the following. If the associated eigenvalue problem (9.1.12) has a positive eigenvalue at all, then simply by scale covariance every positive number is an eigenvalue. Thus, in particular, $\lambda = 1$ is then an eigenvalue and the corresponding eigenfunction is a solution of our field equation.

Most of the subsequent articles on solutions of equations of the form (9.1.1) rely on this idea. We explain a substantial generalisation of this idea in Sect. 9.6 and apply it in Sects. 9.7–9. In these sections we will also learn why translation-invariant global vector field equations with mass zero and critical Sobolev exponent are the most difficult type of equation. In the corresponding case for scalar fields there is a considerable simplification because then the technique of spherically symmetric rearrangement of functions is usually available. In Sect. 9.4 we describe G. Talenti's solution for equation (9.1.3) which is of this type.

It took sometime to learn how to solve vector field equations of type (9.1.1) when the method of spherically symmetric rearrangement of functions does not apply. For the case of translation-invariant global vector field equations (including mass zero) the first results seem to be due to *H. Brezis* and *E. H. Lieb* [9.13] and *P. L. Lions* [9.6]. Here the most important step to learn was how to "concentrate" the elements of a minimising sequence for (9.1.11) on a common compact subset of $\mathbb{R}^d$, since in such a problem the elements u_j of a minimising sequence can easily "disappear towards infinity" for $j \to \infty$, i.e. the weak limit of such a sequence vanishes.

Because of its great importance, the corresponding *"lemma of concentration by translation"* of *Lieb* is proven completely in Sect. 9.2. We think that it is considerably easier to use than the more general *"compactness principle"* of *P. L. Lions* which is explained in great detail in [9.6].

In a more general setting, an abstract solution of this concentration problem for minimising sequences of (9.1.11) was given by one of the authors (*E. Brüning*) in 1986 [9.1], on the basis of an appropriate notion of "symmetry" for the constrained minimisation problem (9.1.11). This solution, which is based on a new continuity criterion for Niemytski operators on $E^{1,2}(\mathbb{R}^d;\mathbb{R}^n)$ (Theorem 9.2.3) also allows some non-translation-invariant problems to be covered easily. Our Sects. 9.6–9 mainly rely on this. In Sect. 9.6 the general setting of a constrained minimisation problem with a scale covariance of general type is explained. A result on the existence of a Lagrange multiplier is proved which, instead of using Fréchet-differentiability of the potential V, uses only Gâteaux-differentiability of V in all directions of $\mathcal{D}$ at all points where V is defined. On an abstract level, a solution of this constrained minimisation problem in terms of necessary and sufficient conditions is presented in Sect. 9.7, while Sect. 9.8 treats some applications, in particular the cases presented in [9.1, 11–13, 15] and some of the examples from [9.6, 16]. An important point of this approach which we want to stress is that it also covers various interesting non-translation-invariant cases (Sect. 9.8.1).

We comment now on the difference between the "positive mass" and the "mass zero" cases. When *W. Strauss* [9.11] considered translation-invariant global scalar field equations with positive mass he worked naturally in the Sobolev space $H^1(\mathbb{R}^d)$, $d \geq 3$; then a minimising sequence for the constrained minimisation problem (9.1.11) is always bounded in $H^1(\mathbb{R}^d)$. However, in the mass zero case for the corresponding minimisation problem, this is no longer true when working with the Sobolev space $H^1(\mathbb{R}^d)$. This is the main reason why we proposed to take the space $E = E^{1,2}(\mathbb{R}^d) \supsetneqq H^1(\mathbb{R}^d)$ as a space of a priori solutions. Then, clearly, minimising sequences for problem (9.1.11) are again bounded in the space of a priori solutions.

The most important aspect of W. Strauss' article [9.11] is the following observation: though the Sobolev space $H^1(\mathbb{R}^d)$ is not compactly embedded in any L^p space, the subspace $H_r^1(\mathbb{R}^d)$ of spherically symmetric elements in $H^1(\mathbb{R}^d)$ is nevertheless compactly embedded into $L^q(\mathbb{R}^d)$ for all $q \in (2, 2^*)$. The main reason for this is that all elements u of a bounded subset M of $H_r^1(\mathbb{R}^d)$ have uniform decay at infinity. This observation has been used in many subsequent articles, for instance in [9.1, 6, 12, 15–19]. We present an independent proof of a slightly extended version of all these results in Sect. 9.8.2 based on the notion of *"concentration of a minimising sequence"* as explained in Sect. 9.7. By the uniform decay property of bounded sets in $H_r^1(\mathbb{R}^d)$ this concentration condition is trivially satisfied.

In favourable situations the variational methods described above yield a weak or distributional solution u of (9.1.1), i.e. an element $u \in E = E^{1,2}$ such that

$$\langle \nabla u, \nabla \varphi \rangle_2 = -\langle u, \Delta \varphi \rangle_2 = \langle \widehat{g}(u), \varphi \rangle_2 \tag{9.1.14}$$

holds for all testfunctions $\varphi \in \mathcal{D}$. But clearly the existence of classical solutions of (9.1.1) is an even more important problem. Therefore, we discuss in Sect. 9.3 some regularity properties of distributional solutions of (9.1.1) for which $\widehat{g}(u) \in L^1_{\text{loc}}$ is assured by appropriate assumptions on g.

To this end we explain and prove in a rather comprehensive way the basic steps of the *"elliptic regularity theory"* for the above type of equation. This will enable us to conclude that in many cases of interest we actually obtain by our variational methods classical solutions of (9.1.1), i.e. functions u which satisfy (9.1.1) pointwise (at least almost everywhere).

Whenever a critical exponent is involved in (9.1.1) there are various problems which are at the moment not fully understood. The special case of (9.1.3) will be discussed in Sect. 9.4. Some results for other global equations are presented in Sect. 9.8 for non-translation-invariant problems. A more comprehensive reference for this is the part "The limit case" of [9.6]. Somewhat surprisingly, even the case of local (i.e. when Ω is a bounded domain in $\mathbb{R}^d$) scalar field equations with critical Sobolev exponent is not fully understood. In order to explain some of the relevant problems we discuss the results of

H. Brezis and *L. Nirenberg* [9.20] obtained in 1983 in some detail in Sect. 9.5. It turns out that the three-dimensional case seems to be considerably harder to solve than the d-dimensional case for $d \geq 4$.

Finally, besides some conclusions, Sect. 9.10 contains several comments on related problems and methods which are not explained in this book but which seem to be quite promising. In addition we suggest some more detailed articles on these kinds of problems for further reading.

9.2 Technical Preliminaries

9.2.1 Some Function Spaces and Their Properties

Consider (9.1.1) over a domain $\Omega \subseteq \mathbb{R}^d$, $d \geq 3$; in Sect. 9.1 we gave a motivation for taking the space $E = E^{1,2}(\Omega; \mathbb{R}^n)$, the completion of $\mathcal{D}(\Omega; \mathbb{R}^n)$ with respect to the energy norm (9.1.7), as the space of a priori solutions. And we also gave some motivation as to why the two-dimensional case requires some other space of a priori solutions. In this section we investigate the basic properties of function spaces of such a type. To this end we consider, somewhat more generally, the following class of spaces.

For $1 \leq p < d$ denote by $E^{1,p}(\Omega; \mathbb{R}^n)$ the completion of $\mathcal{D}(\Omega; \mathbb{R}^n)$ with respect to the norm

$$v_0 \mapsto \|\nabla\varphi\|_p = |\varphi|_{1,p} = \left(\int_\Omega \sum_{j=1}^{n} \sum_{i=1}^{d} \left| \frac{\partial \varphi_j}{\partial x_i}(x) \right|^p dx \right)^{1/p}, \tag{9.2.1}$$

and for $2 \leq d = p$ and some $d \leq q < \infty$ denote by $E_q^1(\Omega; \mathbb{R}^n)$ the completion of $\mathcal{D}(\Omega; \mathbb{R}^n)$ with respect to the norm

$$\varphi \mapsto \|\varphi\| = \|\varphi\|_q + \|\nabla\varphi\|_d, \tag{9.2.2}$$

where $\|\cdot\|_r$, $r = q, d$, denotes the norm of the Lebesgue space $L^r(\Omega)$. Sobolev's inequality says that there is some constant C such that

$$\|\varphi\|_{p^*} \leq C\|\nabla\varphi\|_p, \quad p^* = \frac{pd}{d-p}, \quad 1 \leq p < d, \tag{9.2.3}$$

holds for all $\varphi \in \mathcal{D}(\Omega; \mathbb{R}^n)$. Hence the norms

$$|\cdot|_{1,p} \quad \text{and} \quad |\cdot|_{1,p} + \|\cdot\|_{p^*} = \|\cdot\| \tag{9.2.4}$$

are equivalent and therefore we can suppose that the completion $E^{1,p} = E^{1,p}(\Omega; \mathbb{R}^n)$ is realised as a subspace of $L^{p^*}(\Omega; \mathbb{R}^n)$:

$$E^{1,p}(\Omega; \mathbb{R}^n) = \{u \in L^{p^*}(\Omega; \mathbb{R}^n) \,|\, \partial_i u \in L^p(\Omega; \mathbb{R}^n),\ i = 1, \ldots, d\} \tag{9.2.5}$$

where $\partial_i u$ denotes the weak derivative of u (see Appendix D).

Embedding relations for the second class of spaces $E_q^1(\Omega; \mathbb{R}^n)$, $d \leq q < \infty$, will be discussed later.

In Sect. 9.4 we will determine for $\Omega = \mathbb{R}^d$ the best possible constant S in Sobolev's inequality (9.2.3), i.e.

$$S = S(p,d) = \inf\{c > 0 \mid \|\varphi\|_{p^*} \le c\|\nabla\varphi\|_p \quad \text{for all } \varphi \in \mathcal{D}\}. \tag{9.2.6}$$

Here we recall only that the exponent p^* in inequality (9.2.3) is the only exponent r for which such an inequality holds.

If there is a constant $c > 0$ such that $\|v\|_r \le c\|\nabla v\|_p$ holds for all $v \in \mathcal{D}$, then it holds in particular for all v_σ, $\sigma > 0$, $v_\sigma(\cdot) = v(\sigma\cdot)$, with some fixed $v \in \mathcal{D}$. An elementary calculation shows

$$\|v_\sigma\|_r = \sigma^{-d/r}\|v\|_r\,, \quad \|\nabla v_\sigma\|_p = \sigma^{1-d/p}\|\nabla v\|_p\,, \tag{9.2.7}$$

and by assumption we know that $\sigma^{-d/r}\|v\|_r \le c\sigma^{1-d/p}\|\nabla v\|_p$ for all $\sigma > 0$. Since $v \ne 0$ this inequality holds if, and only if,

$$-d/r = 1 - d/p\,, \quad \text{i.e.} \quad r = p^*\,.$$

Remark 9.2.1. (a) Note that for $2 = d = q$ the space $E_q^1(\Omega, \mathbb{R}^n)$ is just the Sobolev space $H_0^1(\Omega; \mathbb{R}^n)$, $\Omega \subseteq \mathbb{R}^2$.

(b) Later we will use only the space $E^{1,2}(\Omega; \mathbb{R}^n)$ for $d \ge 3$ and the spaces $E_q^1(\Omega; \mathbb{R}^n)$ for $2 = d \le q < \infty$.

Theorem 9.2.1. *The spaces $E^{1,p}(\Omega; \mathbb{R}^n)$, for a domain $\Omega \subseteq \mathbb{R}^d$, $1 < p < d$, and the spaces $E_q^1(\Omega; \mathbb{R}^n)$, $2 \le d \le q < \infty$, are real separable reflexive Banach spaces.*

Proof. We give the proof only for the first class of spaces. For the second class it is essentially the same. It suffices to use some standard method (cf. Appendix D). Introduce a function space

$$F = L^{p^*}(\Omega; \mathbb{R}^n) \oplus L^p(\Omega; \mathbb{R}^n)^d$$

equipped with the norm

$$\|v\|_F = \|v_0\|_{p^*} + \left(\sum_{i=1}^d \|v_i\|_p^p\right)^{1/p} \tag{9.2.8}$$

for $v = (v_0, v_1, \dots, v_d) \in F$, i.e. $v_0 \in L^{p^*}$, $v_i \in L^p$, $i = 1, \dots, d$. Since $1 < p < p^* < \infty$, F is a real separable reflexive Banach space (see, for instance Appendix D).

A linear map $y : E^{1,p}(\Omega; \mathbb{R}^n) \to F$ is well defined by

$$y(u) = (u, \partial_1 u\,, \dots, \partial_d u)$$

and if, for simplicity, the extension of the norm (9.2.4) to $E^{1,p}$ is denoted in the same way, (9.2.4, 5) and (9.2.8) imply for all $u \in E^{1,p}$ that

$$\|y(u)\|_F = \|u\|\,.$$

Thus y is an isometry. By definition of $E^{1,p}$, $y(E^{1,p})$ is a closed subspace of F. It follows by the inverse mapping theorem (Appendix A) that y defines an isomorphism of the Banach space $E^{1,p}$ onto a closed subspace $y(E^{1,p})$ of the separable reflexive real Banach space F. Hence, when equipped with the norm $\|\cdot\| = \|\cdot\|_{p^*} + \|\nabla\cdot\|_p$, $E^{1,p}$ is a separable reflexive real Banach space. But by inequality (9.2.3) this norm is equivalent to the original norm $|\cdot|_{1,p} = \|\nabla\cdot\|_p$ and thus we conclude. □

According to this theorem and Appendix C, it is known that every bounded subset of $E = E^{1,p}$ or of $E = E^1_q$ is relatively weakly compact. But since these spaces are "concrete" function spaces we are in a position to say considerably more about bounded sets. This additional information will be of primary importance in subsequent applications.

Theorem 9.2.2. *For any bounded sequence $(u^j)_{j\in\mathbb{N}}$ in $E = E^{1,p}(\Omega;\mathbb{R}^n)$ there is a subsequence $j(i)$, $i\in\mathbb{N}$, and there is a $u\in E$ such that*

$$u^{j(i)} \to u \quad \textit{for } i\to\infty$$

(a) *almost everywhere on Ω,*
(b) *weakly in E, i.e. $\nabla u^{j(i)} \underset{i\to\infty}{\longrightarrow} \nabla u$ weakly in L^p,*
(c) *weakly in L^{p^*},*
(d) *in $L^r_{\rm loc}(\Omega;\mathbb{R}^n)$ for fixed $r\in[1,p^*)$.*

Proof. (a) Suppose $(u^j)_{j\in\mathbb{N}}$ to be a sequence in E with

$$\sup_j \|\nabla u_j\|_p = C < \infty$$

and suppose that $r\in[1,p^*)$ is fixed. In the first step we show that for any $f\in\mathcal{D}$ the sequence $(fu^j)_{j\in\mathbb{N}}$ is relatively compact in $L^r(\mathbb{R}^d)$.

By Hölder's and Sobolev's inequality this sequence is bounded in $L^r(\mathbb{R}^d)$:

$$\|fu^j\|_r \le \|f\|_s\, \|u^j\|_{p^*} \le \|f\|_s \le SC\,, \quad \frac{1}{s} = \frac{1}{r} - \frac{1}{p^*}\,.$$

Hence by the theorem of Kolmogoroff and M. Riesz (Appendix C) it suffices to show that the translations τ_a, $a\in\mathbb{R}^d$, act uniformly continuously on this sequence. For $r=1$ we know

$$\begin{aligned}\|\tau_a(fu^j) - fu^j\|_1 &= \left\| \int_0^1 dt\, a\tau_{ta}(\nabla(fu^j))\right\|_1 \le |a|\ \|\nabla(fu^j)\|_1 \\ &\le |a|(S\|\nabla f\|_q + \|f\|_{p'})\|\nabla u^j\|_p \le |a|C_f C\end{aligned}$$

where again Hölder's and Sobolev's inequality have been used and where q is the Hölder conjugate exponent to p^*. For $1 < r < p^*$ we use the interpolation inequality

$$\|fu^j\|_r \le \|fu^j\|_1^\theta \|fu^j\|_{p^*}^{1-\theta}, \quad \frac{1}{r} = \theta + \frac{1-\theta}{p^*},$$

to obtain

$$\begin{aligned}\|\tau_a(fu^j) - fu^j\|_r &\le |a|^\theta \|\nabla(fu^j)\|_1^\theta (2\|fu^i\|_{p^*})^{1-\theta} \\ &\le |a|^\theta (C_f C)^\theta (2S)^{1-\theta} \|f\|_\infty^{1-\theta} \|\nabla u^j\|_{p^*}^{1-\theta} \\ &\le |a|^\theta C(f).\end{aligned}$$

Since $\operatorname{supp}(fu^j) \subseteq \operatorname{supp} f$ we conclude that $(fu^j)_{j\in\mathbb{N}}$ is relatively compact in $L^r(\mathbb{R}^d)$.

(b) Now choose some function $\chi \in \mathcal{D}(\mathbb{R}^d)$, $0 \le \chi \le 1$, with $\chi(x) = 1$ for all $x \in \mathbb{R}^d$, $|x| \le 1$, and $\chi(x) = 0$ for all $x \in \mathbb{R}^d$ with $|x| > 2$; then we define a sequence $(\chi_k)_{k\in\mathbb{N}}$ by

$$\chi_k(x) = \chi\Big(\frac{1}{k}x\Big).$$

Clearly $\chi_k \in \mathcal{D}(\mathbb{R}^d)$ and $\chi_k(x) = 1$ for all $|x| \le k$ and $\chi_k(x) = 0$ for all $|x| > 2k$.

By the first step we know that the sequence $(\chi_1 u^j)_{j\in\mathbb{N}}$ is relatively compact in $L^r(\mathbb{R}^d)$. Hence there are a subsequence $j_1(i)$, $i \in \mathbb{N}$, and an element $g_1 \in L^r(\mathbb{R}^d)$ such that

$$\chi_1 u^{j_1(i)} \to g_1 \quad \text{for } i \to \infty$$

in $L^r(\mathbb{R}^d)$ and almost everywhere.

The same argument applies to the sequence $(\chi_2 u^{j_1(i)})_{i\in\mathbb{N}}$. Again there are a subsequence $j_2(i)$, $i \in \mathbb{N}$, of $j_1(i)$, $i \in \mathbb{N}$, and an element $g_2 \in L^r(\mathbb{R}^d)$ such that

$$\chi_2 u^{j_2(i)} \to g_2 \quad \text{for } i \to \infty$$

in $L^r(\mathbb{R}^d)$ and almost everywhere. Clearly $g_1(x) = g_2(x)$ for almost all $|x| \le 1$.

Using this procedure successively, we obtain a sequence of subsequences $(j_k(i))_{i\in\mathbb{N}}$, $k = 1, 2, \ldots,$ with the following properties:

(i) $(j_{k+1}(i))_{i\in\mathbb{N}}$ is a subsequence of $(j_k(i))_{i\in\mathbb{N}}$.

(ii)
$$\chi_k u^{j_k(i)} \to g_k \quad \text{for } i \to \infty$$
in $L^r(\mathbb{R}^d)$ and pointwise almost everywhere.

(iii)
$$g_{k+1}(x) = g_k(x) \quad \text{for almost all } |x| \le k.$$

Denote by N_k a null set in $\mathbb{R}^d$ such that for all $x \notin N_k$

$$\chi_k u^{j_k(i)}(x) \underset{i\to\infty}{\longrightarrow} g_k(x) \quad \text{and} \quad g_{k+1}(x) = g_k(x) \ \text{ if } |x| \le k.$$

Then $N = \bigcup_{k=1}^\infty N_k$ is again a null set in $\mathbb{R}^d$.

By property (iii), a function $u : \mathbb{R}^d \to \mathbb{R}^n$ is well defined by

$$u(x) = g_k(x) \quad \text{if } |x| \le k \text{ and } x \notin N\,,$$
$$u(x) = 0 \quad \text{if } x \in N\,.$$

Consider now the diagonal sequence $j(i) = j_i(i)$, $i \in \mathbb{N}$. For any compact set $K \subset \Omega$ one can find $i_0 \in \mathbb{N}$ such that $\chi_{i_0} \restriction K = 1$. Then for all $i \ge i_0$ and almost every $x \in K$, we know

$$u^{j(i)}(x) = \chi_{i_0}(x) u^{j_i(i)}(x)\,.$$

By construction, the right-hand side converges to g_{i_0} in $L^r(\mathbb{R}^d)$ and pointwise for all $x \in K \setminus N$. Thus it follows for all $x \notin N$ that

$$u^{j(i)}(x) \to u(x)\,, \quad i \to \infty\,,$$

and $u^{j(i)} \to u$ in $L^r_{\text{loc}}(\mathbb{R}^d)$.

(c) $(u^j)_{j\in\mathbb{N}}$ being bounded in the reflexive Banach spaces $E = E^{1,p}(\Omega;\mathbb{R}^n)$ and $L^{p^*}(\Omega;\mathbb{R}^n)$, we can assume in addition that the subsequence $(u^{j(i)})_{j\in\mathbb{N}}$ constructed above converges weakly in E to $v \in E$, let us say, and in L^{p^*} to $w \in L^{p^*}$. If we can show $u = v = w$, the proof is complete.

For any $g \in C^1_0(\Omega,\mathbb{R}^n)$ with compact support K, the following relations hold:

(α)
$$\langle g, w\rangle_2 = \lim_{i\to\infty} \langle g, u^{j(i)}\rangle_2$$
since g belongs to L^q, $1/q + 1/p^* = 1$;

(β)
$$\langle g, u\rangle_2 = \lim_{i\to\infty} \langle g, u^{j(i)}\rangle_2$$
since $g \in L^{r'}$, $1/r' + 1/r = 1$, and $u^{j(i)} \to u$ in L^r_{loc};

(γ)
$$\langle \partial_\nu g, u\rangle_2 = \lim_{i\to\infty} \langle \partial_\nu g, u^{j(i)}\rangle_2$$
for the same reason;

(δ)
$$\lim_{i\to\infty} \langle g, \partial_\nu u^{j(i)}\rangle_2 = \langle g, \partial_\nu v\rangle_2$$
since $g \in L^{p'}$ and $\partial_\nu u^{j(i)} \to \partial_\nu v$ weakly in L^p.

(α) and (β) imply $u = w \in L^{p^*}$, while according to (γ) and (δ),

$$\langle \partial_\nu g, u\rangle_2 = -\langle g, \partial_\nu v\rangle_2$$

holds for all $g \in C^1_0(\Omega;\mathbb{R}^n)$. Hence $u \in L^{p^*}$ has weak derivatives $\partial_\nu u = \partial_\nu v \in L^p$, $\nu = 1,\ldots,d$; thus, $u \in E^{1,p}(\Omega;\mathbb{R}^n)$ and $\nabla u = \nabla v$, and therefore $u - v = \text{const.}$, and the boundary conditions imply $u = v$. □

The corresponding result for the spaces $E^1_q(\Omega;\mathbb{R}^n)$, $2 \le d \le q < \infty$, is prepared for by means of an important embedding theorem for these spaces. Its proof is given in Appendix E.

Theorem 9.2.3. *For all $2 \le d \le q < \infty$ we have*

(a) $$E_q(\mathbb{R}^d;\mathbb{R}^n) \hookrightarrow L^r(\mathbb{R}^d;\mathbb{R}^n) \quad \text{for all } r \ge q$$

with continuous injections expressed by the inequality

$$\|u\|_r^r \le C_r \|u\|_q^q \Big(\frac{1}{d}\|\nabla\|_d\Big)^{r-q} \tag{9.2.9}$$

for all $u \in E_q \equiv E_q(\mathbb{R}^d;\mathbb{R}^n)$.

(b) *In particular, the following inequalities hold for all $u \in E_q$, all $k = 0,1,2,\ldots$:*

$$\|u\|_{r(k)}^{r(k)} \le (\alpha)_k^p \|u\|_q^q \Big(\frac{1}{d}\|\nabla u\|_d\Big)^{kp} \tag{9.2.10}$$

where $r(k) = q + kp$, $\alpha = 1 + q/p$, $1/p = 1 - 1/d$, and $(\alpha)_k = \alpha(\alpha+1)\cdots(\alpha+k-1)$.

Remark 9.2.2. This theorem has some immediate consequences:

(a) For $2 = d \le q' \le q$ one has

$$H^1(\mathbb{R}^2;\mathbb{R}^n) \equiv E_2 \subseteq E_{q'} \subseteq E_q$$

and examples show that $E_{q'} \subsetneqq E_q$ whenever $q' < q$.

(b) For $q = d = 2$ Theorem 9.2.3 provides a simple proof of the well-known Sobolev embeddings

$$H^1 \hookrightarrow L^r(\mathbb{R}^2) \quad \text{for all } r \ge 2$$

with explicitly known embedding constants.

Theorem 9.2.4. *For any bounded sequence $(u^j)_{j\in\mathbb{N}}$ in $E_q = E_q^1(\mathbb{R}^d;\mathbb{R}^n)$, $2 \le d \le q < \infty$, there is a subsequence $j(i)$, $i \in \mathbb{N}$, and there is a $u \in E_q$ such that*

$$u^{j(i)} \to u \quad \text{for } i \to \infty$$

(a) *almost everywhere,*
(b) *weakly in E_q, i.e. $\nabla u^{j(i)} \to \nabla u$ weakly in L^d and $u^{j(i)} \to u$ weakly in L^q,*
(c) *weakly in L^s for any fixed $s \ge q$,*
(d) *in L^r_{loc} for any fixed $r \ge 1$.*

Proof. Relying on Theorem 9.2.3 instead of Sobolev's inequality, the proof is basically the same as that of Theorem 9.2.2. In some parts it is even easier, so it is not repeated here. □

9.2.2 Some Continuity Results for Niemytski Operators

Theorems 9.2.2 and 9.2.4 are typically applied to minimising sequences of a variational problem. Then, in order to proceed, one has to control the potential V according to (9.1.8) on minimising sequences. We prepare for this here by studying continuity properties of Niemytski operators $\widehat{F}$ on the space E of a priori solutions associated with a C-function $F : \Omega \times \mathbb{R}^n \to \mathbb{R}$, $\Omega \subset \mathbb{R}^d$; the notion of convergence on E which is interesting in this context is indicated by conditions (a)–(d) of Theorem 9.2.2 and 9.2.4. We start with quite a general criterion which only relies on pointwise convergence of a given sequence (i.e. condition (a) of Theorem 9.2.2 or 9.2.4). Again we denote by π_A the operator of multiplication by the characteristic function 1_A of a measurable set A and then we write $\pi_A^\perp = 1 - \pi_A$.

For a C-function $F : \Omega \times \mathbb{R}^n \to \mathbb{R}$ and a number $\lambda \geq 0$ we write

$$F^{<\lambda}(x,y) = \begin{cases} F(x,y) & \text{for all } x \in \Omega \text{ and all } y,\ |y| < \lambda\,, \\ 0 & \text{for all } x \in \Omega \text{ and all } y,\ |y| \geq \lambda\,. \end{cases}$$

Theorem 9.2.5. *Suppose $F : \Omega \times \mathbb{R}^n \to \mathbb{R}$ to be a C-function such that for every $\lambda \geq 0$ there is a function $a_\lambda \in L^1_{\text{loc}}(\Omega)$ with*

$$|F(x,y)| \leq a_\lambda(x) \quad \text{for almost every } x \in \Omega \text{ and all } y\,,\ |y| \leq \lambda\,. \tag{9.2.11}$$

If $u_j : \Omega \to \mathbb{R}^1$, $j \in \mathbb{N}$, is a sequence of measurable functions which converges almost everywhere to some function u and for which $\widehat{F}(u_j)$ belongs to $L^1(\Omega)$ for all $j \in \mathbb{N}$, $\widehat{F}(u_j)(x) = F(x, u_j(x))$, then

$$\widehat{F}(u_j) \to \widehat{F}(u) \quad \text{for } j \to \infty \text{ in } L^1(\Omega)$$

if the following conditions of "concentration" (C) and "incomplete uniform decay at infinitely" (IUD) in the L^1-norm sense on compact sets are satisfied:

(C) *For every $\varepsilon > 0$ there is a compact set $K \subset \Omega$ such that*

$$\sup_j \|\pi_K^\perp \widehat{F}(u_j)\|_1 \leq \varepsilon\,. \tag{9.2.12}$$

(IUD) *For every compact set $K \subseteq \Omega$ and every $\varepsilon > 0$ there is $\lambda_0 \in (0, \infty)$ such that for every $\lambda \geq \lambda_0$ there is j_0 with*

$$\sup_{j \geq j_0} \|\pi_K \widehat{F^{\geq\lambda}}(u_j)\|_1 \leq \varepsilon\,. \tag{9.2.13}$$

Proof. (a) First recall that if $(f_j)_{j\in\mathbb{N}}$ is a sequence of measurable functions $\Omega \to \mathbb{R}$ which converges almost everywhere to some function f then $f_j \to f$, $j \to \infty$, in $L^1(\Omega)$ if and only if

(A) for every $\varepsilon > 0$ there is a compact set $K \subset \Omega$ such that

$$\sup_j \|\pi_K^\perp f_j\|_1 \leq \varepsilon$$

and

(B) $\pi_K f_j \to \pi_K f$ in $L^1(\Omega)$ for every compact set $K \subseteq \Omega$.

Then observe that the following information is available:

(i) $$\sup_j |\widehat{F}^{<\lambda}(u_j)| \leq a_\lambda \quad \text{almost everywhere on } \Omega,$$

(ii) $$(\widehat{F}^{<\lambda}(u_j))_{j\in\mathbb{N}} \quad \text{converges almost everywhere on } \Omega.$$

Hence, by dominated convergence for every compact set $K \subseteq \Omega$,

$$(\pi_K \widehat{F}^{<\lambda}(u_j))_{j\in\mathbb{N}} \quad \text{is a Cauchy sequence in } L^1(\Omega)\,. \tag{9.2.14}$$

(b) Now suppose $\widehat{F}(u_j) \to \widehat{F}(u)$ in $L^1(\Omega)$. Condition (A) applied to $f_j = \widehat{F}(u_j)$ and $f = \widehat{F}(u)$ gives just the concentration condition (C). Then, for $K \subset \Omega$ compact, $\lambda > 0$, and $j \in \mathbb{N}$, observe that

$$\|\pi_K \widehat{F}^{\geq\lambda}(u_j)\|_1 \leq \|\pi_K \widehat{F}^{\geq\lambda}(u)\|_1 + \|\pi_K(f_j - f)\|_1 + \|\pi_K(\widehat{F}^{<\lambda}(u_j) - \widehat{F}^{<\lambda}(u))\|_1\,.$$

Therefore, given K and $\varepsilon > 0$ there is, first, $\lambda_0 > 0$ such that

$$\|\pi_K \widehat{F}^{\geq\lambda}(u)\|_1 \leq \varepsilon/3 \quad \text{for all } \lambda \geq \lambda_0$$

and there is j_0 such that for all $j \geq j_0$

$$\|\pi_K(f_j - f)\|_1 \leq \varepsilon/3\,.$$

Then, for fixed $\lambda \geq \lambda_0$, there is by (9.2.14) a $j_1 = j_1(\lambda)$ such that the last term of the above estimate is also dominated by $\varepsilon/3$ hence for $j \geq j_0 \vee j_1$ it follows that

$$\|\pi_K \widehat{F}^{\geq\lambda}(u_j)\| \leq \varepsilon\,,$$

that is, condition (IUD) holds.

(c) Suppose, conversely that conditions (C) and (IUD) hold. Condition (C) gives just (A) for $f_j = \widehat{F}(u_j)$, $j \in \mathbb{N}$. Then we know that $f_j \to f$ for $j \to \infty$ almost everywhere, $f = \widehat{F}(u)$, since F is a C-function. Hence in order to get (B) it suffices to prove that, for every compact set $K \subset \Omega$, $(\pi_K f_j)_{j\in\mathbb{N}}$ is a Cauchy sequence in L^1.

But the following estimate is known:

$$\|\pi_K(f_j - f_i)\|_1 \leq \|\pi_K(\widehat{F}^{<\lambda}(u_j) - \widehat{F}^{<\lambda}(u_i))\|_1 + \|\pi_K \widehat{F}^{\geq\lambda}(u_j)\|_1 + \|\pi_K \widehat{F}^{\geq\lambda}(u_i)\|_1\,.$$

By (IUD), given $\varepsilon > 0$ and $K \subset \Omega$ compact, there are λ and i_0 such that, for all $i, j \geq i_0$, the last two terms are smaller than $\varepsilon/3$. By (9.2.14) there is $j_1 = j_1(\lambda, K)$ such that the first term is smaller than $\varepsilon/3$ for all $i, j \geq j_1$ and this then implies the statement. □

Remark 9.2.3. Theorem 9.2.5 will be used to derive necessary and sufficient conditions for the existence of a solution of the constrained minimisation problem (9.1.11) (Theorem 9.7.1). It is also very useful in many other situations where one wants to ensure L^1-convergence knowing only pointwise convergence almost everywhere.

Another result of this type which allows L^1 convergence to be derived from pointwise convergence almost everywhere and which has proved to be quite useful in variational problems is given by the following lemma.

Lemma 9.2.6. *If $F: \mathbb{R}^n \to \mathbb{R}$ is continuous and satisfies $F(0) = 0$ and if for every $\varepsilon > 0$ there is a constant C_ε such that, for all $a, b \in \mathbb{R}^n$,*

$$|F(a+b) - F(a)| \leq \varepsilon\{|F(a)| + |a|^r\} + C_\varepsilon\{|F(b) + |b|^r\} \tag{9.2.15}$$

holds for some exponent $r \geq 1$, then the Niemytski operator $\widehat{F}$ associated with F has the following continuity property: for any bounded sequence $(u_i)_{i\in\mathbb{N}}$ in $L^r(\Omega; \mathbb{R}^n)$, $\Omega \subseteq \mathbb{R}^d$, which converges almost everywhere to some function u and for which $(\widehat{F}(u_i)_{i\in\mathbb{N}}$ is bounded in L^1, the following holds for every $v \in L^r$ with $\widehat{F}(v) \in L^1$:

(a) $\qquad \widehat{F}(u_i + v) \in L^1, \quad \widehat{F}(u) \in L^1, \quad \widehat{F}(u+v) \in L^1;$

(b) $\quad \widehat{F}(u_i + v) - \widehat{F}(u_i) \to \widehat{F}(u+v) - \widehat{F}(u) \quad$ *for* $\ i \to \infty$ *in* $L^1(\Omega)$.

Proof. $\widehat{F}(u_i + v) \in L^1$ follows directly from our assumptions and inequality (9.2.15). Since $(\widehat{F}(u_i))_{i\in\mathbb{N}}$ converges almost everywhere to $\widehat{F}(u)$ and since $(\widehat{F}(u_i))_{i\in\mathbb{N}}$ is bounded in L^1, Fatou's lemma implies $\widehat{F}(u) \in L^1$. Now $\widehat{F}(u+v) \in L^1$ follows again from (9.2.15). Writing $C = \sup_i\{\|\widehat{F}(u_i)\|_1 + \|u_i\|_r^r\}$, then by (9.2.15), for any measurable subset $A \subset \Omega$ and any $\varepsilon > 0$,

$$\|\pi_A(\widehat{F}(u_i + v) - \widehat{F}(u_i))\|_1 \leq \varepsilon C + C_\varepsilon[\|\pi_A \widehat{F}(v)\|_1 + \|\pi_A v\|_r^r].$$

Therefore, choosing first $\varepsilon > 0$ and then $A \subset \Omega$ appropriately, this estimate implies that for every $\gamma > 0$ there is a compact set $K \subset \Omega$ and a number $\delta > 0$ such that

$$\sup_i \|\pi_K^\perp(\widehat{F}(u_i + v) - \widehat{F}(u_i))\|_1 \leq \gamma,$$

$$\sup_i \|\pi_A(\widehat{F}(u_i + v) - \widehat{F}(u_i))\|_1 \leq \gamma,$$

for all measurable sets $A \subseteq \Omega$, $|A| \leq \delta$. Therefore Vitali's convergence theorem implies statement (b). □

The class of functions satisfying the hypotheses of Lemma 9.2.6 is actually fairly rich.

Lemma 9.2.7. *Any continuous function $F: \mathbb{R}^n \to \mathbb{R}$ of the form*

$$F(y) = f(y)|y|^r \tag{9.2.16}$$

with a bounded function $f \in C^1(\mathbb{R}^n \setminus \{0\})$ satisfying

$$\sup\{|y|\ |\nabla f(y)|\ |\ y \in \mathbb{R}^n \setminus \{0\}\} < \infty$$

has property (9.2.15).

We leave the elementary proof as an exercise and proceed with an immediate corollary.

Corollary 9.2.8. *Suppose $(f_i)_{i\in\mathbb{N}}$ is a bounded sequence in $L^r(\Omega)$ which converges almost everywhere to some function f. Then $f \in L^r(\Omega)$ and*

$$\lim_{i\to\infty} \{\|f_i\|_r^r - \|f - f_i\|_r^r\} = \|f\|_r^r .$$

Proof. By Lemma 9.2.7 we know that Lemma 9.2.6 applies to

$$F(y) = |y|^r$$

Lemma 9.2.6 proves $f \in L^r$ while Lemma 9.2.6 (b) implies

$$\widehat{F}(f_i) - \widehat{F}(f_i - f) \to \widehat{F}(f) - \widehat{F}(f - f) = \widehat{F}(f) \quad \text{in } L^1 .$$

Thus we conclude. □

Remark 9.2.4. (a) Brezis and Lieb seem to have been the first to recognise the importance of condition (9.2.15) in variational problems and to use it accordingly (see for instance [9.13]).

(b) With a different proof, Corollary 9.2.8 is also due to Brezis and Lieb [9.21], in a more general version ($F(y) = |y|^r$ is replaced by some convex function $j(y)$). Our method of proof also shows clearly how to extend this corollary to many other classes of functions.

9.2.3 Some Results on Concentration of Function Sequences

By Theorem 9.2.2 we know that for the constrained minimisation problem (9.1.11) there are a minimising sequence $(u_j)_{j\in\mathbb{N}}$ and an element u in $E = E^{1,2}(\mathbb{R}^d;\mathbb{R}^n)$ such that $u_j \to u$ for $j \to \infty$ in the sense of statements (a)–(d) of this theorem. For a solution of the variational problem (9.1.11) one has to prove that there is a minimising sequence $(u_j)_{j\in\mathbb{N}}$ for which the limit u according to Theorem 9.2.2 is different from zero (see Sect. 9.6 for more details). This requires the existence of a minimising sequence which is concentrated in a certain sense. The meaning of this concentration has already been indicated in the introduction to this chapter and will be explained in detail in a precise version in Sect. 9.7.

This section provides some tools for achieving this concentration. The elements of a minimising sequence for problem (9.1.11) can be concentrated by:

(a) spherically symmetric rearrangement of functions,
(b) translations.

We start by describing the more traditional tool, that is we quickly recall the basic facts about the spherically symmetric rearrangement of functions. For the proofs of these results we recommend the articles of *E. H. Lieb* [9.22] and *G. Talenti* [9.10].

The spherically symmetric rearrangement of functions allows one to associate a spherically symmetric function with a given function in such a way that the values of various integrals over this function remain under control.

For a measurable function $f : \mathbb{R}^d \to \mathbb{R}$ consider its level sets

$$[f > t] = \{x \in \mathbb{R}^d \mid f(x) > t\}$$

for $t \in \mathbb{R}$, and define $\mu(t)$ to be the Lebesgue measure of $[f > t]$.

$$\mu(t) = \mu_f(t) = |[f > t]| \, .$$

μ is a nonincreasing right-continuous function $\mathbb{R} \to \mathbb{R}_+$. The *spherically symmetric rearrangement* f^* of f is defined as the (spherically symmetric) function whose level sets $[f^* > t]$ are balls which have the same measure as the level sets $[f > t]$ of the given function. Thus, if ω_d denotes the volume of the unit ball in $\mathbb{R}^d$, f^* is defined as follows:

$$f^*(x) = f_0(|x|) \quad \text{for all } x \in \mathbb{R}^d \tag{9.2.17}$$

where

$$f_0(r) := \sup\{t \mid \mu_f(t) > \omega_d r^d\} \, .$$

The level sets of f^* are clearly the balls

$$[f^* > t] = \{x \in \mathbb{R}^d \mid |x| < R_t\} \, , \quad R_t = (\omega_d^{-1} \mu_f(t))^{1/d} \, .$$

We now collect the main properties of the mapping $f \mapsto f^*$.

If G is a function on $\mathbb{R}$ such that $G \circ f$ is integrable then $G \circ f^*$ is integrable too and the identity

$$\int_{\mathbb{R}^d} G(f(x)) \, dx = \int_{\mathbb{R}^d} G(f^*(x)) \, dx \tag{9.2.18}$$

is valid.

For all square-integrable functions f, g the following inequality holds:

$$\int_{\mathbb{R}^d} f(x) g(x) \, dx \le \int_{\mathbb{R}^d} f^*(x) g^*(x) \, dx \, . \tag{9.2.19}$$

(For an extension of this inequality see [9.22, 23].) Together with (9.2.18) it implies the continuity of the spherically symmetric rearrangement on $L^2(\mathbb{R}^d)$. For all $f, g \in L^2(\mathbb{R}^d)$,

$$\|f^* - g^*\|_2 \le \|f - g\|_2 \, . \tag{9.2.20}$$

The last important property of the mapping $f \to f^*$ we will use is the following behaviour with respect to weak differentiation.

Whenever f is a real measurable function such that its weak derivative ∇f belongs to $L^p(\mathbb{R}^d)$ for some $1 \leq p < \infty$ then the weak derivative of $|f|^*$ also exists and is p-integrable and satisfies

$$\|\nabla |f|^*\|_p \leq \|\nabla f\|_p \,. \tag{9.2.21}$$

We will typically apply this property for $u \in E^{1,2}(\mathbb{R}^d;\mathbb{R}^n)$ and $p = 2$ and obtain

$$K(|u|^*) = \frac{1}{2}\|\nabla |u|^*\|_2^2 \leq \frac{1}{2}\|\nabla u\|_2^2 = K(u)\,. \tag{9.2.22}$$

In order to illustrate the effect of the mapping $u \mapsto |u|^*$ on $E^{1,2}(\mathbb{R}^d;\mathbb{R}^n)$ we prove a simple result on spherically symmetric functions in this space.

Lemma 9.2.9. *If a function u satisfies $u(x) = u_0(|x|)$ for all $x \in \mathbb{R}^d$ and belongs to $E = E^{1,p}(\mathbb{R}^d;\mathbb{R}^n)$, $1 \leq p < d$, then it is absolutely continuous on $\mathbb{R}^d \setminus \{0\}$ and satisfies*

$$|u(x)| \leq c_{p,d}\|\nabla u\|_p |x|^{1-d/p} \quad \textit{for } x \neq 0\,. \tag{9.2.23}$$

Proof. Since $E \subset L^{p^*}(\mathbb{R}^d)$ we know for such elements that

$$\|u\|_{p^*}^{p^*} = \sigma_d \int_0^\infty |u_0(r)|^{p^*} r^{d-1}\, dr < \infty\,,$$

$$\|\nabla u\|_p^p = \sigma_d \int_0^\infty |u_0'(r)|^p r^{d-1}\, dr < \infty\,,$$

where σ_d denotes the volume of the unit sphere in $\mathbb{R}^d$. Define τ by $p\tau = d-1$ and calculate for $0 < r < r'$

$$u_0(r') - u_0(r) = \int_r^{r'} z^{-\tau} z^\tau u_0'(z)\, dz\,.$$

Hölder's inequality implies

$$|u_0(r') - u_0(r)| \leq \left(\int_r^{r'} z^{-p'z}\, dz\right)^{1/p'} \left(\int_r^{r'} z^{p\tau}|u_0'(z)|^p\, dz\right)^{1/p}.$$

Hence, with $a = p'\tau - 1 = \frac{d-p}{p-1} > 0$ and some constant $c_{p,d}$,

$$|u_0(r') - u_0(r)| \leq [r^{-a} - r'^{-a}]^{1/p'} c_{p,d}\|\nabla u\|_p\,.$$

This estimate implies

$$\lim_{r\to\infty} u_0(r) = 0$$

and thus for all $r > 0$

$$|u_0(r)| \leq c_{p,d}\|\nabla u\|_p r^{-a/p'}\,.$$

This is inequality (9.2.23). □

Corollary 9.2.10. *The elements of a bounded set in* $E_s = \{u \in E^{1,p}(\mathbb{R}^d; \mathbb{R}^n) | u(x) = u_0(|x|)\}$ *have uniform decay at infinity (and are thus essentially concentrated on a common compact set in* $\mathbb{R}^d$*).*

Remark 9.2.5. This uniform decay property of bounded set in E_s can be used to derive results on compact embeddings for subspaces of spherically symmetric functions in $E^{1,p}(\mathbb{R}^d)$ into appropriate Lebesgue spaces. *W. Strauss* seems to have been the first to do this and to apply it [9.11] (for the subspace of spherically symmetric functions of the Sobolev space $H^1(\mathbb{R}^d)$, $d \geq 3$, for translation-invariant scalar field equations of positive mass). Relying on Theorem 9.2.5 we will be able to omit this step because in our applications this uniform decay property will imply immediately the concentration condition (C) of this theorem.

It is well understood why the weak limit of a sequence of functions f_j in $L^r(\mathbb{R}^d)$, $1 < r < \infty$, with $\|f_j\|_r = 1$ for all j, can vanish. If this sequence belongs to $E^{1,p}(\mathbb{R}^d)$ and is also bounded there the weak and pointwise limit f of every subsequence can still vanish according to Theorem 9.2.2. But as we will learn from the following lemma on "concentration by translation", for appropriate exponents r there exist a sequence of translation a_j and a subsequence such that the translated subsequence

$$g_i(\cdot) = f_{j(i)}(\cdot + a_{j(i)}) \tag{9.2.24}$$

has a nonvanishing weak and pointwise limit g. This will usually settle one of the most difficult points of the variational problem (9.1.11).

Lemma 9.2.11. *For a bounded sequence* f_j, $j \in \mathbb{N}$, *in* $E^{1,p}(\mathbb{R}^d)$, $1 \leq p < d$, *assume that for some* $k > 0$

$$\inf\{\|f_j\|_p \,|\, j \in \mathbb{N}\} = k \tag{9.2.25}$$

(where clearly $\|f_j\|_p = +\infty$ *occurs of* $f_j \notin L^p$*). Then there is a constant* $\gamma > 0$, *depending only on* k *and the bound for the* f_j *in* $E^{1,p}(\mathbb{R}^d)$, *and a sequence of translation* a_j, $j \in \mathbb{N}$, *in* $\mathbb{R}^d$ *such that the translated sequence*

$$g_j(\cdot) = f_j(\cdot + a_j)$$

satisfies

$$\gamma < |B \cap \operatorname{supp} g_j| \quad \forall j \in \mathbb{N} \tag{L}$$

where B *is the open unit ball in* $\mathbb{R}^d$ *centered at zero.*

Proof. By assumption we know

$$\sup_{j \in \mathbb{N}} \|\nabla f_j\|_p^p = C < \infty .$$

Since $p < p^*$ we know by Sobolev's inequality $f_j \in L^p_{\text{loc}}$ for all $j \in \mathbb{N}$ but not necessarily $f_j \in L^p$. We define for $j = 1, 2, \ldots$

$$A(f_j) = \|\nabla f_j\|_p^p \cdot \begin{cases} 1 & \text{if } f_j \notin L^p \\ 1 + \|f_j\|_p^{-p} & \text{if } f_j \in L^p \end{cases}$$

Assumption (9.2.25) implies $\sup_j A(f_j) = C_0 < \infty$. Let B be the open unit ball in $\mathbb{R}^d$ centered at $x = 0$ and denote by $B_x = x + B$ the ball B translated by $x \in \mathbb{R}^d$. If we had for all $x \in \mathbb{R}^d$

$$\|\nabla f_j\|^p_{L^p(B_x)} \geq A(f_j)\|f_j\|^p_{L^p(B_x)}$$

we could integrate this inequality and would get by Fubini's theorem

$$\begin{aligned} |B|\ \|\nabla f_j\|_p^p &= \int \|\nabla f_j\|^p_{L^p(B_x)}\, dx \\ &\geq A(f_j) \int \|f_j\|^p_{L^p(B_x)} 1_B(x)\, dx = A(f_j)|B|\ \|f_j\|_p^p\,. \end{aligned}$$

But this is a contradiction to the definition of $A(f_j)$. Therefore for each $j \in \mathbb{N}$ there is a point $a_j \in \mathbb{R}^d$ such that

$$\|\nabla f_j h\|^p_{L^p(Ba_j)} < A(f_j)\|f_j\|^p_{L^p(Ba_j)}$$

Now introduce the translated sequence $g_j(\cdot) = f_j(\cdot + a_j)$. We have just shown that it satisfies for all $j \in \mathbb{N}$

$$\|\nabla g_j\|^p_{L^p(B)} < A(f_j)\|g_j\|^p_{L^p(B)} \leq C_0\|g_j\|^p_{L^p(B)}$$

and thus in particular $\|g_j\|^p_{L^p(B)} > 0$ for all j. Clearly g_j belongs to $E^{1,p}(\mathbb{R}^d)$ and satisfies $\|\nabla f_j\|_p = \|\nabla g_j\|_p$.

Choose an exponent r, $p < r < p^*$. Then Sobolev's embedding theorem yields

$$\begin{aligned} \|g_j\|^p_{L^r(B)} &\leq K_0\{\|g_j\|^p_{L^p(B)} + \|\nabla g_j\|^p_{L^p(B)}\} \\ &< K_0(1 + C_0)\|g_j\|^p_{L^p(B)} = K\|g_j\|^p_{L^p(B)} \end{aligned}$$

for some positive number K. By Hölder's inequality we get $(1/p = 1/r + 1/s)$

$$\|g_j\|_{L^p(B)} \leq \|g_j\|_{L^r(B)}\|1_{B\cap \operatorname{supp} g_j}\|_s \leq \|g_j\|_{L^r(B)}|B \cap \operatorname{supp} g_j|^{1/s}$$

and combining this estimate with the previous one we get the lower bound (L) with $\gamma = K^{-1/(1-p/r)}$. □

Corollary 9.2.12. *For a bounded sequence u_j, $j \in \mathbb{N}$, in $E^{1,p}(\mathbb{R}^d, \mathbb{R}^n)$, $1 \leq p < d$, assume that there exist some positive numbers ε, α such that for all $j \in \mathbb{N}$*

$$|[|u_j| \geq \varepsilon]| \geq \alpha\,. \tag{9.2.26}$$

Then there exist a sequence of translations a_j, $j \in \mathbb{N}$, and a subsequence $j(i)$, $i \in \mathbb{N}$, such that the translated sequence

$$v_i(\cdot) = u_{j(i)}(\cdot + a_{j(i)})$$

has a nonvanishing limit in the sense of conditions (a)–(b) *of Theorem 2.2.*

Proof. Denote $C_0 = \sup_{j \in \mathbb{N}} \|\nabla u_j\|_p^p$ and introduce

$$f_j = \max\{|u_j| - \varepsilon/2, 0\}.$$

It follows $f_j \in E^{1,p}(\mathbb{R}^d)$ and for all $j \in \mathbb{N}$ one has

$$\|\nabla f_j\|_p^p \leq \|\nabla |u_j|\ \|_p^p \leq \|\nabla u_j\|_p^p \leq C_0$$

$$\|f_j\|_p^p \geq \int_{\|\ |u_j| \geq \varepsilon\|} |f_j|^p \, dx \geq \left(\frac{\varepsilon}{2}\right)^p \|[|u_j| \geq \varepsilon]\| \geq \left(\frac{\varepsilon}{2}\right)^p \alpha = k > 0.$$

Thus by Lemma 9.2.11 there are a constant $\gamma = \gamma(k, C_0) > 0$ and a sequence of points $a_j \in \mathbb{R}^d$ such that the translated sequence

$$g_j(x) = f_j(x + a_j) = \max\{|u_j(x + a_j)| - \varepsilon/2, 0\}$$

satisfies for all $j \in \mathbb{N}$

$$\gamma < |B \cap \operatorname{supp} g_j|.$$

Denoting by 1_B the characteristic function of the ball B we get for the function $v_j(\cdot) = u_j(\cdot + a_j)$

$$\int 1_B |v_j(x)| \, dx \geq \int_{\operatorname{supp} g_j} 1_B |v_j(x)| \, dx \geq \int_{\operatorname{supp} g_j} 1_B \frac{\varepsilon}{2} \, dx$$
$$= \frac{\varepsilon}{2} |B \cap \operatorname{supp} g_j| > \frac{\varepsilon}{2} \gamma.$$

Clearly $(v_j)_{j \in \mathbb{N}}$ and $(|v_j|)_{j \in \mathbb{N}}$ are bounded sequences in $E^{1,p}(\mathbb{R}^d; \mathbb{R}^n)$ respectively in $E^{1,p}(\mathbb{R}^d)$. Hence Theorem 9.2.2 assures the existence of a subsequence $j(i)$, $i \in \mathbb{N}$, such that $v_{j(i)}$ and $|v_{j(i)}|$ converge in the sense (a)–(d) of that theorem to $v \in E^{1,p}(\mathbb{R}^d; \mathbb{R}^n)$ respectively to $|v| \in E^{1,p}(\mathbb{R}^d)$. Since in particular $|v_{j(i)}| \to |v|$ for $i \to \infty$ weakly in $L^{p^*}(\mathbb{R}^d)$ and $1_B \in (L^{p^*}(\mathbb{R}^d))'$ it follows

$$\int 1_B |v| \, dx = \lim_{i \to \infty} \int 1_B |v_{j(i)}| \, dx \geq \frac{\varepsilon}{2} \gamma$$

and thus $v \neq 0$. □

Remark 9.2.6 Corollary 9.2.12 is a *lemma of concentration by translations.* In its original form for sequences in the Sobolev space $W^{1,p}(\mathbb{R}^d)$ it is due to *E. H. Lieb* who published it in 1983 as a compactness lemma (Lemma 6 of reference [9.29]). The proof given here by means of Lemma 2.11 is based on an idea of H. Brezis as indicated in reference [9.29].

Lemma 9.2.13. *For any bounded sequence $(f_j)_{j\in\mathbb{N}}$ in $E_q^1(\mathbb{R}^d;\mathbb{R}^n)$, $2\le d\le q<\infty$ satisfying for some positive numbers $\varepsilon>0$, $a>0$*

$$\inf\{\|[|f_j|\ge\varepsilon]\|\ |j\in\mathbb{N}\}=a$$

there are a sequence of translations $(a_j)_{j\in\mathbb{N}}$ and a subsequence $j(i)$, $i\in\mathbb{N}$, such that the translated subsequence $g_i(\cdot)=f_{j(i)}(\cdot+a_{j(i)})$ has a nonvanishing limit in the sense of condition (a)–(d) *of Theorem 9.2.4.*

Proof. We proceed just as in the proofs of Lemma 9.2.11 and Corollary 9.2.12 using however Theorem 9.2.4 instead of Theorem 9.2.2 and Theorem 9.2.3 instead of Sobolev's inequality. □

Remark 9.2.7. Lemma 9.2.13 has an extension. It suffices to assume: The sequence of functions $f_j:\mathbb{R}^d\to\mathbb{R}^n$, $j=1,2,\ldots$, has weak derivatives ∇f_j and satisfies

(i)
$$\sup_j\|\nabla f_j\|_d=C<\infty$$

(ii) for some constants $\varepsilon>0$, $a>0$
$$\inf\{|[|f_j|\ge\varepsilon]||j\in\mathbb{N}\}=a\,.$$

Then the conclusions of Lemma 9.2.13 hold in the sense that the translated subsequence (g_i) converges to some element $g\neq 0$ almost everywhere and $\nabla g_i\to\nabla g$ weakly in $L^d(\mathbb{R}^d)$.

9.2.4 A One-dimensional Variational Problem

The purpose of this section is threefold:

(i) To show in a simple situation how the methods and results developed thus far are typically used in a variational problem;
(ii) to solve a class of nonlinear global differential equations;
(iii) to produce a result for later use.

We begin with some facts about the Sobolev space

$$W^{1,p}(\mathbb{R})=\{f\in L^p(\mathbb{R})\,|\,f'\in L^p(\mathbb{R})\}\quad\text{for some } p\in(1,\infty)\,.$$

Lemma 9.2.14. (a) *Each $f\in W^{1,p}(\mathbb{R})$ is Hölder continuous of exponent $\alpha=1/p'$:*

$$|f(x)-f(y)|\le|x-y|^{1/p'}\|f'\|_p\quad\textit{for all } x,y\in\mathbb{R}\,. \tag{9.2.27}$$

(b) *Symmetric (even) nonincreasing functions f in $W^{1,p}(\mathbb{R})$ are bounded in the following way:*

$$|f(x)|\le\|f\|_{1,p}h(x)\,,\quad h(x)=\min\{1,|x|^{-1/p}\}\,,\quad x\in\mathbb{R}\,. \tag{9.2.28}$$

Proof. (a) follows easily from

$$\left|\int_y^x f'(\xi)\,d\xi\right| \le \|1\|_{L^{p'}([x,y])}\|f'\|_{L^p([x,y])}\,.$$

In order to prove (b) consider the elementary identity

$$f(x) = f(x+\xi) - \int_0^\xi \frac{d}{dt} f(x+t)\,dt$$

and integrate it with respect to ξ over the unit interval. A simple application of Hölder's inequality yields

$$\begin{aligned}|f(x)| &\le \int_0^1 |f(x+\xi)|\,d\xi + \int_0^1 \left(\int_0^\xi |f'(x+t)|\,dt\right) d\xi \\ &\le \left(\int_0^1 |f(x+\xi)|^p\,d\xi\right)^{1/p} + \left(\int_0^1 |f'(x+t)|^p\,dt\right)^{1/p} \le \|f\|_{1,p}\,.\end{aligned}$$

If f is now a symmetric nonincreasing function in $W^{1,p}(\mathbb{R})$, we have, for all $z>0$,

$$\|f\|_p^p = 2\int_0^\infty |f(x)|^p\,dx \ge 2\int_0^z |f(x)|^p\,dx \ge 2\int_0^z |f(x)|^p\,dx = 2z|f(z)|^p\,,$$

and hence

$$|f(z)| \le |z|^{-1/p}\|f\|_p \quad \text{for all } z \ne 0\,,$$

and (9.2.28) follows. □

Theorem 9.2.15. *Suppose $G:\mathbb{R}\to\mathbb{R}$ to be an even continuous function such that $G(0)=0$, $G(y)>0$ somewhere, and for some constant C and some exponent $1<r<\infty$*

$$|G(y)| \le C|y|^r\,. \tag{9.2.29}$$

Then for any p, $1<p<r$, consider the following variational problem. On

$$M = \{f\in W^{1,p}(\mathbb{R})\,|\,\widehat{G}(f)\in L^1(\mathbb{R}), V(f) = \int \widehat{G}(f)\,dx > 0\}$$

define

$$J_p(f) = \|f\|_{1,p}^{-p} V(f)^{p/r}\,. \tag{9.2.30}$$

Then there is a maximising element f_0 for $J_p(f)$ on M:

$$J_p(f_0) = I_p := \sup\{J_p(f)\,|\,f\in M\}\,. \tag{9.2.31}$$

f_0 is a nonnegative nonincreasing absolutely continuous function on $\mathbb{R}$ which decreases at infinity as const.$|x|^{-1/p}$.

If $G\in C^1(\mathbb{R})$, then f_0 is a solution of the following differential equation in $\mathcal{D}'(\mathbb{R})$:

$$[(\operatorname{sign} f')|f'|^{p-1}]' - (\operatorname{sign} f)|f|^{p-1} = -\frac{1}{r} I_p^{-r/p} G'(f)\,. \tag{9.2.32}$$

Proof. (a) Since $G(y) > 0$ somewhere, there are functions in $W^{1,p}(\mathbb{R})$ such that $\widehat{G}(f) \in L^1$ and $V(f) > 0$, for instance $f \in \mathcal{C}_0^1(\mathbb{R})$. Thus M is not empty.

Denote by f^* the spherically symmetric rearrangement of $|f|$, $f \in M$. Then by (9.2.18) and our assumption we know that

$$\widehat{G}(f^*) \in L^1 \quad \text{and} \quad \int \widehat{G}(f^*)\,dx = \int \widehat{G}(|f|)\,dx = \int G(f)\,dx > 0\,,$$

$$\|f^*\|_p = \|f\|_p\,,$$

and by (9.2.21), $\|\frac{d}{dx}f^*\|_p \leq \|f'\|_p$. Hence $\|f^*\|_{1,p} \leq \|f\|_{1,p}$.

Therefore $f^* \in M$ and $J_p(f) \leq J_p(f^*)$. It follows that

$$I_p = \sup\{J_p(f) \mid f \in M^*\}\,, \quad M^* \subsetneqq M\,,$$

if M^* denotes the image of M under the mapping $f \to f^*$.

(b) For $f \in M^*$, estimate (9.2.28) is available. Since the function h in (9.2.28) belongs to $L^r(\mathbb{R})$ for every $r > p$, assumption (9.2.29) easily implies that $J_p(\cdot)$ is bounded on M^* and thus on M. For $f \in M^*$ we have

$$0 < V(f)^{p/r} \leq C^{p/r}\|f\|_r^r \leq C^{p/r}\|f\|_{1,p}^r\|h\|_r^r\,,$$

So we can find a maximising sequence $(f_j)_{j\in\mathbb{N}} \subset M^*$ for problem (9.2.31) satisfying

$$\|f_j\|_{1,p} = 1\,, \quad 0 < V(f_j) \leq V(f_{j+1}) \quad \text{for all } j$$

and

$$I_p = \lim_{j\to\infty} J_p(f_j) = \lim_{j\to\infty} V(f_j)^{p/r}\,.$$

Just as in Theorem 9.2.2, one can prove that there exist a subsequence $j(i)$, $i \in \mathbb{N}$, and an element $f \in W^{1,p}$ such that $f_{j(i)} \to f$ for $i \to \infty$ weakly in $W^{1,p}(\mathbb{R})$ and almost everywhere on $\mathbb{R}$.

So we can and will assume that the above minimising sequence already has these properties.

(c) The proof that $(\widehat{G}(f_j))_{j\in\mathbb{N}}$ converges in $L^1(\mathbb{R})$ to $\widehat{G}(f)$ is now quite easy by Theorem 9.2.5. We simply check the conditions (C) and (IUD). By inequality (9.2.28), for any compact set $K \subset \mathbb{R}$, we have for all j:

$$\|\pi_K^\perp\widehat{G}(f_j)\|_1 \leq \|\pi_K^\perp C|f_j|^r\|_1 \leq C\|f_j\|_{1,p}^r\|\pi_K^\perp|h|^r\|_1 \leq C\|\pi_K^\perp h\|_r^r\,.$$

Since $h \in L^r(\mathbb{R})$ condition (C) follows. By inequality (9.2.28) and the normalisation $\|f_j\|_{1,p} = 1$, the sequence $(f_j)_{j\in\mathbb{N}}$ is uniformly bounded: $|f_j(x)| \leq 1$ for all j and all x. Hence for $\lambda > 1$ all the sets $[\lambda \leq |f_j|]$ are empty and condition (IUD) trivially holds.

Therefore Theorem 9.2.5 implies $\widehat{G}(f_j) \to \widehat{G}(f)$ for $j \to \infty$ in L^1; thus, $\widehat{G}(f) \in L^1(\mathbb{R})$ and $V(f_j) \to V(f)$. It follows that $V(f) \geq V(f_1) > 0$ and therefore that $f \in M$. But f occurs also as the pointwise limit almost everywhere of the sequence $(f_j)_{j\in\mathbb{N}} \subset M^*$. Hence f is a nonnegative nonincreasing symmetric function in $W^{1,p}(\mathbb{R})$, and therefore by Lemma 9.2.14 f is also absolutely continuous and decreases at infinity as const.$|x|^{-1/p}$. Since f belongs to M we know $J_p(f) \le I_p$. As a weak limit of functions in $W^{1,p}(\mathbb{R})$, f satisfies furthermore

$$\|f\|_{1,p} \le \lim_{j\to\infty} \inf \|f_j\|_{1,p} = 1\,,$$

and by construction, the identity

$$I_p = \lim_{j\to\infty} V(f_j)^{p/r} = V(f)^{p/r}$$

holds. This implies $\|f\|_{1,p} = 1$ and $I_p = J_p(f)$, i.e. f is a maximising element for problem (9.2.31). Corollary 9.2.8 finally proves that $f_j \to f$ in $W^{1,p}(\mathbb{R})$.

(d) If the function G is differentiable then it is not hard to show (or to use Proposition 9.6.4) that at every $f \in M$ the functional V has linear continuous Gâteaux derivatives $V'(f;\varphi)$ in all directions $\varphi \in \mathcal{D}(\mathbb{R})$ given by

$$V'(f;\varphi) = \int \widehat{G}'(f)\varphi\, dx = \langle \widehat{G}'(f), \varphi\rangle_2\,.$$

The example of Sect. 9.2.3 shows that the functional $F_p(f) = \|f\|_{1,p}^p$ also has linear continuous Gâteaux derivatives at $f \in M$ in all directions $\varphi \in \mathcal{D}$ and that these are given by

$$F_p'(f;\varphi) = p\langle(\operatorname{sign} f)|f|^{p-1}, \varphi\rangle_2 + p\langle(\operatorname{sign} f)|f'|^{p-1}, \varphi'\rangle_2\,.$$

It follows that the functional $J_p(\cdot)$ has such derivatives at $f \in M$:

$$J_p'(f;\varphi) = \frac{V(f)^{p/r-1}}{F_p(f)}\left\{\frac{p}{r}V'(f;\varphi) - \frac{V(f)}{F_p(f)}F_p'(f;\varphi)\right\}.$$

At the maximising function f_0 of problem (9.2.31) this derivative vanishes. Hence

$$\frac{p}{r}V'(f_0;\varphi) - \frac{V(f_0)}{F_p(f_0)}F_p'(f_0;\varphi) = 0$$

for all $\varphi \in \mathcal{D}(\mathbb{R})$. Using

$$V(f_0) = F_p(f_0)I_p^{r/p}$$

Equation (9.2.32) follows and proves the theorem. □

Remark 9.2.8 For $p = 2$, (9.2.32) simplifies considerably:

$$f'' - f = -aG'(f) \tag{9.2.32'}$$

with some positive constant a. We have solved this equation on all of $\mathbb{R}$ in Theorem 9.2.15 for even C^1 functions G on $\mathbb{R}$ satisfying $G(0) = 0$ and $|G(y)| \le C|y|^r$ for some exponent $r > 2$ and some constant C.

9.3 Some Properties of Weak Solutions of Semilinear Elliptic Equations

A weak solution of

$$-\Delta u(x) = g(x, u(x)) \tag{9.3.1}$$

is, by definition, a distribution $u \in \mathcal{D}'(\mathbb{R}^d; \mathbb{R}^n)$ such that

$$\begin{aligned} &\text{(i)} && \widehat{g}(u) \in L^1_{\text{loc}}(\mathbb{R}^d; \mathbb{R}^n), \\ &\text{(ii)} && \langle u, -\Delta\varphi \rangle = \int \varphi(x)\widehat{g}(u)(x)\,dx, \quad \text{for all } \varphi \in \mathcal{D}(\mathbb{R}^d; \mathbb{R}^n). \end{aligned} \tag{9.3.2}$$

In Sects. 9.7 and 9.8 we will prove the existence of a weak solution of (9.3.1) in the variational approach for $d \geq 3$. There, weak solutions of (9.3.1) in the space $E = E^{1,2}(\mathbb{R}^d; \mathbb{R}^n)$ are obtained under appropriate assumption on g. In Sect. 9.9 the existence of solutions for the two-dimensional case is discussed. We also comment on how to obtain regularity properties of weak solutions for the case $d = 2$.

In this section we study weak solutions in $E = E^{1,2}(\mathbb{R}^d; \mathbb{R}^n)$ for $d \geq 3$. For such solutions (9.3.2) can be rewritten as

$$\langle \nabla u, \nabla\varphi \rangle_2 = \langle \widehat{g}(u), \varphi \rangle_2 \tag{9.3.3}$$

when $\langle \cdot, \cdot \rangle_2$ denotes the scalar product in the real Hilbert space $L^2(\mathbb{R}^d; \mathbb{R}^n)$.

First of all we derive regularity properties of weak solutions. We use arguments from elliptic regularity theory [9.25]. The standard version of this theory [see [Ref. 9.25, Sect. 8], for instance) would proceed by assuming $\widehat{g}(u) \in L^2_{\text{loc}}$ to prove $u \in W^{2,2}_{\text{loc}}$ in the first step. Since we start with $u \in E \subset L^{2^*}(\mathbb{R}^d; \mathbb{R}^n)$, appropriate growth restrictions on g would imply directly $\widehat{g}(u) \in L^2_{\text{loc}}$. However, for the situation of global equations, we know the Green's function of $-\Delta$ or more conveniently of $-\Delta + 1$ explicitly. This knowledge can be used in the first step to prove $u \in L^\infty$ under weaker assumptions on g relying on Kato's inequality for vector fields and some properties of the Green's function J of $-\Delta + 1$ (Appendix E). Then $\widehat{g}(u) \in L^2_{\text{loc}}$ follows easily and we can proceed as indicated above.

In the second part we use these regularity properties of weak solutions to prove some integral identities (the so-called Pokozaev's identities or "virial relations") which every weak solution has to satisfy under the assumptions considered. See Sect. 9.7 for another, much more direct, proof of one of these identities.

9.3.1 Regularity of Weak Solutions

Here we give a comprehensive explanation of that part of *elliptic regularity theory for* (9.3.1) which will be used later. The first lemma on differentiability properties of weak solutions is a version of [Ref. 25, Theorem 8.8].

Lemma 9.3.1. *Suppose $u \in E = E^{1,2}(\mathbb{R}^d;\mathbb{R}^n)$ is a weak solution of* (9.3.1) *such that $\widehat{g}(u) \in L^2_{\mathrm{loc}}$. Then it follows that $u \in W^{2,2}_{\mathrm{loc}}(\mathbb{R}^d;\mathbb{R}^n)$.*

Proof. (a) Since $u \in E^{1,2}(\mathbb{R}^d;\mathbb{R}^n)$ is assumed we only have to show that all the weak derivatives $\partial_i(\nabla u)$, $i = 1,\dots,d$ of $\nabla u \in L^2$ are square-integrable over compact sets. To achieve this we intend to apply the characterisation of the existence of weak derivatives in L^2 as given in Appendix D.1 and accordingly we will derive appropriate bounds for the difference quotients of ∇u.

If for a weak solution u of (9.3.1) in $E = E^{1,2}(\mathbb{R}^d;\mathbb{R}^n)$ the function $\widehat{g}(u)$ is locally square-integrable, then (9.3.3) can be extended from $\mathcal{D}(\mathbb{R}^d;\mathbb{R}^n)$ to all elements $v \in W^{1,2}(\mathbb{R}^d;\mathbb{R}^n)$ with compact support by continuity with respect to φ in the topology of $W^{1,2}(\mathbb{R}^d;\mathbb{R}^n)$.

$$\langle \nabla u, \nabla v\rangle_2 = \langle \widehat{g}(u), v\rangle_2$$

for all $v \in W^{1,2}(\mathbb{R}^d;\mathbb{R}^n)$, $\operatorname{supp} v$ compact.

(b) For any $\chi \in \mathcal{D}(\mathbb{R}^d)$ with $0 \le \chi \le 1$ and $0 < |h|$, define the localised difference quotients

$$v = \chi^2 \Delta^h u\,, \quad \Delta^h u = (\Delta^h u_1,\dots,\Delta^h u_n)\,, \quad \Delta^h u_j = (\Delta^h_1 u_j,\dots,\Delta^h_d u_j)\,.$$

Since $u \in L^{d^*}$ and χ has compact support, these function v are square-integrable and have a compact support contained in $K = \operatorname{supp}\chi$. Their weak derivatives are

$$\nabla v = 2\chi\nabla\chi\Delta^h u + \chi^2\nabla\Delta^h u\,, \quad 0 < |h|\,.$$

Since $\nabla u \in L^2(\mathbb{R}^d)$ and

$$\nabla(\Delta^h u) = \Delta^h(\nabla u)\,, \quad 0 < |h|$$

it follows that ∇v is square-integrable and thus $v \in W^{1,2}$, $\operatorname{supp} v \subseteq K$.

(c) Now consider the family of functions

$$\Delta^{-h} v\,, \quad v = \chi^2\Delta^h u \quad \text{as above,}\ \ 0 < |h| \le 1\,.$$

By step (b) we know that $\Delta^{-h} v \in W^{1,2}(\mathbb{R}^d;\mathbb{R}^n)$ and $\operatorname{supp}\Delta^{-h}v \subseteq K_1 = \{x \in \mathbb{R}^d \mid \operatorname{dist}(x,K) \le 1\}$ for all $0 < |h| \le 1$. Hence, according to step (a), for $0 < |h| \le 1$,

$$\begin{aligned}
\langle \widehat{g}(u), \Delta^{-h}v\rangle_2 &= \langle \nabla u, \nabla(\Delta^{-h}v)\rangle_2 = \langle \nabla u, \Delta^{-h}\nabla v\rangle_2 = -\langle \Delta^h(\nabla u), \nabla v\rangle_2 \\
&= -\langle \Delta^h(\nabla u), 2\chi\nabla\chi\Delta^h u + \chi^2\nabla\Delta^h u\rangle \\
&= -2\langle \chi\Delta^h(\nabla u), \nabla\chi\Delta^h u\rangle_2 - \|\chi\Delta^h(\nabla u)\|_2^2
\end{aligned}$$

and thus

$$\begin{aligned}
A^2 \equiv \|\chi\Delta^h(\nabla u)\|_2^2 &\le 2\|\chi\Delta^h(\nabla u)\|_2\|\nabla\chi\Delta^h u\|_2 + |\langle \widehat{g}(u), \Delta^{-h}v\rangle_2| \\
&\le 2A\|\nabla\chi\|_\infty\|\Delta^h u\|_{L^2(K_1)} + \|\widehat{g}(u)\|_{L^2(K_1)}\|\Delta^{-h}v\|_{L^2(K_1)}
\end{aligned}$$

for all $0 < |h| \leq 1$. By Appendix D.1 we know that

$$\|\Delta^h u\|_{L^2(K_1)} \leq \|\nabla u\|_2 \quad \text{and} \quad \|\Delta^{-h} v\|_{L^2(K_1)} \leq \|\nabla v\|_2$$

and by step (b) $\|\nabla v\|$ is estimated by

$$2\|\chi\nabla\chi\Delta^h u\|_2 + \|\chi^2\Delta^h(\nabla u)\|_2 \leq 2\|\chi\nabla\chi\|_\infty\|\nabla u\|_2 + \|\chi\Delta^h(\nabla u)\|_2 \,.$$

With the abbreviations

$$B = 2\|\chi\nabla\chi\|_\infty\|\nabla u\|_2 \quad \text{and} \quad C = \|\widehat{g}(u)\|_{L^2(K_1)}$$

the above estimate for A^2 reads

$$A^2 \leq (B+C)A + BC \leq \frac{1}{2}(B+C)^2 + \frac{1}{2}A^2 BC \,.$$

This implies

$$A^2 \leq (B+C)^2 + 2BC \leq 2(B+C)^2$$

and therefore

$$A \leq \sqrt{2}(B+C) \,.$$

Inserting the abbreviations yields

$$\|\chi\Delta^h(\nabla u\|_2 \leq \sqrt{2}\{2\|\chi\nabla\chi\|_\infty\|\nabla u\|_2 + \|\widehat{g}(u)\|_{L^2(K_1)}\} \equiv C_0$$

for all $0 < |h| \leq 1$ and all $\chi \in \mathcal{D}(\mathbb{R}^d)$ with $0 \leq \chi \leq 1$ and $K = \operatorname{supp}\chi$.

(d) Now, given a bounded Ω and a subset $\Omega' \Subset \Omega$, open set choose in the above estimate $\chi \in \mathcal{D}$ such that it satisfies in addition

$$\chi \restriction_{\overline{\Omega'}} = 1 \,.$$

Then this estimate implies for all $0 < |h| \leq 1$ that

$$\|\Delta^k(\nabla u)\|_{L^2(\Omega')} \leq \|\chi\Delta^h(\nabla u)\|_2 \leq C_0$$

where the constant C_0 is defined as above. Another apllication of the characterisation of the existence of weak derivatives (Appendix D.1) implies

$$\|\partial_i(\nabla u)\|_{L^2(\Omega')} \leq C_0 < \infty \,, \quad i = 1, \ldots, d \,,$$

and this proves the lemma. □

The following lemmas prepare for proving that under quite natural assumptions on g every weak solution of (9.3.1) in E is essentially bounded.

Let us begin by recalling *Young's inequality*

$$\|h * f\|_r \leq \|h\|_q \|f\|_s \,, \tag{9.3.4}$$

i.e. the closure of Ω' is compact and contained in Ω, which holds for all $r, q, s \in [1, +\infty]$ with $1 + 1/r = 1/q + 1/s$, and the *generalised Young inequality*,

$$\|h * f\|_r \leq C_{r,s}\|h\|_{q,w}\|f\|_s\,, \tag{9.3.5}$$

which holds for all $r, q, s \in (1, +\infty)$ with $1 + 1/r = 1/q + 1/s$.

Lemma 9.3.2. *Denote by J the Bessel potential of order 2 on $\mathbb{R}^d$, $d \geq 3$. If $h \in L^q(\mathbb{R}^d)$ for some $q > d/2$ then*

(a) $$J * h \in L^\infty\,,$$

(b) $$(J * h)(x) \to 0 \quad \textit{for } |x| \to \infty\,.$$

Proof. The conjugate exponent q', $1/q + 1/q' = 1$, satisfies $1 \leq q' < 2^*/2$ if $q > d/2$. From Appendix E we know that $J \in L^{q'}(\mathbb{R}^d)$, and hence Young's inequality proves (a).

In order to prove (b) we write

$$(J*h)(x) = \int_{|x-\xi|\leq 1} J(x-\xi)h(\xi)\,d\xi + \int_{|x-\xi|>1} J(x-\xi)h(\xi)\,d\xi \equiv I_1(x)+I_2(x) \tag{9.3.6}$$

and use the bounds (E.5) for J and obtain by Hölder's inequality

$$|I_1(x)| \leq C\int_{|\xi-x|\leq 1} |x-\xi|^{2-d}|h(\xi|\,d\xi \leq C'\|\chi_{B_1(x)}h\|_q$$

where $B_1(x)$ is the ball with centre x and radius 1. Clearly $h \in L^q$ implies that $\|\chi_{B_1(x)}h\|_q \to 0$ for $|x| \to \infty$. Similarly the relevant estimates for I_2 are

$$\begin{aligned} |I_2(x)| &\leq C\int_{|x-\xi|>1} |x-\xi|^{2-d}\mathrm{e}^{-|x-\xi|/2}|h(\xi)|\,d\xi \\ &\leq C'\left(\int_{|x-\xi|>1} \mathrm{e}^{-q|x-\xi|/4}|h(\xi)|^q\,d\xi\right)^{1/q}, \end{aligned}$$

and by dominated convergence $I_2(x) \to 0$ for $|x| \to \infty$ follows. This proves (b). □

For a weak solution u of (9.3.1) in E we know that $v = |u| \in L^{2^*}$. So Lemma 9.3.2 can be applied directly to such a function v only if $2^* > d/2$, for example if $d < 6$. In the general case the argument is more complicated and we lead up to it via the following lemma.

Lemma 9.3.3. *Suppose $B \subset \mathbb{R}^d$ to be a measurable set of finite (Lebesgue) measure and $p \in [1, \infty]$ some exponent. Suppose furthermore that $K : L^p(B) \to L^p(B)$ is some continuous linear operator with*

(a) $\|K\|_{p,p} < 1$;

(b) *K is monotone with respect to the cone of nonnegative functions;*

(c) *there is some exponent q, $p < q < \infty$, such that*

$$KL^q(B) \subseteq L^q(B) \quad \text{and} \quad \|K\|_{q,q} < 1\,.$$

Then if some $g \in L^p(B)$ satisfies the inequality

$$0 \leq g \leq h + Kg$$

for some $h \in L^q(B)$, it follows that $g \in L^q(B)$.

Proof. By assumption we know that $L^q(B) \subset L^p(B)$. By the monotonicity of K we can iterate the above inequality and obtain for all $n \in \mathbb{N}$

$$0 \leq g \leq \sum_{j=0}^{n} K^j h + K^{n+1} g\,. \tag{9.3.6$'$}$$

Since $\|K\|_{p,p} < 1$ we know $K^{n+1}g \to 0$ in $L^p(B)$ for $n \to \infty$.

Since K maps $L^q(B)$ into itself we know

$$h_n = \sum_{j=0}^{n} K^j h \in L^q(B)$$

for all n. Then $\|K\|_{q,q} < 1$ implies

$$h_n \underset{n\to\infty}{\longrightarrow} \sum_{j=0}^{\infty} K^j h = (1-K)^{-1} h \in L^q(B)\,.$$

Thus in (9.3.6$'$) we can take the limit $n \to \infty$ and obtain

$$0 \leq g \leq (1-K)^{-1} h$$

almost everywhere on B and hence $g \in L^q(B)$ follows. □

Next we prove an integral inequality for weak solutions of (9.3.1).

Lemma 9.3.4. *Assume g to be a C-function on $\mathbb{R}^d \times \mathbb{R}^n$ such that for every $y \in \mathbb{R}^n$ and almost every $x \in \mathbb{R}^d$*

$$|g(x,y)| \leq A_0(x) + A_1(x)|y|^{r_1} + A_2(x)|y|^{r_2} \tag{9.3.7}$$

with

$$A_0 \in L^1_{\text{loc}}\,, \quad A_i \in L^{q_i}_{\text{loc}}\,, \quad \frac{1}{q_i} + \frac{r_i}{2^*} = 1\,, \quad r_1 \leq r_2 \leq 2^*\,.$$

Then for every weak solution u of (9.3.1) in $E = E^{1,2}(\mathbb{R}^d; \mathbb{R}^n)$ the following estimate holds for all $0 \leq \psi \in \mathcal{D}(\mathbb{R}^d)$:

$$\langle \nabla\psi, \nabla|u|\rangle_2 \leq \left\langle \psi, \frac{u}{|u|}\widehat{g}(u)\right\rangle_2 \equiv \int \psi(x)\frac{u}{|u|}\widehat{g}(u)(x)\,dx \tag{9.3.8}$$

where $\frac{u}{|u|}(x)$ is defined to be zero for $u(x) = 0$.

Proof. (1) The first step is to provide some approximations. For $u \in E$ and $\varepsilon > 0$ define $v_\varepsilon = [|u|^2 + \varepsilon^2]^{1/2}$. Using the basic rules for weak derivatives one proves

$$\begin{gathered} v_\varepsilon \underset{\varepsilon \searrow 0}{\longrightarrow} |u| \quad \text{in} \quad W^{1,2}_{\text{loc}} \quad \text{almost everywhere}, \\ \partial_j |u| = \tfrac{u}{|u|} \partial_j u \,. \end{gathered} \tag{9.3.9}$$

For fixed $\varepsilon > 0$ we calculate

$$\sum_i \nabla \Big(\frac{u_i}{v_\varepsilon} \Big) \nabla u_i = \frac{1}{v_\varepsilon} |\nabla u|^2 - \frac{1}{v_\varepsilon} \sum_j \Big(\frac{u}{v_\varepsilon} \partial_j u \Big) \Big(\frac{u}{v_\varepsilon} \partial_j u \Big) \geq 0 \tag{9.3.10}$$

since $|\frac{u}{v_\varepsilon}| < 1$. Hence for $0 \leq \psi \in \mathcal{D}(\mathbb{R}^d)$ we get

$$\nabla \psi \cdot \nabla v_\varepsilon = \sum_i \nabla \Big(\psi \frac{u_i}{v_\varepsilon} \Big) \nabla u_i - \psi \sum_i \nabla \Big(\frac{u_i}{v_\varepsilon} \Big) \nabla u_i \leq \sum_i \nabla \Big(\psi \frac{u_i}{v_\varepsilon} \Big) \nabla u_i \,. \tag{9.3.11}$$

By definition of E we know that for $u \in E$ there is a sequence $\varphi^\nu \in \mathcal{D}(\mathbb{R}^d; \mathbb{R}^n) = \mathcal{D}$, $\nu \in \mathbb{N}$, such that

$$\begin{aligned} &\text{(a)} \ \varphi^{(\nu)} \underset{\nu \to \infty}{\longrightarrow} u \text{ in } L^{2*} \text{ and almost everywhere,} \\ &\text{(b)} \ \partial_j \varphi^{(\nu)} \longrightarrow \partial_j u \text{ in } L^2 \text{ and almost everywhere} \end{aligned} \tag{9.3.12}$$

For such a sequence and fixed $\varepsilon > 0$ define

$$\psi_\varepsilon^{(\nu)} = \frac{\varphi^{(\nu)}}{v_\varepsilon^{(\nu)}} \,, \quad v_\varepsilon^{(\nu)} = [\varepsilon^2 + |\varphi^{(\nu)}|^2]^{1/2} \,. \tag{9.3.13}$$

It follows that $\psi_\varepsilon^{(\nu)} \in \mathcal{D}$, $\nu \in \mathbb{N}$, and

$$\begin{aligned} &\text{(a)} \ \psi_\varepsilon^{(\nu)} \underset{\nu \to \infty}{\longrightarrow} \frac{u}{v_\varepsilon} \text{ almost everywhere and in } L^q_{\text{loc}},\ 1 \leq q < \infty, \\ &\text{(b)} \ \partial_j \psi_\varepsilon^{(\nu)} \underset{\nu \to \infty}{\longrightarrow} \partial_j \Big(\frac{u}{v_\varepsilon} \Big) = \frac{\partial_j u}{v_\varepsilon} - \frac{u}{v_\varepsilon} \Big(\frac{u}{v_\varepsilon} \partial_j u \Big) \text{ almost everywhere and in } L^2. \end{aligned} \tag{9.3.14}$$

(2) Now we apply the above approximation to a weak solution $u \in E$ of (9.3.1). Assumption (9.3.7) easily implies $\widehat{g}(u) \in L^1_{\text{loc}}$. Hence we can apply (9.3.3) for

$$\varphi = \psi \psi_\varepsilon^{(\nu)} \,, \quad \nu = 1, 2, \dots \,, \quad \varepsilon > 0 \,,$$

with some fixed $0 \leq \psi \in \mathcal{D}(\mathbb{R}^d)$ and obtain

$$\langle \nabla u, \nabla (\psi \psi_\varepsilon^{(\nu)}) \rangle_2 = \langle \widehat{g}(u), \psi \psi_\varepsilon^{(\nu)} \rangle_2 \,.$$

The limit $\nu \to \infty$ of this equation exists according to (9.3.14) and is given by

$$\Big\langle \nabla u, \nabla \Big(\psi \frac{u}{v_\varepsilon} \Big) \Big\rangle_2 = \Big\langle \widehat{g}(u), \psi \frac{u}{v_\varepsilon} \Big\rangle_2 \,. \tag{9.3.15}$$

Now we observe that inequality (9.3.11) implies

$$\langle \nabla\psi, \nabla v_\varepsilon \rangle_2 \le \left\langle \nabla\left(\psi \frac{u}{v_\varepsilon}\right), \nabla u \right\rangle_2 . \tag{9.3.16}$$

Thus

$$\langle \nabla\psi, \nabla v_\varepsilon \rangle_2 \le \left\langle \widehat{g}(u), \psi \frac{u}{v_\varepsilon} \right\rangle_2 .$$

Finally, by (9.3.9) we can take the limit $\varepsilon \searrow 0$ in this inequality and obtain (9.3.8). □

Remark 9.3.1. Note that the proof of this lemma contains a version of *Kato's inequality* [9.26, 27] *for vector fields*:

$$-\Delta|u| \le \frac{u}{|u|}(-\Delta u) \tag{9.3.17}$$

with the following interpretation according to (9.3.16):

$$\langle \nabla\psi, \nabla|u| \rangle_2 \le \left\langle \nabla\left(\psi \frac{u}{|u|}\right), \nabla u \right\rangle_2 \tag{9.3.17'}$$

for all $0 \le \psi \in \mathcal{D}(\mathbb{R}^d)$.

Lemma 9.3.5 *Under the assumptions of Lemma* 9.3.4 *and the additional assumption*

$$|yg(x,y))| \le a_0(x)|y| + a_1(x)|y|^{p_1+1} + a_2(x)|y|^{p_2+1} \tag{9.3.18}$$

for all $y \in \mathbb{R}^n$ and almost every $x \in \mathbb{R}^d$ with

$$a_i \in L^{s_i}, \quad \alpha' < s_i \le +\infty ,$$

and for $i = 1, 2$:

$$\frac{p_i - 1}{2^*} + \frac{1}{s_i} \le \frac{1}{\alpha'} \tag{9.3.19}$$

($\alpha' = d/2 =$ conjugate exponent of $\alpha = 2^*/2$), every weak solution u of (9.3.1) *in E satisfies almost everywhere on $\mathbb{R}^d$*

$$|u| \le (J * \widehat{f})(u) \tag{9.3.20}$$

with

$$\widehat{f}(u)(x) := |u(x)| + \frac{u}{|u|}\widehat{g}(u)(x) \tag{9.3.20}$$

and

$$J \text{ being the kernel of } (-\Delta + 1)^{-1} .$$

Proof. For all $0 \le \varphi \in \mathcal{D}(\mathbb{R}^d)$ we have by Lemma 9.3.4 and definition (9.3.21)

$$\langle \nabla\varphi, \nabla|u| \rangle_2 + \langle \varphi, |u| \rangle_2 \le \langle \varphi, \widehat{f}(u) \rangle_2 . \tag{9.3.22}$$

Pointwise positivity of the Bessel potential J implies

$$0 \le \varphi := J * \psi = (-\Delta + 1)^{-1}\psi \tag{9.3.23}$$

whenever $0 \le \psi \in \mathcal{D}(\mathbb{R}^d)$.

Clearly the functions φ given by (9.3.23) are $\mathcal{C}^\infty$ but have no compact support. So in order to prove that nevertheless they can be inserted into (9.3.22) we have to use an approximation.

For fixed $\chi \in \mathcal{D}(\mathbb{R}^d)$, $0 \le \chi \le 1$, $\chi(x) = 1$ for $|x| \le 1$ and $\chi(x) = 0$ for $|x| > 2$, define $\chi_j(x) = \chi(x/j)$, $j = 1, 2, \ldots$, and

$$\varphi_j = \chi_j \varphi = \chi_j(J * \psi), \quad j \in \mathbb{N}.$$

Clearly these functions belong to $\mathcal{D}(\mathbb{R}^d)$ and are nonnegative. Hence (9.3.22) applies for all j:

$$\langle \nabla \varphi_j, \nabla |u| \rangle_2 + \langle \varphi_j, |u| \rangle_2 \le \langle \varphi_j, \widehat{f}(u) \rangle_2 . \tag{9.3.24}$$

By Young's inequality and the L^p properties of J, we know that $\varphi = J * \psi$ and $J * \partial_j \psi$ belong to $L^r(\mathbb{R}^d)$ for all $1 \le r \le +\infty$. It follows that

$$\varphi_j \to \varphi \quad \text{for } j \to \infty \text{ in } L^r \text{ for all } r \in [1, +\infty]$$

and

$$\nabla \varphi_j \to \nabla \varphi \quad \text{for } j \to \infty \text{ in } L^2 .$$

Hence the limit of the left-hand side of inequality (9.3.24) is

$$\langle \nabla \varphi, \nabla |u| \rangle_2 + \langle \varphi, |u| \rangle_2 = \langle (-\Delta + 1)\varphi, |u| \rangle_2 = \langle \psi, |u| \rangle_2 . \tag{9.3.25}$$

In order to determine the limit of the right-hand side of inequality (9.3.24) it suffices to ensure that $\varphi \widehat{f}(u)$ belongs to $L^1(\mathbb{R}^d)$. By (9.3.21) and (9.3.18) we know that

$$|\widehat{f}(u)| \le a_0 + |u| + a_1 |u|^{p_1} + a_2 |u|^{p_2} . \tag{9.3.26}$$

Since $\varphi \in L^r(\mathbb{R}^d)$ for all $1 \le r \le +\infty$, $\varphi \widehat{f}(u) \in L^1(\mathbb{R}^d)$ follows form $a_0 \in L^{s_0}$ with some $s_0 \in [1, \infty]$ and $a_i \in L^{s_i}$ with $p_i/2^* + 1/s_i \le 1$ for $i = 1, 2$ by Hölder's inequality. These conditions are implied by (9.3.19). Thus, dominated convergence proves

$$\langle \varphi_j, \widehat{f}(u) \rangle_2 \underset{j \to \infty}{\longrightarrow} \langle \varphi, \widehat{f}(u) \rangle_2 = \int (J * \psi) \widehat{f}(u)\, dx . \tag{9.3.27}$$

If we knew

$$J * \widehat{f}(u) \in L^1_{\mathrm{loc}} \tag{9.3.28}$$

Fubini's theorem would imply

$$\langle J * \psi, \widehat{f}(u) \rangle_2 = \langle \psi, J * \widehat{f}(u) \rangle_2 \tag{9.3.29}$$

and therefore

$$\langle \psi, |u| \rangle_2 \le \langle \psi, J * \widehat{f}(u) \rangle_2 \tag{9.3.30}$$

for arbitrary $0 \le \psi \in \mathcal{D}(\mathbb{R}^d)$. This implies inequality (9.3.20).

Hence, the last step of the proof is to ensure condition (9.3.28). We will do this in a way which will also be used in the proof of the next theorem. By pointwise positivity of J and (9.3.26) we have

$$|J * \widehat{f}(u)| \leq J * |\widehat{f}(u)| \leq J * a_0 + J * (|u| + a_1|u|^{p_1} + a_2|u|^{p_2}). \qquad (9.3.31)$$

The contributions from the small and the large values of $|u|$ are treated separately. To this end we split up $|u|$ as

$$|u| = [|u| \leq \lambda]|u| + [\lambda < |u|]|u|$$

where $[v \leq \lambda]$ denotes the set $\{x \in \mathbb{R}^d \,|\, v(x) \leq \lambda\}$ and, at the same time, the characteristic function of this set, since the meaning will always be clear from the context.

Our assumption $a_i \in L^{s_i}$ with $\alpha' < s_i$, implies

$$J * a_i \in L^\infty .$$

Therefore, we continue estimate (9.3.31) by

$$|J * \widehat{f}(u)| \leq C_\lambda + J * (A_\lambda(v)v), \quad v = |u| \qquad (9.3.32)$$

with

$$C_\lambda = \|J * a_0\|_\infty + \lambda\|J\|_1 \lambda^{p_1}\|J * a_1\|_\infty + \lambda^{p_2}\|J * a_2\|_\infty$$

and

$$A_\lambda(v) = [\lambda < v]\{1 + a_1 v^{p_1-1} + a_2 v^{p_2-1}\}. \qquad (9.3.33)$$

Since $v = |u| \in L^p(\mathbb{R}^d)$, $p = 2^*$, we know for the Lebesgue measure $|[\lambda < v]|$ of the set $[\lambda < v]$ for $\lambda > 0$ that

$$|[\lambda < v]| \leq \lambda^{-p}\|v\|_p^p < \infty \qquad (9.3.34)$$

and, in addition,

$$\|[\lambda < v]v\|_p \to 0 \quad \text{for } \lambda \to +\infty .$$

The second term in (9.3.32) is controlled by the generalised Young inequality and then by Hölder's inequality. For $\gamma > \alpha = 2^*/2$ with $v_\lambda = [\lambda < v]v$,

$$\|J * (A_\lambda(v)v)\|_\gamma \leq C_{\gamma,\rho}\|J\|_{\alpha,w}\|A_\lambda(v)\|_r\|v_\lambda\|_\sigma \qquad (9.3.35)$$

if

$$1 + \frac{1}{\gamma} = \frac{1}{\alpha} + \frac{1}{\rho} \quad \text{and} \quad \frac{1}{\rho} = \frac{1}{r} + \frac{1}{\sigma}$$

with $\gamma, \rho, \alpha \in (1, \infty)$ and $1 \leq \sigma \leq 2^*$.

By (9.3.33) and Hölder's inequality we know that

$$\begin{aligned}\|A_\lambda(v)\|_r \leq \|[\lambda < v]\|_r &+ \|[\lambda < v]a_1\|_{r_1}\|v_\lambda\|_p^{p_1-1}\\ &+ \|[\lambda < v]a_2\|_{r_2}\|v_\lambda\|_p^{p_2-1}\end{aligned} \qquad (9.3.36)$$

with $\frac{1}{r} = \frac{1}{r_i} + \frac{p_i - 1}{p}$ for suitable r_i.

A simple way to ensure the finiteness of $\|A_\lambda(v)\|_r$ is to choose $r = \alpha'$ and then, by (9.3.35), $\sigma = \gamma$. Then by our assumption (9.3.19) we know that

$$\frac{1}{r_i} = \frac{1}{\alpha'} - \frac{p_i - 1}{p} \geq \frac{1}{s_i}.$$

Hence it follows that

$$\|[\lambda < v]a_i\|_{r_i} \leq C_\lambda \|a_i\|_{s_i} < \infty \tag{9.3.37}$$

and

$$\|[\lambda < v]a_i\|_{r_i} \to 0 \quad \text{for } \lambda \to \infty.$$

This, in turn, implies by (9.3.34) and (9.3.36):

$$\|A_\lambda(|u|)\|_{\alpha'} \to 0 \quad \text{for } \lambda \to \infty. \tag{9.3.38}$$

Therefore the estimate (9.3.32) says in particular that statement (9.3.28) holds. This proves the lemma. □

Theorem 9.3.6. *Suppose that $g : \mathbb{R}^d \times \mathbb{R}^n \to \mathbb{R}^n$ is a C-function satisfying conditions* (9.3.7), (9.3.18), *and* (9.3.19). *In* (9.3.7) *suppose in addition that $A_i \in L^2_{\text{loc}}$, $i = 0, 1, 2$. Then every weak solution u of*

$$-\Delta u(x) = g(x, u(x))$$

in $E = E^{1,2}(\mathbb{R}^d; \mathbb{R}^n)$, $d \geq 3$, $n \geq 1$, has the following properties:

(a) $u \in L^\infty$,
(b) $u(x) \to 0$, *for* $|x| \to \infty$,
(c) $u \in W^{2,2}_{\text{loc}}$.

Proof. According to Lemma 9.3.6 we know the estimate (9.3.20) for $v = |u|$, if $u \in E$ is a weak solution of (9.3.1). Therefore, by inequality (9.3.32),

$$v = |u| \leq C_\lambda + J * (A_\lambda(v) v_\lambda) \tag{9.3.39}$$

holds for $\lambda > 0$. If $q > p$ is given, apply (9.3.35) for $\gamma = p$ and $\gamma = q$. Now choose some $\lambda > 0$ according to (9.3.38) such that

$$C_{\gamma,\rho} \|J\|_{\alpha,W} \|A_\lambda(v)\|_{\alpha'} < 1 \quad \text{for } \gamma = p \text{ and } \gamma = q. \tag{9.3.40}$$

Given $B \subset \mathbb{R}^d$, $|B| < \infty$, define $B_\lambda = B \cup [\lambda < v]$. It follows that $|B_\lambda| < \infty$. Denote by χ the characteristic function of B_λ and for $g \in L^p(B_\lambda)$ define

$$Kg = \chi J * (A_\lambda g), \quad A_\lambda = A_\lambda(v). \tag{9.3.41}$$

Inequality (9.3.35) implies for $\gamma = p$ and $p = q$ (and $r = \alpha'$)

$$\|Kg\|_{L^\gamma(B_\lambda)} \leq C_{\gamma,\rho} \|J\|_{\alpha,w} \|A_\lambda\|_{\alpha'} \|g\|_{L^p(B_\lambda)}.$$

Hence K is a linear operator on $L^p(B_\lambda)$ with norm $\|K\|_{p,p} < 1$ and this operator leaves the subspace $L^q(B_\lambda)$ invariant. Its norm with respect to this subspace is denoted by $\|K\|_{q,q}$. By (9.3.40) it satisfies $\|K\|_{q,q} < 1$.

If we multiply estimate (9.3.39) by χ we obtain

$$\chi v \leq C_\lambda \chi + \chi(J * A_\lambda \chi v) = C_\lambda \chi + K(\chi v),$$

that is, inequality (9.3.6) with $h = C_\lambda \chi \in L^q(B_\lambda)$ and $g = \chi v \in L^p(B_\lambda)$. Since the operator K defined by (9.3.41) is clearly monotone with respect to the cone of nonnegative functions, Lemma 9.3.3 applies and yields $g \in L^q(B_\lambda)$. Since $B \subset \mathrm{I\!R}^d$, $|B| < \infty$, was arbitrary this shows that

$$|u| \in L^q_{\mathrm{loc}}(\mathrm{I\!R}^d), \quad 1 \leq q < \infty, \tag{9.3.42}$$

in the sense $\|\, |u| \,\|_{L^q(B)} < \infty$ for all measurable sets $B \subset \mathrm{I\!R}^d$ with $|B| < \infty$.

An elementary calculation yields for a weak solution u of (9.3.1)

$$(-\Delta + 1)|u|^2 = -2u \cdot \Delta u - 2|\nabla u|^2 + |u|^2 = 2u \cdot \widehat{g}(u) + |u|^2 - 2|\nabla u|^2 .$$

Since $|\nabla u|^2 \in L^1$, its convolution with J is well defined. We are going to show that $2u \cdot \widehat{g}(u) + |u|^2$ admits convolution with J. This implies that the above equation admits the convolution with J and thus implies by monotonicity of J and assumption (9.3.18):

$$\begin{aligned} |u|^2 &= J * \{2u \cdot \widehat{g}(u) + |u|^2 - 2|\nabla u|^2\} \\ &\leq J * \{2u \cdot \widehat{g}(u) + |u|^2\} \\ &\leq J * \{2a_0|u| + 2a_1|u|^{p_1+1} + 2a_2|u|^{p_2+1} + |u|^2\} \\ &\leq \lambda J * \{(2a_0 + 2a_1\lambda^{p_1} + 2a_2\lambda^{p_2} + \lambda) + J * h_\lambda \end{aligned} \tag{9.3.43}$$

where

$$h_\lambda = 2a_0 v_\lambda + 2a_1 v_\lambda^{p_1+1} + 2a_2 v_\lambda^{p_2+1} + v_\lambda^2, \quad v_\lambda = [\lambda < |u|]|u| .$$

Since $|[\lambda < |u|]| < \infty$ for all $\lambda > 0$, statement (9.3.42) applies and proves $v_\lambda \in L^q(\mathrm{I\!R}^d)$ for all $1 \leq q < \infty$. This in turn easily implies by assumption (9.3.19) and Hölder's inequality that $h_\lambda \in L^{q_0}(\mathrm{I\!R}^d)$ for some $q_0 > \alpha' = d/2$. Therefore, Lemma 9.3.2 proves $J * h_\lambda \in L^\infty$ and $J * h_\lambda(x) \to 0$ for $|x| \to \infty$. Again, by (9.3.19) we have

$$\begin{aligned} &J * \{2a_0 + 2a_1\lambda^{p_1} + 2a_2\lambda^{p_2} + \lambda\} \\ &\qquad \leq 2\|J * a_0\|_\infty + 2\lambda^{p_1}\|J * a_1\|_\infty + 2\lambda^{p_2}\|J * q_2\|_\infty + \lambda\|J\|_1 = C_\lambda , \end{aligned}$$

as in Lemma 9.3.5, and thus (9.3.43) implies

$$|u|^2 \leq \lambda C_\lambda + J * h \tag{9.3.44}$$

and it follows that $|u| \in L^\infty$ and

$$\lim_{|x|\to\infty} |u(x)|^2 \leq \lambda C_\lambda .$$

But we are allowed to choose λ arbitrarily small; hence $|u(x)| \to 0$ for $|x| \to \infty$; and this proves parts (a) and (b). Finally, part (c) follows by Lemma 9.3.1 since $|u| \in L^\infty$ and $A_i \in L^2_{\text{loc}}$, $i = 0, 1, 2$, immediately imply $\widehat{g}(u) \in L^2_{\text{loc}}$.

□

Clearly not all the assumptions used in Theorem 9.3.6 are independent. Therefore we present a simplified version with all assumptions stated explicitly.

Corollary 9.3.7. *Suppose that* $g : \mathbb{R}^d \times \mathbb{R}^n \to \mathbb{R}^n$, $d \geq 3$, $n \geq 1$, *is a* C*-function satisfying in addition either* (A) *or* (B). *Then the conclusions of Theorem* 3.6 *hold.*

(A) *For all* $y \in \mathbb{R}^n$ *and almost every* $x \in \mathbb{R}^d$

$$|g(x,y)| \leq A_0(x) + A_2(x)|y|^{2^*} \tag{9.3.45}$$

with $A_0 \in L^2_{\text{loc}}$, $A_2 \in L^\infty_{\text{loc}}$, *and*

$$|y \cdot g(x,y)| \leq a_0(x)|y| + a_2(x)|y|^{p_2+1}$$

with $a_i \in L^{s_i}$, $d/2 < s_i \leq +\infty$, $1 < p_2 \leq 2^* - 1$, $\frac{p_2-1}{2^*} + \frac{1}{s_2} \leq \frac{2}{d}$.

(B) g *is independent of* $x \in \mathbb{R}^d$ *(translation-invariant case) and for all* $y \in \mathbb{R}^n$, *for some constants* C_i,

$$\begin{aligned} |g(y)| &\leq C_0 + C_1|y|^{2^*}, \\ |y \cdot g(y)| &\leq C_0|y| + C_1|y|^{2^*}. \end{aligned} \tag{9.3.46}$$

Proof. (A) and (B) imply all the assumptions of Theorem 9.3.6. □

We mention without proof a result about support and decay properties of a weak solution of (9.3.1) for the *translation-invariant case* (9.13).

If in addition to assumption (B) of Corollary 9.3.7, there are constants $C > 0$, $\delta > 0$ and an exponent $r \in [1,2]$ such that

$$yg(y) \leq -C|y|^r \quad \text{for all } y \in \mathbb{R}^n \text{ with } |y| \leq \delta,$$

then in the case when $r = 2$, every weak solution $u \in E$ decays exponentially for $|x| \to \infty$, while in the case $1 \leq r < 2$ the solution u has compact support.

Remark 9.3.2. Since according to (9.3.9–14) "sufficiently good" approximations are available, the results of Theorem 9.3.6 and Corollary 9.3.7 can be extended to the case where the nonlinear term g also depends on ∇u, i.e. to equations of the form

$$-\Delta u(x) = g(x, u(x), \nabla u(x)).$$

We indicate briefly some cases where this is possible. Suppose g splits into $g_0 + g_1$ where g_0 can be treated by Corollary 9.3.7 or Theorem 9.3.6 and where the following conditions are supposed for g_1:

$$|\widehat{g}_1(u)| \le c_1 |\nabla u|^{\alpha_1} |u|^{\beta_1} + c_2 |\nabla u|^{\alpha_2} |u|^{\beta_2} ,$$
$$|u \cdot \widehat{g}_1(u)| \le c_3 |\nabla u|^{r_1} |u|^{p_1+1} + c_4 |\nabla u|^{r_2} |u|^{p_2+1} ,$$

with constants $c_i \ge 0$ and exponents α_i, β_i, r_i, p_i satisfying the following restrictions:

$$\alpha_i, r_i \in [0,2] , \quad r_i < \frac{4}{d} , \quad \alpha_1 < \alpha_2 , \quad \beta_1 < \beta_2 ,$$

$$\frac{\alpha_i}{2} + \frac{\beta_i}{2^*} \le 1 , \quad 1 \le p_i \le 1 + 2^* \left\{ \frac{2}{d} - \frac{r_i}{2} \right\} .$$

These additional terms can be treated as in Theorem 9.3.6 simply by identifying in (9.3.7):

$$A_i = |\nabla u|^{\alpha_i}$$

and in (9.3.18, 19)

$$a_i = |\nabla u|^{r_i}$$

and using $|\nabla u| \in L^2$.

For instance additional terms g_1 on $\mathbb{R}^3$ are allowed if they satisfy

$$|\widehat{g}_1(u)| \le C |\nabla u| \, |u|^2$$

since then $2^* = 6$.

9.3.2 Pohozaev's Identities

By Theorem 9.3.6 and Corollary 9.3.7 we know explicit conditions on g which imply

$$\widehat{g}(u) \in L^1_{\text{loc}} \quad \text{and} \quad \Delta u \in L^2_{\text{loc}} \tag{9.3.47}$$

for a weak solution $u \in E$ of (9.3.1). Indeed, by these results it follows that $u \in W^{2,2}_{\text{loc}} \cap L^\infty$ and hence (9.3.47). Then it is clear that a solution satisfies (9.3.1) not only in the sense of (9.3.2) but also in the sense of pointwise equality almost everywhere:

$$-\Delta u(x) = g(x, u(x)) \quad \text{for almost every } x \in \mathbb{R}^d . \tag{9.3.48}$$

If (9.3.1) is considered on some open bounded subset $\Omega \subset \mathbb{R}^d$ then (9.3.48) follows almost everywhere on Ω since Theorem 9.3.6 is of local nature.

As *Pohozaev* has observed, the information provided by (9.3.47) and (9.3.48) can be used to derive some integral identities as very useful necessary conditions for solutions of (9.3.1) [9.28].

Proposition 9.3.8. *If $u \in E$ is a weak solution of* (9.3.1) *such that* (9.3.47) *and*

$$u \cdot \widehat{g}(u) \in L^1(\mathbb{R}^d) \tag{9.3.49}$$

hold, then this solution satisfies

$$\int u \cdot \widehat{g}(u)\, dx = \int |\nabla u|^2\, dx = 2K(u) \tag{9.3.50}$$

If (9.3.1) *is considered over a region $\Omega \subsetneqq \mathbb{R}^d$ with smooth boundary $\partial\Omega$ then the boundary condition*

$$u \restriction \partial\Omega = 0 \tag{9.3.51}$$

is assumed in the sense that u is an element of $E^{1,2}(\Omega; \mathbb{R}^n)$ (Sect. 9.1).

Proof. Choose a fixed $\varphi \in \mathcal{D}$, $0 \le \varphi \le 1$, such that $\varphi(x) = 1$ for all $|x| \le 1$ and $\varphi(x) = 0$ for all $|x| > 2$. Then introduce $\varphi_k \in \mathcal{D}$ by $\varphi_k(x) = \varphi(x/k)$ for all $x \in \mathbb{R}^d$ and all $k \in \mathbb{N}$. It follows that

$$\varphi_k \to 1\,, \quad \nabla\varphi_k \to 0\,, \quad \text{for } k \to \infty\,, \tag{9.3.52}$$

uniformly on compact sets, and

$$\sup_{x\in\mathbb{R}^d} \sup_{k\in\mathbb{N}} |(x\nabla\varphi_k(x)| = c < \infty\,, \tag{9.3.53}$$

$$\|\nabla\varphi_k\|_d = \|\nabla\varphi\|_d \quad \text{for all } k \in \mathbb{N}\,. \tag{9.3.54}$$

Since $\Delta u \in L^2_{\text{loc}}$ is known, Hölder's inequality implies for $k \in \mathbb{N}$ that

$$\|\Delta u\, u\varphi_k\|_1 \le \|\pi_{B(k)}\Delta u\|_2\, \|u\|_{2^*} \|\varphi_k\|_d < \infty$$

if $B(k)$ denotes the compact support of φ_k. Hence we are allowed to multiply (9.3.48) by $u\varphi_k$ and to integrate.

$$\begin{aligned} \int \varphi_k u \cdot \widehat{g}(u)\, dx &= \int (-\Delta u) \cdot u\varphi_k\, dx = \int \nabla u \cdot \nabla(u\varphi_k)\, dx - \int \operatorname{div}\{u\varphi_k \nabla u\}\, dx \\ &= \int \varphi_k |\nabla u|^2\, dx + \int u \cdot \nabla u \cdot \nabla\varphi_k\, dx \end{aligned} \tag{9.3.55}$$

where in the case of a bounded region Ω the boundary condition (9.3.51) is used and where we integrate only over Ω.

For any measurable subset $A \subset \mathbb{R}^d$ Hölder's inequality implies the following uniform estimate for the last integral in (9.3.55) if (9.3.54) is used:

$$\|\pi_A u \cdot \nabla u \cdot \nabla\varphi_k\|_1 \le \|\pi_A u\|_{2^*} \|\pi_A \nabla u\|_2 \|\nabla\varphi\|_d\,, \quad k \in \mathbb{N}\,,$$

where the dot indicates matrix multiplication. Since $u \in L^{2^*}$ and $\nabla u \in L^2$, this estimate proves that

$$\sup_k \|\pi_K^{\perp} u \cdot \nabla u \cdot \nabla\varphi_k\|_1 \to 0 \quad \text{for } K \subset \mathbb{R}^d \text{ compact, } K \nearrow \mathbb{R}^d\,,$$

$$\sup_k \|\pi_A u \cdot \nabla u \cdot \nabla\varphi_k\|_1 \to 0 \quad \text{for } A \subset \mathbb{R}^d \text{ measurable, } |A| \to 0\,.$$

Therefore, by pointwise convergence of the integrand according to (9.3.52), Vitali's convergence theorem allows us to conclude that

$$\int u \cdot \nabla u \cdot \nabla \varphi_k \, dx \to 0 \quad \text{for } k \to \infty \,.$$

According to our assumptions and (9.3.52) the convergence of the other two integrals follows easily by dominated convergence and thus, in the limit $k \to \infty$ of (9.3.55), we obtain (9.3.50). □

In the variational approach, one solves (9.3.1) under the additional assumption that the nonlinearity g has a "potential" $G : \mathbb{R}^d \times \mathbb{R}^n \to \mathbb{R}$ such that

$$\begin{aligned} &G(x,0) = 0 \quad \text{for almost every } x \in \mathbb{R}^d ; \quad \text{and for almost every } x \in \mathbb{R}^d , \\ &y \mapsto G(x,y) \quad \text{is continously differentiable on } \mathbb{R} \setminus \{0\} \text{ and} \\ &g(x,y) = \operatorname{grad}_y G(x,y) \quad \text{for } y \neq 0 \,. \end{aligned} \tag{9.3.56}$$

Further growth restrictions on G and g implying the existence of a weak solution are formulated later. For the following discussion we have to assume in addition that the following chain rule for weak differentiation,

$$\frac{\partial}{\partial x_i} G(x, u(x)) = G_{,x_i}(x, u(x)) + g(x, u(x)) \partial_i u(x) \,, \tag{9.3.57}$$

holds for G and $u \in E$. Some comments on the validity of this equation are given at the end of this section.

Proposition 9.3.9. *Suppose $u \in E$ is a weak solution of* (9.3.1) *satisfying* (9.3.47) *and*

$$\widehat{G}(u), x \cdot \widehat{G}_{,x}(u) \in L^1(\mathbb{R}^d) \,. \tag{9.3.58}$$

Suppose furthermore that the potential G of g according to (9.3.56) *satisfies* (9.3.57) *and that in the case of a bounded region Ω with smooth boundary $\partial\Omega$ the boundary condition* (9.3.51) *holds. Then this solution satisfies the following integral identities:*

$$(d-2)K(u) - d \cdot V(u) - \int_\Omega x \cdot \widehat{G}_{,x}(u) \, dx = \int_{\partial\Omega} \left\{ \frac{1}{2} x (\nabla u)^2 - \nabla u \cdot (x \cdot \nabla u) \right\} d\underline{o} \tag{9.3.59}$$

and

$$(d-2)K(u) = d \cdot V(u) + \int_{\mathbb{R}^d} x \cdot \widehat{G}_{,x}(u) \, dx \,. \tag{9.3.60}$$

Proof. If $u \in E$ satisfies $\Delta u \in L^2_{\text{loc}}$, then for every fixed $\varphi \in \mathcal{D}$ a simple calculation using the basic rules for weak differentiation shows that the following identity holds almost everywhere:

$$-\Delta u\varphi(x\cdot\nabla u)=(1-d/2)\varphi(\nabla u)^2+(x\cdot\nabla u)(\nabla\varphi\cdot\nabla u)-\frac{1}{2}(x\cdot\nabla\varphi)(\nabla u)^2$$
$$+\operatorname{div}\left\{\frac{1}{2}\varphi x(\nabla u)^2-\varphi\nabla u(x\cdot\nabla u)\right\}. \qquad (9.3.61)$$

As in the proof of Proposition 9.3.8 one verifies that all terms in this equation belong to $L^1(\mathbb{R}^d)$.

In the same sense, one proves by another elementary calculation that

$$\varphi(x\cdot\nabla u)\widehat{g}(u)=d\cdot\varphi\cdot\widehat{G}(u)-x\cdot\widehat{G}_{,x}(u)\varphi-\widehat{G}(u)(x\cdot\nabla\varphi)$$
$$+\operatorname{div}\{x\varphi(x)\widehat{G}(u)\}. \qquad (9.3.62)$$

By (9.3.48), the left-hand sides of (9.3.61) and (9.3.62) are equal for a weak solution u, and thus the right-hand sides also agree.

Now we treat the local case of a bounded-region $\Omega\subset\mathbb{R}^d$ with smooth boundary $\partial\Omega$ and the global case separately. In the local case we can take $\varphi\in\mathcal{D}$ such that $\varphi=1$ on Ω, and by integration over Ω it follows that

$$(d/2-1)\int_\Omega(\nabla u)^2\,dx-d\cdot\int_\Omega\widehat{G}(u)\,dx-\int_\Omega x\cdot\widehat{G},x(u)\,dx$$
$$=\int_{\partial\Omega}\left\{\frac{1}{2}x(\nabla u)^2-\nabla u\cdot(x\cdot\nabla u)-x\widehat{G}(u)\right\}\cdot d\underline{o},$$

since $\widehat{G}(u), x\cdot\widehat{G}_{,x}(u)\in L^1(\Omega)$ is assumed. Hence, using the boundary condition (9.3.51), assumption (9.3.56), and the notation introduced in Sect. 9.1, equation (9.3.59) follows.

In the global case of solutions on $\Omega=\mathbb{R}^d$, denote the integral of the left-hand side of (9.3.61) by $L(\varphi)$. Since $\varphi\in\mathcal{D}$, the integral of the divergence term vanishes and thus

$$L(\varphi)=(1-d/2)\int_{\mathbb{R}^d}\varphi(\nabla u)^2\,dx+\int_{\mathbb{R}^d}\left\{(x\cdot\nabla u)(\nabla\varphi\cdot\nabla u)-\frac{1}{2}(x\cdot\nabla\varphi)(\nabla u)^2\right\}dx.$$

Similarly, the integral of the right-hand side of (9.3.62) is denoted by $R(\varphi)$; then, as above,

$$R(\varphi)=-d\int_{\mathbb{R}^d}\varphi\widehat{G}(u)\,dx-\int_{\mathbb{R}^d}\varphi x\cdot\widehat{G}_{,x}(u)\,dx-\int_{\mathbb{R}^d}\widehat{G}(u)(x\cdot\nabla\varphi)\,dx,$$

and by (9.3.48), (9.3.61) and (9.3.62) we know that

$$L(\varphi)=R(\varphi)\quad\text{for all }\varphi\in\mathcal{D}. \qquad (9.3.63)$$

This identity holds in particular for all test functions φ_k, $k\in\mathbb{N}$, introduced in the proof of Proposition 9.3.8. Since $\nabla u\in L^2$ is known, dominated convergence implies by (9.3.52) and (9.3.53) that

$$L(\varphi_k)\to(2-d)K(u)\quad\text{for }k\to\infty.$$

Similarly, $\widehat{G}(u), x \cdot \widehat{G}_{,x}(u) \in L^1$ imply

$$R(\varphi_k) \to -d \cdot V(u) - \int_{\mathbb{R}^d} x \cdot \widehat{G}_{,x}(u)\, dx \quad \text{for } k \to \infty$$

and therefore, by (9.3.63), equation (9.3.60) follows. □

Under the hypotheses of Proposition 9.3.9 the following simplifications occur.

Corollary 9.3.10. *For a bounded region Ω with smooth boundary $\partial\Omega$, denote by ν the outer normal unit vector on $\partial\Omega$. Then, for a weak solution of* (9.3.1) *satisfying the boundary condition*

$$u \restriction \partial\Omega = 0 \quad \text{and} \quad \nabla u_i \restriction \partial\Omega \text{ is proportional to } \nu\,, \tag{9.3.64}$$

(9.3.59) *simplifies to*

$$(d-2)K(u) - d \cdot V(u) - \int_\Omega x \cdot \widehat{G}_{,x}(u)\, dx = -\frac{1}{2}\int_{\partial\Omega} (x\cdot\nu)(\nabla u)^2 do\,. \tag{9.3.65}$$

In particular, in the translation-invariant case ($G_{,x} \equiv 0$),

$$(d-2)K(u) + \frac{1}{2}\int_{\partial\Omega} (x\cdot\nu)(\nabla u)^2 do = d \cdot V(u)\,. \tag{9.3.66}$$

In the global translation-invariant case a weak solution of (9.3.1) *satisfies*

$$(d-2)K(u) = d \cdot V(u)\,. \tag{9.3.67}$$

Remark 9.3.3. (a) The relations (9.3.50), (9.3.66) and (9.3.67) are known as *Pohozaev identities.* Note that relation (9.3.67) is nothing but the well-known *virial equation.*

(b) In Sect. 9.6 we will show in the context of general scale-covariant functionals K and V that the virial equation always holds for certain weak solutions ("minimal-action solution") of (9.3.1) under much weaker differentiability assumptions on G.

(c) In a form of substantial generality the proof of the chain rule (9.3.57) is by no means easy. We leave it as an exercise for the reader. Some results are available for the translation-invariant case. Section 7.4 of [9.25] treats the case where weakest assumptions on u but correspondingly stronger assumptions on G and g are used. Brezis and Lieb [9.13] indicate a proof for $u \in E \cap L^\infty_{\text{loc}}$ under considerably weaker assumptions on G and g ($G \in C^1(\mathbb{R}^n \setminus \{0\})$, if $g \in L^\infty_{\text{loc}}$ is known.

9.4 Best Constant in Sobolev Inequality

For $1 < p < d$ there is, according to Sect. 9.2, a constant $C_{p,d}$ such that for all $u \in E^{1,p}(\mathbb{R}^d; \mathbb{R}^n)$

$$\|u\|_{p^*} \leq C_{p,d}\|\nabla u\|_p$$

holds. Here we consider the case $n = 1$ and ask what is the best possible choice for the constant in this inequality. Equivalently, what is the value of

$$S = \inf\{C \mid C > 0, \|u\|_{p^*} \leq C\|\nabla u\|_p \quad \text{for all } u \in E^{1,p}(\mathbb{R}^d)\} \ ?$$

$S \equiv S_{d,p}$ is called the *best* (or *sharp*) *constant in Sobolev inequality.* A third formulation of this question is as follows. For $u \in E \equiv E^{1,p}(\mathbb{R}^d)$, $u \neq 0$, consider

$$R_p(u) = \frac{\|u\|_{p^*}^p}{\|\nabla u\|_p^p} \,. \tag{9.4.1}$$

It is clear that

$$S^p = \sup\{R_p(u) \mid u \in E, u \neq 0\} \,. \tag{9.4.2}$$

Thus one possible way to determine S is to show that the bounded functional R_p has a maximising function u_p and to calculate S as

$$S = R_p(u_p)^{1/p} \,. \tag{9.4.3}$$

In this way we arrive at a typical and important variational problem. First we give the results and then discuss some parts of the proof due to *G. Talenti* [9.10] and *E. H. Lieb* [9.20].

Theorem 9.4.1. (a) *There is a maximising function u_p for R_p:*

$$R_p(u_p) = \sup\{R_p(u) \mid u \in E, u \neq 0\} \,.$$

(b) *Up to normalisation and translation, every maximising function u_p has the form*

$$u_p(x) = (a + |x|^{p'})^{1-d/p} \,, \quad a > 0 \,, \quad \frac{1}{p'} + \frac{1}{p} = 1 \,. \tag{9.4.4}$$

(c) *The value of the best Sobolev constant is*

$$S = S_{d,p} = \pi^{-1/2} d^{-1/p} \left(\frac{p-1}{d-p}\right)^{1-1/p} \left\{\frac{\Gamma(1+d/2)\Gamma(d)}{\Gamma(d/p)\Gamma(1+d-d/p)}\right\}^{1/d} \,. \tag{9.4.5}$$

Our discussion of the proof will be complete only for the case $p = 2$ which is considerably easier than the others. This has been observed by E. H. Lieb. In the general case the proof uses methods which have not been developed in this book and therefore we have to refer the reader to [9.10] for details.

For a subset F of $\dot{E} = \{u \in E \,|\, u \neq 0\}$ let us write

$$I(F) = \sup\{R_p(u) \,|\, u \in F\}.$$

Clearly

$$I(F) \leq I(\dot{E}) = S^p.$$

The first step of the proof consists in finding a suitable subset F of $\dot{E}$ such that

$$I(F) = I(\dot{E}). \tag{9.4.6}$$

For this to be satisfied it is necessary and sufficient that for every $\varepsilon > 0$ and every $u \in \dot{E}$ there exist a $v \in F$ such that

$$R_p(u) - \varepsilon \leq R_p(v). \tag{9.4.7}$$

We verify this condition for

$$F = \{u \in W^{1,p}_{sd}(\mathbb{R}^d) \cap L^\infty(\mathbb{R}^d) \,|\, u \neq 0, u \geq 0\}. \tag{9.4.8}$$

Here $W^{1,p}_{sd}(\mathbb{R}^d)$ denotes the image of $W^{1,p}(\mathbb{R}^d)$ under spherically symmetric rearrangement. By Sobolev's inequality R_p is a continuous function on $\dot{E}$. Since $W^{1,p}(\mathbb{R}^d) \cap L^\infty(\mathbb{R}^d)$ is dense in E, given $u \in \dot{E}$ and $\varepsilon > 0$ there is a $u_\varepsilon \in W^{1,p}(\mathbb{R}^d) \cap L^\infty(\mathbb{R}^d)$, $u_\varepsilon \neq 0$, such that

$$R_p(u) - \varepsilon \leq R_p(u_\varepsilon).$$

From the results on the spherically symmetric rearrangement of functions it is known that, for all $f \in \dot{E} \cap L^p(\mathbb{R}^d)$,

$$\|f\|_p = \|\,|f|^*\|_p, \quad \|f\|_q = \|\,|f|^*\|_q, \quad q = p^*,$$

and

$$\|\nabla |f|^*\|_p \leq \|\nabla f\|_p$$

hold. It follows that $|f|^* \in W^{1,p}_{sd}(\mathbb{R}^d)$, $|f|^* \neq 0$, and

$$R_p(f) \leq R_p(|f|^*).$$

Therefore, if we define $v = |u_\varepsilon|^*$, we get $v \in F$ and

$$R_p(u_\varepsilon) \leq R_p(v),$$

and this implies condition (9.4.7) for this choice of the subset F. We conclude that

$$S^p = \sup\{R_p(u) \,|\, u \in F\}. \tag{9.4.9}$$

Elements in F are of the form $u(x) = u_0(|x|)$ with a decreasing and bounded function $u_0 : \mathbb{R}_+ \to \mathbb{R}_+$ satisfying

$$\left.\begin{aligned} \sigma_d \int_0^\infty dr\, r^{d-1} |u_0(r)|^\sigma &= \|u\|_\sigma^\sigma \quad \text{for } \sigma = p \quad \text{and} \quad \sigma = p^*, \\ \sigma_d \int_0^\infty dr\, r^{d-1} |u_0'(r)|^p &= \|\nabla u\|_p^p \end{aligned}\right\} \tag{9.4.10}$$

if σ_d denotes the area of the unit sphere in $\mathbb{R}^d$. It follows that

$$R_p(u) = \sigma_d^{p/q-1} Q_p(u_0),$$

$$Q_p(u_0) = \left\{ \int_0^\infty dr\, r^{d-1} |u_0(r)|^q \right\}^{p/q} \left\{ \int_0^\infty dr\, r^{d-1} |u_0'(r)|^p \right\}^{-1}. \tag{9.4.11}$$

The proof in [9.10] now proceeds by showing that the functions $u_{p,o}$ given by (9.4.4) are maximising for Q_p. This proof relies on classical methods of the calculus of variations (Mayer fields of extremals, the Hilbert invariant integral, and the Weierstrass E function) which have not been developed here. This then proves part (a) and part (b) of the theorem. The functions u_p of (9.4.4) being known, the proof of part (c) is a straightforward calculation according to (9.4.11).

For $p = 2$ the proof is considerably simplified by introducing logarithmic radial coordinates which reduce the functional Q_p to a functional J_p for which it is easy to determine a maximising function.

Logarithmic radial coordinates are defined by

$$z = \tau \ln r \quad \text{for} \quad r > 0 \quad \text{or} \quad r = e^{z/\tau} \quad \text{for} \quad z \in \mathbb{R} \tag{9.4.12}$$

for $\tau > 0$. For $u \in W_s^{1,p}(\mathbb{R}^d)$, $u(x) = u_0(|x|)$, define $f = \phi u$ by

$$f(z) = e^z u_0(e^{z/\tau}). \tag{9.4.13}$$

Then, choosing

$$\tau = \frac{d}{p^*}, \tag{9.4.14}$$

we get

$$\begin{aligned} \|u\|_q^q &= \tau^{-1} \sigma_d \|\phi u\|_q^q, \\ \|\nabla u\|_p^p &= \tau^{p-1} \sigma_d \|(\phi u)' - \phi u\|_p^p, \end{aligned} \tag{9.4.15}$$

where, on the right-hand side, $\|\cdot\|_q$ and $\|\cdot\|_p$ are the norms of $L^q(\mathbb{R})$ and of $L^p(\mathbb{R})$.

If $u \in F$ then, as in Sects. 9.2.3 and 9.2.4, the decay estimate

$$|u_0(r)| \leq \min\{\|u\|_\infty, c_d \|u\|_p r^{-d/p}\}$$

is available. It implies for $f = \phi u$ the estimate

$$|f(z)| \leq e^z \min\{\|u\|_\infty, c_d \|u\|_p e^{-\frac{p^*}{p} z}\} \tag{9.4.16}$$

and, in particular, $f(z) \to 0$ for $|z| \to \infty$.

It follows that $f \in L^p(\mathbb{R})$ and, by (9.4.13), also, that $f' \in L^p(\mathbb{R})$. Therefore, $f \in W^{1,p}(\mathbb{R})$. Furthermore, since $u \in F$ implies $f = \phi u \geq 0$, we get

$$|f(z')^p - f(z)^p| \leq \left(\int_z^{z'} |f'(p)|^p \, dp\right)^{1/p} \left(\int_z^{z'} |f(p)|^p \, dp\right)^{1/p'} \leq \|f\|^p_{W^{1,p}(\mathbb{R})} \tag{9.4.17}$$

and thus $\|f\|_\infty \leq \|f\|_{W^{1,p}}$. In the case $p = 2$, (9.4.16) implies

$$2\int_{-\infty}^{\infty} dz \, f(z) f'(z) = \int_{-\infty}^{\infty} dz \frac{d}{dz} f(z)^2 = 0\,,$$

and therefore

$$\|f' - f\|_2^2 = \|f'\|_2^2 + \|f\|_2^2 = \|f\|^2_{W^{1,2}}\,.$$

Combining this with (9.4.15) yields

$$R_2(u) = C_{2,d} \frac{\|\phi u\|_q^2}{\|\phi u\|_{W^{1,2}}}\,, \qquad u \in F\,, \tag{9.4.18}$$

where $C_{p,d} = \sigma_d^{p/q-1} \tau^{p/d-p}$.

But for the functional occurring on the right-hand side of (9.3.18), it is easy to find a maximising function if we use Theorem 9.2.15. To this end we introduce the functional J_2 on $W^{1,2}(\mathbb{R})$ by

$$J_2(f) = \frac{\|f\|^2_{2^*}}{\|f\|^2_{1,2}}\,, \qquad f \in W^{1,2}(\mathbb{R})\,, \quad f \neq 0\,.$$

Then (9.4.18) reads

$$R_2(u) = C_{2,d} J_2(\phi u)\,, \quad u \in F\,. \tag{9.4.19}$$

Conversely, for $f \in W^{1,2}(\mathbb{R})$, $f \neq 0$, we define a function $u \equiv \phi^{-1} f$ on $\mathbb{R}^d$ by $u(x) = u_0(|x|)$, with

$$u_0(r) = r^{-\tau} f(\tau \ln r)\,, \quad r > 0\,, \quad \tau 2^* = d\,.$$

As above, it follows that

$$\|u\|^{2^*}_{2^*} = \tau^{-1} \sigma_d \|f\|^{2^*}_{2^*} \quad \text{and} \quad \|\nabla u\|_2^2 = \tau \sigma_d \|f\|^2_{1,2}\,.$$

Hence $u \in E^{1,2}(\mathbb{R}^d)$ and

$$R_2(u) = C_{2,d} J_2(f)\,. \tag{9.4.20}$$

Now (9.4.19) and (9.4.20) imply easily

$$S_2^2 = \sup\{R_2(u) \,|\, u \in f\} = C_{2,d} I_2\,, \tag{9.4.21}$$

$$I_2 = \sup\{J_2(f) \,|\, f \in W^{1,2}(\mathbb{R}), f \neq 0\}\,. \tag{9.4.22}$$

The variational problem (9.4.22) is solved by Theorem 9.2.15 if we choose there $G(y) = |y|^r$, $r = 2^* > 2 = p$. It implies that there is a nonnegative, nonincreasing, symmetric function f_2 in $W^{1,2}(\mathbb{R})$ such that

$$I_2 = J_2(f_2)$$

holds. This function f_2 is a nonnegative solution of the differential equation

$$f'' - f = -af^{r-1}, \quad a = I_2^{-2^*/2}. \tag{9.4.23}$$

The unique classical solution of this equation is

$$f(z) = A\left(\cosh\left(\frac{z}{\tau}\right)\right)^{-\tau}, \quad A = \left(\frac{r}{2a}\right)^{1/(r-2)}. \tag{9.4.24}$$

Hence, up to normalisation and Euclidean transformation, the maximising function u_2 for R_2 is given by

$$u_2(x) = (1 + |x|^2)^{-\tau} \tag{9.4.25}$$

which is a special case of (9.4.4) for $p = 2$. By evaluating some integrals, the value of $I_2 = J_2(f_2)$ can be calculated and then by (9.4.21) the value of the sharp Sobolev constant S_2 follows in the case $p = 2$.

Remark 9.4.1. The maximising function (9.4.25) satisfies the differential equation

$$-\Delta u = 2\tau d u^{2^*-1}. \tag{9.4.26}$$

With respect to the type of equations that will be considered in Sects. 9.5–8 this equation is a *"limit case"* since the exponent in it equals the Sobolev exponent minus 1. For comments on the physical relevance of this equation see [9.30].

9.5 The Local Case with Critical Sobolev Exponent

In this section we discuss the variational approach to the following type of nonlinear elliptic differential equation:

$$-\Delta u = u^{r-1} + f(\cdot, u) \quad \text{and} \quad u > 0 \text{ on } \Omega,$$

$$u \restriction \partial\Omega = 0.$$

Here Ω is a bounded domain in $\mathbb{R}^d$ with smooth boundary $\partial\Omega$. f is supposed to be a perturbation of lower order than u^{r-1} satisfying $f(x, 0) = 0$ for all $x \in \Omega$. In the case of

$$2 < r < p = 2^* = \frac{2d}{d-2},$$

the methods of Sect. 8.4 (see in particular Theorem 8.4.7 and the following examples) could be used for a variational approach to the "corresponding eigenvalue problem" in the space $W^{1,2}(\Omega)$. These methods then apply quite independently of geometrical and topological properties of Ω, and in particular quite uniformly in $d \geq 3$. Here we want to point out that, on the contrary, the existence of solutions depends appreciably on geometrical and topological properties of Ω, in particular on the dimension $d \geq 3$ if

$$r \text{ equals the } \textit{critical Sobolev exponent } p = 2^* .$$

If $r = p$ then the above problem turns out to be much harder than the corresponding one for $r < p$, and it was solved only in 1983 by *H. Brezis* and *L. Nirenberg* [9.20]. In the general form above it still requires methods [9.20] which have not been developed in this book. Therefore, we discuss only the simplest version of this problem in which, however, the most important and striking features are already revealed.

Thus we consider the following problem. Let $\Omega \subset \mathbb{R}^d$ be a bounded domain with smooth boundary and let λ be a real number. For

$$p = \frac{2d}{d-2}, \quad d \geq 3,$$

we are looking for solutions $u \neq 0$ of

$$\begin{aligned} -\Delta u &= u^{p-1} + \lambda u \quad \text{on } \Omega, \\ u &\geq 0 \quad \text{on } \Omega, \end{aligned} \tag{9.5.1}$$

$$u \restriction \partial\Omega = 0 .$$

In the variational approach to this problem we consider the functional Q_λ defined by

$$Q_\lambda(u) = \|u\|_p^{-2}\{\|\nabla u\|_2^2 - \lambda\|u\|_2^2\} \tag{9.5.2}$$

on $u \in H_0^1(\Omega)$, $u \neq 0$, and we try to find a minimiser $v_\lambda : Q_\lambda(v_\lambda) = S_\lambda$,

$$S_\lambda = \inf\{Q_\lambda(u) \mid u \in H_0^1(\Omega), u \neq 0\} . \tag{9.5.3}$$

This then leads to a solution of (9.5.1) (see Sects. 2.2 and 2.3). The derivative of Q_λ in $u \in H_0^1(\Omega)$, $u \neq 0$, is

$$\begin{aligned} DQ_\lambda(u)(v) = \|u\|_p^{-2} 2\,\mathrm{Re}\{&\langle \nabla u, \nabla v\rangle_2 - \lambda\langle u, v\rangle_2 \\ &+ \|u\|_p^{2-p} Q_\lambda(u)\langle |u|^{p-1}\mathrm{sign}\, u, v\rangle_2\} \end{aligned} \tag{9.5.4}$$

for all $v \in H_0^1(\Omega)$. If $Q_\lambda(v_\lambda) = S_\lambda$ now holds, then $Q_\lambda(|u_\lambda|) = S_\lambda$ also (see [Ref. 9.25, Lemma 7.6]) and thus we may assume $v_\lambda \geq 0$. Then, by (9.5.4) we see that v_λ satisfies

$$-\Delta v_\lambda = \lambda v_\lambda + \|v_\lambda\|_p^{2-p} Q_\lambda(v_\lambda) v_\lambda^{p-1} .$$

As we will show that $S_\lambda = Q_\lambda(v_\lambda) > 0$, we can define

$$u_\lambda = \|v_\lambda\|_p^{-1} S_\lambda^{1/p-2} v_\lambda$$

to obtain a (weak) solution of (9.5.1). A (weak) solution of (9.5.1) is characterised by

$$\langle \nabla u, \nabla v \rangle_2 - \lambda \langle u, v \rangle_2 = \langle u^{p-1}, v \rangle_2 \tag{9.5.5}$$

for all $v \in H_0^1(\Omega)$.

If $\Omega \subset \mathbb{R}^d$ is bounded, the Dirichlet-Laplace operator $-\Delta$ is known to have a discrete spectrum in $L^2(\Omega)$ (Theorem 6.4.1) and the least (first) eigenvalue λ_1 is characterised by

$$\lambda_1 = \inf\{\|u\|_2^{-2} \|\nabla u\|_2^2 \,|\, u \in H_0^1(\Omega), u \neq 0\}\,. \tag{9.5.6}$$

For the corresponding eigenfunction v_1 we may again assume that $v_1 > 0$ in Ω. Equation (9.5.5) implies

$$\langle u^{p-1}, v_1 \rangle_2 = \langle \nabla u, \nabla v_1 \rangle_2 - \lambda \langle u, v_1 \rangle_2 = (\lambda_1 - \lambda) \langle u, v_1 \rangle_2$$

and as we know $\langle u^{p-1}, v_1 \rangle_2 > 0$ and $\langle u, v_1 \rangle_2 > 0$, we can only have a solution $u > 0$, $u \neq 0$ of (9.5.1) if

$$\lambda < \lambda_1\,. \tag{9.5.7}$$

Next we discuss whether condition (9.5.7) is also sufficient for the existence of a solution. For special domains Ω this is actually the case: Kazhdan and Warner have shown that for an annulus Ω there exist radial solutions for every $\lambda < \lambda_1$ [9.31].

For the class of star-shaped domains Ω (with respect to the origin) the following can be said. In this situation *Pohozaev's identity* applies:

$$(1 - d/2) \int_\Omega g(u) u \, dx + d \int_\Omega \cdot G(u) \, dx = \frac{1}{2} \int_{\partial\Omega} (x \cdot \nu) \left(\frac{\partial u}{\partial \nu} \right)^2 ds\,. \tag{9.5.8}$$

In our case

$$g(s) = \lambda s + s^{p-1}\,, \quad G(s) = \int_0^s g(t) \, dt = \frac{1}{2} \lambda s^2 + \frac{1}{p} s^p\,,$$

and it yields for a solution u of (9.5.1)

$$\lambda \|u\|^2 = \frac{1}{2} \int_{\partial\Omega} (x \cdot \nu) \left(\frac{\partial u}{\partial \nu} \right)^2 ds\,. \tag{9.5.9}$$

For a star-shaped domain Ω one has $x \cdot \nu > 0$ almost everywhere on the boundary $\partial\Omega$. When $\lambda < 0$ it follows immediately from (9.5.9) that $u = 0$. When $\lambda = 0$, (9.5.9) implies $\partial u / \partial v = 0$ almost everywhere on $\partial\Omega$ and therefore, by (9.5.1),

$$\int_\Omega u^p \, dx = \int_\Omega -\Delta u \, dx = -\int_{\partial\Omega} \frac{\partial u}{\partial \nu} \, ds = 0$$

and again $u = 0$ follows.

Thus for a star-shaped domain one should expect a solution of (9.5.1) to exist only in the case $\lambda \in (0, \lambda_1)$. And this is actually the case if $d \geq 4$ as we are going to show now.

According to Sect. 9.4, the best constant for the Sobolev embedding $H_0^1(\Omega) \hookrightarrow L^p(\Omega)$ is given by

$$\begin{aligned} S = S_0 &= \inf\{Q_{\lambda=0}(u) \mid u \in H_0^1(\Omega), u \neq 0\} \\ &= \inf\{\|\nabla u\|_2^2 \mid u \in H_0^1(\Omega), \|u\|_p = 1\} \end{aligned} \tag{9.5.10}$$

Using Pohozaev's identity it is not hard to see that in contrast to the results of Sect. 9.4 the infimum in (9.5.10) is never achieved if Ω is bounded [9.20]. The minimising functions for (9.5.10) in the case of $\Omega = \mathbb{R}^d$ are, according to Sect. 9.4,

$$u_\varepsilon(x) = C(\varepsilon + |x|^2)^{1-d/2}, \quad \varepsilon > 0. \tag{9.5.11}$$

Multiplication with a nonnegative function $\varphi \in \mathcal{D}(\Omega)$ which equals 1 in a neighborhood of $0 \in \Omega$ yields a family of functions

$$v_\varepsilon = \varphi \cdot u_\varepsilon, \quad \varepsilon > 0,$$

$$v_\varepsilon \in H_0^1(\Omega) \quad \text{and} \quad v_\varepsilon \neq 0 \quad \text{for } \varepsilon > 0.$$

Some straightforward but lengthy calculations allow one to evaluate the functional Q_λ on these functions v_ε, $\varepsilon > 0$. The result is

$$Q_\lambda(v_\varepsilon) = \begin{cases} S + o(\varepsilon^{d/2-1}) - \lambda \frac{K_3}{K_2}\varepsilon & \text{if } d \geq 5, \\ S + o(\varepsilon) - \lambda \frac{K_3}{K_2}\varepsilon|\log\varepsilon| & \text{if } d = 4, \end{cases} \tag{9.5.12}$$

where (according to (9.5.2, 3, 10) $S = K_1/K_2$, $K_1 = \|\nabla u_1\|_2^2$, $K_2 = \|u_1\|_p^2$ and K_3 is the area of the unit sphere in $\mathbb{R}^d$. As a consequence of this crucial step we have that whenever $d \geq 4$ and $\lambda > 0$, there is $\varepsilon > 0$ such that $Q_\lambda(v_\varepsilon) < S$. This proves:

Lemma 9.5.1. *If $d \geq 4$ and $\lambda > 0$, the infimum S_λ defined by* (9.5.3) *satisfies*

$$S_\lambda < S.$$

Theorem 9.5.2. *If $d \geq 4$ and $\Omega \subset \mathbb{R}^d$ is star shaped with respect to $0 \in \Omega$, then for every $\lambda \in (0, \lambda_1)$, where $\lambda_1 = \lambda_1(\Omega)$ is the least eigenvalue of the Dirichlet-Laplacian $-\Delta$ in $L^2(\Omega)$, there is $u_\lambda \in H_0^1(\Omega)$, $u_\lambda \neq 0$, $u_\lambda \geq 0$ such that*

$$Q_\lambda(u_\lambda) = S_\lambda.$$

u_λ is a positive solution of (9.5.1).

Proof. According to the definition of λ_1 and the assumption $\lambda < \lambda_1$, we know

$$\|\nabla u\|_2^2 - \lambda\|u\|_2^2 \geq (\lambda_1 - \lambda)\|u\|_2^2 \geq 0. \tag{9.5.13}$$

Therefore $Q_\lambda(u)$ is bounded from below by 0 and S_λ is well defined and satisfies $S_\lambda \geq 0$.

There is a minimising sequence $\{u_j\}$ in $H_0^1(\Omega)$ such that

$$S_\lambda = \lim_{j\to\infty} Q_\lambda(u_j) \leq Q_\lambda(u_j) \leq 1\,, \tag{9.5.14 a}$$

$$\|u_j\|_p = 1 \quad \text{for all } j\,. \tag{9.5.14 b}$$

As $\lambda_1 - \lambda > 0$, (9.5.13) and (9.5.14) imply first that $\{u_j\}$ is bounded in $L^2(\Omega)$; then (9.5.14 a) implies that $\{\nabla u_j\}$ is also bounded in $L^2(\Omega)$, that is $\{u_j\}$ is bounded in $H_0^1(\Omega)$. Thus we have a subsequence $(u_i = u_{j(i)})_{i\in\mathbb{N}}$ and $u \in H_0^1(\Omega)$ such that

(a) $$u_i \to u \quad \text{weakly in } H_0^1(\Omega)\,, \tag{9.5.15 a}$$

(b) $$\nabla u_i \to \nabla u \quad \text{weakly in } L^2(\Omega)\,. \tag{9.5.15 b}$$

According to Appendix D, the embedding $H_0^1(\Omega) \hookrightarrow L^2(\Omega)$ is known to be compact. Therefore (9.5.15 a) implies

$$u_i \to u \quad \text{in } L^2(\Omega)\,. \tag{9.5.16}$$

By extracting another subsequence if necessary, we may assume in addition

$$u_i \to u \quad \text{almost everywhere on } \Omega\,. \tag{9.5.17}$$

Fatou's lemma and equations (9.5.14 b) and (9.5.17) now imply

$$\|u\|_p \leq 1\,. \tag{9.5.18}$$

By (9.5.16), the limit of (9.5.14 a) can be evaluated in more detail:

$$S_\lambda = \lim_{i\to\infty} (\|\nabla u_i\|_2^2 - \lambda\|u_i\|_2^2) = \lim_{i\to\infty} \|\nabla u_i\|_2^2 - \lambda\|u\|_2^2\,.$$

According to (9.5.10) and (9.5.14 b), $\|\nabla u_i\|_2^2 \geq S$ is known. It follows that

$$A := \lim_{i\to\infty} \|\nabla u_i\|_2^2 = S_\lambda + \lambda\|u\|_2^2 \geq S \tag{9.5.19}$$

and thus by Lemma 9.5.1, for $0 < \lambda < \lambda_1$,

$$\|u\|_2^2 \geq \lambda^{-1}(S - S_\lambda) > 0\,,$$

and in particular $u \neq 0$.

Let us write $v_i = u - u_i$. Corollary 9.2.8 implies

$$1 = \lim_{i\to\infty} \|u_i\|_p^p = \|u\|_p^p + \lim_{i\to\infty} \|v_i\|_p^p\,. \tag{9.5.20}$$

Hence by (9.5.19) and (9.5.15 b),

$$\lim_{i\to\infty} \|\nabla v_i\|_2^2 = A - \|\nabla u\|_2^2 = S_\lambda - Q_\lambda(u)\|u\|_p^2 \tag{9.5.21}$$

follows. As $u \neq 0$ there is an i_0 such that, for all $i \geq i_0$, $\|v_i\|_p < 1$, and as $2 < p$ we know $\|v_i\|_p^p \leq \|v_i\|_p^2$ for these i and also that $\|u\|_p^p \leq \|u\|_p^2$. Therefore, from (9.5.20) we deduce

$$1 \leq \|u\|_p^2 + \|v_i\|_p^2 + \varepsilon_i \quad \text{for } i \geq i_0 \tag{9.5.22}$$

with $\lim_{i\to\infty} \varepsilon_i = 0$.

By Sobolev's inequality one obtains

$$1 \leq \|u\|_p^2 + \frac{1}{S}\|\nabla v_i\|_2^2 + \varepsilon_i$$

and therefore, in the limit, using (9.5.21),

$$1 \leq \|u\|_p^2 + \frac{1}{S}(A - \|\nabla u\|_2^2),$$

and thus

$$\|u\|_p^2 + S^{-1}S_\lambda \geq 1 + S^{-1}\|u\|_p^2 Q_\lambda(u) \geq 1 + S^{-1}S_\lambda\|u\|_p^2. \tag{9.5.23}$$

Since by Lemma 9.5.1

$$S^{-1}S_\lambda < 1$$

is known, this estimate implies $\|u\|_p^2 \geq 1$. Together with inequality (9.5.18) we conclude that $\|u\|_p = 1$. Then, again, inequality (9.5.23) proves $S_\lambda \geq Q_\lambda(u)$ and therefore $S_\lambda = Q_\lambda(u)$; i.e. u is a minimiser. Now (9.5.20) and (9.5.21) show

$$u_i \to u \quad \text{in } L^p \quad \text{and} \quad \nabla u_i \to \nabla u \quad \text{in } L^2.$$

This also proves by (9.5.16), that $u_i \to u$ in $H_0^1(\Omega)$. □

For $d = 3$, (9.5.1) reads

$$\begin{aligned} -\nabla u &= u^5 + \lambda u \quad \text{on } \Omega, \\ u &\geq 0 \quad \text{on } \Omega \text{ and } u \restriction \partial\Omega = 0. \end{aligned} \tag{9.5.1'}$$

Surprisingly it turns out that the problem of finding solutions for this equation is much more delicate than the corresponding problem for $d \geq 4$. At the moment, a complete answer is known only for the case where Ω is a ball which may then be supposed to the be unit ball with centre at the origin. For more discussion on this problem we refer the reader to [9.20].

Looking at the proof of Theorem 9.5.2 we notice that the dimension d only enters through Lemma 9.5.1, and here, again, the first step where the difference occurs is the corresponding Lemma 9.5.3.

Lemma 9.5.3. *Define S_λ and S according to* (9.5.3) *and* (9.5.10) *for the ball $\Omega = \{x \in \mathbb{R}^3 \mid |x| < 1\}$ in $\mathbb{R}^3$ and denote by $\lambda_1 = \lambda_1(\Omega)$ the least eigenvalue of the Dirichlet-Laplacian in Ω.*

Then for all $\lambda > \lambda_1/4$ it follows that

$$S_\lambda < S.$$

Proof. Once more, the proof consists in evaluating the functional Q_λ on a suitably chosen family of functions U_ε, $\varepsilon > 0$. This time we take

$$U_\varepsilon(x) = \frac{\varphi(|x|)}{(\varepsilon + |x|^2)^{1/2}}, \quad \varepsilon > 0, \tag{9.5.24}$$

where φ is a fixed smooth function satisfying $\varphi(0) = 1$, $\varphi'(0) = 0$, and $\varphi(1) = 0$. Again $x \mapsto (\varepsilon + |x|^2)^{-1/2} = u_\varepsilon(x)$, $\varepsilon > 0$, are minimising functions for (9.5.10) on $\mathbb{R}^3$.

As in the proof of Lemma 9.5.1, a lengthy calculation shows

$$Q_\lambda(v_\varepsilon) = S + \varepsilon^{1/2}\frac{4\pi}{K_2}\left\{\int_0^1 |\varphi'(r)|^2\,dr - \lambda\int_0^1 \varphi^2(r)\,dr\right\} + o(\varepsilon). \tag{9.5.25}$$

Note that the least eigenvalue of the Dirichlet-Laplacian in the ball is known to be

$$\lambda_1 = \pi^2 .$$

The corresponding eigenfunction is $\varphi_1(x) = |x|^{-1}\sin(\pi|x|)$. If we now choose $\varphi(r) = \cos(\pi r/2)$ it satisfies the assumptions given above and (9.5.25) leads for this choice to

$$Q_\lambda(v_\varepsilon) = S + \varepsilon^{1/2}(\pi^2/4 - \lambda)C + o(\varepsilon) \tag{9.5.26}$$

with some positive constant C. Now one concludes by choosing $\varepsilon > 0$ small enough. □

At this step the same information is available as in the case of dimension $d \geq 4$. Indeed we know

$$0 \leq S_\lambda < S \quad \text{for } \lambda \in (\lambda_1/4, \lambda_1).$$

Thus the same arguments apply as in the proof of Theorem 9.5.2 and yield in this case a minimiser $u_\lambda \in H_0^1(\Omega)$, $0 \leq u_\lambda$, $u_\lambda \not\equiv 0$, such that $Q_\lambda(u_\lambda) = S_\lambda > 0$. This function u_λ is a positive solution of (9.5.1′).

We already saw at the beginning that there is no solution for $\lambda \geq \lambda_1$ and $\lambda \leq 0$. The question then arises of what happens for $0 < \lambda \leq \lambda_1/4$? The answer is given by the following striking lemma.

Lemma 9.5.4. *For* $0 < \lambda \leq \lambda_1/4$ *there is no solution of* (9.5.1′).

Proof. The assumption that there is a solution of (9.5.1′) leads for $0 < \lambda \leq \lambda_1/4$ to a contradiction.

According to a result in [9.32], a solution v of (9.5.1′) has to be spherically symmetric: $v(x) \equiv u(|x|)$. Thus u satisfies

$$-u'' - \frac{2}{r}u' = u^5 + \lambda u \quad \text{on } (0,1)$$

and

$$u'(0) = u(1) = 0\,. \tag{9.5.27}$$

Take a smooth function g on $(0,1)$ with $g(0) = 0$, multiply (9.5.27) by $r^2 g u'$ and integrate over $(0,1)$. Do the same with $(r^2 g'/2 - rg)u$, perform some partial integrations and combine. The result is

$$\int_0^1 \Big(\lambda g' + \frac{1}{4} g'''\Big) u^2 r^2 \, dr = \frac{2}{3} \int_0^1 (rg - r^2 g') u^6 \, dr + \frac{1}{2} g(1) |u'(1)|^2 \,. \tag{9.5.28}$$

A special solution of $\lambda g' + g'''/4 = 0$ with $g(1) \geq 0$ is $g(r) = \sin(r\sqrt{\lambda}/2)$. It satisfies in addition

$$rg(r) - r^2 g'(r) > 0 \quad \text{on } (0,1)\,.$$

Thus (9.5.28) yields a contradiction for $u \neq 0$ and the lemma is proven. □

Theorem 9.5.5. *If $\Omega \subset \mathbb{R}^3$ is a ball and λ_1 the least eigenvalue of the Dirichlet-Laplacian in Ω then* (9.5.1′) *has a solution if $\lambda \in (\lambda_1/4, \lambda_1)$.*

Remark 9.5.1. Let $\Omega \subset \mathbb{R}^d$, $d \geq 3$, be a bounded domain. The calculations for Lemmas 9.5.1 and 9.5.3 show for $\lambda \leq 0$:

$$Q_\lambda(U_\varepsilon) = \begin{cases} S + o(\varepsilon) & \text{if } d \geq 5\,, \\ S + o(\varepsilon |\log \varepsilon|) & \text{if } d = 4\,, \\ S + o(\varepsilon^{1/2}) & \text{if } d = 3\,, \end{cases}$$

and therefore $S_\lambda \leq S$. But clearly $S_\lambda \geq S$ for $\lambda \leq 0$ and therefore $S_\lambda = S$ for $\lambda \leq 0$. As we have already remarked, the infimum (9.5.10) is not attained for $\lambda = 0$, so it follows that it is not attained for $\lambda < 0$ either.

9.6 The Constrained Minimisation Method Under Scale Covariance

The constrained minimisation method is explained in its original form in Chap. 4 and applied in Chaps. 6 and 8. Recall that it is used quite often to find solutions of eigenvalue equations of the form

$$F'(u) = \lambda G'(u) \tag{9.6.1}$$

where F' and G' are the Fréchet derivatives of some real differentiable functions F and G on some real Banach space E. A solution of (9.6.1) is obtained by minimising the function F on a level surface of G defining the constraint, that is by proving that for some $c \in \mathbb{R}$ there is $u \in E$ such that

$$F(u) = \inf\{F(v) \,|\, v \in E, G(v) = c\} = I_c \quad \text{and} \quad G(u) = c \tag{9.6.2}$$

holds. If the "minimiser" u is a regular point of the level surface $G^{-1}(c)$ the theorem about the existence of a Lagrange multiplier (Theorem 4.2.3') proves u to be a solution of (9.6.1) where the eigenvalue λ occurs as the Lagrange multiplier.

In this form the constrained minimisation method relies in an essential way on the assumption that the functions F and G are continuously Fréchet-differentiable. But in many applications, for instance those mentioned in Sect. 9.1, this assumption is too restrictive. Therefore we present here a version which relies only on the existence of Gâteaux derivatives in sufficiently many directions. Another basic ingredient of this version is a general scale covariance condition for the functions F and G. This covariance condition allows one in some cases to find a solution of (9.6.1) with eigenvalue $\lambda = 1$, so that many important applications to semilinear equations of the form (9.1.1) become possible.

Suppose $E = E(\mathbb{R}^d, \mathbb{R}^n)$ is a Banach space of functions $\mathbb{R}^d \to \mathbb{R}^n$ in which $\mathcal{D} = \mathcal{D}(\mathbb{R}^d; \mathbb{R}^n)$ is densely and continuously embedded. Here $d \geq 3$ and $n \geq 1$ is assumed. Suppose also that for every $\sigma > 0$ a linear map $T_\sigma : E \to E$ which leaves $\mathcal{D}$ invariant is well defined such that for all $\sigma_i > 0$, $i = 1, 2$,

$$T_{\sigma_1} T_{\sigma_2} = T_{\sigma_1 \sigma_2}, \quad T_1 = \mathrm{id} \tag{9.6.3}$$

holds. Furthermore, we assume that we are given a real function $K : E \to \mathbb{R}$ which is covariant with respect to the group $\{T_\sigma \,|\, \sigma > 0\}$ on E: for all $u \in E$ and all $\sigma > 0$

$$K(T_\sigma u) = \sigma^r K(u) \tag{9.6.4}$$

holds for some fixed real r.

Finally, we assume that a real function V is defined on some subset $D(V)$ of E having the following properties:

$$\begin{aligned} &(\mathrm{a})\ \mathcal{D} \subset D(V), \\ &(\mathrm{b})\ u \in D(V),\ v \in \mathcal{D} \Rightarrow u + v \in D(V), \\ &(\mathrm{c})\ u \in D(V),\ \sigma > 0 \Rightarrow T_\sigma u \in D(V). \end{aligned} \tag{9.6.5}$$

V itself satisfies the following *covariance condition*: For some fixed real s

$$V(T_\sigma u) = \sigma^s V(u) \tag{9.6.6}$$

holds for all $u \in D(V)$ and all $\sigma > 0$.

Remark 9.6.1. We will use explicitly the groups $\{T_\sigma \,|\, \sigma > 0\}$:

(a) $T_\sigma u = \sigma^\beta u$, $u \in E$, for some $\beta \in \mathbb{R}$,

(b) $T_\sigma u(\cdot) = u(\sigma \cdot)$, $u \in E$,

(c) $T_\sigma u(\cdot) = \sigma^\alpha u(\sigma \cdot)$, $u \in E$, for some $\alpha \in \mathbb{R}$.

Now consider the following minimisation problem

$$I_c = \inf\{K(u) \mid u \in D(V), V(u) = c\} \tag{9.6.7}$$

for some $c > 0$. Clearly, if the functions K and V satisfy the above covariance conditions, then we have

$$I_{\sigma c} = \sigma^{r/s} I_s, \quad \sigma > 0. \tag{9.6.8}$$

Hence it suffices to consider the case $c = 1$ and, accordingly, we do so and write henceforth $I = I_1$.

The following proposition describes how to obtain the existence of a Lagrange multiplier and to determine its value in a situation where on one side only Gâteaux-differentiability is available but on the other side *scale-covariance* is assumed.

Proposition 9.6.1. *Suppose that K and V are functions on a Banach space E satisfying the covariance conditions (9.6.4–6) as described above. If the constrained minimisation problem*

$$I = \inf\{K(v) \mid v \in D(V), V(v) = 1\} \tag{9.6.9}$$

has a solution $u \in D(V)$, i.e. if there is

$$u \in D(V) \quad \textit{with } V(u) = 1 \textit{ and } K(u) = I \tag{9.6.10}$$

and if K and V have linear continuous Gâteaux derivatives $K'(u;v)$ and $V'(u;v)$ at u in all directions $v \in \mathcal{D}$, then the following holds:

(a)
$$K'(u;v) = \gamma I V'(u;v) \quad \textit{for all } v \in \mathcal{D} \tag{9.6.11}$$

if $s \neq 0$ and $\gamma = r/s$.

(b) *K and V have linear continuous Gâteaux derivatives at every $T_\sigma u$, $\sigma > 0$, and*

$$K'(T_\sigma u; v) = \gamma I \sigma^{r-s} V'(T_\sigma u; v) \tag{9.6.12}$$

holds for all $v \in \mathcal{D}$ and $\sigma > 0$.

(c) *If $\gamma I > 0$ and $s - r \neq 0$ then for $\overline{\sigma} = (\gamma I)^{1/(s-r)}$ the "scaled minimiser" $\overline{u} := T_{\overline{\sigma}} u$ satisfies in $\mathcal{D}'$:*

$$K'(\overline{u}; \cdot) = V'(\overline{u}; \cdot). \tag{9.6.13}$$

Proof. For arbitrary but fixed $v \in \mathcal{D}$, we know by covariance (9.6.6) and the definition of $b = V'(u;v)$ that for every $\sigma > 0$

$$V(T_\sigma u + \varepsilon T_\sigma v) = \sigma^s \{1 + \varepsilon b + o(\varepsilon)\}$$

holds (see Sect. 2.3). There is an $\varepsilon_0 > 0$ such that $1 + \varepsilon b + o(\varepsilon) > 0$ for all $|\varepsilon| < \varepsilon_0$. For such ε there is a $\sigma = \sigma(\varepsilon)$ such that

$$V(T_\sigma u + \varepsilon T_\sigma v) = 1\,, \quad \text{i.e.} \quad \sigma = \{1 + \varepsilon b + o(\varepsilon)\}^{-1/s}\,.$$

By (9.6.4) and the definition of $a = K'(u; v)$ according to Sect. 2.4, it follows from (9.6.9) that

$$I \leq K(T_\sigma u + \varepsilon T_\sigma v) = \sigma^r K(u + \varepsilon v) = K(u) + \varepsilon\{a - \gamma I b\} + o(\varepsilon)\,.$$

This inequality holds for all $|\varepsilon| \leq \varepsilon_0$ if, and only if, $a - \gamma I b = 0$. But this is (9.6.11).

In order to prove (b) we write for $\sigma > 0$

$$K(T_\sigma u + \varepsilon v) = \sigma^r K(u + \varepsilon T_\sigma^{-1} v)$$

and proceed as above to prove

$$K'(T_\sigma u; v) = \sigma^r K'(u; T_\sigma^{-1} v)$$

and in the same way

$$V'(T_\sigma u; v) = \sigma^s V'(u; T_\sigma^{-1} v)\,.$$

This is possible since by assumption $\mathcal{D}$ is invariant under the group $\{T_\sigma \,|\, \sigma > 0\}$. Equation (9.6.11) now proves equation (9.6.12). Part (c) is a simple special case of part (b). □

In our applications of this proposition the following additional information is available:

$$\begin{array}{lll} \text{(a)} & K(u) \geq 0 \quad \text{for all } u \in E \text{ and } K(u) = 0 \text{ iff } u = 0\,. & \\ \text{(b)} & V(0) = 0 \quad \text{and} \quad V(u) > 0 \quad \text{for some } u \in D(V)\,. & \end{array} \tag{9.6.14}$$

Then the constrained minimisation problem (9.6.9) is well-posed and $I \geq 0$ follows. But for applications of Proposition 9.6.1 one needs to know when $I > 0$. The following lemma gives a useful translation of this condition. Later, this lemma will also be used to derive restrictions on V ensuring $I > 0$.

Lemma 9.6.2. *Assume that the function K and V satisfy* (9.6.4–6) *and* (9.6.14).

(a) *I, defined by* (9.6.9) *is positive if, and only if, S, defined by*

$$S = \sup\{R(v) \,|\, v \in D(V), V(v) > 0\}\,, \quad R(v) = K(v)^{-1} V(v)^\gamma \tag{9.6.15}$$

is finite; then the equation

$$IS = 1 \tag{9.6.16}$$

holds.

(b) *If $I > 0$ then there is a minimiser for the constrained minimisation problem* (9.6.9) *if, and only if, there is a maximising element for the problem* (9.6.15).

Proof. By definition we have

$$S \geq \sup\{R(v) \,|\, v \in D(V), V(v) = 1\} = \sup\{K(v)^{-1} \,|\, v \in D(V), V(v) = 1\},$$

and hence $S \geq 1/I$. If $v \in D(V)$ satisfies $V(v) > 0$ there is $\sigma > 0$ such that

$$V(T_\sigma v) = \sigma^s V(v) = 1$$

and therefore

$$I \leq K(T_\sigma v) = \sigma^r K(v) = V(v)^{-\gamma} K(v) = R(v)^{-1}.$$

Hence $R(v) \leq I^{-1}$ and thus

$$S \leq 1/I.$$

This implies part (a). Statement (b) is a simple consequence of (9.6.10) and part (a). □

By covariance, the function W defined on $D(V)$ by

$$W(u) = K(u) - V(u)$$

is in general not bounded, either from above or from below. The following result shows that on certain subsets W nevertheless can have a minimising element.

Proposition 9.6.3. *For functions K and V satisfying* (9.6.4–6) *and* (9.6.14) *assume*

$$0 < I \quad \textit{and} \quad \gamma = r/s \in (0,1). \tag{9.6.17}$$

Then the following identities hold for the solution $\overline{u}$ of (9.6.16) *as defined by part* (c) *of Proposition* 9.6.1:

$$V(\overline{u}) = \gamma K(\overline{u}) \tag{9.6.18}$$

$$W(\overline{u}) = \inf\{W(v) \,|\, v \in D(V), \quad v \neq 0, \quad V(v) = \gamma K(v)\} \tag{9.6.19}$$

Proof. By part (c) of Proposition 9.6.1 one has

$$\overline{u} = T_{\overline{\sigma}} u, \quad (\overline{\sigma})^{s-r} = \gamma I,$$

where u satisfies (9.6.10). By covariance of V and K, (9.6.18) easily follows.

Now suppose that $v \in D(V)$ satisfies $V(v) = \gamma K(v)$ and $v \neq 0$. Then $V(v) > 0$ and there is a $\sigma > 0$ such that $\sigma^s = V(v)$. Define $v_0 = T_\sigma^{-1} v$ and write the equation $V(v) = \gamma K(v)$ in terms of v_0. This yields $\sigma^s = \gamma\sigma^r K(v_0)$ and therefore $\sigma^{s-r} = \gamma K(v_0)$. $W(v)$ can now be expressed in terms of $K(v_0)$:

$$W(v) = (1-\gamma)K(v) = (1-\gamma)\sigma^r K(v_0) = (1-\gamma)[\gamma K(v_0)]^{r/(s-r)} K(v_0) .$$

Similarly we get by (9.6.18) and (9.6.10)

$$W(\overline{u}) = (1-\gamma)\overline{\sigma}^r K(u) = (1-\gamma)[\gamma I]^{r/(s-r)} I .$$

Since $V(v_0) = 1$ we know $I \leq K(v_0)$ and obtain $W(\overline{u}) \leq W(v)$. Hence (9.6.19) follows. □

Remark 9.6.2. In Sect. 9.3.2 we proved the virial identity (9.6.18) for a weak solution of (9.1.1) using some regularity assumptions. There the functions K and V have the interpretation of "kinetic energy" and "potential energy". Thus, in this context the funtion W can be interpreted as an "action". Therefore, by Proposition 9.6.3 we learn that the solution of (9.1.1) obtained by the constrained minimisation method according to part (c) of Proposition 9.6.1 is a *"minimal-action solution"* which satisfies the virial identity (9.6.18) in a situation where these regularity properties are not known but where a certain scale covariance is assumed.

In order to prove the existence of a weak solution of (9.1.1) by the application of Proposition 9.6.1, two conditions have to be ensured:

(A) Existence of a minimiser for problem (9.6.9).
(B) Existence of linear continuous Gâteaux derivatives for the functions K and V at a minimiser, in all directions $\varphi \in \mathcal{D}$.

Condition (A) presents the main difficulty. It will be treated in Sects. 9.7 and 9.8. For condition (B) we observe that by its definition the function K is always continuoulsy Fréchet-differentiable. According to Sects. 9.1 and 9.2 the function K is defined on $E = E^{1,2}(\mathbb{R}^d; \mathbb{R}^n)$ by

$$K(u) = \frac{1}{2}\|\nabla u\|_2^2$$

and therefore, for all $u, v \in E$,

$$K'(u)(v) = \langle \nabla u, \nabla v \rangle_2 = K'(u; v) . \tag{9.6.20}$$

There are many ways of ensuring the differentiability of the function V,

$$V(u) = \int G(x, u(x))\, dx , \quad u \in D(V) . \tag{9.6.21}$$

Some simple ways are discussed in Sect. 9.2. Some more refined versions are given in [9.1] and [9.13], but they do not cover the cases we intend to discuss.

So, we present a result which seems to be rather close to the most general result for which one can expect linear continuous Gâteaux derivatives to exist.

Proposition 9.6.4. *If the pair of functions (g, G) satisfies the hypotheses* (H_1) *and* (H_2) *and if in addition the following growth conditions*

$$|g(x,y)| \leq b(x)\{|y|^{p-1} + |y|^{q-1}\}, \tag{9.6.22}$$

$$|G(x,y)| \leq Cb(x)\{|y|^p + |y|^q\} \tag{9.6.23}$$

hold for almost every $x \in \mathbb{R}^d$ and all $y \in \mathbb{R}^n$ for some exponents $1 \leq p < q < +\infty$, some function $b \in L^s_{\mathrm{loc}}$, $s \geq q$, and some constant $C < \infty$, then the function V has linear continuous G-derivatives at every $u \in D(V) \cap L^q_{\mathrm{loc}}$ in all directions $\varphi \in \mathcal{D}$, given by

$$V'(u;\varphi) = \int g(x,u(x))\varphi(x)\,dx \equiv \int \widehat{g}(u)\varphi\,dx\,. \tag{9.6.24}$$

Furthermore, $u \in D(V) \cap L^q_{\mathrm{loc}}$ and $\varphi \in \mathcal{D}$ imply $u + \varphi \in D(V)$.

Proof. For a given pair $u \in D(V) \cap L^q_{\mathrm{loc}}$, $\varphi \in \mathcal{D}$, introduce the measureable sets

$$M_0 = [u \neq 0] \cap [\varphi \neq 0]\,, \quad M_t = M_0 \cap [u + t\varphi \neq 0]\,, \quad t \in \mathbb{R}\,, \quad |t| \leq 1\,,$$

and denote by χ_t the characteristic function of M_t. Obviously

$$\lim_{t\to 0} \chi_t = \chi_0 = \text{the characteristic function of } M_0\,.$$

in the sense of pointwise convergence. Then observe the following decomposition:

$$1 = [u = 0] + \chi_{M_0\setminus M_t} + \chi_t + \chi$$

where χ is the characteristic function of $[\varphi = 0] \setminus [u = 0]$. On $M_0 \setminus M_t$ we know $u = -t\varphi \neq 0$, and hence we get

$$\widehat{G}(u + t\varphi) - \widehat{G}(u) = [u = 0]\widehat{G}(t\varphi) - \chi_{M_0\setminus M_t}\widehat{G}(-t\varphi) + \chi_t\{\widehat{G}(u + t\varphi) - \widehat{G}(u)\}\,.$$

Condition (9.2.25) easily implies $\widehat{G}(t\varphi) \in L^1$ with

$$\|\widehat{G}(t\varphi)\|_1 = o(t)\,.$$

On M_t the differentiability of G can be used: for every $x \in M_t$ there is a $\tau = \tau(x,t) \in [0,1]$ such that

$$(\widehat{G}(u + t\varphi) - \widehat{G}(u))(x) = t(\widehat{g}(u)\varphi)(x) + t([\widehat{g}(u + \tau t\varphi) - \widehat{g}(u)]\varphi)(x)\,.$$

By condition (9.6.22) and Hölder's inequality one has ($K = \operatorname{supp}\varphi$)

$$\int |\widehat{g}(u)\varphi|\,dx \leq \|\varphi\|_\infty\{\|\pi_K b\|_p \|\pi_K u\|_p^{p-1} + \|\pi_K b\|_q \|\pi_K u\|_q^{q-1}\} \tag{9.6.25}$$

which is finite by assumption; hence $\widehat{g}(u)\varphi \in L^1$. Similarly $\widehat{g}(u+t\tau\varphi)\varphi \in L^1$ follows. This then implies $\widehat{G}(u+t\varphi) - \widehat{G}(u) \in L^1$ for $|t| \leq 1$ and thus $\widehat{G}(u+\varphi) \in L^1$ since $u \in D(V)$.

Now we show that $F_t = \widehat{g}(u+t\tau\varphi)\varphi$ converges in L^1 to $\widehat{g}(u)\varphi$ by application of Vitali's convergence theorem. Since F_t has a compact support contained in $K = \operatorname{supp}\varphi$ for all t and since $F_t \to \widehat{g}(u)\varphi$ for $t \to 0$ almost everywhere, it suffices to show that ($A \subset K$ measurable)

$$\sup_{|t|\leq 1} \|\pi_A F_t\|_1 \to 0 \quad \text{for } |A| \to 0\,.$$

By assumption (9.6.22) the following estimate is available:

$$\begin{aligned}
\|\pi_A F_t\|_1 &\leq \|\pi_A b|\varphi|\,\|_p \|\pi_K|u+t\tau\varphi|\,\|_p^{p-1} \\
&\quad + \|\pi_A b|\varphi|\,\|_q \|\pi_K|u+t\tau\varphi|\,\|_q^{q-1} \\
&\leq \|\pi_A b|\varphi|\,\|_p \{\|\pi_K u\|_p + \|\varphi\|_p\}^{p-1} \\
&\quad + \|\pi_A b|\varphi|\,\|_q \{\|\pi_K u\|_q + \|\varphi\|_q\}^{q-1}
\end{aligned}$$

Since $b \in L^s_{\mathrm{loc}}$ for some $s \geq q > p$ is assumed, we conclude, hence,

$$t \int [\widehat{g}(u+t\tau\varphi) - \widehat{g}(u)]\varphi \, dx = o(t)\,.$$

By collecting together the various terms we obtain

$$\begin{aligned}
V(u+t\varphi) - V(u) &= \int [u=0]\widehat{G}(t\varphi)\,dx - \int_{M_0\setminus M_t} \widehat{G}(-t\varphi)\,dx \\
&\quad + t\int \chi_t \widehat{g}(u)\varphi\,dx + t\int \chi_t[\widehat{g}(u+t\tau\varphi) - \widehat{g}(u)]\varphi\,dx \\
&= t\int \chi_t \widehat{g}(u)\varphi\,dx + o(t)\,.
\end{aligned}$$

Hence

$$\lim_{t\to 0} \frac{1}{t}(V(u+t\varphi) - V(u)) = \int \chi_0 \widehat{g}(u)\varphi\,dx = \int \widehat{g}(u)\varphi\,dx\,,$$

and (9.6.22) implies that $\varphi \mapsto \int \widehat{g}(u)\varphi\,dx$ is a linear continuous function on $\mathcal{D}$. Thus V has a linear continuous Gâteaux derivative at $u \in D(V) \cap L^q_{\mathrm{loc}}$ given by (9.6.24). □

Remark 9.6.3. (a) Note that the statement of Proposition 9.6.4 remains valid if assumption (9.6.23) is replaced by

$$|G(x,y)| \leq a(x)\{|y|^r + |y|^s\}$$

for almost all $x \in \mathbb{R}^d$ and all $y \in \mathbb{R}^n$ for some exponents $1 < r \leq s < \infty$ and some function $a \in L^\alpha_{\mathrm{loc}}$, $\alpha \geq 1$.

(b) In order to prove the existence of a minimiser for the problem (9.6.9) some hypotheses about the behaviour of $G(\cdot,y)$ near $y=0$ and near $|y|=\infty$ have to be used. Under these assumptions $u\in L^q_{\text{loc}}$ follows for all $u\in D(V)$ so that then Proposition 9.6.4 proves the existence of linear continuous Gâteaux-derivatives at every $u\in D(V)$ (see Proposition 9.7.4).

9.7 Existence of a Minimiser I: Some General Results

9.7.1 Symmetries

The first part of this section points out a major difficulty in the constrained minimisation problem for functions of the form (9.1.5) and (9.1.8). In the main part some necessary and sufficient conditions for the existence of a minimiser are presented. One of these conditions is the "concentration condition" (C). It is discussed in the third part of this section and a possible way of realising this condition is indicated. Finally, growth restrictions on translation-invariant potentials $y\mapsto G(y)$ are derived within a fairly general class of potentials in order that the constrained minimisation problem for the functions (K,V) given by (9.1.5) and (9.1.8) may be treated along the lines of Sect. 9.6.

Suppose that the functions (K,V) satisfy condition (9.6.14) and the co-variance relations (9.6.4–6) for some choice of the group $\{T_\sigma,\sigma>0\}$. By our general assumptions (H_1) and (H_2) the constrained minimisation problem

$$I=\inf\{K(v)\,|\,v\in D(V),V(v)=1\} \tag{9.7.1}$$

is well defined. Therefore $I\geq 0$ and there are minimising sequences $\underline{v}=(v_j)_{j\in\mathbb{N}}$ for (9.7.1), that is there is

$$v_j\in D(V)\,,\quad V(v_j)=1\,,\quad \text{and}\quad I=\lim_{j\to\infty}K(v_j)\,. \tag{9.7.2}$$

Such a sequence is bounded in $E=E^{1,2}(\mathbb{R}^d;\mathbb{R}^n)$ and hence the results of Sect. 9.2 apply. Thus we arrive at a minimising sequence $\underline{u}=(u_i)_{i\in\mathbb{N}}$ and an element $u\in E$ such that

$$u_i\to u\quad \text{for } i\to\infty \tag{9.7.3}$$

weakly in E and almost everywhere on $\mathbb{R}^d$. Clearly this limit $u\in E$ appears as a natural candidate for a minimiser, if $u\neq 0$, because then, by Theorem 9.2.2 b and weak lower semicontinuity of the norm on E,

$$0<K(u)\leq I \tag{9.7.4}$$

is known. However, in general it is not easy to prove $u\neq 0$. This difficulty occurs typically if the "symmetry group" $S=S(K,V)$ of the constrained minimisation problem (9.7.1) is "large", as we are going to explain now.

In this context, any map $\phi : D(V) \to D(V)$ satisfying

$$\begin{aligned} &\text{(a)} \qquad K(\phi v) \le K(v) \,, \\ &\text{(b)} \qquad V(\phi v) = V(v) \,, \end{aligned} \tag{9.7.5}$$

for all $v \in D(V)$ is called a *"symmetry"* for the pair of functions (K, V); and the set of all such maps under usual composition of maps forms the *"symmetry group"* $S = S(K, V)$. The following two facts are easy to see:

(1) If $\underline{v} = (v_j)_{j \in \mathbb{N}}$ is any minimising sequence for (9.7.1) and if $\underline{\phi} = (\phi_j)_{j \in \mathbb{N}}$ is any sequence in S, then

$$\underline{\phi}(\underline{v}) = (\phi_j v_j)_{j \in \mathbb{N}} \tag{9.7.6}$$

is another minimising sequence for (9.7.1).

(2) If there is a minimiser $\overline{u}$ for the problem (9.7.1) then the set of all minimisers for this problem is at least

$$S\overline{u} = \{u = \phi\overline{v} \mid \phi \in S\} \,.$$

Thus, whenever the symmetry group S is nontrivial, a solution of the constrained minimisation problem (9.7.1) requires to study the complicated factor space of all minimising sequences modulo symmetries in the sense of relation (9.7.6). In Theorem 9.7.1 we will learn that the existence proof of a minimiser for (9.7.1) is equivalent to the proof of existence of very special elements in this factor space. Thus, for example, if the potential G does not depend on the space variable $x \in \mathbb{R}^d$, then it is easy to see that the full Euclidean group $E(d)$ of $\mathbb{R}^d$ defines symmetries for (K, V) by

$$\phi u = u \circ \varphi \quad \text{if } \varphi \in E(d) \,.$$

In particular, if in this situation $(u_j)_{j \in \mathbb{N}}$ is a minimising sequence for (9.7.1) which converges (strongly in E) to some $u \in E$, $u \neq 0$, and if $(a_j)_{j \in \mathbb{N}}$ is a sequence of points in $\mathbb{R}^d$ such that $|a_j| \to \infty$ for $j \to \infty$, then the "translated sequence"

$$v_j(\cdot) = u_j(\cdot + a_j) \,, \quad j \in \mathbb{N} \,,$$

is another minimising sequence for (9.7.1). However, the weak limit of any subsequence of this sequence vanishes.

This discussion indicates the following: A "good" minimising sequence for (9.7.1) is a minimising sequence which, at least after a transformation by symmetries according to (9.7.6), contains a subsequence which converges in E to some nontrivial element. In this sense, the existence of a minimiser is characterised by the existence of a "good" minimising sequence.

9.7.2. Necessary and Sufficient Conditions

Clearly, there are many ways of formulating necessary and sufficient conditions for the existence of a minimiser for the constrained minimisation problem (9.7.1). We present here a set of such conditions which rely on the continuity of the "Niemytski operator"

$$u \mapsto \widehat{G}(u), \quad \widehat{G}(u)(x) = G(x, u(x)), \tag{9.7.7}$$

associated with the "potential" G as a map from $D(V)$ into $L^1(\mathbb{R}^d)$. The relevant continuity condition is given in Theorem 9.2.5.

Theorem 9.7.1. *Under the general assumptions* (H_1) *and* (H_2) *on the potential* G *define the functions* K *and* V *on* $E = E^{1,2}(\mathbb{R}^d; \mathbb{R}^n)$ *by equations* (9.1.5) *and* (9.1.8). *Suppose that* K *and* V *satisfy the covariance relations* (9.6.4–6) *with* $\gamma = r/s \in (0,1)$.

Then the constrained minimisation problem (9.7.1) *for* (K, V) *has a minimiser if, and only if, there is a minimising sequence* $\underline{u} = (u_j)_{j\in\mathbb{N}}$ *for* (9.7.1) *satisfying the concentration condition* (C) *and the condition* (IUD) *of "incomplete uniform decay at infinity in the* L^1*-norm sense" for* $F = G_+$ *i.e.*

(C) *For every* $\varepsilon > 0$ *there is a compact set* $K \subset \mathbb{R}^d$ *such that*

$$\sup_j \|\pi_K^{\perp} \widehat{G}_+(u_j)\|_1 \leq \varepsilon. \tag{9.7.8}$$

(IUD) *For every compact set* $K \subset \mathbb{R}^d$ *and every* $\varepsilon > 0$ *there is* $\lambda_0 \in (0, \infty)$ *such that for every* $\lambda \geq \lambda_0$ *there is* j_0 *with*

$$\|\pi_K \widehat{G}_+^{\geq\lambda}(u_j)\|_1 \leq \varepsilon \tag{9.7.9}$$

for all $j \geq j_0$.

Proof. If there is a minimiser for (9.7.1) then clearly there is a minimising sequence $(u_j)_{j\in\mathbb{N}} \subset D(V)$ such that

$$\widehat{G}(u_j) \to \widehat{G}(u_j) \quad \text{in } L^1.$$

It follows that $\widehat{G}_+(u_j) \to \widehat{G}_+(u_j)$ in L^1 for the Niemytski operator $\widehat{G}_+$ defined by the positive part of G. Now Theorem 9.2.5 implies that for this sequence and G_+ the conditions (C) and (IUD) are satisfied.

Conversely, suppose that there is a minimising sequence for which (9.7.3) and conditions (C) and (IUD) hold. Then by Theorem 9.2.5 we know that

$$\widehat{G}_+(u_j) \to \widehat{G}_+(u) \quad \text{in } L^1(\mathbb{R}^d),$$

and hence that $V_+(u_j) \to V_+(u)$ for $j \to \infty$, and thus that $V_-(u_j) = V_+(u_j) - 1 \to V_+(u) - 1$ for $j \to \infty$. By Fatou's lemma we deduce $\widehat{G}_-(u) \in L^1$ and

$$V_-(u) = \int \widehat{G}_-(u)\,dx \le \lim_{j\to\infty} V_-(u_j) = V_+(u) - 1\,.$$

Hence $u \in D(V)$ and $1 \le V_+(u) - V_-(u) = V(u)$. Therefore $u \ne 0$ and (9.7.3) implies (9.7.4).

Suppose $V(u) > 1$. Then according to the covariance relation (9.6.6) there is $\sigma > 0$, $0 < \sigma^s < 1$, such that

$$V(T_\sigma u) = \sigma^s V(u) = 1\,.$$

On the other hand, by (9.6.4), (9.7.1) and (9.7.4),

$$0 < K(u) \le I \le K(T_\sigma u) = \sigma^r K(u)$$

and thus $1 \le \sigma^r$.

However, by assumption, $r = \gamma s$ with $0 < \gamma < 1$, and we arrive at a contradiction. This proves $V(u) = 1$, and thus $K(u) \ge I$, and finally, by (9.7.4), $K(u) = I$. Hence the weak limit u of relation (9.7.3) is a minimiser, and it follows that u is the limit in E of the sequence $(u_i)_{i\in\mathbb{N}}$. □

Remark 9.7.1. (a) There are two cases where the condition (IUD) is easily satisfied:

(i) the minimising sequence $(u_j)_{j\in\mathbb{N}}$ is uniformly bounded:

$$|u_j(x)| \le \lambda_0 \quad \text{for all } j \in \mathbb{N} \text{ and almost every } x \in \mathbb{R}^d\,;$$

(ii) there is $\lambda_0 < \infty$ such that $G_+^{\ge\lambda_0} = 0$.

In all other cases (IUD) is realised by growth restrictions on $G_+(\cdot, y)$ for $|y| \to \infty$. Concrete examples for all three cases will follow in Sect. 9.8.

(b) In order to ensure the concentration condition (C) a clever choice of the minimising sequence has to be made. This is particularly difficult if the symmetry group $S(K,V)$ is large, as indicated in Sect. 9.7.1. But sometimes one can show:

(i) there is a subset $\mathcal{F} \subset D(V)$ such that all $u \in \mathcal{F}$ satisfy condition (C);
(ii) there is a minimising sequence in $\mathcal{F}$.

Such a subset $\mathcal{F}$ is appropriately called a "minimising subset". This notion is introduced and discussed in Sect. 9.7.4. Concrete examples of minimising subsets are given in Sect. 9.8.2.

9.7.3 The Concentration Condition

At least on an abstract level (and also for some concrete examples) it is not difficult to decide when the choice of a minimising sequence satisfying the concentration condition is possible. This is decided by the value of the "concentration functional" $F(\underline{v})$ for a minimising sequence $\underline{v} = (v_j)_{j\in\mathbb{N}}$, as we are going to describe now.

For a minimising sequence $\underline{v}$ of problem (9.7.1) assume

$$\sup_j V_+(v_j) = \Lambda < \infty . \tag{9.7.10}$$

Then we define a sequence of functions $F_j : \mathbb{R}_+ \times S \to \mathbb{R}_+$ by

$$F_j(r, \phi) = \|\pi^{\perp}_{B(r)} \widehat{G}_+(\phi u_j)\|_1 \tag{9.7.11}$$

where $B(r)$ denotes as usual the ball of radius $r > 0$ and centre $x = 0$ in $\mathbb{R}^d$. Next we define

$$F_j(r) = \inf_{\phi \in S} F_j(r, \phi) . \tag{9.7.12}$$

This sequence of functions $F_j : \mathbb{R}_+ \to \mathbb{R}_+$ has the following properties:

(a) For each $j : r \mapsto F_j(r)$ is nonincreasing and

$$F_j(r) \to 0 \quad \text{for } r \to \infty .$$

(b) For all $j \in \mathbb{N}$ and all $r \in \mathbb{R}_+ : F_j(r) \leq \Lambda$.

Hence by a classical theorem of Helly, for instance in [9.33], there are a subsequence $(F_{j(i)})_{i\in\mathbb{N}}$ and a nonincreasing function $F : \mathbb{R}_+ \to [0, \Lambda]$ such that for all $r \in \mathbb{R}_+$

$$F_{j(i)}(r) \to F(r) \quad \text{for } i \to \infty . \tag{9.7.13}$$

Thus

$$F_\infty = \lim_{r\to\infty} F(r) \tag{9.7.14}$$

is finite and

$$F_\infty \in [0, \Lambda] .$$

The *concentration functional* F for the problem (9.7.1) is defined on all minimising sequences $\underline{v}$ for (9.7.1) satisfying condition (9.7.10) by

$$F(\underline{v}) = F_\infty , \quad F_\infty \text{ according to (9.7.11–14).} \tag{9.7.15}$$

The relevant characterisation is given in the following lemma.

Lemma 9.7.2. *Suppose that a minimising sequence $\underline{v} = (v_j)_{j\in\mathbb{N}}$ for the constrained minimisation problem* (9.7.1) *satisfies* (9.7.10). *Then there is a sequence $\underline{\phi} = (\phi_j)_{j\in\mathbb{N}} \subset S(K, V)$ of symmetries for (K, V) such that the transformed minimising sequence $\underline{u} = (u_j = \phi_j v_j)_{j\in\mathbb{N}}$ satisfies the concentration condition* (C) *if, and only if,*

$$F(\underline{v}) = 0 . \tag{9.7.16}$$

Proof. Clearly it suffices to consider in condition (C) only the family of closed balls $B(r)$, $r > 0$.

First, suppose that condition (C) holds for the minimising sequence $u_j = \phi_j v_j$, $\phi_j \in S$, $j \in \mathbb{N}$, where $\underline{v} = (v_j)_{j\in\mathbb{N}}$ is some given minimising sequence. Now, given $\varepsilon > 0$, choose $R > 0$ according to condition (C). Then for all j, by definition of F_j, $F_j(R) \leq F_j(R, \phi_j)$, and hence $F(R) = \lim_{i\to\infty} F_{j(i)}(R) \leq \varepsilon$. Therefore,

$$F(\underline{v}) = \lim_{r\to\infty} F(r) \leq F(R) \leq \varepsilon$$

and thus $F(\underline{v}) = 0$.

Conversely, suppose that for every sequence $\underline{\phi} = (\phi_j)_{j\in\mathbb{N}} \subset S$ there is an $\varepsilon_0 > 0$ such that for every $r > 0$ there is $j_0 = j_0(r)$ with

$$F_{j_0}(r, \phi_{j_0}) \geq \varepsilon_0 \,. \tag{9.7.17}$$

Fix $R > 1$; by definition of $F_j(R)$ there are $\phi_j \in S$ and $\varepsilon_j > 0$ such that

$$F_j(R, \phi_j) \leq F_j(R) + \varepsilon_j \tag{9.7.18}$$

for all $j \in \mathbb{N}$ where $\varepsilon_j \to 0$ for $j \to \infty$. For the sequence $(\phi_j)_{j\in\mathbb{N}}$, there is, by assumption (9.7.17), a positive number $\varepsilon_0 > 0$ such that for a sequence $(R_k)_{k\in\mathbb{N}}$ with $R_k \geq R$ and $R_k \to \infty$ for $k \to \infty$ there is $j(k) = j_0(R_k)$ such that

$$F_{j(k)}(R_k, \phi_{j(k)}) \geq \varepsilon_0 \,. \tag{9.7.19}$$

Suppose there is a k_0 such that $j(k) = j(k_0)$ for all $k \geq k_0$. Then for all $k \geq k_0$,

$$\varepsilon_0 \leq F_{j(k)}(R_k, \phi_{j(k)}) = F_{j(k_0)}(R_k; \phi_{j(k_0)}) \to 0 \quad \text{for } k \to \infty$$

by dominated convergence, since $R_k \to \infty$ for $k \to \infty$. By this contradiction we see that there is a subsequence $k(i)$, $i \in \mathbb{N}$, such that $j(k(i)) \to \infty$ for $i \to \infty$.

If we now combine (9.7.18) and (9.7.19) for this subsequence, we get

$$\varepsilon_0 \leq F_{j(k(i))}(R_{k(i)}) + \varepsilon_{j(k(i))}\,, \qquad i \in \mathbb{N}\,.$$

Given $r \geq R$ there is i_0 such that $R_{k(i)} \geq r$ for all $i \geq i_0$; thus

$$\varepsilon_0 \leq F_{j(k(i))}(r) + \varepsilon_{j(k(i))}$$

for all $i \geq i_0$, or in the limit $i \to \infty$ we have $\varepsilon_0 \leq F(r)$. This implies $\varepsilon_0 \leq F(\underline{v})$. Hence, if the concentration condition does not hold for the given minimising sequence $\underline{v}$ transformed by any sequence $\underline{\phi}$ of symmetries for (K, V), then the value of the concentration function is strictly positive: $F(\underline{v}) > 0$. Thus we conclude. □

Remark 9.7.2. (a) By its definition the concentration functional has values in $[0, \Lambda]$, Λ given by (9.7.10). Lemma 9.7.2 characterises the value $F(\underline{v}) = 0$ as the favourable case in the sense of Theorem 9.7.2. So it remains to consider the values $0 < F(\underline{v}) < \Lambda$ and $F(\underline{v}) = \Lambda$. We comment on their meaning.

(b) If $F(\underline{v}) = \Lambda$ holds, then it is not hard to show that, for all $R > 0$,

$$\lim_{j\to\infty} \sup_{\phi\in S} \|\pi_{B(R)} \widehat{G}_+(\phi v_j)\|_1 = 0\,, \tag{9.7.20}$$

i.e. sequences with $F(\underline{v}) = \Lambda$ finally leave every ball uniformly with respect to all symmetries.

(c) If $0 < F(\underline{v}) < \Lambda$ then at least intuitively the sequence $\underline{v}$ "contains" two parts. One part finally leaves every ball while the other part, after applying a suitable sequence of symmetries, remains localised in the sense of condition (C). Then, in favourable circumstances, given $\varepsilon > 0$, one can show (cf. [9.6]) that the sequence $\underline{v}$ splits in the following sense: there are two sequences $\underline{v}^i$, $i = 1, 2$, and $\lambda_i > 0$, λ_i independent of $\varepsilon > 0$, such that, for all j,

$$\begin{gathered} K(v_j^1 + K(v_j^2) \leq K(v_j) + \varepsilon\,, \\ V(v_j) - \varepsilon \leq V(v_j^1) + V(v_j^2) \leq V(v_j) + \varepsilon\,, \\ V(v_j^i) \to \lambda_i \quad \text{for } j \to \infty\,. \end{gathered} \tag{9.7.21}$$

And this implies

$$I_{\lambda_1} + I_{\lambda_2} \leq I + \varepsilon \quad \text{and} \quad 1 - \varepsilon \leq \lambda_1 + \lambda_2$$

where I_λ is defined by relation (9.6.7). Since λ_1, λ_2 are independent of $\varepsilon > 0$, one obtains finally

$$I_{\lambda_1} + I_{\lambda_2} \leq I \quad \text{and} \quad 1 \leq \lambda_1 + \lambda_2\,. \tag{9.7.22}$$

But by covariance we know according to (9.6.8) that $I_\lambda = \lambda^\gamma I$, $0 < \gamma < 1$, so that for $I > 0$ we arrive at the contradiction

$$\lambda_1^\gamma + \lambda_2^\gamma \leq 1 \leq (\lambda_1 + \lambda_2)^\gamma\,, \quad \lambda_i > 0\,.$$

(d) At least on an intuitive level, it is thus clear how minimising sequences have to be chosen. The choice has to be such as to avoid relation (9.7.20), because if (9.7.20) holds then the weak and pointwise limit u according to relation (9.7.3) vanishes. The occurrence of splitting in the sense of relations (9.7.21) is ruled out by the covariance assumption on V.

(e) A complete proof of these statements is in general (and also in many concrete examples) rather long, and many technical details have to be taken into account. So we have not done it here. Using similar concentration functionals, a detailed discussion along these lines with many applications can be found in [9.6].

9.7.4 Minimising Subsets

Whether the concentration condition is satisfied or not depends on the special choice of the minimising sequence $\underline{v}$. In some cases it might be possible to choose a minimising sequence within a certain subset $\mathcal{F}$ of $D(V)$, the elements of which are "concentrated" by the very definition of $\mathcal{F}$. But clearly one has to ensure that

$$I(\mathcal{F}) = \inf\{K(v) \,|\, v \in \mathcal{F}, V(v) = 1\} \tag{9.7.23}$$

equals the infimum I as defined in relation (9.7.1).

A subset $\mathcal{F} \subset D(V)$ is called a *minimising subset* for the *constrained minimisation problem* (9.7.1) if, and only if,

$$I(\mathcal{F}) = I$$

holds. The following lemma gives two simple characterisations of a minimising subset which are quite useful in later applications.

Lemma 9.7.3. *Suppose that the functions K and V satisfy condition* (9.6.14) *and the covariance relations* (9.6.4–6) *with respect to some group* $\{T_\sigma \mid \sigma > 0\}$.

Then a subset $\mathcal{F} \in D(V)$ which is invariant under this group satisfies $I(\mathcal{F}) = I$ if, and only if,

(a) *for every $u \in D(V)$, $V(u) = 1$, and every $\varepsilon > 0$ there is $v \in \mathcal{F}$ such that*

$$K(v) \leq K(u) + \varepsilon \quad \textit{and} \quad 1 \leq V(v)$$

or

(b) *for every $u \in D(V)$, $V(u) > 0$, and every $\varepsilon > 0$ there is $v \in \mathcal{F}$ such that*

$$R(u) - \varepsilon \leq R(v)$$

where the function R is defined in (9.6.15).

Proof. According to their definitions we know $I \leq I(\mathcal{F})$. Then clearly the condition stated in (a) with $V(v) = 1$ instead of $V(v) \geq 1$ is necessary.

Suppose, conversely, that condition (a) holds and that u and v are chosen according to this condition. Define $\sigma > 0$ by

$$1 = V(T_\sigma v) = \sigma^s V(v) .$$

Hence $0 < \sigma \leq 1$. It follows that

$$I(\mathcal{F}) \leq K(T_\sigma v) = \sigma^r K(v) \leq K(v) \leq K(u) + \varepsilon$$

and we deduce $I(\mathcal{F}) \leq I$, and thus $I(\mathcal{F}) = I$. Therefore condition (a) is sufficient.

If we introduce, in analogy with Lemma 9.6.2,

$$S(\mathcal{F}) = \sup\{R(v) \mid v \in \mathcal{F}, V(v) > 0\}$$

then condition (b) can be reformulated as $S \leq S(\mathcal{F})$. Since, however, by definition $S(\mathcal{F}) \leq S$, condition (b) is equivalent to $S(\mathcal{F}) = S$.

Now observe that by Lemma 9.6.2 $I(\mathcal{F})$ and $S(\mathcal{F})$ satisfy the same relation as I and S, i.e.

$$I(\mathcal{F}) S(\mathcal{F}) = 1 = IS .$$

Thus condition (b) is necessary and sufficient. □

Remark 9.7.3. (a) Condition (b) can be applied if some subset $\mathcal{F} \subset D(V)$ is known to be dense in $D(V)$ with respect to some topology in which R is continuous.

(b) If ϕ_0 is a symmetry for the functions (K, V) such that

$$T_\sigma \phi_0 D(V) \subseteq \phi_0 D(V) \quad \text{for all } \sigma > 0$$

then it is easy to show that

$$\mathcal{F} = \phi_0 D(V)$$

is a minimising subset for the problem (9.7.1).

(c) In Sect. 9.8, in concrete examples, minimising subsets are given explicitly along the lines of the above remark.

9.7.5 Growth Restrictions on the Potential

As we have learned in Sect. 9.6, the infimum I defined in (9.7.1) has to be positive if the constrained minimisation method of Proposition 9.6.1 is to be applied. In this section, for the translation-invariant case, we deduce, within a fairly general class $\mathcal{G}$ of potentials, growth restrictions on $G \in \mathcal{G}$ from the assumption $I > 0$.

For a translation-invariant potential, i.e. if the function V is of the form

$$V(u) = \int G(u(x))\, dx\,, \quad u \in D(V)\,,$$

the group $\{T_\sigma \,|\, \sigma > 0\}$ can always be chosen to be

$$T_\sigma u(\cdot) = u(\sigma \cdot)\,. \tag{9.7.24}$$

The covariance relations are then

$$\begin{aligned} V(T_\sigma u) &= \sigma^s V(u)\,, \quad u \in D(V)\,, \quad s = -d\,, \\ K(T_\sigma u) &= \sigma^r K(u)\,, \quad u \in E = E^{1,2}(\mathbb{R}^d; \mathbb{R}^n)\,, \quad r = 2 - d\,, \end{aligned} \tag{9.7.25}$$

and clearly $\gamma = r/s = 1 - 2/d \in (0,1)$ since $d \geq 3$.

Let us say that a continuous function $G : \mathbb{R}^n \to \mathbb{R}$ has *polynomial behaviour* at $y = 0$ (at infinity) if and only if there is an exponent $p \geq 1$ ($q \geq 1$) such that

$$|y|^{-p} G(y) \quad \text{is bounded on } \{y \,|\, |y| \leq 1\}$$

($|y|^{-q} G(y)$ is bounded on $\{y \,|\, |y| \geq \lambda_0\}$ for some $\lambda_0 > 1$).

With this definition we introduce the following class of potentials:

$$\mathcal{G} = \{G \in \mathcal{C}(\mathbb{R}^n; \mathbb{R}) \,|\, G \text{ has polynomial behaviour at } y = 0 \text{ and at infinity}\}\,. \tag{9.7.26}$$

The following proposition shows that within $\mathcal{G}$ the condition $I > 0$ (I according to (9.7.1)) can be translated into growth restrictions on G.

Proposition 9.7.4. *Suppose $G : \mathbb{R}^n \to \mathbb{R}$ to be continuous, $G(0) = 0$, $G(y) > 0$ somewhere and define the function K on E and V on $D(V) \subset E$ by (9.1.5) and (9.1.8). Then the constrained minimisation problem (9.7.1) for K and V is well defined and the functions K and V satisfy condition (9.6.14) and the covariance relations (9.6.4–6) with respect to the group (9.7.24) with exponents r and s given in (9.7.25). In addition, the following holds:*

(a) *If $G_+(y) \le C|y|^{2^*}$ for all $y \in \mathbb{R}^n$ and some $C \in \mathbb{R}_+$ then $I > 0$.* (9.7.27)

(b) *If $G \in \mathcal{G}$ and $I > 0$ then (9.7.27) holds.*

Proof. The first part is obvious. So we prove (a) and (b).

For $u \in D(V)$ with $V(u) > 0$ one has, by inequality (9.7.27),

$$V(u) \le V_+(u) \le C\|u\|_{2^*}^{2^*} .$$

Hence, by Sobolev's inequality, $V(u)^\gamma \le C^\gamma K(u)$ and thus, by Lemma 9.6.2, $I > 0$.

In order to prove (b), observe that for $G \in \mathcal{G}$ and $u \in D(V)$ the following basic estimate is available ($0 < \delta < \lambda < \infty$):

$$\begin{aligned} C_1\|[|u| < \delta]u\|_p^p + \int |G^{[\delta,\lambda]}(u)|\, dx + C_2\|[\lambda < |u|]u\|_q^q &\le \int |G(u)|\, dx \\ \le C_3\|[|u| < \delta]u\|_p^p + \int |G^{[\delta,\lambda]}(u)|\, dx + C_4\|[\lambda < |u|]u\|_q^q & \quad (9.7.28) \end{aligned}$$

where the exponent p characterises the polynomial behaviour of G at $y = 0$ while the exponent q does the same at infinity. Thus $u \in D(V)$ implies

$$\|[|u| < \delta]u\|_p < \infty \quad \text{and} \quad \|[\lambda < |u|]u\|_q < \infty \qquad (9.7.29)$$

for all $0 < \delta,\ \lambda < \infty$. In particular it follows that if $1 \le p < 2^*$ then $u \in L^p$ and if $2^* < q < \infty$ then $u \in L^q$.

If $I > 0$ is supposed, then by Lemma 9.6.2,

$$R(v) = K(v)^{-1}V(v)^\gamma$$

is known to be bounded on $\{v \in D(V) \,|\, V(v) > 0\}$.

Now we study this function on the orbits of the group $\{\widehat{T}_\sigma \,|\, \sigma > 0\}$:

$$\widehat{T}u(\cdot) = \lambda_0 u(\sigma\cdot)\,, \quad \lambda_0 = \sigma^{d/2^*}\,, \quad \sigma > 0\,. \qquad (9.7.30)$$

For this group one has $K(\widehat{T}_\sigma u) = K(u)$ for all $\sigma > 0$. Therefore, for $u \in D(V)$ and $V(u) > 0$, we introduce

$$\Sigma = \Sigma(u) = \{\sigma > 0 \,|\, \widehat{T}_\sigma u \in D(V), V(\widehat{T}_\sigma u) > 0\}$$

and we know by the assumption $I > 0$ that

$$\sup\{V(\widehat{T}_\sigma u)\,|\,\sigma \in \Sigma(u)\} \leq (SK(u))^{1/\gamma}\,, \tag{9.7.31}$$

where $0 < S < \infty$ is given by Lemma 9.6.2.

To proceed we observe that for all $\sigma > 0$, $0 < \delta \leq \lambda < \infty$,

$$\|G^{[\sigma,\lambda]}(\widehat{T}_\sigma u)\|_1 \leq C_{\delta,\lambda}\delta^{-2^*}\|u\|_{2^*}^{2^*} \tag{9.7.32}$$

holds with $C_{\delta,\lambda} = \sup\{|G(y)|\,|\,\delta \leq |y| \leq \lambda\}$. Thus, if we apply inequality (9.7.28) to $\widehat{T}_\sigma u$ for $\sigma > 0$ and $u \in D(V)$, $V(u) > 0$, we see that we have to control

$$J_p(\sigma) = \|[|\widehat{T}_\sigma u| < \delta]\widehat{T}_\sigma u\|_p^p \quad \text{and} \quad I_q(\sigma) = \|[\lambda < |\widehat{T}_\sigma u|]\widehat{T}_\sigma u\|_q^q \tag{9.7.33}$$

and this can easily be done. First, it follows that

$$u \in D(V) \Rightarrow \widehat{T}_\sigma u \in D(V) \quad \text{for all } \sigma > 0\,.$$

Then one proves that (9.7.31) requires $2^* \leq p < \infty$ if $G(y)$ is positive near $y = 0$ while there is no restriction on $p \geq 1$ if $G(y)$ is negative near $y = 0$.

Similarly, the growth at infinity is restricted. If G is positive near infinity then (9.7.31) requires $1 \leq q \leq 2^*$. If, however, G is negative near infinity, then all $q \in [1, \infty)$ are allowed. Therefore $I > 0$ implies $G_+(y) \leq C|y|^{2^*}$. □

9.8 Existence of a Minimiser II: Some Examples

In this section we show, for some classes of examples, that the constrained minimisation problem (9.7.1) actually has a solution if appropriate growth restrictions on the potential are taken into account in addition to the general assumptions of Theorem 9.7.1.

Section 9.8.1 treats a class of examples where, by assumption on the growth properties of

$$(x, y) \longmapsto G(x, y)\,,$$

the potential G cannot be independent of $x \in \mathbb{R}^d$. In particular, a case with spherical symmetry is discussed.

The following two sections are concerned with the case where the potential does not depend on $x \in \mathbb{R}^d$ (translation-invariant case). If spherical symmetry is available or if the problem can be reduced to such a situation by spherically symmetric rearrangement, the concentration condition is easily realised. This is explained in Sect. 9.8.2.

In the general translation-invariant case this concentration condition is much harder to realise. In Sect. 9.8.3 we explain the elegant solution due to Brezis and Lieb [9.13] where the concentration condition is obtained indirectly by a clever combination of the special scale covariance properties of the problem and their lemma of "concentration by translation".

9.8.1 Some Non-translation-invariant Cases

First of all, the existence of a minimiser is proved under fairly general assumptions on the potential. This is then explained via some more concrete examples. Here the concentration condition results from natural restrictions on the potential.

Proposition 9.8.1. *Suppose that $G : \mathbb{R}^d \times \mathbb{R}^n \to \mathbb{R}$ is some C-function which is strictly positive on some set of positive measure. Define the functions K and V by (9.1.5) and (9.1.8) and suppose the covariance relations (9.6.4–6) to hold for some group $\{T_\sigma \mid \sigma > 0\}$ with $\gamma = r/s \in (0,1)$. Suppose furthermore:*

(a) *there is a function $h \in L^q \cap L^\alpha_{\mathrm{loc}}$, $\alpha > q$, $1/q = 1 - p/2^*$, $p\gamma = 2$, such that for all $y \in \mathbb{R}^n$ and almost every $x \in \mathbb{R}^d$,*

$$G_+(x,y) \le h(x)|y|^p\,, \tag{9.8.1}$$

or

(b) *there are functions $h \in L^\infty$ with*
   (i) $\|\pi^\perp_{B(R)} h\|_\infty \to$ *for* $R \to \infty$,
   (ii) $b_\lambda(\cdot) = \sup_{|y| \ge \lambda} |y|^{-2^*} G_+(\cdot, y)$,
   satisfying $b_\lambda \to 0$ in L^∞_{loc} for $\lambda \to \infty$, such that for all $y \in \mathbb{R}^n$ and almost every $x \in \mathbb{R}^d$,

$$G_+(x,y) \le h(x)|y|^{2^*} \quad \text{and} \quad \gamma 2^* = 2\,. \tag{9.8.2}$$

Then the constrained minimisation problem (9.7.1) for (K,V) has a nontrivial minimiser.

Proof. The estimates (9.8.1) and (9.8.2) imply

$$V_+(u) \le \|h\|_q \|u\|^p_{2^*}\,.$$

Hence, by Sobolev's inequality, since $\gamma p = 2$ in both cases,

$$V_+(u)^\gamma \le C K(u)\,.$$

Therefore Lemma 9.6.2 proves $I > 0$.

The function a_λ of hypothesis (H_2) is just $a_\lambda = \lambda^p h$; thus $a_\lambda \in L^1_{\mathrm{loc}}$ and (H_2) is satisfied.

Now take a minimising sequence $(u_j)_{j \in \mathbb{N}}$ for the problem (9.7.1). Then for a compact set $K \subset \mathbb{R}^d$ and $\lambda > 0$, we have, if (a) is assumed,

$$\|\pi^\perp_K \widehat{G}_+(u_j)\|_1 \le \|\pi^\perp_K h\|_q \|u_j\|^p_{2^*}\,,$$

$$\|\pi_K \widehat{G}^{\ge\lambda}_+(u_j)\|_1 \le \|\pi_K h\|_\alpha \|\pi_K u_j\|^p_{2^*} \|\pi_K [\lambda < |u_j|]\|_\rho\,,$$

with $1/q = 1/\alpha + 1/\rho$, i.e. $\rho \in (1, \infty)$. Since

$$\|[\lambda \le |u_j|]\|_\rho \le C\lambda^{-2^*/\rho}, \quad 0 < \lambda < \infty, \tag{9.8.3}$$

is known, the conditions (C) and (IUD) follow easily.

If (b) is assumed, the concentration condition (C) follows in the same way from (i). For (IUD) we note that, by (ii),

$$\|\pi_K \widehat{G}_+^{\geq\lambda}(u_j)\|_1 \le \|\pi_K b_\lambda\|_\infty \|u_j\|_{2^*}^{2^*}$$

holds. This implies (IUD).

Therefore, by Theorem 9.7.1, the constrained minimisation problem for the functions K and V has a nontrivial minimiser. □

Note that in this proposition no further information about the minimising sequence is taken into account. If some additional information about minimising sequences is available, then the hypotheses on the potential can be relaxed (see Example 9.8.2).

Example 9.8.1. As a potential we take

$$G(x, y) = Q(x)|y|^p, \quad x \in \mathbb{R}^d, \ y \in \mathbb{R}^n, \quad 2 < p < 2^*, \tag{9.8.4}$$

with a measurable function $Q : \mathbb{R}^d \to \mathbb{R}$ such that

$$Q_+ \in L^q \cap L^\alpha_{\text{loc}}, \quad \alpha > q, \quad 1/q = 1 - p/2^* \quad \text{and } Q_+ > 0 \text{ on some ball}.$$

The covariance relations (9.6.4–6) are satisfied for the group $\{T_\sigma \,|\, \sigma > 0\}$ on $E = E^{1,2}(\mathbb{R}^d; \mathbb{R}^n)$ defined by

$$T_\sigma u = \sigma u, \quad \sigma > 0, \quad u \in E. \tag{9.8.5}$$

These relations are

$$K(T_\sigma u) = \sigma^2 K(u), \quad V(T_\sigma u) = \sigma^p V(u).$$

Hence, $\gamma = 2/p \in (0, 1)$ and Proposition 9.8.1 applies.

This example has an easy extension to the case where $K(u) = \frac{1}{2}\|\nabla u\|_2^2$ is replaced by

$$K_1(u) = \frac{1}{2}\|\nabla u\|_2^2 + \frac{1}{2}\|au\|_2^2, \tag{9.8.6}$$

a being an L^∞ function on $\mathbb{R}^d$ with values in the real vector space of $n \times n$ matrices. This is so because

(i) K_1 has the same covariance properties with respect to the group (9.8.5),
(ii) $K(u) \le K_1(u)$ for all $u \in E$.

Example 9.8.2. In the Sobolev space $H^1(\mathbb{R}^d)$, the constrained minimisation problem (9.7.1) is considered for the functions

$$K(u) = \frac{1}{2}\|\nabla u\|_2^2 + \frac{1}{2}\|Au\|_2^2 \,, \tag{9.8.7}$$

$$V(u) = \int Q(|x|)|u(x)|^p \, dx \,, \quad 2 < p \le 2^* \,.$$

Here A is some linear operator in $L^2(\mathbb{R}^d)$ which depends only on $r = |x|$ and for which $H^1(\mathbb{R}^d)$ is contained in the domain of A. Furthermore,

$$\|Au\|_2^2 \ge a\|u\|_2^2 \quad \text{for all } u \in H^1(\mathbb{R}^d)$$

with some $a > 0$ is assumed.

Our assumptions on Q are the following: Q is a measurable function $\mathbb{R}_+ \to \mathbb{R}$ which is positive on a set of positive measure. If $2 < p < 2^*$ then $Q_+ \in L^q_{\text{loc}}(\mathbb{R}^d)$, $1/q < 1 - p/2^*$, and with $R_d = d - 1$ we have

$$\Sigma := \int_{R_d}^{\infty} r^l Q_+(r) \, dr < \infty \,, \quad l = d - 1 - p\alpha \,, \quad 2\alpha = d - 1 \,. \tag{9.8.8}$$

If $p = 2^*$ then suppose

$$Q(r) \le 0 \quad \text{for } 0 \le r \le R_d \tag{9.8.9}$$

and the above integrability condition for $p = 2^*$.

As by assumption, radial symmetry is available, the problem is treated in the subspace $E_0 = H^1_r(\mathbb{R}^d)$ of radial functions in $H^1(\mathbb{R}^d)$. Then by the radial lemma of Strauss [9.11] (cf. Lemma 9.2.9) the following estimate is available for $u \in E_0$:

$$|u(x)| \le C_d \|u\|_{H^1} |x|^{-\alpha} \quad \text{for } |x| \ge R_d \,. \tag{9.8.10}$$

If this bound is used to estimate $V_+(u)$ we obtain by (9.8.8), using Hölder's inequality and the abbreviation $B = B(R_d)$,

$$V_+(u) \le \|\pi_B Q_+\|_{p_1} \|\pi_B u\|_p^p + C\Sigma \|u\|_{H^1}^p$$

where $1/p_1 = 1 - p/2^*$. Then Sobolev's inequality implies $V_+(u) \le C\|u\|_{H_1}^p$.

The group $\{T_\sigma \,|\, \sigma > 0\}$ and the covariance relations are the same as in Example 9.8.1. Hence $\gamma = 2/p$ and $V_+(u)^\gamma \le C^\gamma K(u)$ follows. This implies $I > 0$.

Clearly, Hypothesis (H_3) is satisfied under our assumptions. Now we check the conditions (C) and (IUD) for a minimising sequence $(u_j)_{j \in \mathbb{N}} \subset E_0$:

$$u_j \in D(V) \subset E_0 \,, \quad V(u_j) = 1 \,, \quad K(u_j) \to I \quad \text{for } j \to \infty \,.$$

Then $(u_j)_{j \in \mathbb{N}}$ is bounded in $H^1(\mathbb{R}^d)$ so that we may assume that this sequence converges to some $u \in H^1(\mathbb{R}^d)$, weakly in $H^1(\mathbb{R}^d)$ and almost everywhere. It follows that $u \in E_0$.

For $|x| \geq R > R_d$ we use the decay estimate (9.8.10) and obtain

$$\|\pi_{B(R)}^{\perp} \widehat{G}_+(u_j)\|_1 \leq C \int_R^{\infty} r^l Q_+(r)\, dr\,,$$
$$C = \sup_j \{C_d \|u_j\|_{H^1}\}^p < \infty\,.$$

The integrability assumption (9.8.8) now implies the concentration condition.

For the condition (IUD) we distinguish between the cases when $2 < p < 2^*$ and $p = 2^*$.

The decay estimate (9.8.10) implies for all $|x| \geq R$, $R \geq R_d$, and all i,

$$|u_i(x)| \leq C_1 |x|^{-\alpha} \leq C_1 R_d^{-1} =: \lambda_0 < \infty\,.$$

Hence, for all i and $\lambda > \lambda_0$,

$$[\lambda < |u_i|] \cap B^{\perp} = \emptyset\,,$$

and, if $2 < p < 2^*$, then by assumption

$$\|\pi_{B(R)} \widehat{G}_+^{\geq \lambda}(u_i)\|_1 \leq \|\pi_B Q_+\|_q \|u_i\|_{2^*}^p \|[\lambda < |u_i|]\|_\rho$$

where $\rho \in [1, \infty)$ is defined by $1/q + 1/\rho = 1 - p/2^*$. Now inequality (9.8.3) implies the condition (IUD) for this case.

For the case $p = 2^*$, $\pi_B Q_+$ vanishes by assumption. Thus, again, (IUD) holds since on B^c the sequence $(u_j)_{j \in \mathbb{N}}$ is uniformly bounded. Now Theorem 9.7.1 proves that u is a minimiser for this constrained minimisation problem.

9.8.2 Spherically Symmetric Cases

These cases are those for which the constrained minimisation problem (9.7.1) can be reduced to the corresponding problem with respect to spherically symmetric functions u. This is typically the case for functions u with values in $\mathbb{R}$ ($n = 1$) and translation-invariant even potentials G because then, by the results on the spherically symmetric rearrangement of functions (Sect. 9.2.3), we know that the map ϕ,

$$u \longmapsto \phi u = |u|^* = \text{the spherically symmetric rearrangement of } |u|, \quad u \in E = E^{1,2}(\mathbb{R}^d) \tag{9.8.11}$$

is a symmetry of the functions K and V defined by

$$K(u) = \frac{1}{2}\|\nabla u\|_2^2\,, \quad V(u) = \int G(u(x))\, dx\,, \quad u \in D(V) \subset E^{1,2}(\mathbb{R}^d)\,. \tag{9.8.12}$$

This means, for all $u \in D(V)$, that

$$V(\phi u) = V(u) \quad \text{and} \quad K(\phi u) \leq K(u)\,. \tag{9.8.13}$$

By definition, ϕu is of the form

$$\begin{aligned} &\phi u(x) = v(|x|)\,, \\ &\qquad 0 \le v\,, \quad v \text{ nonincreasing on } \mathbb{R}_+\,, \quad v' \in L^2(\mathbb{R}_+, r^{d-1}\,dr)\,. \end{aligned} \tag{9.8.14}$$

The corresponding class of functions is denoted by E_0:

$$E_0 = \{u \in E^{1,2}(\mathbb{R}^d)\,|\,u(x) = v(|x|), \quad v \text{ satisfies (9.8.14)}\}\,.$$

Then by Lemma 9.7.3 (a) we know that $D_0(V) = D(V) \cap E_0$ is a minimising subset for the constrained minimisation problem (9.7.1) for the function K and V of (9.8.12). Hence it suffices to consider the following reduced constrained minimisation problem:

$$I = \inf\{K(u)\,|\,u \in D_0(V), V(u) = 1\}\,. \tag{9.8.15}$$

This kind of minimisation problem has been considered many times under various assumptions [9.1, 6, 11, 12, 15]. Based on Theorem 9.7.1 we present a simplified proof of the existence of a minimiser for the problem (9.8.15) under slightly more general assumptions on the potential.

Proposition 9.8.2. *Suppose $G : \mathbb{R} \to \mathbb{R}$ to be an even continuous function satisfying*

(a) $$G(0) = 0 \quad \textit{and} \quad G(y) > 0 \textit{ somewhere}\,, \tag{9.8.16 a}$$

(b) $$G_+(y) = o(|y|^{2^*}) \quad \textit{for } |y| \to 0 \textit{ and } |y| \to \infty\,. \tag{9.8.16 b}$$

Then the constrained minimisation problem (9.7.1) for the functions K and V of (9.8.12) has a nontrivial minimiser which is a nonnegative spherically symmetric nonincreasing function in $D_0(V)$.

Proof. In the case of a translation-invariant potential G we can always use the group $\{T_\sigma\,|\,\sigma > 0\}$ defined by

$$T_\sigma u(\cdot) = u(\sigma\cdot)\,, \quad \sigma > 0\,, \quad u \in E = E^{1,2}(\mathbb{R}^d)\,. \tag{9.8.17}$$

The corresponding covariance relations are $(u \in D(V))$

$$\begin{aligned} K(T_\sigma u) &= \sigma^r K(u)\,, \quad r = 2 - d\,, \\ V(T_\sigma u) &= \sigma^s V(u)\,, \quad s = -d\,, \end{aligned} \tag{9.8.18}$$

and hence $\gamma = r/s = 2/2^* \in (0,1)$.

Assumption (9.8.16 b) implies $G_+(y) \le C|y|^{2^*}$ for all y and some constant $C \in \mathbb{R}_+$; hence $V_+(u)^\gamma \le C'K(u)$ and Lemma (9.6.2) proves $I > 0$. Since our general assumption (H_2) of Theorem 9.7.1 is trivially satisfied we are left with verifying the conditions (C) and (IUD). The basic observation which makes this easy is that for $u \in E_0$ one knows

$$|u(x)| \leq C\|\nabla u\|_2 |x|^{-d/2^*} \quad \text{for all } x \neq 0\,. \tag{9.8.19}$$

The proof of (9.8.19) consists of combining Sobolev's inequality with the simple estimate

$$\|u\|_{2^*}^{2^*} \geq \frac{\omega_d}{d} v(R)^{2^*} R^d \quad \text{for all } R > 0$$

according to (9.8.14), where ω_d denotes the volume of the unit sphere in $\mathbb{R}^d$.

Applied to a minimising sequence $(u_j)_{j\in\mathbb{N}}$ for the problem (9.8.15) this estimate can be reformulated as

$$\sup_{x\in\mathbb{R}^d} \sup_{j\in\mathbb{N}} \pi_{B(R)}^{\perp} |u_j(x)| = \gamma(R) \tag{9.8.20}$$

where $\gamma(R) < \infty$ for $0 < R < \infty$ is explicitly known but we only use $\gamma(R) \to 0$ for $R \to \infty$. Clearly (9.8.20) expresses a strong form of a concentration for a minimising sequence.

Now we show how the assumption about the behaviour of the potential near $y = 0$ together with (9.8.20) imply concentration in the sense of condition (C) of Theorem 9.7.1. According to (9.8.16 b) we know

$$G_+^{<\delta}(y) \leq a(\delta)|y|^{2^*}[|y| < \delta] \quad \text{with } a(\delta) \to 0 \text{ for } \delta \to 0\,.$$

Hence we obtain for $R > 0$

$$\|\pi_{B(R)}^{\perp}\widehat{G}_+(u_j)\|_1 = \|\pi_{B(R)}^{\perp}\widehat{G}_+^{<\gamma(R)}(u_j)\|_1 \leq a(\gamma(R))\|u_j\|_{2^*}^{2^*}$$

and condition (C) follows immediately since $\{\|u_j\|_{2^*}\}$ is bounded by Sobolev's inequality.

Similarly, condition (IUD) follows from the assumed behaviour of $G_+(y)$ for $|y| \to \infty$. An appropriate reformulation of (9.8.16 b) is

$$G_+^{\geq\lambda}(y) \leq b(\lambda)|y|^{2^*}[\lambda \leq |y|] \quad \text{with } b(\lambda) \to 0 \text{ for } \lambda \to \infty\,.$$

Therefore, for a compact set $K \subset \mathbb{R}^d$ and $0 < \lambda < \infty$, we have

$$\|\pi_K \widehat{G}_+^{\geq\lambda}(u_j)\|_1 \leq b(\lambda)\|u_j\|_{2^*}^{2^*}$$

and we conclude as above.

Therefore, by Theorem 9.7.1, there is a minimiser u for our constrained minimisation problem. Since u appears as the pointwise limit of a sequence of functions in E_0, u itself belongs to E_0. This then proves the proposition. □

Now we indicate how the above results can be extended to spherically symmetric vector field equations, i.e. for equations of the form

$$\begin{aligned} -\Delta u = g(u)\,, \quad & u \in E^{1,2}(\mathbb{R}^d;\mathbb{R}^n) \equiv E\,, \quad d \geq 3\,, \quad n > 1\,, \\ g(y) = g_0(|y|)y\,, \quad & y \in \mathbb{R}^n\,, \quad g_0 : R_+ \to \mathbb{R} \text{ continuous}\,. \end{aligned} \tag{9.8.21}$$

Here we define G_0 on $\mathbb{R}_+$ by $G_0(s) = \int_0^s g_0(t)\,dt$ and extend it as an even function on $\mathbb{R}$ and use

$$G(y) = G_0(|y|)$$

as the potential to arrive at the associated constrained minimisation problem for this potential.

Denote by E_1 the space $E^{1,2}(\mathbb{R}^d;\mathbb{R})$ and define $V_0(v)$ on

$$D(V_0) = \{v \in E_1 \mid G_0(v) \in L^1\}$$

as usual and consider the constrained minimisation problem

$$I_0 = \inf\{K_0(v) \mid v \in D(V_0), V_0(v) = 1\} \tag{9.8.22}$$

where for $v \in E_1$, $K_0(v) = \frac{1}{2}\|\nabla v\|_2^2$. Using properties of weak derivatives (as for instance in [9.25]) one can easily prove that

$$u \in E\,, \quad v := |u| \;\Rightarrow\; v \in E_1 \quad \text{and} \quad K_0(v) \leq K(u)\,. \tag{9.8.23}$$

Since by definition $V(u) = V_0(|u|)$, $u \in D(V)$, is known, we deduce

$$I_0 \leq I\,.$$

But since E_1 can be considered as a subset of E the converse inequality is trivial; hence

$$I = I_0$$

and it follows that the constrained minimisation problem for the functions K and V in E has a minimiser if, and only if, problem (9.8.22) has a solution. Problem (9.8.22) can now be treated by Proposition 9.8.2.

9.8.3 The Translation-invariant Case Without Spherical Symmetry

In the general case of (9.8.21), i.e. in the case of a vector field equation without spherical symmetry, one cannot use the spherically symmetric rearrangement of functions as we did in the previous section. Then concentration by spherically symmetric rearrangement of functions as in inequality (9.8.19) is not available. We really have to take into account the symmetry group $S = S(K,V)$ for the function K of "kinetic energy" and the function V of "potential energy". Recall that in the case of a translation-invariant potential G this symmetry group always contains at least the full Euclidean group of $\mathbb{R}^d$. Hence all the difficulties mentioned in Sect. 9.7 can occur in the associated constrained minimisation problem. Up to now these difficulties have prevented a direct proof of the concentration condition. Then, as indicated in Remark 9.7.2, one could proceed by excluding the other possible values of the concentration functional F in a way similar to what has been done in [9.7]. But this causes many technical complications, and the proof is considerably less direct. The most direct proof of the existence of a minimiser in this case seems to be that of *Brezis and Lieb* [9.13]. So we prefer to present their proof. This proof relies on a clever use of the special covariance relations (9.8.18) in connection with their "lemma of concentration by translation" (Corollary 9.2.12) in order to realise the concentration condition in two steps. The condition (IUD) is realised in the same way as in the previous section by growth restrictions on the potential.

Proposition 9.8.3. *Under the following assumptions on the potential G:*

$$G : \mathbb{R}^n \to \mathbb{R}, \quad \textit{continuous}, \ G(0) = 0, \ G(u) > 0 \ \textit{somewhere}, \tag{9.8.24}$$

$$G_+(y) = o(|y|^{2^*}) \quad \textit{for } |y| \to 0 \textit{ and } |y| \to \infty, \tag{9.8.25}$$

For every $\varepsilon > 0$ there is $C_\varepsilon \in \mathbb{R}_+$ such that for all $a, b \in \mathbb{R}^n$

$$|G(a+b) - G(a)| \le \varepsilon\{|G(a)| + |a|^{2^*}\} + C_\varepsilon\{|G(b)| + |b|^{2^*}\}, \tag{9.8.26}$$

the associated constrained minimisation problem (9.7.1) for the functions K and V has a minimiser.

Proof. (a) The covariance group $\{T_\sigma \mid \sigma > 0\}$ is defined on $E = E^{1,2}(\mathbb{R}^d; \mathbb{R}^n)$ by the same formula as in (9.8.17). Then the covariance relations (9.8.18) hold. Since by assumption (9.8.25) $G_+(y) \le C|y|^{2^*}$ is known, Proposition 9.7.4 (a) proves $I > 0$. Furthermore, conditions (9.6.14) are satisfied because of assumption (9.8.24). Therefore the general frame for the constrained minimisation method as explained in Sect. 9.6 is realised. In particular there is a minimising sequence

$$\begin{gathered} v_j \in D(V), \quad V(v_j) = 1, \\ \lim_{j\to\infty} K(v_j) = I = \inf\{K(v) \mid v \in D(V), V(v) = 1\} > 0. \end{gathered} \tag{9.8.27}$$

According to assumption (9.8.25) we estimate in the following way, for $0 < \delta < \lambda < \infty$

$$\begin{aligned} 1 = V(v_j) \le V_+(v_j) &\le a(\delta)\|[|v_j| < \delta]v_j\|_{2^*}^{2^*} + b(\lambda)\|[\lambda < |v_j|]v_j\|_{2^*}^{2^*} \\ &\quad + \int G_+(v_j)[\delta \le |v_j| \le \lambda]\, dx \\ &\le (a(\delta) + b(\lambda))\|v_j\|_{2^*}^{2^*} + C_{\delta,\lambda}|[\delta \le |v_j| \le \lambda]| \end{aligned}$$

where $C_{\delta,\lambda} = \sup\{G_+(u) \mid \delta \le |y| \le \lambda\}$ is finite. Since $C = \sup_j \|v_j\|_{2^*}^{2^*}$ is also finite we can find $0 < \delta < \lambda < \infty$ such that $(a(\delta) + b(\lambda))C \le 1/2$ because $a(\delta) \to 0$ for $\delta \to 0$ and $b(\lambda) \to 0$ for $\lambda \to \infty$ by assumption (9.8.25). Then it follows that

$$\frac{1}{2} \le C_{\delta,\lambda}|[\delta \le |v_j| \le \lambda]|$$

and therefore the lemma of concentration by translations applies in the form of Corollary 9.2.12. It proves the existence of a minimising sequence $(u_i)_{i\in\mathbb{N}}$

which has a nonvanishing limit u in the sense of conditions (a)–(d) of Theorem 9.2.2:

$$u_i \to u \quad \text{for } i \to \infty \text{ weakly in } E \text{ and almost everywhere on } \mathbb{R}^d, \qquad u \neq 0. \tag{9.8.28}$$

(b) Statement (9.8.28) allows one to apply Lemma 9.2.6 with $r = 2^*$ and $F = G$. Hence it follows that $u \in D(V)$,

$$V(u_i + v) - V(u_i) \to V(u+v) - V(u) \quad \text{for } i \to \infty \tag{9.8.29}$$

for all $v \in D(V)$. Similarly we get $v = u_\sigma - u \in D(V)$, $u_\sigma = T_\sigma u$ and thus by (9.8.29)

$$V(u_i + v) \to 1 + V(u_\sigma) - V(u) = 1 + (\sigma^{-d} - 1)V(u)$$

for $i \to \infty$ and all $\sigma > 0$. There is $\varepsilon_0 > 0$ such that for all $|\sigma - 1| \le \varepsilon_0$

$$|(\sigma^{-d} - 1)V(u)| \le \frac{1}{2}.$$

Therefore there is $i_0 \in \mathbb{N}$ such that for all $i \ge i_0$ and all $|\sigma - 1| < \varepsilon_0$

$$V(u_i + v) > 0.$$

Lemma 9.6.2 implies $IV(u_i + v)^\gamma \le K(u_i + v)$ for all $i \ge i_0$ and all these $\sigma > 0$. Since $(u_i)_{i \in \mathbb{N}}$ is a minimising sequence which converges weakly to u, the right-hand side of this inequality also converges:

$$\begin{aligned}\lim_{i\to\infty} K(u_i + v) &= I + K(v) + \langle \nabla u, \nabla v \rangle_2 = I - K(u) + K(u+v) \\ &= I + (-1 + \sigma^{2-d})K(u),\end{aligned}$$

where the covariance relations (9.8.16) have been used. So for the limit $i \to \infty$ of the above inequality we obtain

$$I[1 + (\sigma^{-d} - 1)V(u)]^\gamma \le I + (-1 + \sigma^{2-d})K(u)$$

or with $\sigma = 1 + \varepsilon$, $|\varepsilon| < \varepsilon_0$, $\varepsilon \to 0$,

$$I - \gamma d \varepsilon I V(u) + o(\varepsilon) \le I + (2-d)\varepsilon K(u) + o(\varepsilon).$$

Hence, if we use $\gamma d = d - 2$,

$$K(u) = IV(u). \tag{9.8.30}$$

(c) In part (a) of the proof we have seen that $u \neq 0$ and $I > 0$. Therefore (9.8.30) implies $V(u) > 0$. This allows us to apply Lemma 9.6.2 to obtain also

$$K(u) \ge IV(u)^\gamma.$$

Hence $V(u) \geq 1$, since $\gamma \in (0,1)$. Thus (9.8.30) implies $K(u) \geq I$ and together with inequality (9.7.4), $K(u) = I$ follows. Then again, by (9.8.30) we finally get $V(u) = 1$. This proves u to be a minimiser. □

Remark 9.8.1. (a) Clearly, if $|G(y)| \leq C|y|^{2^*}$ for all $y \in \mathbb{R}^n$ or if G belongs to the class $\mathcal{G}$ of potentials introduced in Sect. 9.7.5, then the potential G satisfied condition (9.8.26) (Lemma 9.2.7).

(b) Originally, Brezis and Lieb proved Proposition 9.8.3 under a slightly weaker assumption than (9.8.26). On the right-hand side of inequality (9.8.26) they admit an additional constant depending on ε. The only difference in the proof then is that $u_\sigma - u$ is approximated by $v_k \in D(V)$, $\operatorname{supp} v_k$ compact. For $v \in D(V)$ with compact support, relation (9.8.29) can be proved under this weaker hypothesis.

9.9 Nonlinear Field Equations in Two Dimensions

Some of the specific complications which occur in the case of two-dimensional field equations have already been pointed out in Sect. 9.1. Another problem is connected with the special scale covariance of the "kinetic energy" functional K and the "potential energy" functional V, in this case according to (9.1.13):

$$K(u_\sigma) = K(u)\,, \quad V(u_\sigma) = \sigma^{-2} V(u)\,, \quad u_\sigma(\cdot) = u(\sigma \cdot) \quad \text{for all } \sigma > 0\,. \tag{9.9.1}$$

and all $u \in D(V)$. This implies that the constrained minimisation problem makes no sense in the form (9.1.11). It took some time to realise that instead, the following constrained minimisation problem (9.9.3) is quite useful for solving a two-dimensional field equation of the form

$$\begin{gathered} -\Delta u = g(u)\,, \\ u : \mathbb{R}^2 \to \mathbb{R}^n\,, \quad g : \mathbb{R}^n \to \mathbb{R}^n\,, \quad n \geq 1\,. \end{gathered} \tag{9.9.2}$$

This minimisation problem reads:

$$I = \inf\{K(u) \,|\, u \in D(V), u \neq 0, V(u) \geq 0\}\,. \tag{9.9.3}$$

Global field equations in two variables have been solved only relatively recently using the constrained minimisation problem (9.9.3). In 1983 *H. Berestycki et al.* [9.17] considered the positive mass case for scalar field equations in two variables with a *"strong" nonlinearity* g (i.e. a continuous function g which is allowed to grow at infinity like an exponential function) and indicated the existence of infinitely many solutions in the subspace $H_s^1(\mathbb{R}^2)$ of spherically symmetric functions in the Sobolev space $H^1(\mathbb{R}^2)$.

In previous sections we have already explained why the positive mass case for scalar field equations is considerably simpler than the mass zero case for vector fields (i.e. $n > 1$). The existence of a "ground-state" solution for this case was proved in 1984 by H. Brezis and E. H. Lieb [9.13] under the assumption of a polynomially bounded nonlinearity g and a quite general assumption about the behaviour of g near $y = 0$. Their space of a priori solutions E is well adopted to the behaviour of g at $y = 0$ but at the expense of E not being a Banach space of functions on $\mathbb{R}^2$.

Finally in [9.29], *two-dimensional mass zero vector field equations* have been solved under the assumption of a strong nonlinearity g whose behaviour at $y = 0$ is characterised by some exponent $q \geq 2$. Accordingly the space of a priori solutions E is chosen to be

$$E = E_q \equiv E_q^1(\mathbb{R}^d; \mathbb{R}^n).$$

Here we follow [9.29] and briefly discuss later the results of [9.13].

9.9.1 Some Properties of Niemytski Operators on E_q

First of all some results about Niemytski operators $\widehat{F}$ on E_q associated with a continuous F have to be developed.

Lemma 9.9.1. *Suppose F is a continuous function $\mathbb{R}^n \to \mathbb{R}$ such that for all $y \in \mathbb{R}^n$*

$$|F(y)| \leq C|y|^q \sum_{k=0}^{\infty} a_k |y|^{pk/s} \tag{9.9.4}$$

with some exponent q, $d \leq q < \infty$, some constant $C > 0$, some number $s \in [1, \infty)$, $1/p = 1 - 1/d$, and some coefficients $a_k \geq 0$ satisfying

$$\limsup_{k\to\infty} |b_k|^{1/k} = \frac{1}{R} < \infty, \quad b_k = a_k (\alpha)_k^{p/s} \tag{9.9.5}$$

where $(\alpha)_k$ is defined in Theorem 9.2.3. Then the Niemytski operator $\widehat{F}$ associated with F maps the subset

$$Z_R = \left\{ u \in E_q \,\middle|\, \frac{1}{d}\|\nabla u\|_d < R^{s/p} \right\}$$

into $L^1(\mathbb{R}^d)$ and satisfies

$$\|\widehat{F}(u)\|_1 \leq C\|u\|_q^q \sum_{k=0}^{\infty} b_k \left[\frac{1}{d}\|\nabla u\|_d\right]^{pk/s} < \infty . \tag{9.9.6}$$

Proof. For $u \in Z_R$ the following estimate is available by Theorem 9.2.3:

$$\begin{aligned}
\|\widehat{F}(u)\|_1 &\leq \int C \sum_{k=0}^{\infty} a_k |u|^{q+pk/s} dx \leq C \sum_{k=0}^{\infty} a_k \| \, |u|^{q/s} |u|^{r(k)/s} \|_1 \\
&\leq C\|u\|_q^{q/s'} \sum_{k=0}^{\infty} a_k \|u\|_{r(k)}^{r(k)/s} \leq C\|u\|_q^q \sum_{k=0}^{\infty} b_k \left[\frac{1}{d}\|\nabla u\|_d\right]^{pk/s} .
\end{aligned}$$

since $1/d\|\nabla u\|_d < R^{s/p}$ is assumed, the series converges and thus proves the lemma. □

Lemma 9.9.2. *Under the assumptions of Lemma 9.9.1 with $R = \infty$, the Niemytski operator $\widehat{F}$ is sequentially continuous as a map from E_q equipped with the weak topology into L^1_{loc}, i.e. if $u_i \to u$ weakly in E_q then $\widehat{F}(u_i) \to \widehat{F}(u)$ in $L^1_{\mathrm{loc}}(\mathbb{R}^d)$.*

Proof. If a sequence $(u_i)_{i\in\mathbb{N}}$ converges weakly in E_q to some element $u \in E_q$ it is strongly bounded:

$$\sup_i \|u_i\| = C < \infty\,,$$

and we may assume that conditions (a)–(d) of Theorem 9.2.4 hold.

By continuity of F and condition (a), it follows that $\widehat{F}(u_i) \to \widehat{F}(u)$ for $i \to \infty$ for almost every $x \in \mathbb{R}^d$. Thus the statement follows from Vitali's convergence theorem if we can show

$$\sup_i \|\pi_A \widehat{F}(u_i)\|_1 = H(|A|) \to 0 \quad \text{for } |A| \to 0$$

for any measurable subset $A \subset \mathbb{R}^d$, $|A| < \infty$.

As in the proof of the previous lemma, we have for such a set A,

$$\|\pi_A \widehat{F}(u_i)\|_1 \le \|\pi_A u_i\|_q^{q/s'} C \sum_{k=0}^{\infty} a_k \|u_i\|_{r(k)}^{r(k)/s}\,.$$

If we observe that

$$\|\pi_A u_i\|_q \le |A|^{1/\beta} \|u_i\|_{r(1)}\,, \quad \frac{1}{\beta} = \frac{1}{q} - \frac{1}{q+p} > 0$$

holds, we get by Theorem 9.2.3

$$\|\pi_A \widehat{F}(u_i)\|_1 \le |A|^{q/\beta s'} \|u_i\|_{r(1)}^{q/s'} C \|u_i\|_q^{q/s} \sum_{k=0}^{\infty} b_k \left(\frac{1}{d}\|\nabla u_i\|_d\right)^{pk/s}$$

and thus by boundedness of $(u_i)_{i\in\mathbb{N}}$ in E_q and again by Theorem 9.2.3,

$$\sup_i \|\pi_A \widehat{F}(u_i)\|_1 \le \mathrm{const}|A|^{q/\beta s'}$$

since the series converges everywhere. Thus we conclude. □

Remark 9.9.1. Lemma 9.9.2 has an obvious modification for the case $0 < R < \infty$.

Some examples will illustrate the growth restrictions used above.

Example 9.9.2. Suppose $F : \mathbb{R}^n \to \mathbb{R}$ is a continuous function which satisfies

$$|F(y)| \leq C(a|y|^\gamma)^{n_\gamma} \sum_{k=0}^{\infty} \frac{a^k |y|^{\gamma k}}{(n_\gamma + k)!} = C\left\{ \mathrm{e}^{a|y|^\gamma} - \sum_{k=0}^{n_\gamma - 1} \frac{(a|y|^\gamma)^k}{k!} \right\}$$

with some constants $q > 0$, $C > 0$, some exponent γ, $0 < \gamma < \infty$, and $q = \gamma n_\gamma \geq d$.

Example 9.9.3. The continuous function $F : \mathbb{R}^n \to \mathbb{R}$ is supposed to satisfy for some constants $a, b > 0$, and some exponents $0 < \gamma < \infty$, $d \leq q < \infty$

$$|F(y)| \leq b|y|^q \cosh(a|y|^\gamma)\,.$$

In both cases we get for the radius R_γ of convergence, according to definition (9.9.5),

$$R_\gamma = \begin{cases} \infty & \text{for } 0 < \gamma < 1\,, \\ \text{finite} & \text{for } \gamma = 1\,, \\ 0 & \text{for } \gamma > 1\,. \end{cases}$$

Hence these lemmas can be conveniently applied for $0 < \gamma < 1$, i.e. if F is bounded at infinity by some exponential of the form

$$\mathrm{e}^{a|y|^\gamma}\,.$$

For the case of a linear exponential (i.e. $\gamma = 1$) more complications arise. This case is not yet worked out in detail.

For $q = 2$, the space E_q coincides with the usual Sobolev space $H^1 = H^1(\mathbb{R}^2; \mathbb{R}^n)$. It is well known that the Fourier transform can be used to characterise conveniently elements of L^p spaces by integrability conditions on their Fourier transforms only for $p = 2$. Relying on [Ref. 9.34, Theorem 5.3], this fact can be used to give a simple proof of the following inequalities:

$$\|u\|_r \leq \alpha(r)\|u\|_{1,2}\,, \quad u \in H^1\,, \quad \|u\|_{1,2}^2 = \|u\|_2^2 + \|\nabla u\|_2^2 \tag{9.9.7}$$

for all $r \geq 2$ with

$$\alpha(r) = 2^{-3(1/2-1/r)}\pi^{-1/2+3/2}(r-2)^{1/2-1/r} \leq \text{const.}r^{1/2}\,. \tag{9.9.8}$$

Notice that Theorem 9.2.3 implies in this case the weaker estimate

$$\|u\|_r \leq \text{const.}r\|u\|_{1,2}$$

Using the sharper estimate (9.9.7, 8) we can reconsider our examples and obtain easily for the radius of convergence:

$$R_\gamma = \begin{cases} \infty & \text{for } 0 < \gamma < 2\,, \\ \text{finite} & \text{for } \gamma = 2\,, \\ 0 & \text{for } \gamma > 2\,. \end{cases}$$

If Lemma 9.9.1 is applied to the function

$$F(u) = |y|^q \cosh(a|y|^\gamma), \quad a > 0, \quad q > 2, \quad 0 < \gamma \leq 1,$$

the following estimate results for all $u \in E_q$, $u \neq 0$, and all $\beta > 0$ with $A = (\beta a/2)(1 + q/2) < 1$ and $\gamma = 1$:

$$\int_{\mathbb{R}^2} |u|^q \left\{ e^{a\beta(|u|/\|\nabla u\|_2)^\gamma} + e^{-a\beta(|u|/\|\nabla u\|_2)^\gamma} \right\} dx \leq 2\|u\|_q^q (1 - A)^{-1} \quad (9.9.9)$$

while for $0 < \gamma < 1$ this integral is finite for all $\beta > 0$ since then $R_\gamma = \infty$. In the case $q = 2$ the same kind of inequality is obtained for all $0 < \gamma \leq 2$.

Inequality (9.9.9) can be considered as a *global version* of the *Moser-Trudinger inequality* [9.25, 35] which holds over bounded domains.

9.9.2 Solution of Some Two-Dimensional Vector Field Equations

First we list our hypotheses on the nonlinearity g:

(H_0) $g : \mathbb{R}^n \to \mathbb{R}^n$ is continuous, $g(0) = 0$, such that $g(y) = \operatorname{grad} G(y)$ for $y \neq 0$ for some potential G, and for some $\delta > 0 : g(x) \neq 0$ for all $0 < |y| < \delta$.

(H_1) $G : \mathbb{R}^n \to \mathbb{R}$ is continuous, of class C^1 in $\mathbb{R}^n \setminus \{0\}$, $G(0) = 0$, and $G(y) > 0$ somewhere.

(H_2) G admits a decomposition $G = G_+ - G_-$ into nonnegative continuous functions G_+, G_-, $G_\pm(0) = 0$, such that there exists an exponent $q \geq 2$ and a constant $b > 0$ with

$$b|y|^q \leq G_-(y) \quad \text{for all } y \in \mathbb{R}^n$$

and

$$G_+(y) = o(|y|^q) \quad \text{for } |y| \to 0 .$$

(H_3) $y \longmapsto |y|\,|g(y)|$ satisfies the growth restrictions as expressed by relations (9.9.4, 5) with $R = \infty$.

(H_3') $G(\cdot)$ satisfies the growth restrictions of assumption (H_3).

Remark 9.9.4. (a) Clearly assumptions (H_0), (H_1), and (H_3) imply assumption (H_3').

(b) Our assumptions imply in particular that $G_+(y) \leq C|y|^q$ near $y = 0$. If $0 \leq C < b$ then $G(y) < 0$ for sufficiently small $y \neq 0$ and our assumption about the behaviour of the nonlinearity g at $y = 0$ is a particular case of the assumption used in [9.13].

(c) According to our examples discussed in Sect. 9.9.1, the following behaviour is allowed for F, $F = G$ or $F(\cdot) = |\cdot|\,|g(\cdot)|$:

$$|F(y)| \leq C|y|^q e^{a|y|^\gamma} \quad \text{for } |y| \to \infty$$

with some constants $a, C > 0$, some exponents $q > 2$ and $0 < \gamma < 1$, or $q = 2$ and $0 < \gamma < 2$.

According to these assumptions we decide to look for solutions in the Banach space $E_q = E^1_q(\mathbb{R}^2;\mathbb{R}^n)$ where the exponent $q \ge 2$ is given by assumptions (H_2) and (H_3). We prepare for the main results of this section by two additional lemmas.

Lemma 9.9.3. *Under the assumptions* (H_0), (H_1) *and* (H_3) *we define the functional* V *on*

$$D(V) = \{u \in E_q \mid G(u(\cdot)) \in L^1(\mathbb{R}^2)\}$$

by

$$V(u) = \int G(u(x))\,dx\,. \tag{9.9.10}$$

At every $u \in D(V)$ V *has linear continuous Gâteaux derivatives* $V'(u;v)$ *in all directions* $v \in C_0^\infty(\mathbb{R}^2;\mathbb{R}^n)$ *given by*

$$V'(u;v) = \int \widehat{g}(u)v\,dx \equiv \langle \widehat{g}(u), v\rangle_2 \tag{9.9.11}$$

Proof. The standard versions of such a differentiability result assume some polynomial bounds for g and its potential G. So they do not apply directly. However if we take into account the basic inequalities from Theorem 9.2.3, it is not hard to show that the proof of Proposition 9.6.4 can be extended to the present case if we manage to control the contributions of the large values of u to $\langle \widehat{g}(u), v\rangle_2$. These contributions are

$$S_\lambda = \|\widehat{g}(u)v[\lambda \le |u|]\|_1\,, \quad \lambda \to \infty\,.$$

They are dominated by

$$\|\widehat{g}(u)\|_2\|v\|_\infty\|[\lambda \le |u|]\|_2\,.$$

By Lemma 9.9.6, $\widehat{g}(u)$ is square-integrable for all $u \in E_q$. Since $\|[\lambda \le |u|]\| \le \lambda^{-q}\|u\|_q^q$ holds for every $\lambda > 0$, we conclude $S_\lambda \to 0$ for $\lambda \to \infty$. This suffices to allow us to proceed as for Proposition 9.6.4. □

Lemma 9.9.4. *Suppose* g *satisfies assumptions* (H_0)–(H_2). *If for some* $u \in E_q$

$$\widehat{g}(u) = 0$$

holds in the sense of distributions, then the function u *vanishes:* $u = 0$.

Proof. By our assumptions on g we know $\widehat{g}(u) \in L^1_{\text{loc}}$ for every $u \in E_q$. Hence $\widehat{g}(u) = 0$ almost everywhere on $\mathbb{R}^2$. By our assumption about the zero of g, i.e.

$$g^{-1}(0) \subseteq \{0\} \cup \{y \in \mathbb{R}^n \mid |y| \ge \delta\}\,,$$

this equation can hold only if $|u|$ has a jump of height at least δ. But in Appendix E it is shown that weakly differentiable functions have not finite jumps. Thus $u = 0$ follows. □

Theorem 9.9.5. *If the assumptions* (H_0)–(H_3) *are satisfied by the nonlinearity* g, *then the global vector field equation in two variables* (9.9.2) *has a nontrivial solution in* E_q.

Proof. (a) By our assumptions the following minimisation problem is well defined:

$$I = \inf\{K(v) \mid v \in D(V), v \neq 0, V(v) \geq 0\}$$

and there exists a minimising sequence

$$f_j \in D(V), \quad f_j \neq 0, \quad V(f_j) \geq 0, \quad I = \lim_{j \to \infty} K(f_j). \tag{9.9.12}$$

Because of the covariance properties (9.9.1) we may assume in addition that

$$\|f_j\|_q = 1 \quad \text{for all } j \in \mathbb{N}. \tag{9.9.13}$$

By assumptions (H_2), (H_3) this normalisation together with the bounds for $\|\pi_A \widehat{G}_+(f_j)\|_1$ given in the proof of Lemma 9.9.2 implies as in the proof of Proposition 9.8.3 the hypothesis of the concentration Lemma 9.2.13. Hence by this lemma there is a sequence of translations $(a_j)_{j\in\mathbb{N}}$ and there is a subsequence $j(i)$, $i \in \mathbb{N}$, such that the translated sequence $u_i(\cdot) = f_{j(i)}(\cdot + a_{j(i)})$ has a nonvanishing limit $u \in E_q$ in the sense of Theorem 9.2.4, conditions (a)–(d). Clearly $\|\nabla u_i\|_2 = \|\nabla f_{j(i)}\|_2$ and $\|u_i\|_q = 1$ holds for all $i \in \mathbb{N}$.

(b) Inequality (9.9.6) of Lemma 9.9.1 implies for all $i \in \mathbb{N}$ that

$$\|\widehat{G}(u_i)\|_1 \leq C\|u_i\|_q^q \sum_{k=0}^{\infty} b_k \left[\frac{1}{2}\|\nabla u_i\|_2\right]^{2k/s} \leq C^1.$$

Since $\widehat{G}(u_i) \underset{i\to\infty}{\longrightarrow} \widehat{G}(u)$ almost everywhere on $\mathbb{R}^2$, Fatou's lemma proves $\widehat{G}(u) \in L^1(\mathbb{R}^2)$, i.e. $u \in D(V)$.

If $V(u) \geq 0$ were also known, the limit function u would be a nontrivial minimiser. In the case of scalar fields and an even potential it is easy to prove this (see later remark). Here, instead, it is shown directly that a suitably scaled version of u is a weak solution of our equation (without showing $V(u) \geq 0$ first).

(c) Take a fixed $v \in C_0^\infty(\mathbb{R}^2; \mathbb{R}^n)$ with compact support K and apply Lemma 9.9.2 to the sequences $(\pi_K u_i)_{i\in\mathbb{N}}$ and $(\pi_K u_i + v)_{i\in\mathbb{N}}$. Since $\pi_K \widehat{G}(u_i + v) = \widehat{G}(\pi_K u_i + v)$ this lemma implies

$$\widehat{G}(\pi_K u_i) \to \widehat{G}(\pi_K u) \quad \text{and} \quad \widehat{G}(\pi_K u_i + v) \to \widehat{G}(\pi_K u + v)$$

in $L^1(\mathbb{R}^2)$ for $i \to \infty$. Hence

$$\widehat{G}(\pi_K u_i + v) - \widehat{G}(\pi_K u_i) \underset{i\to\infty}{\longrightarrow} \widehat{G}(\pi_K u + v) - \widehat{G}(\pi_K u) \quad \text{in } L^1(\mathbb{R}^2). \tag{9.9.14}$$

However, $\widehat{G}(\pi_K u + v) - \widehat{G}(\pi_K u) = \widehat{G}(u+v) - \widehat{G}(u)$, and therefore (9.9.14) proves

$$V(u_i + v) - V(u_i) \underset{i\to\infty}{\longrightarrow} V(u+v) - V(v). \tag{9.9.15}$$

(d) In the next step we show that for any $v \in C_0^\infty$

$$V'(u;v) > 0 \quad \text{implies} \quad K'(u;v) = \langle \nabla u, \nabla v \rangle_2 \geq 0 \tag{9.9.16}$$

holds for the limit function u of (a).

According to Lemma 9.9.3 we have

$$V(u+tv) - V(u) = tV'(u;v) + o(t).$$

Hence, if $V'(u,v) > 0$, there is t_0 such that for all $0 < t < t_0$

$$V(u+tv) - V(u) > 0,$$

and therefore by (9.9.15) there is $i_0 = i_0(t)$ such that for all $i \geq i_0$

$$V(u_i + tv) - V(u_i) > 0.$$

Since by construction $V(u_i) \geq 0$ is known, it follows that $V(u_i + tv) > 0$ and therefore that

$$K(u_i + tv) \geq I \quad \text{for } i \geq i_0.$$

By weak convergence of $u_i \to u$ and (9.9.12) we deduce

$$I \leq \lim_{i\to\infty} K(u_i + tv) = I + tK'(u_i;v) + t^2(K(v).$$

Since this inequality holds for all $0 < t < t_0$ statement (9.9.16) follows.

(e) Since $u \neq 0$ is known, there is $v_0 \in C_0^\infty$ such that $K'(u;v_0) = -1$. Statement (9.9.16) implies

$$V'(u;v_0) = -\lambda \leq 0.$$

For arbitrary $v \in C_0^\infty$ and $t \in \mathbb{R}$ define $v(t) = tv_0 + K'(u;v)v_0 + v$; it follows that $K'(u;v(t)) = -t$ and thus $t > 0$ implies

$$0 \leq V'(u;v(t)) = V'(u;v) - \lambda K'(u;v) - t\lambda$$

or, in the limit $t \searrow 0$,

$$0 \leq V'(u;v) - \lambda K'(u;v).$$

This inequality also applies to $-v$ and thus proves

$$0 = V'(u;v) - \lambda K'(u;v) \quad \text{for all} \quad v \in C_0^\infty \tag{9.9.17}$$

with some $\lambda \geq 0$. If $\lambda = 0$, then by Lemma 9.9.3, $\widehat{g}(u) = 0$ in the sense of distributions and therefore by Lemma 9.9.4 $u = 0$. This contradiction proves $\lambda > 0$ and we may rescale the function according to

$$\overline{u} = u_\sigma\,, \quad \sigma = \sqrt{\lambda}\,,$$

to obtain a nontrivial weak solution of our equation:

$$V'(\overline{u}; v) = K'(\overline{u}, v) \quad \text{for all} \quad v \in C_0^\infty\,. \tag{9.9.18}$$

□

Remark 9.9.5. In the case of scalar fields and an even potential G, spherically symmetric rearrangement of functions can be used. Thus we can assume that the elements of a minimising sequence $(f_i)_{i\in\mathbb{N}}$ according to (9.9.12) and (9.9.13) are spherically symmetric nonincreasing functions in E_q. As we did earlier, we can show for such functions the following decay estimate:

$$\begin{aligned} |f_i(x)| \le \|f_i\| h(x)\,, \quad h(x) = \min\{1, |x|^{-2/q}\}C\,, \\ \|f_i\| = \|f_i\|_q + \|\nabla f_i\|_2 \end{aligned} \tag{9.9.19}$$

Thus, bounded sequences in E_q are uniformly bounded on $\mathbb{R}^2$, and hence the condition (IUD) of Theorem 9.2.5 is easily realised for such a minimising sequence.

If assumption (H_2) is supplemented by the condition

$$G_+(y) = o(|y|^q) \quad \text{for} \quad |y| \to 0 \tag{9.9.20}$$

then as in Sect. 9.8 the concentration condition (C) of Theorem 9.2.5 follows from (9.9.19). Theorem 9.2.5 now implies

$$\widehat{G}_+(u_i) \to \widehat{G}_+(u) \quad \text{in } L^1 \text{ for } \ i \to \infty$$

for the sequence (u_i) and its limit u acccording to step (1) of the proof. Therefore, by Fatou's lemma we get

$$V-(u) = \int \widehat{G}_-(u)\,dx \le \liminf_{i\to\infty} \int \widehat{G}_-(u_i)\,dx \le \liminf_{i\to\infty} \int \widehat{G}_+(u_i)\,dx = V_+(u)\,.$$

Hence $V(u) = V_+(u) - V_-(u) \ge 0$, and thus $K(u) \ge I$. Since by weak convergence of the sequence $(u_i)_{i\in\mathbb{N}}$ we also know that $K(u) \le I$, this limit u is a nontrivial minimiser of the variational problem (9.9.3).

Now, without step (c), the implication (9.9.16) follows. Therefore, as above, we obtain a weak solution of (9.9.2).

Recall from Lemma 9.3.1, that the basic step of the elliptic regularity theory applies and proves for a weak solution u of (9.9.2) that it actually belongs to $W^{2,2}_{\mathrm{loc}} \cap E_q$ if $\widehat{g}(u) \in L^2_{\mathrm{loc}}$ is known. Then such a solution solves (9.9.2) in the sense of equality almost everywhere. The following lemma provides the necessary information.

Lemma 9.9.6. *If the continuous function g satisfies hypothesis (H_3) then for any $u \in E^1_q(\mathbb{R}^2; \mathbb{R}^n)$ the function $\widehat{g}(u)$ is square-integrable on $\mathbb{R}^2$.*

Proof. By (H_3) one has

$$\int_{\mathbb{R}^2} |\widehat{g}(u)|^2 \, dx \leq \sum_{k,l=0}^{\infty} a_k a_l \, \| \, |u|^{2(q-1)+2k/s+2l/s} \|_1 \, .$$

If $q \geq 2$ then $q' = 2(q-1) \geq q$ so that we can apply the estimates of Theorem 9.2.3 for $r'_k = q' + 2k$, i.e. $r'_{2k} = 2(r_k - 1)$, where $r_k = r(k) = q + 2k$. Hence, with $\alpha' = 1 + q'/2 = q$, we have

$$\|u\|_{r'_{2k}}^{r'_{2k}} \leq (\alpha')_{2k}^2 \|u\|_{q'}^{q'} \left(\frac{1}{2}\|\nabla u\|_2\right)^{2(2k)}$$

and thus by Hölder's inequality and Theorem 9.2.3

$$\begin{aligned} \| \, |u|^{2(q-1)+2k/s+2l/s} \|_1 &\leq \| \, |u|^{q-1+2k/s} \|_2 \, \| \, |u|^{q-1+2l/s} \|_2 \\ &\leq \|u\|_{q'}^{q'/s} \|u\|_{r'_{2k}}^{r'_{2k}/s} \|u\|_{r'_{2l}}^{r'_{2l}/s} \leq \|u\|_{q'}^{q'} (\alpha')_{2k}^{1/s} (\alpha')_{2l}^{1/s} A^{k+l} \end{aligned}$$

where $A = (\frac{1}{2}\|\nabla u\|_2)^{2/s}$. This proves finally

$$\|\widehat{g}(u)\|_2 \leq \|u\|_{2(q-1)}^{q-1} \sum_{k=0}^{\infty} a_k (\alpha')_{2k}^{1/s} A^k < \infty$$

since $\lim_{k \to \infty} \sup |a_k (\alpha)_k^{2/s}|^{1/k} = 0$ implies $\lim_{k \to \infty} \sup |a_k (\alpha')_{2k}^{1/s}|^{1/k} = 0$, so that the above series converges for every $A \in \mathbb{C}$. □

Remark 9.9.6. In [9.13] Brezis and Lieb take as a space of a priori solutions of (9.9.2) the space E:

$$E = \{u \in L^1_{\text{loc}}(\mathbb{R}^2; \mathbb{R}^n) \mid \nabla u \in L^2(\mathbb{R}^2), |[|u| > \lambda]| < \infty \text{ for all } \lambda > 0\} \, .$$

Accordingly they normalise their minimising sequence $(f_j)_{j \in \mathbb{N}}$ for problem (9.9.3) by

$$|[|f_j| > \varepsilon]| = 1 \quad \text{for all } j \in \mathbb{N}$$

where $\varepsilon > 0$ is given by the assumption that

$$G(y) < 0 \quad \text{for } 0 < |y| < \varepsilon \, .$$

Thus the concentration by translation lemma easily applies.

Now, in its main steps, the strategy for the rest of their proof is that given above. However, there are some complications connected with differentiability properties of the functional V due to the quite general behaviour of g near $y = 0$, and there are some simplifications due to the assumption of a polynomial bound for g.

9.10 Conclusion and Comments

9.10.1 Conclusion

In this chapter we considered equations of the form

$$-\Delta u(x) = g(x, u(x)),$$
$$u : \Omega \to \mathbb{R}^n, \quad \Omega \subseteq \mathbb{R}^d \text{ a domain}, \quad d \geq 2,\ n \geq 1. \tag{9.10.1}$$

Besides the standard assumptions (H_0)–(H_3) we used the nontrivial restriction of a scale-covariance for the nonlinearity g, i.e. the existence of a representation of $\mathbb{R}_+$ on the space E of a priori solutions by continuous linear operators T_σ, $\sigma > 0$, which leave $\mathcal{D} = \mathcal{D}(\Omega, \mathbb{R}^n)$ invariant such that the equations

$$\begin{aligned} -\Delta(T_\sigma u) &= \sigma^r(-\Delta u) \circ T_\sigma^{-1}, \\ g(\cdot, T_\sigma u) &= \sigma^s g(\cdot, u) \circ T_\sigma^{-1}, \end{aligned} \tag{9.10.2}$$

hold in the sense of distributions for all $\sigma > 0$ with $\gamma = r/s \in (0,1)$. Since the existence of linear continuous Gâteaux derivatives in all directions of $\mathcal{D}$ has been established in sufficient generality by Proposition 9.6.4, the problem of proving the existence of weak solutions of (9.10.1) is reduced to proving the existence of a minimiser for the associated constrained minimisation problem. This is done in Sects. 9.4–9 under various assumptions. These assumptions also ensure that the regularity results of Sect. 9.3 apply so that we actually get "classical" solutions of (9.10.1) by solving the associated variational problem.

In order to illustrate the range of these results in more concrete terms we discuss some simple examples which are covered by our general assumptions.

Whenever the function Q satisfies conditions (9.8.4) then by the results of Sect. 9.8 the equation

$$-\Delta u(x) = Q(x)|u|^{p-2}(x)u(x) \tag{9.10.3}$$

has, for $2 < p < 2^*$, a nontrivial solution in $E = E^{1,2}(\mathbb{R}^d; \mathbb{R}^n)$, $d \geq 3$, $n \geq 1$. The same considerations apply to spherically symmetric equations of the form

$$(-\Delta u)(x) + (A_r^* A_r u)(x) = Q(|x|)|u(x)|^{p-2}u(x) \tag{9.10.4}$$

and prove the existence of a spherically symmetric solution $u : \mathbb{R}^d \to \mathbb{R}^n$ if the operator A_r and the function Q have the properties specified in (9.9.7–9) for $2 < p \leq 2^*$. Relying on a different method (the "Mountain-Pass Lemma", cf. Sect. 9.10.2) this type of equation has also been considered in [9.36] for $n = 1$, $2 < p < 2^*$, and $Q \geq 0$.

The results of Sect. 9.8.2 imply in particular that the following type of spherically symmetric vector field equations,

$$-\Delta u(x) = g_0(|u(x)|)u(x),$$
$$u \in E^{1,2}(\mathbb{R}^d; \mathbb{R}^n), \quad 3 \leq d, \quad 1 \leq n, \tag{9.10.5}$$
$$g_0 : \mathbb{R}_+ \to \mathbb{R} \text{ continuous}, \quad g_0(t) > 0 \text{ somewhere},$$

have solutions if, for instance, the following assumptions about the behaviour of $g_0(t)$ for $t \to 0$ and for $t \to \infty$ are made:

(a) $$g_0(t) = at^\alpha \quad \text{for } t \to 0\,, \quad g_0(t) = bt^\beta \quad \text{for } t \to \infty$$
with $a, b > 0$ and $0 < \beta < 2^* - 2 < \alpha$;

(b) $$g_0(t) = -at^\alpha \quad \text{for } t \to 0\,, \quad g_0(t) = -bt^\beta \quad \text{for } t \to \infty$$
with $a, b > 0$ and $\alpha, \beta > -1$; or

(c) any combination of the behaviour of g_0 at $t = 0$ and at $t = +\infty$ as specified in (a) and (b).

Many other explicit examples are presented in [9.1] and [9.6].

Finally, we mention a simple class of examples of nonlinear field equations in two variables which have a nontrivial solution according to our results in Sect. 9.9. As a potential G we take, for instance,

$$G(y) = |y|^q\{Ae^{a|y|^\alpha} - \beta e^{-b|y|^\beta} - C\}\,, \quad y \in \mathbb{R}^n \tag{9.10.6}$$

with $A, B, a, b > 0$, $C \geq 1$ and $\alpha, \beta \in (0,1)$ for $2 < q < \infty$ and $\alpha, \beta \in (0,2)$ for $q = 2$.

Then, by Theorem 9.9.5, the equation

$$-\Delta u = g(u)\,, \quad g(y) = \operatorname{grad} G(y)\,, \quad y \neq 0\,, \quad g(0) = 0\,, \tag{9.10.6'}$$

has a nontrivial solution u in $E = E_q^1(\mathbb{R}^2; \mathbb{R}^n)$. For the proof we simply observe that with the choice

$$G_-(y) = |y|^q\{C + Be^{-b|y|^\beta}\}\,, \quad G_+(y) = |y|^q Ae^{a|y|^\alpha}$$

the hypotheses (H_0)–(H_3') are satisfied.

Throughout Chap. 9 we have used the functional V of "potential energy" to define the constraint while the functional K of "kinetic energy" was minimised on a level surface of V. From the point of view of Sect. 8.4 however, it would have been much more natural to take the functional K to define the constraint since the level surfaces of K in $E^{1,2}(\mathbb{R}^d; \mathbb{R}^n)$ are "good" in the sense of Sect. 8.4 as they are "sphere-like" (Theorem 8.13), while on the other hand not much is known about the level surfaces of V. Nevertheless, the results of this chapter show that working with the level surface $V^{-1}(1)$ is quite efficient in proving the existence of a weak solution for a fairly general class of nonlinearities.

In the "dual version" the constraint is defined by the functional K. For $E = E^{1,2}(\mathbb{R}^d; \mathbb{R}^n)$, $d \geq 3$, the one-sphere $S_1(E)$ of E is just the level surface $\{u \in E \,|\, K(u) = 1/2\}$ of K. The associated variational problem in the dual version then is the constrained maximisation problem

$$J = \sup\{V(f) \,|\, f \in D(V) \cap S_1(E)\}\,.$$

Thus it corresponds just to the reformulation of the constrained minimisation problem given in Lemma 9.6.2 with $S = J^\gamma$.

9.10.2 Generalisations

Here we indicate some immediate generalisations. These generalisations are concerned with more general second-order differential operators than the Laplacian on $\mathbb{R}^d$.

Suppose that for $d \geq 3$ and $n \geq 1$, F is a real function on $\mathbb{R}^n \times \mathbb{R}^{nd}$ of class C^1 which has the following properties: there is some $p \in (1, d)$ and there are some constants $0 < a \leq A < \infty$ such that

$$\begin{aligned} &\text{(i)} && F(y, \sigma z) = \sigma^p F(y, z), \\ &\text{(ii)} && a|z|^p \leq F(y, z) \leq A|z|^p \end{aligned} \tag{9.10.7}$$

hold for all $y \in \mathbb{R}^n$, all $z \in \mathbb{R}^{nd}$, and all $\sigma > 0$. Then a functional K on the space

$$E = E^{1,p}(\mathbb{R}^d; \mathbb{R}^n)$$

of Sect. 9.2 is well defined by

$$K(u) = \int_{\mathbb{R}^d} F(u(x), \nabla u(x))\, dx\,. \tag{9.10.8}$$

It satisfies

$$a\|\nabla u\|_p^p \leq K(u) \leq A\|\nabla u\|_p^p \quad \text{and} \quad K(u_\sigma) = \sigma^{p-d} K(u) \tag{9.10.9}$$

for all $u \in E$ and all $\sigma > 0$, $u_\sigma(\cdot) = u(\sigma \cdot)$. The functional V is defined as usual by a potential G on $D(V) \subseteq E$. Then the associated constrained minimisation problem is

$$I = \inf\{K(u) \,|\, u \in D(V), V(u) = 1\} \tag{9.10.10}$$

and we assume that K and V have linear continuous Gâteaux derivatives in all directions of $\mathcal{D}$.

Since minimising sequences for this problem are bounded in E, there are a minimising sequence $(u_i)_{i \in \mathbb{N}}$ and an element $u \in E$ which are related by conditions (a)–(d) of Theorem 9.2.2, i.e. in particular $u_i \to u$ weakly in E and pointwise almost everywhere.

If $I > 0$ and if the following two continuity conditions,

$$K(u_i + v) - K(u_i) \longrightarrow K(u + v) - K(u), \tag{9.10.11}$$

$$i \to \infty$$

$$V(u_i + v) - V(u_i) \longrightarrow V(u + v) - V(u), \tag{9.10.12}$$

hold for every $v \in \mathcal{D}$ then it follows as in Sect. 9.6 that

$$K'(u; v) = \gamma I V'(u; v), \quad v \in \mathcal{D}, \tag{9.10.13}$$

i.e. a suitably scaled version $\overline{u} = u_\sigma$ of u is a weak solution of

$$\begin{aligned} -\operatorname{div}(a(u, \nabla u)) + b(u, \nabla u) &= g(u), \\ u \in E^{1,p}(\mathbb{R}^d; \mathbb{R}^n), \quad d \geq 3, \quad &n \geq 1, \end{aligned} \tag{9.10.14}$$

where $a(y, z) = \frac{\partial F}{\partial z}(y, z)$, $b(y, z) = \frac{\partial F}{\partial y}(y, z)$, $g(y) = \frac{\partial G}{\partial y}(y)$.

However it is not yet known that $u \neq 0$! If, for instance by Lemma 9.2.11 and its corollary, $u \neq 0$ can be proved, the existence of a nontrivial weak solution $\overline{u} = u_\sigma$ follows.

Notice that the function u need not be a minimiser of the above constrained minimisation problem! As in Sects. 9.7, 8 u can be proved to be a minimiser if, for instance, conditions (9.10.11) and (9.10.12) can be extended to all $v \in D(V)$, because then it follows that $K(u) = IV(u)$.

Clearly the new and nontrivial problem here is to prove the continuity property (9.10.11) for the functional K under appropriate restrictions on F. Up to now this has been known only for a relatively simple class of examples for F.

Another direction for generalisations is connected with the choice of the space of a priori solutions. It is well known that different kinds of problems in the theory of (nonlinear) partial differential equations usually need different kinds of basic function spaces as spaces of a priori solutions for a successful treatment [9.5, 7, 8, 25, 37, 38]. (Recall that the various function spaces were introduced and investigated in order to be able to study such problems.) In order to limit the amount of necessary preparation in all of Chap. 9 we used only the spaces

$$E^{1,p}(\Omega;\mathbb{R}^n), \quad \Omega \subseteq \mathbb{R}^d, \quad d \geq 3, \quad 1 < p < d, \quad n \geq 1$$

and

$$E_q^1(\Omega;\mathbb{R}^n), \quad \Omega \subseteq \mathbb{R}^2, \quad 2 \leq q < \infty, \quad n \geq 1,$$

or the corresponding Sobolev spaces as subspaces. And we learned that for many interesting and important semilinear elliptic equations we could find solutions in these spaces.

Clearly one would expect this too for the kind of equations considered in this chapter, but with a different kind of nonlinearity a different kind of function space as a space of a priori solutions is needed, for instance an appropriate Sobolev-Orlicz space [9.3] if the nonlinearity g is allowed to grow more rapidly than a polynomial. But it seems that at the moment not much is known in this direction.

9.10.3 Comments

The following kinds of variational problems have been the concern of this chapter:

(1) variational problems with "lack of compactness";
(2) such problems for "indefinite functionals";
(3) such problems for functionals which are not Fréchet-differentiable on their natural domain of definition.

"Lack of compactness" means the following: for a domain $\Omega \subseteq \mathbb{R}^d$ denote by $E(\Omega)$ the space of functions in which the problem at hand has to be considered naturally. Then for $E(\Omega)$ there are no compact embeddings available because

(i) Ω is bounded and in the functionals which have to be considered the critical Sobolev exponent occurs (as in the problems of Sect. 9.5);
(ii) Ω equals $\mathbb{R}^d$ (as in Sects. 9.6–9);
(iii) Ω equals $\mathbb{R}^d$ and the critical Sobolev exponent is involved (as in Sect. 9.4).

"Indefinite functionals" are functionals which are not bounded on their natural domain of definition, either from above or from below. Such functionals do not have critical points of "minimum" or "maximum type". So more sophisticated methods than minimisation or maximisation have to be used in order to detect critical points. Usually these problems also show lack of compactness, in particular if they are considered over $\Omega = \mathbb{R}^d$.

In recent years there has been considerable progress in many aspects of these problems. Since our treatment in this chapter starts on a relatively elementary level it is clear that we could explain only some of the recent solutions of such problems. Many other results and some important methods had to be omitted. Therefore, we want to indicate briefly some of these methods and maybe, more importantly to make some suggestions for further reading.

H. Brezis [9.39] gives a detailed account with many applications of variational problems with lack of compactness. Our treatment of variational problems for indefinite functionals relies on the simplifying assumption of a general scale-covariance (which in many important applications is known) but allows functionals which are not Fréchet-differentiable on their natural domain of definition to be treated. So we explained the "constrained minimisation method" for functionals which have Gâteaux derivatives in sufficiently many directions at all points where they are defined (Sects. 9.6–9.9). There are other methods for this kind of problems which do not rely on scale-covariance. A good reference for this is P. H. Rabinowitz' article [9.42]. The main abstract tool of these methods is the *"Mountain-Pass Lemma"* of Ambrosetti-Rabinowitz [9.39–42]. The use of this lemma is usually quite involved and therefore in applications to concrete problems it is sometimes less powerful than the constrained minimisation method. This can, for instance, be seen, if one compares the results for Examples 9.8.1, 9.8.2 with those [9.36] (see also (9.10.3)).

It is quite obvious that in the context of Sects. 9.6–9.9 we can in general not expect to have uniqueness of the solution because a 'large' symmetry group is involved (see Sect. 9.7). But in Sect. 9.8.2, where nonnegative spherically symmetric solutions have been determined the question of uniqueness seems to be more sensitive within this class of solutions. And indeed some uniqueness results for this class are known [9.43]. For some further results we refer the reader to [9.1, 46]. The existence of symmetric solution is usually proved by studying the associated radial equation and using methods from the theory of ordinary differential equations.

On the other hand, it is important and interesting to know about the existence of many solutions. In the case of a bounded domain Ω of $\mathbb{R}^d$ we saw in Sect. 8.4 how the Ljusternik-Schnirelman theory of critical points in

Browder's formulation can be used to prove the existence of infinitely many eigensolutions. But clearly it is not at all obvious how to extend these arguments to the case of equations over the whole space $\mathbb{R}^d$ due to the reasons mentioned at the beginning of this section.

Nevertheless, *H. Berestycki* and *P. L. Lions* in [Ref. 9.1, part II] and *M. Struwe* in [9.18.19] succeeded in adapting the above mentioned part of critical point theory in such a way that if applied to the case of global scalar field equations their results allow the existence of infinitely many distinct spherically symmetric positive classical solutions to be proved.

M. Struwe's approach [9.18.19] provides a critical point theory for functionals which are only Gâteaux-differentiable in sufficiently many directions at all points where they are defined naturally but which are not necessarily Fréchet-differentiable like the functionals in Sect. 8.4. Adopted to this weaker notion of differentiability, a *"generalised Palais-Smale condition"* is formulated and applied successfully as in Sect. 8.4.

For the case of global vector-field equations the "dual version" mentioned in Sect. 10.2 might be a good starting print to investigate the existence of infinitely many solutions in such a frame. But this has not yet been done.

9.11 Complementary Remarks

If we compare the results and strategy of this section with those of the corresponding preceding sections we notice an important difference.

The results of Sect. 8.4 rely in an essential way on the use of the Palais-Smale condition (PS) (part B of Sect. 8.3) or the Smale condition (S) (Sect. 7.2) which in turn are based on the existence of appropriate compact embeddings for the space of a priori solutions E. In the present section we discuss problems in which critical Sobolev exponents are involved and the corresonding embedding $E = H_0^1(\Omega) \to L^p(\Omega)$, $p = 2^*$, is, as we know, not compact. Thus the most important aspect of the results of Brezis and Nirenberg is that they have shown how to overcome this *lack of compactness*. Naturally this was the starting point of further investigations. Accordingly we discuss briefly some of the works this paper by Brezis and Nirenberg has stimulated.

As slight generalisation of problem (5.1) is the problem

$$\begin{aligned} -\Delta u - \lambda u &= u|u|^{2^*-2} \quad \text{in } \Omega \\ u &= 0 \quad \text{on } \partial\Omega\,. \end{aligned} \tag{9.11.1}$$

Weak solutions of this boundary value problem can be obtained as critical points of the functional

$$I_\lambda(u) = \frac{1}{2}\int_\Omega \{|\nabla u|^2 - \lambda u^2\}\,dx - \frac{1}{2^*}\int_\Omega |u|^{2^*}\,dx \tag{9.11.2}$$

on $E = H_0^1(\Omega)$. This functional is related to the functional Q_λ of definition (5.2) by the equation

$$I_\lambda(u) = \frac{1}{2}\|u\|_{2^*}^2 Q_\lambda(u) = \frac{1}{2^{2^*}}\|u\|_{2^*}^{2^*}$$

so that an elementary argument shows that

$$Q_\lambda(u) < S \quad \text{iff} \quad I_\lambda(u) < \frac{1}{d}S^{d/2}\,.$$

Thus if we look at the proof of Theorem 5.2 we can read off a proof of the following "local PS-type condition":

If for $\lambda > 0$ one knows for a sequence u_j, $j \in \mathbb{N}$, in $E = H_0^1(\Omega)$ that

$$\begin{aligned} I_\lambda(u_j) &\underset{j\to\infty}{\longrightarrow} C < \frac{1}{d}S^{d/2} \quad \text{and} \\ I'_\lambda(u_j) &\underset{j\to\infty}{\longrightarrow} 0 \quad \text{strongly in } E'\,, \end{aligned} \tag{9.11.3}$$

then there is a subsequence which converges strongly in E.

In this form another proof is presented in [9.45]. There it is also shown that for $C = (1/d)S^{d/2}$ there exists a sequence in E satisfying the above local PS-assumption which is not relatively compact. What is then the origin of this "noncompactness" for $C \geq (1/d)S^{d/2}$?

This question has found a fairly comprehensive answer in [9.46], where it is shown that apart from "jumps of the topological type of the admissible functions" compactness is preserved globally, that is for all $C \in \mathbb{R}$ (Proposition 2.1 of [9.46]).

Proposition. *Let $d \geq 3$, $\lambda \in \mathbb{R}$ be given and suppose that a sequence u_j in $E = H_0^1(\Omega)$ satisfies*

$$I_\lambda(u_j) \leq C\,, \quad I'_\lambda(u_j) \to 0\,, \quad j \to \infty\,,\ \textit{strongly in } E'\,.$$

Then there exists a solution u^0 of (9.11.1), a finite number k of solutions $u^1, \ldots, u^k$ of the equation

$$-\Delta u = u|u|^{2^*-2} \quad \textit{in } \mathbb{R}^d\,, \quad u(x) \to 0 \quad \textit{for } |x| \to \infty \tag{9.11.4}$$

as described in Sect. 9.4, a sequence of points $x_j^1, \ldots, x_j^k \in \mathbb{R}^d$ and a sequence of radii r_j^i, $r_j^i > 0$ such that for some subsequence for $j \to \infty$

$$\begin{aligned} u_j^0 &= u_j \to u^0 \quad \textit{weakly in } H_0^1(\Omega)\,, \\ u_j^i &= (u_j^{i-1} - u^{i-1})_{r_j^i, x_j^i} \to u^i \quad \textit{weakly in } H_0^1(\mathbb{R}^d)\,, \end{aligned}$$

for $i = 1, \ldots, k$,

$$\|u_j\|_E^2 \to \sum_{i=0}^{k} \|u^i\|_E^2\,,$$

$$I_\lambda(u_j) \to I_\lambda(u^0) + \sum_{i=1}^{k} I_\lambda^*(u^i),$$

where $u_{r,y}(x) \coloneqq r^{(d-2)/2} u(r(x-y))$ *and* $I_\lambda^*(u) = \frac{1}{2}\int_{\mathbb{R}^n} |\nabla u|^2\, dx - \frac{1}{2^*}\int_{\mathbb{R}^n} |u|^{2^*}\, dx$.

An essential ingredient of the proof is the rescaling argument.

With this improved picture of how compactness can fail further results on the existence of (infinitely many) solutions of equation (9.11.1) follow.

To describe these, denote by $\lambda_j = \lambda_j(\Omega)$, $j \in \mathbb{N}$, the sequence of eigenvalues of $-\Delta : H_0^1(\Omega) \to H_0^1(\Omega)'$.

In reference [9.45] it is shown that from any eigenvalue λ_j there exists a "bifurcation" and there exists a number $\lambda_j^* < \lambda_j$ such that for all λ in (λ_j^*, λ_j) the number of nontrivial solutions of equation (9.11.1) is at least twice the multiplicity of λ_j. This is true for any $d \geq 3$. A further improvement of the original result (Theorem 5.2) of Brezis-Nirenberg is that for $d \geq 4$ equation (9.11.1) as a nontrivial solution for all $\lambda > 0$ ([9.47]).

A simplified proof and a slight extension of this result based on "dual variational methods" in combination with the Mountain Pass Lemma was given in 1986 by Ambrosetti and Struwe [9.49].

Finally in reference [9.48] the existence of infinitely many solutions of equation (9.11.1) has been proved; the more precise statement is that for $d \geq 6$ and all $\lambda \in (0, \lambda_1)$ there are at least two pairs of nontrivial solutions, while for $d \geq 7$, if Ω is some open ball, for all $\lambda \in (0, \lambda_1)$ equation (9.11.1), for any $k \in \mathbb{N}$, always has a pair of radial solutions $u_k^\pm$ with exactly k nodes in $(0, R]$ and $u_k^-(0) < 0 < u_k^-(0)$.

10. Thomas-Fermi Theory

10.1 General Remarks

In this chapter we wish to study those problems from Thomas-Fermi theory whose solutions depend almost exclusively on the methods of the calculus of variations. Questions of validity and physical interpretation are discussed by Thirring [10.1] and Lieb and Simon [10.2] and in references quoted therein. Justification for statements which we do not prove below can be found in [10.1, 3].

The starting point of Thomas-Fermi (TF) theory is the problem of determining the ground-state energy of a quantum mechanical many-electron system. One tries to solve this linear problem by minimising a certain nonlinear functional $\mathcal{E}$ approximating the quantum mechanical ground-state energy subject to a constraint. This so-called *Thomas-Fermi energy functional* $\mathcal{E}$ is defined as

$$\mathcal{E}(\rho) \equiv \mathcal{E}(\rho, V) = K(\rho) - A_V(\rho) + R(\rho)\,, \tag{10.1.1}$$

$$K(\rho) = \frac{3}{5}\int \rho(x)^{5/3}\,dx\,,$$

$$A_V(\rho) = \int V(x)\rho(x)\,d^3x\,,$$

$$R(\rho) = \frac{1}{2}\int \rho(x)\rho(y)|x-y|^{-1}\,d^3x\,d^3y\,, \tag{10.1.1'}$$

and the constraint is

$$\int \rho\, d^3x = N\,, \quad \rho > 0\,.$$

In this chapter we will assume $\int = \int_{\mathbb{R}^3}$. (We have actually chosen an appropriate normalisation for the form (10.1) of $\mathcal{E}$ by utilising the scaling property.)

Many people tried in vain for a long time to solve this minimisation problem in accordance with the "old" calculus of variations. They constructed for this purpose the equations characterising the stationary value of this functional and then tried to solve them. These equations, known as the *Thomas-Fermi equations*, are given as

$$\rho^{2/3} = \max\{\phi_\rho - \phi_0, 0\}\,,$$

$$\phi_\rho = V - W * \rho\,, \quad W(x) = \frac{1}{|x|}\,,$$

$$\int \rho\, d^3x = N\,, \tag{10.1.2}$$

where ϕ_0 is a certain constant. Numerical calculations indicated that these equations could be solved uniquely; however, a direct proof of solvability was obtained by Brezis only after the variational solution by Lieb, Simon, and Thirring.

In 1973 *Lieb* and *Simon* solved the TF equations by minimising the TF energy functional and by showing that the Euler equations for this problem are the TF equations.

The form of the term $K(\rho)$ for the kinetic energy in TF theory suggests to formulate this problem in the Banach space $X = L^{5/3}(\mathbb{R}^3)$.

The interpretation of the functions $\rho : \mathbb{R}^3 \to \mathbb{R}$ as densities and the constraint $\int \rho\, d^3x = N < +\infty$ require the following domain of definition M for the functional:

$$M = \{f \in L^{5/3}(\mathbb{R}^3) \,|\, f \geq 0,\ \|f\|_1 < +\infty\}\,. \tag{10.1.3}$$

Thus M is the convex cone of the nonnegative functions in $L^{5/3} \cap L^1$. The fact that M does not possess an inner point in $L^{5/3} \cap L^1$ gives rise to complications.

For physical applications V represents the potential due to the attractive forces of the nuclei. For nuclei at the positions $x_1, \ldots, x_k \in \mathbb{R}^3$ with nuclear charge $z_j > 0$, V has the explicit form

$$V(x) = \sum_{j=1}^{k} \frac{z_j}{|x - x_j|}\,. \tag{10.1.4}$$

However, we wish to underline the fact that most of our arguments remain valid for a general class of potentials.

A *quantum mechanical N-body system* is called *stable* if and only if the corresponding Hamilton operator H_N satisfies the bound $H_N \geq cN\mathbb{1}$ where $c > -\infty$. Applying TF theory, Lieb and Thirring have considerably simplified the original proof for the stability of atoms and matter (as put forward) by Dyson and Lenard. They also succeeded in improving the numerical value of the lower bound by a factor of 10^{-14}. Here the Fermi statistics of the electrons plays a crucial role. Correction terms for the TF energy functional were introduced by Dirac in 1930 and by von Weizsäcker in 1935. The corresponding functionals are

$$\mathcal{E}^{\mathrm{TFD}}(\rho) = \mathcal{E}(\rho) - c\int \rho(x)^{4/3}\, dx\,, \quad \mathcal{E}^{\mathrm{TFW}}(\rho) = \mathcal{E}(\rho) + \delta \int [\nabla \rho^{1/2}(x)]^2]\, dx\,.$$

Recently Benguria, Brezis, Lieb and others started the mathematical investigation of these functionals. For a review of these methods and results, see [10.4].

Before approaching the minimisation problem it will be useful to recall some results from the theory of L^p spaces which we would like to summarise in the following short section. This summary has been written as a preparation for the application of Theorem 1.2.5.

10.2 Some Results from the Theory of L^p Spaces $(1 \leq p \leq \infty)$

Let $L^p = L^p(\mathbb{R}^n)$ be, as usual, the vector space of all (equivalence classes of) Lebesgue-measurable real-valued functions on $\mathbb{R}^n$ which are integrable for the pth power and for which

$$\|f\|_p = \left[\int_{\mathbb{R}^n} |f(x)|^p \, dx \right]^{1/p}$$

is finite ($p < +\infty$). $L^\infty = L^\infty(\mathbb{R}^n)$ is the vector space of all (equivalence classes of) real-valued Lebesgue-measurable functions on $\mathbb{R}^n$ which are bounded almost everywhere, that is to say

$$\|f\|_\infty = \operatorname*{ess\,sup}_{x \in \mathbb{R}^n} |f(x)| < +\infty .$$

We begin with some important inequalities. A good reference for further reading is [10.5].

Lemma 10.2.1. *For $1 \leq q_j \leq +\infty$ the following holds:*

(a) *If $q_1 < q_2$ and $q_1 < p < q_2$ then*

$$\|f\|_p \leq \|f\|_{q_1}^{\alpha} \|f\|_{q_2}^{1-\alpha} ,$$

where $0 < \alpha < 1$ is defined by

$$\frac{1}{p} = \frac{\alpha}{q_1} + \frac{1-\alpha}{q_2} ,$$

and therefore

$$L^{q_1} \cap L^{q_2} = \bigcap_{q_1 \leq p \leq q_2} L^p .$$

(b) *From $f_j \in L^{q_j}$, $j = 1, 2$, it follows that $f_1 * f_2 \in L^{q_3}$ where*

$$1 + \frac{1}{q_3} = \frac{1}{q_1} + \frac{1}{q_2}$$

and

$$\|f_1 * f_2\|_{q_3} \leq \|f_1\|_{q_1} \|f_2\|_{q_2}$$

(Young's inequality).

Proof. Part (a) can be deduced easily from Hölder's inequality

$$\|f_1 f_2\|_1 \le \|f_1\|_{q_1} \|f_2\|_{q_2}, \quad \frac{1}{q_1} + \frac{1}{q_2} = 1;$$

for the proof of (b) see [10.5]. □

The following lemma exhibits clearly why we prefer to work in the Banach space $L^{5/3}$ for TF theory instead of L^1 or $L^1 \cap L^{5/3}$.

Lemma 10.2.2. *For $1 < p < \infty$ the following holds:*

(a) *L^p is a reflexive Banach space and the dual space of L^p is (isomorphic to) L^q with $1/q + 1/p = 1$.*
(b) *Bounded sets in L^p are weakly relatively compact. Every weakly countably compact subset of L^p is weakly compact and weakly sequentially compact.*

Lemma 10.2.3. *For $1 < p < \infty$ it follows that*

(a) *$f \to \|f\|_p$ is a weakly lower semicontinuous function on L^p.*
(b) *For $1 \le q < \infty$ the function $f \to \|f\|_q$ on $L^p \cap L^q$ is lower semicontinuous with respect to the weak topology of L^p; consequently, sets of the form $\{f \in L^p \mid \|f\|_q \le R < +\infty\}$ are weakly closed in L^p.*
(c) *The weak convergence in L^p and the boundedness with respect to the norm $\|\cdot\|_p + \|\cdot\|_q$ with $1 \le q < p$ imply weak convergence in L^r for all $r \in (q, p)$.*

Proof. First note that

$$\frac{1}{p'} + \frac{1}{p} = 1, \quad \frac{1}{q'} + \frac{1}{q} = 1,$$

$$\|f\|_p = \sup\{\langle g, f\rangle_2 \mid g \in L^{p'}, \|g\|_{p'} = 1\},$$

$$\|f\|_q = \sup\{\langle g, f\rangle_2 \mid g \in L^{p'} \cap L^{q'}, \|g\|_{q'} = 1\};$$

for $L^{p'} \cap L^{q'}$ is dense in $L^{q'}$. Here we have set

$$\langle g, f\rangle_2 = \int_{\mathbb{R}^n} g(x) f(x)\, dx.$$

Since the supremum of continuous functions is lower semicontinuous (Appendix B), (a) and (b) follow immediately.

If a net $\{f_\alpha\}_{\alpha \in A} \subset L^p$ converges weakly to f in L^p and $\|f_\alpha\|_p + \|f_\alpha\|_q \le c$, $\forall \alpha \in A$, is satisfied, then it follows from (b) that $\|f\|_p + \|f\|_q \le c$. It remains to be shown that $(\langle g, f - f_\alpha\rangle_2)_{\alpha \in A}$ in $\mathbb{R}$ converges to zero for all $g \in L^{r'}$, $1/r + 1/r' = 1$, $q < r < p$. Since $L^{r'} \cap L^{p'}$ is dense in $L^{r'}$, for every $\varepsilon > 0$ and arbitrary $g \in L^{r'}$ there exists a g_ε in $L^{r'} \cap L^{p'}$ with $\|g - g_\varepsilon\|_r \le \varepsilon/4c$. From Theorem 10.4.2 it follows for all $\alpha \in A$ that

$$\begin{aligned}|\langle g, f - f_\alpha \rangle_2| &\leq |\langle g_\varepsilon, f - f_\alpha \rangle_2| + |\langle g - g_\varepsilon, f - f_\alpha \rangle_2| \\ &\leq |\langle g_\varepsilon, f - f_\alpha \rangle_2| + \|g - g_\varepsilon\|_{r'} [\|f\|_r + \|f_\alpha\|_1] \\ &\leq |\langle g_\varepsilon, f - f_\alpha \rangle_2| + \frac{\varepsilon}{2},\end{aligned}$$

and this establishes our assertion since $g_\varepsilon \in L^{p'}$. □

10.3 Minimisation of the Thomas-Fermi Energy Functional

We prepare here for the application of the version Theorem 1.2.5 of the fundamental theorem of the calculus of variations to the TF energy functional. We first prove a number of properties for this functional.

Lemma 10.3.1. (a) *The function*

$$W : \mathbb{R}^3 \to \mathbb{R}, \quad W(x) = \frac{1}{|x|}, \quad x \neq 0,$$

can be expressed as

$$W = W_1 + W_2, W_j \in L^{p_j}(\mathbb{R}^3), \quad 1 \leq p_1 < 3 < p_2 \leq \infty,$$

e.g.

$$W_1(x) = \begin{cases} \frac{1}{|x|}, & |x| \leq R, \quad x \neq 0, \\ 0, & |x| > R > 0. \end{cases}$$

(b)

$$(f, g) \to \langle f, g \rangle \equiv \langle f, W * g \rangle_2 = \int_{\mathbb{R}^3} \frac{f(x)g(y)}{|x - y|} \, dx \, dy$$

defines a scalar product on $L^1 \cap L^{5/3}$*; furthermore,*

$$|\langle f, g \rangle| \leq \|f\|_{q_2^1} \|W_1\|_{q_2^1} \|g\|_{q_3^1} + \|f\|_{q_1^2} \|W_2\|_{q_2^2} \|g\|_{q_3^2}$$

holds for all $\{q_j^i \mid i = 1, 2, j = 1, 2, 3\}$ *with*

$$2 = \sum_{j=1}^{3} \frac{1}{q_j^i}, \quad i = 1, 2 \quad \textit{and} \quad q_1^i, q_3^i \in [1, 5/3], \quad q_2^1 < 3 < q_2^2.$$

Thus $\langle , \rangle$ *is in particular continuous with respect to the norm*

$$\| \cdot \|_{5/3} + \| \cdot \|_r, \quad 1 \leq r < 5/3.$$

Proof. The proof of (a) is obtained by an elementary calculation. The estimate in (b) follows from a simple combination of Hölder's and Young's inequalities and utilising (a). This means that $\langle , \rangle$ is a well-defined bilinear form on $L^1 \cap L^{5/3}$, which is symmetric and possesses the required continuity stated above; Fourier transformation gives the representation

$$\langle f,g\rangle = c\int \frac{\overline{\tilde{f}(p)}\,\tilde{g}(p)}{p^2}\,d^3p\,,\quad c>0$$

and thus is a scalar product. □

Lemma 10.2.1 (a) shows that we can admit all potentials

$$V\in \operatorname{lin}\bigcup_{q\in[(s/2),\infty]} L^q(\mathbb{R}^3)\quad \text{in } A(\rho)=\langle V,\rho\rangle_2$$

while maintaining the domain of definition M according to (10.1.3). To be more precise, let us deal with the case

$$V=V_1+V_2\,,\quad V_1\in L^{5/3}(\mathbb{R}^3)\,,\quad V_2\in L^q(\mathbb{R}^3)\,,\quad \frac{5}{2}<q<\infty\,. \tag{10.3.1}$$

($V_2\in L^\infty$ is also allowed and is even simpler to deal with in some respects.) For potentials of the form (10.3.1) it follows for all $f\in L^1\cap L^{5/3}$ that

$$\begin{aligned}|A(f)| = |\langle V,f\rangle_2| &\le \|V_1\|_{5/2}\|f\|_{5/3}+\|V_2\|_q\|f\|_{q'}\\ &\le \|V_1\|_{5/2}\|f\|_{5/3}+\|V_2\|_q\|f\|_1^\alpha\|f\|_{5/3}^{1-\alpha}\,,\end{aligned} \tag{10.3.2}$$

$$\frac{1}{q'}+\frac{1}{q}=1\,,\quad 0<\alpha<1\,,$$

with

$$\frac{1}{q'}=\alpha+\frac{1-\alpha}{5/3}$$

and according to Lemma 10.2.1 (a),

$$\alpha = 1-\frac{5/2}{q}\,.$$

The most important properties of the TF energy functional are summarised in the following theorem.

Theorem 10.3.2. Properties of the TF Energy Functional. *For potentials V of the form* (10.3.1) *the* TF *energy functional $\mathcal{E}=\mathcal{E}(\cdot,V)$ is well defined on M and possesses the following properties:*

(a) *$\mathcal{E}$ is strictly convex.*
(b) *$\mathcal{E}$ is continuous with respect to the norm*

$$\|\cdot\|_{5/3}+\|\cdot\|_r\quad \textit{with } r=r(a)\in[1,5/3]\,.$$

On all sets

$$M_\lambda=\{\rho\in M\,|\,\|\rho\|_1\le\lambda\}\,,\quad \lambda>0 \tag{10.3.3}$$

$\mathcal{E}$ is

(c) *bounded below*
(d) *coercive: $\mathcal{E}(\rho)\to+\infty$ if $\|\rho\|_{5/3}\to+\infty$.*
(e) *weakly sequentially lower semicontinuous.*

Proof. (a) Setting

$$q_1' = 5/3\,, \quad q_3' = r_1\,, \quad q_1^2 = q_3^2 = r_2\,,$$

in Lemma 10.3.1 (b), we obtain, utilising (10.3.2), for

$$\mathcal{E}(\rho) = K(\rho) - H(\rho) + \frac{1}{2}\langle \rho, \rho\rangle\,, \tag{10.3.4}$$

the bound

$$|\mathcal{E}(\rho)| \le K(\rho) + \|V_1\|_{5/2}\|\rho\|_{5/3} + \|V_2\|_q\|\rho\|_{q'} + \frac{1}{2}[\|\rho\|_{5/3}\|W_1\|_{q_2^1}\|\rho\|_{r_1} + \|W_2\|_{q_3^2}\|\rho\|_{r_2}^2]\,.$$

Here the conditions

$$2 = \frac{3}{5} + \frac{1}{q_2^1} + \frac{1}{r_1}\,, \quad q_2^1 < 3\,,$$

$$2 = \frac{2}{r_2} + \frac{1}{q_2^2}\,, \quad 3 < q_2^2\,,$$

have to be observed. If we have $1 \le r_j \le \frac{5}{3}$ then the functional $\mathcal{E}$ is well defined on M because of Lemma 10.1 (a), since $q \in (\frac{5}{2}, +\infty)$ implies $q' \in (1, \frac{5}{3})$. Again by Lemma 10.1 (a) continuity of the functional $\mathcal{E}$ with respect to the norm $\|\cdot\|_{5/3} + \|\cdot\|_r$ follows, if we can assure $1 < r_j \le q' < \frac{5}{3}$ and if we then set $r = \min\{r_1, r_2\}$. But this is easily done: For

$$\frac{1}{\frac{2}{5} + \frac{1}{q}} < q_2^1 < \frac{5}{2}$$

one gets $\frac{3}{5} < \frac{1}{q'} = 1 - \frac{1}{q} < \frac{7}{5} - \frac{1}{q_2^1} = \frac{1}{r_1} < 1$; and $2q_2^2 > \max\{q, 6\}$ implies

$$\frac{3}{5} < \frac{1}{q'} = 1 - \frac{1}{q} < 1 - \frac{1}{2q_2^2} = \frac{1}{r_2} < 1\,.$$

If we now estimate $\|\rho\|_{q'}$ and $\|\rho\|_{r_j}$ by $\|\rho\|_r$ and $\|\rho\|_{5/3}$. Using Lemma 10.1 (a), continuity of $\mathcal{E}$ follows indeed.

(b) Since $x \mapsto x^{5/3}$ is strictly convex on $[0, \infty)$, the functional $\rho \mapsto K(\rho) = \frac{3}{5}(\|\rho\|_{5/3})^{5/3}$ is strictly convex on M. $R(\rho) = \frac{1}{2}\langle \rho, \rho\rangle$ is strictly convex since it is a scalar product. Hence $\rho \mapsto \mathcal{E}(\rho)$ is strictly convex on M.

(c) Estimate (10.3.2) implies for $\rho \in M_\lambda$

$$\mathcal{E}(\rho) = K(\rho) - A(\rho) + \frac{1}{2}\langle \rho, \rho\rangle \ge \frac{3}{5}\|\rho\|_{5/3}^{5/3} - |A(\rho)|$$

$$\ge \frac{3}{5}\|\rho\|_{5/3}^{5/3} - \|V_1\|_{5/2}\|\rho\|_{5/3} - \|V_2\|_q \lambda^{1-5/(2q)}\|\rho\|_{5/3}^{5/(2q)}\,;$$

therefore the functional $\mathcal{E}$ is bounded from below on M_λ and is coercive since $\frac{5}{2} < q$.

(d) A weakly convergent sequence in $L^{5/3}$ is bounded in norm; thus Lemma 10.3 (c) says that a sequence $\{\rho_n\}_{n\in\mathbb{N}} \subset M_\lambda$ converges weakly in all the spaces L^r, $1 < r < \frac{5}{3}$, whenever it converges weakly in $L^{5/3}$. Hence

$$\rho \mapsto A(\rho) = \langle V_1, \rho\rangle_2 + \langle V_2, \rho\rangle_2 \,, \quad V_1 \in L^{5/2} \,, \quad V_2 \in L^q \,, \quad \frac{5}{2} < q < +\infty$$

is a weakly sequentially continuous function on M_λ.

Since the functional $R(\rho) = \frac{1}{2}\langle \rho, \rho\rangle$ derives from a scalar product it suffices to show, by Lemma 1.1 (a) and Lemma 10.40, that for arbitrary but fixed $\rho_0 \in M_\lambda$ the functional $\rho \mapsto \langle \rho, \rho_0\rangle$ is weakly sequentially continuous. But we have

$$\langle \rho, \rho_0\rangle = \langle \rho, W_1 * \rho_0\rangle + \langle \rho, W_2 * \rho_0\rangle_2$$

where for instance $W_1 \in L^{5/2}$, $W_2 \in L^4$ so that Lemma 10.1 (b) implies $W_1 * \rho_0 \in L^{5/2}$ and $W_2 * \rho_0 \in L^4$. Therefore $\rho \mapsto \langle \rho, W_1 * \rho_0\rangle_2$ is weakly continuous in $L^{5/3}$, and by Lemma 10.3 (c) it follows in addition that $\rho \mapsto \langle \rho, W_2 * \rho_0\rangle_2$ is weakly sequentially continuous since for sequences weak convergence in $L^{5/3}$ implies again weak convergence in $L^{4/3}$ and $L^4 \cong (L^{4/3})'$. □

Now, according to Lemma 10.2.3 (b) all sets of the form $M_\lambda = \{\rho \in M \,|\, \|\rho\|_1 \leq \lambda\}$, $\lambda > 0$, are weakly closed in $L^{5/3}$. Theorem 10.3.2 therefore establishes the fact that the second version of the assumption of Theorem 1.2.5 is satisfied. Further, since $\mathcal{E}$ is strictly convex on M the question of uniqueness is also settled by the Theorem 1.1.3 and we obtain the next theorem.

Theorem 10.3.3. *For all potentials* $V = V_1 + V_2$ *of the form* (10.3.1) *and all* $\lambda \geq 0$, *there exists exactly one* $\rho_{\lambda,V} \in M_\lambda$ *such that*

$$\mathcal{E}(\rho_{\lambda,V}; V) = \inf_{\rho\in M_\lambda} \mathcal{E}(\rho; V) \,. \tag{10.3.5}$$

We have advanced a long way towards the solution of the real problem, namely determining

$$\inf\left\{\mathcal{E}(\rho; V) \,|\, \rho \in M_\lambda, \|\rho\|_1 = \int \rho \, dx = \lambda\right\}.$$

The roundabout way of determining first the infimum on all of M_λ seems to be necessary since, although M_λ is weakly closed in $L^{5/3}$,

$$\left\{\rho \in M_\lambda \,|\, \|\rho\|_1 = \int \rho \, dx = \lambda\right\}$$

is not (Appendix C).

The strategy of solving the minimisation problem

$$E(\lambda) = \inf\left\{\mathcal{E}(\rho; V) \,|\, \rho \in M, \|\rho\|_1 = \int \rho \, dx = \lambda\right\} = \mathcal{E}(\rho_{\lambda,V}) \tag{10.3.6}$$

where

$$\rho_{\lambda,V} \in M\,, \quad \int \rho_{\lambda,V}\,dx = \lambda = \|\rho_{\lambda,V}\|_1\,,$$

with the help of Theorem 10.3.3 consists in showing first (Theorem 10.3.4)

$$\inf\{\mathcal{E}(\rho)\,|\,\rho \in M_\lambda\} = \inf\left\{\mathcal{E}(\rho)\,|\,\rho \in M, \int \rho\,dx = \lambda\right\}$$

and then demonstrating for certain potentials of the form (10.3.1) under which conditions the density $\rho_{\lambda,V} \in M_\lambda$, which is uniquely determined by Theorem 10.3.3, will necessarily satisfy

$$\int \rho_{\lambda,V}\,dx = \|\lambda_{\lambda,V}\|_1 = \lambda\,.$$

Theorem 10.3.4. *For all potentials* V *of the form* (10.3.1) *and all* $\lambda \geq 0$ *the following holds:*

$$\inf\{\mathcal{E}(\rho;V)\,|\,\rho \in M_\lambda\} = \inf\{\mathcal{E}(\rho;V)\,|\,\rho \in M, \|\rho\|_1 = \lambda\}\,.$$

Proof. Obviously, it is sufficient to show

$$\inf\{\mathcal{E}(\rho)\,|\,\rho \in M_\lambda\} \geq \inf\{\mathcal{E}(\rho)\,|\,\rho \in M, \|\rho\|_1 = \lambda\}\,.$$

Since $\mathcal{D}(\mathbb{R}^3)$ is dense in $L^1 \cap L^{5/3}$, one has, according to Theorem 10.3.3,

$$\mathcal{E}(\rho_{\lambda,V};V) = \inf\{\mathcal{E}(\rho;V)\,|\,\rho \in M_\lambda \cap \mathcal{D}(\mathbb{R}^3)\}\,.$$

We may prove the desired inequality by showing that for every $\varepsilon > 0$ and every $\rho \in \mathcal{D}(\mathbb{R}^3) \cap M_\lambda$ there exists a $\varphi_\varepsilon \in M$, $\|\rho_\varepsilon\|_1 = \lambda$, such that

$$|\mathcal{E}(\rho;V) - \mathcal{E}(\rho_\varepsilon;V)| < \varepsilon$$

holds. Now let $\rho \in M_\lambda \cap \mathcal{D}(\mathbb{R}^3)$ be given. For every $n \in \mathbb{N}$ there exists then a Lebesgue-measurable set $A_n \subset \mathbb{R}^3$ with

(i) $$A_n \cap \operatorname{supp}\rho = \emptyset\,,$$

(ii) $$|A_n| = \int_{A_n} dx = n(\lambda - \|\rho\|_1)\,,$$

since ρ has a compact support. Let χ_n be the characteristic function of the set A_n. Then for $\rho_n = \rho + \chi_n/n$ one has

(i) $$0 \leq \rho_n \in L^1 \cap L^{5/3}\,,$$

(ii) $$\|\rho_n\|_1 = \|\rho\|_1 + \frac{1}{n}|A_n| = \lambda\,,$$

and thus $\rho_\lambda \in M$ and $\|\rho_n\|_1 = \lambda$, $\forall n$. Moreover, this sequence approximates the given function ρ in the $(\|\cdot\|_{5/3} + \|\cdot\|_r)$-norm $(r > 1)$ according to Theorem 10.3.2:

$$\begin{aligned}\|\rho - \rho_n\|_{5/3} + \|\rho - \rho_n\|_r &= \frac{1}{n}\|\chi_n\|_{5/3} + \frac{1}{n}\|\chi_n\|_r \\ &= \frac{1}{n}[n(\lambda - \|\rho\|_1)]^{3/5} + \frac{1}{n}[n(\lambda - \|\rho\|_1)]^{1/r} \underset{n\to+\infty}{\longrightarrow} 0\,.\end{aligned}$$

Since $\mathcal{E}$ is continuous with respect to this norm according to Theorem 10.3.2, there exists a n_ε for every given $\varepsilon > 0$, such that for all $n \geq n_\varepsilon$ one has

$$|\mathcal{E}(\rho; V) - \mathcal{E}(\rho_n; V)| < \varepsilon\,, \qquad \square$$

In conclusion, we would like to discuss some of the consequences which follow from Theorems 10.3.3 and 10.8.4 in connection with the strict convexity of $\mathcal{E}$. These theorems state that for a fixed potential V of the form (10.3.1) we can define a function $E(\cdot) = E(\cdot; V)$ on $[0, \infty)$ by

$$E(\lambda) = \inf\{\mathcal{E}(\rho, V) \,|\, \rho \in M, \|\rho\|_1 = \lambda\} = \min\{\mathcal{E}(\rho) \,|\, \rho \in M_\lambda\} = \mathcal{E}(\rho_\lambda)\,.$$

Here $\rho_\lambda \in M_\lambda$ is uniquely determined and satisfies

$$0 \leq \lambda_1 = \|\rho_\lambda\|_1 \leq \lambda\,.$$

Let us first consider a few simple properties of $E(\lambda)$.

Lemma 10.3.5. (a) *$E(\cdot)$ is convex and monotonically decreasing on $[0, \infty)$.*
(b) *If the density $\rho_\lambda \in M$, which minimises $\mathcal{E}$ on M_λ according to Theorem 10.3.3, satisfies $\|\rho_\lambda\|_1 = \lambda_1 < \lambda$, then $E(\cdot)$ is constant on $[\lambda_1, \infty)$.*

Proof. (a) Since $\mathcal{E}(\cdot)$ is strictly convex, one obtains immediately the convexity of $E(\cdot)$ as the infimum of $\mathcal{E}(\cdot)$ on a convex set. Further, $\lambda_2 > \lambda_1$ implies that $M_{\lambda_2} \supset M_{\lambda_1}$; it is thus obvious that $E(\cdot)$ decreases monotonically.
(b) If $\lambda_1 = \|\rho_\lambda\|_1 < \lambda$ holds, then

$$E(\lambda_1) \leq \mathcal{E}(\rho_\lambda) = E(\lambda)$$

follows and therefore from (a) we get $E(\lambda_1) = E(\lambda)$. Again by utilising (a) we establish the fact that $E(\cdot)$ is constant not only on $[\lambda_1, \lambda]$ but on all of $[\lambda_1, \infty)$. $\square$

With the help of these simple considerations and some additional assumptions about the potential, which are natural, we can describe the solution of the minimisation problem (10.1.1) almost completely.

For a potential V of the form (10.3.1) which satisfies $V(x) \leq 0$ almost everywhere, we have

$$-A_V(\rho) = \langle -V, \rho \rangle_2 \geq 0\,, \quad \forall \rho \in M\,.$$

For potentials of this type the minimisation problem (10.1.1) is, however, trivial to solve, namely, one has

$$\mathcal{E}(\rho) \geq 0 = \mathcal{E}(0)\,, \quad \forall \rho \in M\,.$$

Thus the only interesting potentials V are which are not almost everywhere negative.

Theorem 10.3.6. *Let V be a potential of the form (10.3.1) and let V be positive almost everywhere on an open nonempty subset $U \subseteq \mathbb{R}^3$. Then there exists one and only one $\lambda_0 = \lambda_0(V)$, $0 < \lambda_0 < \infty$, with the following properties:*

(i) *for all $\lambda \in [0, \lambda_0]$ one has*

$$E(\lambda) = \mathcal{E}(\rho_\lambda) \quad \textit{where } \rho_\lambda \in M \textit{ and } \|\rho_\lambda\|_1 = \lambda\,;$$

(ii) *$E(\cdot)$ is constant on $[\lambda_0, \infty)$.*

Proof. (a) Under these conditions for the potential V there exists a continuous function $\rho \geq 0$ with compact support in U such that

$$\langle V, \rho \rangle_2 = \int_U V \rho \, d^3x = \alpha > 0$$

holds. For all $t > 0$ this ρ satisfies

$$\mathcal{E}(t\rho) = t^{5/3} K(\rho) - t\alpha + t^2 R(\rho)\,,$$

and thus for sufficiently small $t > 0$, $\mathcal{E}(t\rho)$ will be negative.

(b) Now let $\lambda > 0$ and $\rho_\lambda \in M_\lambda$ be determined according to Theorem 10.3.3 such that

$$E(\lambda) = \min\{\mathcal{E}(\rho) \mid \rho \in M_\lambda\} = \mathcal{E}(\rho_\lambda)$$

holds. Then $\lambda_1 = \|\rho_\lambda\|_1$ is also uniquely determined and satisfies $0 \leq \lambda_1 \leq \lambda$. If we had $\lambda_1 = 0$, then we would also have $\rho_\lambda = 0 \in M_\lambda$ and thus $\mathcal{E}(\rho_\lambda) = 0$. By scaling, it follows further from (a) that $\mathcal{E}(\rho)$ can also assume negative values on M_λ for $\lambda > 0$. This is a contradiction to the minimisation property of ρ_λ. Therefore we have $0 < \lambda_1$ and thus $\rho_1 = \rho_\lambda$ minimises the TF energy functional $\mathcal{E}$ on $\{\rho \in M \mid \|\rho\|_1 = \lambda_1\}$.

(c) We now show that for all $\lambda \in (0, \lambda_1)$, $\mathcal{E}(\cdot)$ will be minimised by a $\rho \in M$, $\|\rho\|_1 = \lambda$ on M_λ. For this purpose consider an arbitrary λ with $0 < \lambda < \lambda_1$ and let $\rho \in M_\lambda$ be the uniquely determined element satisfying $E(\lambda) = \mathcal{E}(\rho_\lambda)$. If we had $\lambda' = \|\rho\|_1 < \lambda$, then $E(\cdot)$ would be constant on $[\lambda', \infty)$ by Lemma 10.3.5 (b), and according to (a), $\lambda' > 0$. It follows therefore that

$$E(\lambda') = E(\lambda) = E(\lambda_1)$$

and thus

$$\mathcal{E}(\rho) = \mathcal{E}(\rho_1)$$

which means that both ρ and ρ_1 would minimise $\mathcal{E}$ on M_{λ_1}. This is a contradiction since

$$\|\rho\|_1 = \lambda' < \lambda_1 = \|\rho_1\|_1 \,.$$

Therefore one has $\|\rho\|_1 = \lambda$, as asserted above.

(d) If $\lambda_1 < \lambda$ (where λ and λ_1 are as in (b)), then it follows that $E(\cdot)$ is constant on $[\lambda_1, \infty)$, and then by (c) $\lambda_0(V) = \lambda_1$ possesses the required properties. If $\lambda_1 = \lambda$ holds in (b), then consider the minimisation problem on $M_{\lambda'}$ for $\lambda' > \lambda$. If for all $\lambda' > \lambda$ the minimising ρ_λ satisfies the condition $\|\rho_{\lambda'}\| = \lambda'$, then set $\lambda_0(V) = +\infty$. There exists then a $\lambda' > \lambda$ for which $\lambda \leq \lambda_1' = \|\rho_{\lambda'}\| < \lambda'$ holds. According to (c), the conditions (i) and (ii) are then valid on $[0, \lambda_1']$ and $[\lambda_1', \infty)$, respectively. In this case we set $\lambda_0(V) = \lambda_1'$.

□

Remark 10.3.1. Brezis [10.6] has used more detailed considerations to determine the value and an interpretation of $\lambda_0(V)$.

Corollary 10.3.7. *Under the assumptions of Theorem* 10.3.6, *there exists a* $\lambda_0(V) > 0$ *such that* $E(\cdot)$ *is strictly convex on* $[0, \lambda_0(V)]$ *and monotonically strictly decreasing on* $[0, \lambda_0(V)]$. *Then* ρ_{λ_0} *determined by Theorem* 10.3.6 *minimises* $\mathcal{E}(\rho)$ *on all of* M.

Proof. Clearly we choose $\lambda_0 = \lambda_0(V)$ as in Theorem 10.3.6. Now let $0 < \lambda_1 < \lambda_2 < \lambda_0$ and $0 < t < 1$. Then one has

$$E(t\lambda_1 + (1-t)\lambda_2) \leq \mathcal{E}(t\rho_{\lambda_1} + (1-\lambda)\rho_{\lambda_2}),$$

if ρ_{λ_j}, $j = 1, 2$, is the minimising ρ of $\mathcal{E}$ in $\{\rho \in M \mid \|\rho\| = \lambda_j\}$. Since $\mathcal{E}$ is strictly convex, one obtains

$$E(t\lambda_1 + (1-\lambda)t_2) < t\mathcal{E}(\rho_{\lambda_1}) + (1-t)\mathcal{E}(\rho_{\lambda_2}) = tE(\lambda_1) + (1-t)E(\lambda_2) \,.$$

The assertion follows from Lemma 10.3.5 (a). Since $E(\lambda)$ is constant on $[\lambda_0, \infty)$ and has the value $E(\lambda_0) = \mathcal{E}(\rho_{\lambda_0})$, the second assertion follows automatically.

□

10.4 Thomas-Fermi Equations and the Minimisation Problem for the TF Functional

The objective of this section is to prove the equivalence of solving the minimisation problem and solving the TF equations. This is essentially a variational problem with constraints. Therefore we could try to apply the Theorem of

Ljusternik (Theorem 4.2.3) to the variational problem (10.8.6). For this, one would have to regard the TF energy functional $\mathcal{E}$ of the form

$$\mathcal{E}(\rho) = \frac{3}{5}\|\rho\|_{5/3}^{5/3} - \langle V, \rho\rangle_2 + \frac{1}{2}\langle \rho, \rho\rangle$$

as a function on $X = L^1 \cap L^{5/3}$ and further take into account the constraints

(i) $$\|\rho\|_1 = \int \rho\, dx = \lambda\,,$$

(ii) $$\rho \geq 0\,.$$

The first constraint can be reformulated as a regular mapping $\psi : X \to \mathbb{R}$ in the sense of the Theorem of Ljusternik (see Chap. 4):

$$\psi(\rho) = \int \rho\, dx$$

with the Fréchet derivative

$$\psi'(\rho)(h) = \int h\, dx\,, \quad \forall \rho, h \in X\,.$$

The second constraint can be incorporated only by modifying the domain of definition X to M as given in 10.1.3.

We are easily able to calculate the Fréchet derivative of the function

$$\mathcal{E} : X \to \mathbb{R} :$$

$$\rho \in X \Rightarrow |\rho|^{2/3} \in L^{5/2} \Rightarrow \operatorname{sign}\rho(x)|\rho(x)|^{2/3} \in L^{5/2}\,.$$

Consequently,

$$\begin{aligned} h \to \mathcal{E}'(\rho)(h) &= \int \operatorname{sign}\rho(x)|\rho(x)|^{2/3} h(x)\, dx - \langle V, h\rangle_2 + \langle \rho, h\rangle \\ &= \langle \operatorname{sign}\rho|\rho|^{2/3} - V + W * \rho, h\rangle_2 \end{aligned}$$

is in fact a continuous linear mapping $X \to \mathbb{R}$ with respect to the norm $\|\cdot\|_{5/3} + \|\cdot\|_1$ on X.

If we consider only the first constraint Ljusternik's Theorem states: for a minimising point ρ_0 of $\mathcal{E}$ there exists a constant, say $-\phi_0$, such that

$$\mathcal{E}'(\rho_0) = -\phi_0 \psi'(\rho_0)$$

holds, which means

$$\operatorname{sign}(\rho_0)|\rho_0|^{2/3} - V + W * \rho_0 = -\phi_0\,. \tag{10.4.1}$$

If one knew $\rho_0 \geq 0$, then the TF equations (10.1.2) would in fact have followed.

If we now take into account the second constraint by regarding the TF energy functional only as a function on $M = \{f \in X \mid f \geq 0\}$, then $\mathcal{E}$ is not Fréchet-differentiable, but possesses at every point $\rho \in M$ only the one-sided Gâteaux derivative in directions $h \in M$:

$$\frac{d^+}{dt}\mathcal{E}(\rho + th)\Big|_{t=0} = \lim_{t\to 0}\frac{1}{t}[\mathcal{E}(\rho + th) - \mathcal{E}(\rho)] = \langle \delta\mathcal{E}/\delta\rho, h\rangle_2\,, \tag{10.4.2}$$

where for potentials of the form (10.3.1), one has by Lemma 10.2.1

$$\frac{\delta \mathcal{E}}{\delta \rho} = \rho^{2/3} - V + W * \rho \in L^{5/2} + L^q + L^\infty , \quad \forall \rho \in M . \tag{10.4.3}$$

The following theorem confirms the above heuristic considerations and leads to the following formulation.

Theorem 10.4.1. *Let V be a potential of the form* (10.1.3). *For $\rho_0 \in M$ with $\|\rho_0\|_1 = \int \rho_0 \, dx = \lambda$ the following holds:*

$$\mathcal{E}(\rho_0) = \mathcal{E}(\rho_0; V) = \inf \left\{ \mathcal{E}(\rho; V) \,|\, \rho \in M, \int \rho \, dx = \lambda \right\}$$

if, and only if, $\rho_0 \in M$ is a solution of the Thomas-Fermi equations (10.1.2) *and satisfies $\int \rho_0 \, dx = \lambda$.*

Proof. If $\rho_0 \in M$ minimises the TF energy functional $\mathcal{E}$ on M subject to the constraint $\int \rho \, dx = \lambda$, then it follows that

$$0 \le \lim_{t \to 0} \frac{1}{t} [\mathcal{E}(\rho_0 + t(\rho - \rho_0)) - \mathcal{E}(\rho_0)] = \frac{d^+}{dt} \mathcal{E}(t\rho + (1-t)\rho_0) \Big|_{t=0}$$

and therefore, from (10.4.2),

$$0 \le \left\langle \frac{\delta \mathcal{E}}{\delta \rho_0}, \rho - \rho_0 \right\rangle_2 , \quad \forall \rho \in M , \quad \int \rho \, dx = \lambda , \tag{10.4.4}$$

Taking into account the characterisation

$$\begin{aligned} \mathcal{F} &= \left\{ \rho - \rho_0 \,|\, \rho \in M, \int \rho \, dx = \lambda \right\} \\ &= \left\{ f \in L^1 \cap L^{5/3} \,\middle|\, \int f \, dx = 0, \; f \upharpoonright [\rho_0 > 0]^c > 0 \right\} \end{aligned}$$

and the fact that for the subset

$$\mathcal{F}_0 = \{ f \in \mathcal{F} \,|\, f \upharpoonright [\rho_0 > 0]^c = 0 \}$$

the relation $-\mathcal{F}_0 = \mathcal{F}_0$ holds, then it follows that

$$0 = \left\langle \frac{\delta \mathcal{E}}{\delta \rho_0}, f \right\rangle_2 , \quad \forall f \in \mathcal{F}_0 . \tag{10.4.5}$$

This equation enables us to show that

$$\frac{\delta \mathcal{E}}{\delta \rho_0} \upharpoonright [\rho_0 > 0] = \text{const. almost everywhere.}$$

According to (10.4.3) $\delta\mathcal{E}/\delta\rho_0$ is bounded almost everywhere on $\mathbb{R}^3$. This means that $[\rho_0 > 0]$ possesses the following representation:

$$[\rho_0 > 0] = \bigcup_{n=1}^{\infty} C_n \,, \quad C_n \subseteq C_{n+1} \,, \quad |C_n| < +\infty$$

for all $n \in N$ and $0 < |C_n|$ for $n \geq n_0$ say, where, for example,

$$C_n = \Big\{ x \in \mathbb{R}^3 \;\Big|\; \frac{1}{n} \leq \rho_0(x) \leq n \text{ almost everywhere, } \left|\frac{\delta\mathcal{E}}{\delta\rho_0(x)}\right| \\ \leq n \text{ almost everywhere, } |x| \leq n \Big\}$$

has been chosen.

Since $\delta\mathcal{E}/\delta\rho_0$ is bounded almost everywhere on C_n and also since $L^1 \cap L^{5/3}$ is dense in L^1, the functional

$$f \to \left\langle \frac{\delta\mathcal{E}}{\delta\rho_0}, f \right\rangle_2 \quad \text{on} \quad \{ f \in \mathcal{F}_0 \,|\, \operatorname{supp} f \subseteq C_n \}$$

possesses a unique extension to a continuous linear functional

$$f \to \left\langle \frac{\delta\mathcal{E}}{\delta\rho_0}, f \right\rangle_2 \quad \text{on} \quad \left\{ f \in L^1(C_n) \,\Big|\, \int f \, dx = 0 \right\}.$$

According to (10.4.5) this must then be the null functional. We therefore have

$$\left\langle \frac{\delta\mathcal{E}}{\delta\rho_0}, f \right\rangle_2 = 0 \,, \quad \forall f \in L^1(C_n) \,, \quad \int f \, dx = 0 \,.$$

Now for arbitrary $g \in L^1(C_n)$ and $n \geq n_0$,

$$f_g \equiv g - \frac{1}{|C_n|} \int g \, dx \, \chi_{C_n}$$

is an element of $L^1(C_n)$ which satisfies $\int f_g \, dx = 0$. It follows then that

$$\left\langle \frac{\delta\mathcal{E}}{\delta\rho_0}, g \right\rangle_2 = \int g \, dx \left\langle \frac{\delta\mathcal{E}}{\delta\rho_0}, \frac{1}{|C_n|} \chi_{C_n} \right\rangle_2 \,, \quad \forall g \in L^1(C_n) \,.$$

The dual space of $L^1(C_n)$ is known to be $L^\infty(C_n)$. Consequently, it follows that in $L^\infty(C_n)$

$$\frac{\delta\mathcal{E}}{\delta\rho_0} \restriction C_n = \left\langle \frac{\delta\mathcal{E}}{\delta\rho_0}, \frac{1}{|C_n|} \chi_{C_n} \right\rangle = \alpha_n = \text{constant} \,.$$

$C_n \subseteq C_{n+1}$ implies $\alpha_n = \alpha_{n+1}$. There exists then a constant independent of n, say $-\phi_0$, such that

$$\frac{\delta \mathcal{E}}{\delta \rho_0(x)} = -\phi_0 \quad \text{almost everywhere on } [\rho_0 > 0] \tag{10.4.6}$$

holds. The relation (10.4.4) states further that

$$\left\langle \frac{\delta \mathcal{E}}{\delta \rho_0}, \rho \right\rangle_2 \geq \left\langle \frac{\delta \mathcal{E}}{\delta \rho_0}, \rho_0 \right\rangle_2 = -\phi_0 \lambda ,$$

and therefore

$$\left\langle \frac{\delta \mathcal{E}}{\delta \rho_0} + \phi_0, f \right\rangle_2 \geq 0 , \quad \forall f \in M , \quad \int \rho \, dx = \lambda .$$

By scaling with $\|\rho\|_1$ one obtains

$$\left\langle \frac{\delta \mathcal{E}}{\delta \rho_0} + \phi_0, \rho \right\rangle_2 \geq 0 , \quad \forall \rho \in M ,$$

and therefore, by well-known arguments,

$$\frac{\delta \mathcal{E}}{\delta \rho_0} + \phi_0 \geq 0 \quad \text{almost everywhere on } \mathbb{R}^3 . \tag{10.4.7}$$

Taking (10.4.3) into account one could also express (10.1.6, 7) as

$$\rho_0^{2/3} = \max\{\phi_{\rho_0} - \phi_0, 0\} ,$$

$$\phi_{\rho_0} = V - W * \rho_0 ,$$

which are the TF equations (10.1.2) with $\int \rho_0 \, dx = \lambda$.

Conversely, let $\rho_0 \in M$ be a solution of the TF equations. It follows then that

$$\frac{\delta \mathcal{E}}{\delta \rho_0} = \rho_0^{2/3} - \phi_{\rho_0} = \begin{cases} -\phi_0 & \text{on } [\rho_0 > 0] , \\ \geq -\phi_0 & \text{on } [\rho_0 > 0]^c , \end{cases}$$

and therefore, according to (10.4.2), for all $\rho \in M$, $\int \rho \, dx = \lambda$:

$$\frac{d^+}{dt} \mathcal{E}(t\rho + (1-t)\rho_0)\Big|_{t=0} = \left\langle \frac{\delta \mathcal{E}}{\delta \rho_0}, \rho - \rho_0 \right\rangle_2 \geq \langle -\phi_0, \rho - \rho_0 \rangle_2 = 0 .$$

Since $\mathcal{E}$ is strictly convex according to Theorem 10.3.2 we have

$$0 \leq \frac{d^+}{dt} \mathcal{E}(t\rho + (1-t)\rho_0)\Big|_{t=0} \leq \mathcal{E}(\rho) - \mathcal{E}(\rho_0)$$

and thus

$$\mathcal{E}(\rho_0) = \min \left\{ \mathcal{E}(\rho) \,\Big|\, \rho \in M, \int \rho \, dx = \lambda \right\} ,$$

which means that a solution of the TF equations (10.1.2) also minimises the TF energy functional (with respect to the same “particle number” λ). □

Remark 10.4.1. As a consequence of the strict convexity of $\mathcal{E}$ this theorem shows that for a fixed "particle number" λ the TF equations have at most one solution, even when we allow for different constants ϕ_0, $\phi_0' \dots$ in these equations.

The following considerations for the differentiability on $E(\lambda)$ and the value of the derivative $(dE/d\lambda)(\lambda)$, which are based mainly on the convexity of $\mathcal{E}(\cdot)$, are naturally of interest for the case of potentials for which $\lambda_0(V) > 0$ can be proved. Since, however, no special property of the potential enters into the proof, we could just as well express this result in the general setting. If we assume $\lambda_0(V) > 0$, then it implies that for $0 < \lambda < \lambda_0$ the uniqueness of the solution of the TF equations is assured (Theorem 10.4.1). These equations contain a certain parameter $\phi_0 = \phi_0(\lambda) = \phi_0(\lambda, V)$ which admits an interpretation through the following theorem.

Theorem 10.4.2. *Let V be a potential of the form* (10.3.1) *for which $\lambda_0 = \lambda_0(V) > 0$ holds. Then $E(\lambda)$ is differentiable on $[0, \lambda_0]$ with the derivative*

$$\frac{dE}{d\lambda}(\lambda) = -\phi_0(\lambda)\,, \quad \forall \lambda \in [0, \lambda_0]\,. \tag{10.4.8}$$

Further, one has $\phi_0(\lambda) > 0$ for $\lambda \in [0, \lambda_0)$ and $\phi_0(\lambda_0) = 0$.

Proof. We prove the differentiability of $E(\lambda)$ by showing that at a point $\lambda_1 \in (0, \lambda_0)$ the right-hand and the left-hand derivatives exist and are equal. Let us take $\lambda_1 \in (0, \lambda_0)$. According to Theorem 10.3.6 and Theorem 10.3.5, there exists precisely one solution ρ_1 of the TF equations with $\int \rho_1 \, dx = \lambda_1$ and $\mathcal{E}(\rho_1) = E(\lambda_1)$. For any other arbitrary $\lambda \in (0, \infty)$ it follows as a consequence of the convexity of $\mathcal{E}$ for all $\rho \in M$ with $\int \rho \, dx = \lambda$ that

$$\mathcal{E}(\rho) - \mathcal{E}(\rho_1) \geq \frac{d^+}{dt}\mathcal{E}(t\rho + (1-t)\rho_1) = \left\langle \frac{\delta \mathcal{E}}{\delta \rho_1}, \rho - \rho_1 \right\rangle_2 \geq -\phi_0(\lambda_1)(\lambda - \lambda_1)$$

because of (10.4.2) and because ρ_1 solves the TF equations. We then have

$$E(\lambda) - E(\lambda_1) \geq -\phi_0(\lambda_1)(\lambda - \lambda_1)\,, \tag{10.4.9}$$

and therefore

$$\frac{d^+E}{d\lambda}(\lambda_1) \geq -\phi_0(\lambda_1) \quad \text{and} \quad \frac{d^-E}{d\lambda}(\lambda_1) \leq -\phi_0(\lambda_1)\,.$$

Since we have established the fact that $\mathcal{E}$ is Gâteaux-differentiable along the directions in M, it follows for $|t| < 1$ that

$$\begin{aligned} \mathcal{E}(\rho_1 + t\rho_1) &= \mathcal{E}(\rho_1) + t\left\langle \frac{\delta \mathcal{E}}{\delta \rho_1}, \rho_1 \right\rangle_2 + o(t) \\ &= E(\lambda_1) - t\phi_0(\lambda_1)\lambda_1 + o(t)\,, \end{aligned}$$

and thus

$$E(\lambda_1 + t\lambda_1) \leq E(\lambda_1) - \lambda_1 t\phi_0(\lambda_1) + o(t), \tag{10.4.10}$$

which immediately gives

$$\frac{d^+E}{d\lambda}(\lambda_1) \leq -\phi_0(\lambda_1) \quad \text{and} \quad \frac{d^-E}{d\lambda}(\lambda_1) \geq -\phi_0(\lambda_1).$$

All the four inequalities taken together show that $(dE/d\lambda)(\lambda_1)$ exists, and one has

$$\frac{dE}{d\lambda}(\lambda_1) = -\phi_0(\lambda_1), \quad \forall\lambda_1 \in (0, \lambda_0).$$

Since $E(\cdot)$ is strictly monotonically decreasing on $[0, \lambda_0]$, for $\lambda \in [0, \lambda_0)$ it follows that $\phi_0(\lambda) > 0$.

The relations (10.4.9, 10) obviously is also valid at the point λ_0. For $\lambda > \lambda_0$, combining (10.4.9) with Theorem 10.3.6, one gets

$$0 \geq -\phi_0(\lambda_0)(\lambda - \lambda_0),$$

and corresponding to (10.1.10) for $1 > t > 0$

$$0 \geq -t\lambda\phi_0(\lambda_0).$$

It follows then that $\phi_0(\lambda_0) = 0$.

The minimisation property of ρ_0 on M is obvious on the basis of Theorem 10.3.6, as well as the monotonicity property of $\phi_0(\lambda)$ as the derivative of a strictly concave function. □

Remark 10.4.2. Utilising (10.4.8), $\phi_0(\lambda)$ can be interpreted as the *chemical potential* of the system, as it is the negative of the derivative of the energy with respect to the particle number.

10.5 Solution of TF Equations for Potentials of the Form $V(x) = \sum_{j=1}^{k} \frac{z_j}{|x - x_j|}$

As seen in the previous section we have concluded the variational part of the proof for the uniqueness of the solution of the TF equations. Potentials of the form (10.1.4), which is to say

$$V(x) = \sum_{j=1}^{k} \frac{z_j}{|x - x_j|}, \quad z_j > 0,$$

satisfy the assumptions of Theorem 10.3.6 by Lemma 10.3.1 (a). We therefore know on the basis of Theorems 10.3.6, 10.3.7, 10.4.1, 10.4.3 that there exists a $\lambda_0 = \lambda_0(V) > 0$ with the following properties: for $0 \le \lambda < \lambda_0$ the TF equations

$$\rho^{2/3} = \max\{\phi_\rho - \phi_0, 0\},$$

$$\phi_\rho = V - W * \rho, \quad \int \rho\, dx = \lambda,$$

are uniquely solvable by a certain $\rho_\lambda \in M$ which minimises the TF energy functional $\mathcal{E}(\cdot) = \mathcal{E}(\cdot; V)$ namely

$$\mathcal{E}(\rho_\lambda) = \min\left\{\mathcal{E}(\rho) \,\Big|\, \rho \in M, \int \rho\, dx = \lambda\right\} = E(\lambda). \tag{10.5.1}$$

One has

$$\phi_0 = \phi_0(\lambda) = -\frac{dE}{d\lambda}(\lambda) > 0 \quad \text{for } 0 \le \lambda < \lambda_0$$

and $\phi_0(\lambda_0) = 0$.

For completeness, we would like to determine the explicit form of $\lambda_0(V)$ for the potentials of the form (10.1.4). For orientation let us first consider how arbitrary potentials of the form (10.3.1) are related to the admissible particle number $\lambda = \int \rho\, dx$ of a solution of the TF equations. If, for example, $\phi_\rho = V - W * \rho$ holds, then by averaging over the unit sphere S^2 in $\mathbb{R}^3$ and utilising

$$\frac{1}{4\pi}\int_{S^2} \frac{d\omega}{|r\omega - y|} = \frac{1}{\max\{r, |y|\}},$$

we have

$$\widehat{\phi}_\rho(r) = \frac{1}{4\pi}\int_{S^2} d\omega\, \phi(r\omega) = \widehat{V}(r) - \int \frac{\rho(y)\, dy}{\max\{r, |y|\}},$$

and therefore

$$\widehat{V}(r) - \frac{\lambda}{r} \le \widehat{\phi}_\rho(r) \le \widehat{V}(r) - \frac{1}{r}\int_{|y|\le r} \rho(y)\, dy \tag{10.5.2}$$

or

$$\lambda = \lim_{r\to+\infty} r[\widehat{V}(r) - \widehat{\phi}_\rho(r)].$$

For potentials of the form (10.1.4) one knows $\widehat{V}(r)$ explicitly:

$$\widehat{V}(r) = \sum_{j=1}^{k} z_j \int_{S^2} \frac{d\omega}{4\pi}\, \frac{1}{|r\omega - x_j|} = \sum_{j=1}^{k} \frac{z_j}{\max\{r, |x_j|\}};$$

thus

$$\lim_{r\to\infty} r\widehat{V}(r) = \sum_{j=1}^{k} z_j = Z. \tag{10.5.3}$$

Therefore, a control on the asymptotic behaviour of the TF potentials ϕ_ρ is necessary.

Lemma 10.5.1. *Let V be given as*

$$V(x) = \sum_{j=1}^{k} z_j |x - x_j|^{-1}, \quad z_j > 0.$$

(a) *For all $\rho \in M$, $\phi_\rho = V - W * \rho$ is a continuous function on $\dot{\mathbb{R}}^3 = \mathbb{R}^3 \setminus \{x_1, x_2, \ldots, x_k\}$ with $\phi_\rho(x) \to 0$ for $|x| \to +\infty$.*

(b) *Furthermore, if $\rho \in M$ is a solution of the* TF *equations, which means that if $\rho^{2/3} = \max\{\phi_\rho - \phi_0, 0\}$ is also valid for a certain $\phi_0 > 0$, then it follows that $\phi_\rho(x) \geq 0$, $\forall x \in \mathbb{R}^3$.*

Proof. (a) Since the convolution $f_1 * f_2$ of two functions

$$f_j \in L^{p_j}, \quad 1 < p_j < +\infty, \quad 1/p_1 + 1/p_2 = 1,$$

is a continuous function which vanishes at infinity, the assertion follows from the decomposition

$$W * \rho = W_1 * \rho + W_2 * \rho, \quad W_1 \in L^{5/2}, \quad W_2 \in L^4$$

and taking into account Lemma 10.2.1 (a): this implies

$$\rho \in M \Rightarrow \rho \in L^p, \quad \forall p \in [1, 5/3]$$

and the corresponding properties of V.

(b) We prove the assertion indirectly by showing that the assumption that the set

$$A = [\phi_\rho < 0] = \{x \in \mathbb{R}^3 \mid \phi_\rho(x) < 0\}$$

is nonempty leads to a contradiction.

First note that A is an open set in $\mathbb{R}^3$ since obviously $A \subset \dot{\mathbb{R}}^3$ holds and since according to (a) ϕ_ρ is continuous on $\dot{\mathbb{R}}^3$. $\phi_0 > 0$ implies $\phi_\rho - \phi_0 < 0$ on A and therefore $\rho \restriction A = 0$. We thus have

$$\Delta \phi_\rho = \Delta V + 4\pi\rho = 0$$

on A. ϕ_ρ is then a harmonic function on A which vanishes on the boundary of A by (a). The maximum principle for harmonic functions shows the contradiction $\phi_\rho \restriction A = 0$. □

Theorem 10.5.2. *For potentials of the form*

$$V(x) = \sum_{j=1}^{k} z_j |x - x_j|^{-1}, \quad z_j > 0,$$

$\lambda_0(V)$, which is uniquely determined by Theorem 10.3.6, has the value

$$\lambda_0(V) = \sum_{j=1}^{k} z_j = Z, \tag{10.5.4}$$

Proof. (a) For $\lambda_0 = \lambda(V) > Z$ there exists, according to Theorems 10.3.6 and 10.4.1 solutions ρ_λ of the TF equations corresponding to the particle number $\lambda = \int \rho_\lambda \, dx$ for all $\lambda \in (Z, \lambda_0)$. For sufficiently large r, then,

$$V(r) = -\frac{1}{r} \int_{|x| \leq r} \rho_\lambda(y) \, dy \quad \left(\approx \frac{Z - \lambda}{r} \right)$$

will be negative according to (10.5.3) which is a contradiction to Lemma 10.5.1 (b) and 10.5.2. Therefore we have $\lambda_0 \leq Z$.

(b) By Theorem 10.4.2 the solution of the TF equations for the particle number λ_0 has the chemical potential $\phi_0(\lambda_0) = 0$: this means that

$$\rho_0^{2/3} = \max\{\phi_{\rho_0}, 0\} ,$$

$$\phi_{\rho_0} = V - W * \rho_0$$

and

$$\lambda_0 = \int \rho_0 \, dx = 4\pi \int_0^R dr \, r^2 \widehat{\rho}_0(r) + 4\pi \int_R^{+\infty} dr \, r^2 \widehat{\rho}_0(r) .$$

From Hölder's inequality it follows that

$$\widehat{\rho}_0(r) \geq \widehat{\phi}_{\rho_0}(r)^{3/2} ,$$

and from (10.5.2, 3) $\widehat{\phi}_{\rho_0}$ satisfies the bound

$$\widehat{\phi}_0(r) \geq \frac{Z - \lambda_0}{r} ,$$

for sufficiently large r so that the assumption $\lambda_0 < Z$ leads to the divergence of $\int_R^{+\infty} dr \, r^2 \widehat{\rho}_0(r)$ and therefore to a contradiction. Consequently, one has $\lambda_0 = Z$. □

Remark 10.5.1. (a) It is shown in [10.2] that all potentials of the form

$$V(x) = \sum_{j=1}^{k} z_j |x - x_j|^{-1} + V_0(x) ,$$

where $z_j > 0$, $V_0 \in L^{5/2}$, $\operatorname{supp} V_0$ is compact and $V_0(x) \leq 0$, possess the same $\lambda_0(V) = \sum_{j=1}^{k} z_j = Z$, and for the case $V_0(x) \geq 0$ it is certainly possible that $\lambda_0(V) > Z$.

(b) From Lemma 10.5.1 and Theorem 10.5.2 one can easily derive the properties of the solutions of the TF equations for the particle number $\int \rho_\lambda \, dx = \lambda \leq Z$:

(i) ρ_λ is continuous on $\dot{\mathbb{R}}^3$.
(ii) For $0 < \lambda < Z$ (that is to say for an ionised molecule) ρ_λ has a compact support.

(iii) For $\lambda = Z$, $\rho_Z(x) \to 0$ holds for $|x| \to +\infty$.
(iv) For $x \to x_j$, ρ_λ satisfies

$$\rho_\lambda(x) = \frac{z_j^{2/3}}{|x - x_j|^{3/2}} + \frac{\gamma_j}{|x - x_j|} + \mathrm{O}(|x - x_j|^{1/2}), \quad j = 1, 2, \ldots, k.$$

10.6 Remarks on Recent Developments in Thomas-Fermi and Related Theories

One way of classifying our applications of the direct methods of the calculus of variations is by the space E in which we decide to look for a solution of the problem at hand. Accordingly, we presented applications to classical mechanics (and a preliminary version of classical field theories) in which the space E of a priori solutions was a Banach space of once continuously differentiable functions on a compact space (Chap. 5). Then, in Chaps. 6–8 we used for E a "local" Sobolev space $W^{1,p}(G)$, $1 < p < \infty$, in order to treat linear and semilinear elliptic boundary and eigenvalue problems over a bounded domain $G \subset \mathbb{R}^d$, $d \geq 3$. Chapter 9 covered the existence of solutions of systems of global semilinear elliptic equations and presented several applications of the direct methods of the calculus of variations in the spaces $E = E^{1,2}(\mathbb{R}^d; \mathbb{R}^n)$, $d \geq 2$, $n = 1, 2, \ldots$ This final Chap. 10 illustrates the successful applications of these methods to a recent important minimisation problem in a "global" Lebesgue space $E = L^p(\mathbb{R}^d)$, $1 < p < \infty$, $d \geq 3$. The purpose of this chapter is thus not to give a comprehensive treatment of the Thomas-Fermi theory but, as mentioned at the beginning of the introduction to this chapter, to give only a presentation of the variational aspects of the solution of this theory.

Clearly this theory, together with its applications and extensions, is very interesting in itself and accordingly deserves a comprehensive treatment of its own. Fortunately, an encyclopaedic, detailed exposition of the whole theory up to 1981 is available ([10.40]).

We think that the solution of the Thomas-Fermi equations (10.1.2) is a good example of the power and simplicity of the direct methods of the calculus of variations. In order to really appreciate the achievements of *E. H. Lieb* and *B. Simon* in [10.2] one has to recall some facts about the history of Thomas-Fermi theory. The Thomas-Fermi equation was suggested as early as 1927 [10.7, 8]. Since then it has attracted much attention and stimulated a lot of research. Though approximate numerical calculations suggested that this equation had a (unique) solution, it was not until 1969, that E. Hille succeeded in proving the existence of a solution for the spherically symmetric case. A "direct proof" of the solvability of the Thomas-Fermi equation for the general case was given by *H. Brezis* in 1978 after the proof by E. H. Lieb and B. Simon in 1977. And it turned out that the proof by variational methods is far simpler.

The power and simplicity of the direct methods as exemplified in [10.2] encouraged many researchers to try to get on this basis a more realistic description of the physical world of atoms, molecules, and large systems. There are several possibilities for improving or extending the original Thomas-Fermi theory which was shown in [10.2] to be asymptotically exact in the limit $Z \to \infty$ in a nonrelativistic setting. The main directions of these investigations are to include

(a) temperature dependence,
(b) nonrelativistic corrections to the kinetic energy functional in the Thomas-Fermi energy functional $\mathcal{E}$,
(c) relativistic corrections to the kinetic energy functional,
(d) effects of magnetic fields,

and thus many interesting new applications of the direct methods of the calculus of variations have been stimulated in various function spaces E. To give a full account of all these recent developments would fill another book. Therefore, we can only suggest some references for further reading.

Temperature-dependent Thomas-Fermi theory is essentially as old as the Thomas-Fermi theory itself and has taken a somewhat independent line of development [10.9, 10].

Important early suggestions for corrections to the Thomas-Fermi energy functional are due to F. Dyson and C. F. von Weizsäcker. They suggested using the functionals

$$\mathcal{E}^{\mathrm{TFD}}(\rho) = \mathcal{E}(\rho) - c \int \rho(x)^{4/3}\, dx$$

and

$$\mathcal{E}^{\mathrm{TFW}}(\rho) = \mathcal{E}(\rho) + \delta \int [\nabla \rho^{1/2}(x)]^2\, dx\,.$$

The latter functional has been studied by methods similar to those in Thomas-Fermi theory and similar conclusions have been drawn [10.11]. Notice, however, the additional complications due to the gradient term which requires to work in a global Sobolev space instead of a global Lebesgue space.

The effects of a relativistic kinetic energy term have been investigated in [10.12, 13]. Finally, a complicated minimisation problem of a different type arises if one includes magnetic fields. Nevertheless, the direct methods of the calculus of variation have again been powerful enough to prove the stability of Coulomb systems with magnetic fields [10.14–16].

Just as the proof of the existence of a solution of the Thomas-Fermi theory was a rather important step in the proof of the stability of matter in [10.3, 17], some of these extended and refined versions of Thomas-Fermi theory allow important conclusions about the stability of atoms, molecules, and stars under more realistic conditions. A particularly beautiful presentation of the main physical and mathematical aspects of this problem has been given recently in [10.18].

Appendix A. Banach Spaces

In the appendices A, B and C we summarise the fundamental concepts and results we require from the theory of Banach spaces. For the proofs, we shall refer to well-known textbooks, and we give a limited selection at the end. We assume that the reader is familiar with the fundamental topological concepts (topology, neighborhood, basis of neighborhoods, open, contact point, etc.) and give full proofs only for a few results from the theory of Banach spaces which are of central significance for our applications.

As a rule, we base our considerations on a real vector space. In order to be able to deal with topological concepts such as open sets, limit points, convergent sequences, Cauchy sequences, etc. in a vector space X, such as in the Euclidean space $\mathbb{R}^n$, we will assume that a generalised "concept of length" or a "seminorm" p is given in X. A *seminorm* is a function $p : X \to [0, \infty)$ with the properties $p(\lambda x) = |\lambda| p(x)$ and $p(x+y) \leq p(x) + p(y)$ for all $\lambda \in \mathbb{R}$ and all $x, y \in X$. If $p(x) = 0$ always implies $x = 0$, a seminorm is called a *norm*, which is written as $p(x) = \|x\|$. A vector space X together with a norm $\|\cdot\|$ defined for its elements is called a *normed space.* In a normed space $(X, \|\cdot\|$, we obtain a topology, and hence all topological concepts, by defining the following sets as open sets. A subset $A \subset X$ is called *open* if, for every $x \in A$, there exists an $r = r_x > 0$ such that the *open ball*

$$B_r(x) = \{y \in X \mid \|y - x\| < r\}$$

with centre x and radius r belongs to $A : B_r(x) \subset A$. In a normed space $(X, \|\cdot\|)$ a *sequence* $(x_n)_{n \in \mathbb{N}}$ *converges to a point* $x \in X$ if, and only if, for every $\varepsilon > 0$ there exists an n_ε such that $\|x_n - x\| < \varepsilon$ holds for all $n \geq n_\varepsilon$. A sequence $(x_n)_{n \in \mathbb{N}}$ is called a *Cauchy sequence* if, and only if, for every $\varepsilon > 0$ there exists an n_ε such that for all $n, m \geq n_\varepsilon$, $\|x_n - x_m\| < \varepsilon$ holds. A normed space in which every Cauchy sequence converges is said to be *complete*, or a *Banach space.* A Banach space is said to be *separable* if there exists a countable dense subset of it.

Two norms $\|\cdot\|_1$ and $\|\cdot\|_2$ on a vector space X are said to be *equivalent* if there exist positive numbers c, c' such that $\|x\|_1 \leq c\|x\|_2$ and $\|x\|_2 \leq c'\|x\|_1$ hold for all $x \in X$. Equivalent norms generate the same topology. A subset M of a normed space $(X, \|\cdot\|)$ is said to be *bounded* if the norm is bounded on M:

$\sup\{\|x\| \mid x \in M\} < \infty$. Equivalent norms have identical bounded sets. Complete normed spaces, which can be regarded as (infinite-dimensional) analogous of the Euclidean spaces $\mathbb{R}^n$, can be obtained in the following manner: let there be a *scalar product* $\langle\cdot,\cdot\rangle$ defined on a vector space X. This is a function $\langle\cdot,\cdot\rangle : X \times X \to \mathbb{R}$ with the following properties:

$$\langle x, \lambda_1 y_1 + \lambda_2 y_2\rangle = \lambda_1\langle x, y_1\rangle + \lambda_2\langle x, y_2\rangle\,, \quad \langle x, y\rangle = \langle y, x\rangle$$

for all $x, y_1, y_2, y \in X$ and all $\lambda_j \in \mathbb{R}$, $\langle x, x\rangle \geq 0$, and $\langle x, x\rangle = 0$ for all $x \in X$ only when $x = 0$. A vector space X with a scalar product $\langle\cdot\,,\cdot\rangle$ is called a *pre-Hilbert* space. In a pre-Hilbert space $(X, \langle\cdot\,,\cdot\rangle)$ one has a norm defined by $x \to \|x\| = +\sqrt{\langle x, x\rangle}$, called the *norm induced by the scalar product.* If the normed space $(X, \|\cdot\| = \sqrt{\langle\cdot\,,\cdot\rangle})$ is complete, then $(X, \langle\cdot\,,\cdot\rangle)$ is called a *Hilbert space.*

The *Theorem of Jordan and von Neumann* characterises those norms which are induced by a scalar product. The characteristic of these norms is the parallelogram law

$$\|x_1 + x_2\|^2 + \|x_1 - x_2\|^2 = 2\|x_1\|^2 + 2\|x_2\|^2$$

for all $x_j \in X$.

The *Cauchy-Schwarz* inequality $|\langle x, y\rangle| \leq \|x\| \cdot \|y\|$ for all $x, y \in X$ holds in every pre-Hilbert space.

The *Projection Theorem* is valid in every Hilbert space just as in any finite-dimensional Euclidean space: let M be a subspace (which means a closed vector subspace) of a Hilbert space $\mathcal{H}$. Then every vector $x \in \mathcal{H}$ can be uniquely represented as $x = u_x + v_x$ where $u_x \in M$ and $v_x \in M^\perp = \{x \in \mathcal{H} \mid \langle x, y\rangle = 0,\ \forall y \in M\}$, and one has

$$\|v_x\| = c(x, M) = \inf_{u \in M} \|x - u\| = \|x - u_x\|\,.$$

We then write $\mathcal{H} = M \oplus M^\perp$ and say that $\mathcal{H}$ is the *direct orthogonal sum* of M and $M^\perp$. If the orthogonal subspace $M^\perp$ of M is finite dimensional, then we say that M has *finite codimension.*

We now come to a few examples.

Example A.1. *The Euclidean Spaces* $\mathbb{R}^n$, $n \in \mathbb{N}$. One can introduce the following norms in the vector space $\mathbb{R}^n$: for an arbitrary number $p \in [1, \infty)$ and $x \in \mathbb{R}^n$ one sets

$$\|x\|_p = \left(\sum_{i=1}^{n} |x_i|^p\right)^{1/p}, \quad \|x\|_\infty = \sup_{1 \leq i \leq n} |x_i|\,.$$

It follows from the inequalities $\|x\|_\infty \leq \|x\|_p \leq n^{1/p}\|x\|_\infty$, $x \in \mathbb{R}^n$, that all these norms are equivalent. $(\mathbb{R}^n, \|\cdot\|_p)$ is a Banach space for $1 \leq p \leq +\infty$. It is only for $p = 2$ that $(\mathbb{R}^n, \|\cdot\|_p)$ is a Hilbert space with the scalar product

$$\langle x, y \rangle = \sum_{i=1}^{n} x_i y_i \,.$$

Example A.2. *The Sequence Spaces* $l^p(\mathbb{R})$, $1 \le p \le \infty$. Let $l^p(\mathbb{R})$ denote the vector space of all sequences $x = (x_n)_{n\in\mathbb{N}}$ of real numbers x_n for which

$$\sum_{n=1}^{\infty} |x_n|^p < \infty\,, \quad 1 \le p < \infty \quad \text{and} \quad \sup_{n\in\mathbb{N}} |x_n| < \infty \quad \text{for } p = +\infty$$

hold. Together with the norms

$$\|x\|_p = \left(\sum_{n=1}^{\infty} |x_n|^p\right)^{1/p}, \quad 1 \le p < \infty\,,$$

$$\|x\|_\infty = \sup_{n\in\mathbb{N}} |x_n|\,,$$

respectively, the sequences spaces $(l^p(\mathbb{R}), \|\cdot\|_p)$ are complete normed spaces. In this case different values of p give rise to nonequivalent norms. Once again, only the sequence space $l^2(\mathbb{R})$ is a Hilbert space. Its scalar product is $\langle x, y \rangle = \sum_{n=1}^{\infty} x_n y_n$. For $1 \le p \le +\infty$ the *dual exponent* p' corresponding to p is defined by the equation $1/p' + 1/p = 1$. For all $x \in l^p(\mathbb{R})$ and all $y \in l^{p'}(\mathbb{R})$, one has $x \cdot y = (x_n y_n)_{n\in\mathbb{N}} \in l^1(\mathbb{R})$ and

$$\|x \cdot y\|_1 = \sum_{n=1}^{\infty} |x_n y_n| \le \|x\|_p \|y\|_{p'} \text{ (Hölder's Inequality)}\,.$$

Example A.3. *Spaces of Continuous Functions.* Let K denote a compact topological space (see also Appendix C) and $\mathcal{C}(K;\mathbb{R}) = \mathcal{C}(K)$, be the set of all continuous functions $f : K \to \mathbb{R}$. One knows that $\mathcal{C}(K)$ is a real vector space in a natural way and $\|f\|_K = \sup_{x\in K} |f(x)|$ defines a norm on $\mathcal{C}(K)$. $\mathcal{C}(K)$ is complete with respect to this norm, and is thus a Banach space but not a Hilbert space.

Example A.4. *Spaces of Differentiable Functions.* For a compact interval $I = [a, b]$, let $\mathcal{C}^1(I;\mathbb{R}^n)$ be the space of all n-tuples of real functions continuously differentiable in (a, b), which together with their derivatives possess continuous extensions to the boundary points $t = a$ and $t = b$. The real vector space $\mathcal{C}^1(I, \mathbb{R}^n)$ is a Banach space with respect to the norm

$$\|q\|_{1,I} = \sup_{t\in I}\{|q(t)|, |\dot{q}(t)|\}\,.$$

For $q(t) = (q_1(t)\,, \dots, q_n(t))$, $q_i \in \mathcal{C}^1(I, \mathbb{R})$, let

$$\dot{q}(t) = (\dot{q}_1(t)\,, \dots, \dot{q}_n(t))\,, \quad \dot{q}_i(t) = \frac{dq_i}{dt}(t)$$

and

$$|q(t)| = \left(\sum_{i=1}^{n} |q_i(t)|^2 \right)^{1/2} .$$

A norm equivalent to $\| \cdot \|_{1,I}$ is given by the following:

$$\|q\|'_{1,I} = \sup_{t \in I} \sup_{1 \le i \le n} \{ |q_i(t)|, |\dot{q}_i(t)| \} .$$

Example A.5. *The Lebesgue Spaces* $L^p(\mathbb{R}^n)$. For every real number $p \ge 1$ let $\mathcal{L}^p(\mathbb{R}^n) = \{ f : \mathbb{R}^n \to \mathbb{R} \,|\, f \text{ measurable, } |f| \text{ summable} \}$. $\mathcal{L}^p(\mathbb{R}^n)$ is a real vector space and $f \to \|f\|_p$ where

$$\|f\|_p = \left(\int_{\mathbb{R}^n} |f(x)|^p \, dx \right)^{1/p}$$

is a seminorm on $\mathcal{L}^p(\mathbb{R}^n)$. One has $\|f\|_p = 0 \Leftrightarrow f = 0$ almost everywhere. Denoting by N the subspace of all functions from $\mathcal{L}^p(\mathbb{R}^n)$ which vanish almost everywhere, the quotient space

$$L^p(\mathbb{R}^n) = \mathcal{L}^p(\mathbb{R}^n)/N$$

together with the norm induced by $\| \cdot \|_p$ (which is again denoted by $\| \cdot \|_p$) is a Banach space (*Theorem of Riesz-Fischer*).

Similarly one defines for $p = +\infty$,

$$\mathcal{L}^\infty(\mathbb{R}^n) = \{ f : \mathbb{R}^n \to \mathbb{R} \,|\, f \text{ measurable and bounded almost everywhere} \}$$
$$\|f\|_\infty = \inf\{ c \,|\, c > 0 \text{, and apart from a null set } |f(x)| \le c \text{ holds} \}$$

and, further,

$$L^\infty(\mathbb{R}^n) = \mathcal{L}^\infty(\mathbb{R}^n)/N .$$

Then $L^\infty(\mathbb{R}^n)$ is also a Banach space.

Just as in the case of sequence spaces, *Hölder's inequality* holds here also, which means that if $f \in L^p(\mathbb{R}^n)$, $g \in L^{p'}(\mathbb{R}^n)$ with p' being the dual exponent to $p \ge 1$, then $f \cdot g \in L^1(\mathbb{R}^n)$, and one has

$$\|f \cdot g\|_1 \le \|f\|_p \|g\|_{p'} .$$

Once again we obtain a Hilbert space only for $p = 2$. $L^2(\mathbb{R}^n)$ is a Hilbert space with the scalar product

$$\langle f, g \rangle_2 = \int_{\mathbb{R}^n} f(x) g(x) \, dx .$$

The norm thereby induced is $\| \cdot \|_2$. (In our notation we do not make explicit the distinction between the equivalence classes and the representatives from the equivalence classes.)

Replacing the space $\mathbb{R}^n$ by a measurable set $G \subseteq \mathbb{R}^n$ one obtains the Banach spaces $L^p(G)$, $1 \leq p \leq +\infty$, of Lebesgue. The Banach spaces $L^p(G)$ with $1 \leq p < \infty$ are separable.

For real vector spaces X_1 and X_2, let $L(X_1, X_2)$ be the collection of all linear maps or operators from X_1 to X_2. $L(X_1, X_2)$ is again a vector space in a naturally way. If $X_2 = \mathbb{R}$, then $X_1^* = L(X_1, \mathbb{R})$ is called the *algebraic dual space* of X_1. For normed spaces, the vector subspace $\mathcal{L}(X_1, X_2)$ of continuous linear maps from X_1 to X_2 is what we are mainly interested in. It can be shown that a linear map $T: X_1 \to X_2$ is continuous only when it is *bounded*, which means that it maps bounded subsets of X_1 onto the bounded subsets of X_2. A nonlinear map which is bounded is not necessarily continuous.

The vector space $\mathcal{L}(X_1, X_2)$ becomes a normed space by defining

$$T \to \|T\| = \sup\{\|Tx\|_2 \mid \|x\|_1 \leq 1\}.$$

Here $\|\cdot\|_i$ denotes the norm for the space X_i. If X_2 is a Banach space, then $(\mathcal{L}(X_1, X_2), \|\cdot\|)$ is also a Banach space. In particular, the space $\mathcal{L}(X, \mathbb{R})$ of all continuous linear functionals of a normed space $(X, \|\cdot\|)$ in $\mathbb{R}$ is a Banach space. The norm for the (topologically) *dual space* $X' = \mathcal{L}(X, \mathbb{R})$ of X is given by

$$\|T\|' = \sup\{|Tx| \mid \|x\| \leq 1\}.$$

An operator $T \in \mathcal{L}(X_1, X_2)$ is called a *homeomorphism* from X_1 onto X_2 if T is bijective and T^{-1} is continuous. If there exists a homeomorhism from X_1 onto X_2, then the normed spaces X_1 and X_2 are said to be *homeomorphic.* If X_1 and X_2 are Banach spaces, then every bijective operator $T \in \mathcal{L}(X_1, X_2)$ is a homeomorphism (*theorem of the inverse operator*).

The (topologically) dual space $(X')'$ of a dual space X' is called the *bidual space* X'' of the normed space X, and thus $X'' = \mathcal{L}(X', \mathbb{R})$. It can easily be seen that X is embedded in X'' in a natural way due to the canonical map

$$J: X \to X'', \quad J(x)(f) = f(x) = \langle f, x \rangle, \quad f \in X', \quad x \in X.$$

Here $\langle \cdot, \cdot \rangle$ expresses the *duality* between X' and X, which means that the bilinear map $X' \times X \to \mathbb{R}$ is defined by $\langle f, x \rangle = f(x)$. Since the canonical embedding J of X in X'' is isometric, which means that J is linear, and since $\|J(x)\|'' = \|x\|$, $\forall x \in X$, it is always injective and continuous. If X is a Banach space and J is surjective, then J is a homeomorphism from X onto X''. In this case the Banach space X is *reflexive.* A reflexive Banach space is thus characterised by $J(X) = X''$. In this case one assumes $J(X)$ and X to be identical. Then one can state with Plessner that either $X = X''$ holds, or that all the spaces X', X'', X''', ... are different.

A Banach space X is reflexive if and only if its dual space X' is reflexive. All finite-dimensional Banach spaces and all Hilbert spaces are *examples of reflexive Banach spaces.* For the case of Hilbert spaces this follows from the Theorem of Riesz-Fréchet: *Theorem of Riesz-Fréchet*: consider an element T

of the dual space $\mathcal{H}'$ of a Hilbert space $\mathcal{H}$. Then there exists a unique $u \in \mathcal{H}$ such that for all $x \in \mathcal{H}$, $T(x) = \langle u, x \rangle$ and consequently $\|T\|' = \|u\|$ holds.

Every closed subspace of a reflexive Banach space is itself a reflexive Banach space.

Most of the Banach spaces used in applications are *uniformly convex*, which means that corresponding to every $\varepsilon \in (0,2)$ there exists a $\delta = \delta_\varepsilon(0,1)$ such that for all $r > 0$ the following holds:

$$x, y \in X\,, \quad \|x\| \leq r\,, \quad \|y\| \leq r \quad \text{and} \quad \|x-y\| \geq \varepsilon r \;\Rightarrow\; \|x+y\| \leq 2(1-\delta)r\,.$$

The uniform convexity is a metric property. Uniformly convex Banach spaces possess the *approximation property*: for a closed *convex* subset $K \subseteq X$ ($x_i \in K$, $0 \leq \lambda \leq 1 \Rightarrow \lambda x_1 + (1-\lambda)x_2 \in K$) every $x \in X$ can be 'approximated optimally' by a $u_x \in K$ in the sense that

$$d(x, K) = \inf_{y \in K} \|x - y\| = \|x - u_x\|$$

holds. These Banach spaces are therefore of great importance for approximation theory.

Every uniformly convex Banach space is reflexive (*Theorem of Milman*). Important examples of uniformly convex spaces are the Lebesgue spaces $L^p(G)$, $G \subseteq \mathbb{R}^n$ measurable, $1 < p < +\infty$ (*Theorem of Clarkson*). The reflexivity for these spaces also follows directly from the explicit construction of the dual and bidual spaces. For $1 < p < \infty$ one has, namely, the results that $(L^p(G))'$ and $L^{p'}(G)$, with $1/p + 1/p' = 1$, are isometrically isomorphic by virtue of the canonical mapping

$$j : L^{p'}(G) \to (L^p(G))'\,, \quad j(f)(g) = \int_G f(x)g(x)\,dx\,,$$

for all $g \in L^p(G)$ and all $f \in L^{p'}(G)$.

Similarly, $L^\infty(G)$ is isometrically isomorphic to $(L^1(G))'$; however, $L^1(G)$ is not isomorphic to $(L^\infty(G))'$.

Until now the topology induced by the norm of a normed space has always been taken as the basis of our considerations. This topology is called the *norm topology* or also the *strong topology*. Another important topology on a normed space X is the *weak topology* $\sigma = \sigma(X, X')$. This is generated by the seminorms $\{q_f \mid f \in X'\}$ where q_f is defined as $q_f(x) = |f(x)|$ for $x \in X$. *Weak neighborhoods* of a point x have thus the typical form

$$U^\varepsilon_{f_1 \dots f_n} = \{x' \in X \mid q_{f_i}(x' - x) < \varepsilon,\, i = 1, 2, \dots, n\}\,, \quad f_i \in X'\,, \quad \varepsilon > 0\,.$$

The weak topology $\sigma(X, X')$ is the coarsest locally convex topology on X with respect to which all $f \in X'$ are continuous.

A sequence $\{x_n\}_{n \in \mathbb{N}}$ *converges weakly* to a point $x \in X$ if for every $f \in X'$ one has $\lim_{n\to\infty} f(x_n) = f(x)$. We write, for short,

$$x = w - \lim_{n\to\infty} x \quad \text{or} \quad x_n \xrightarrow[n\to\infty]{w} x\,.$$

The concept of a *weak Cauchy sequence* is defined in a similar fashion. A Banach space X is said to be *weakly complete* if every weak Cauchy sequence in X converges weakly. More precisely, one should express this as *weakly sequentially complete.* Every Hilbert space is weakly complete.

It should be clear by now that *strong convergence* (which means convergence with respect to the strong topology) always implies *weak convergence.* But the converse of this is not always true. However, for finite-dimensional spaces the weak and strong topologies agree.

A set $M \subset X$ is said to be *weakly bounded* if it is bounded with respect to the weak topology $\sigma(X, X')$. A set $M \subset X$ is weakly bounded if, and only if, it is 'pointwise' bounded, which means the sets $f(M) = \{f(x) \mid x \in M\}$ are bounded in $\mathbb{R}$ for all $f \in X'$. Obviously the strongly bounded sets are weakly bounded. The converse is also true, the proof of which being based on a fundamental result of functional analysis, the *principle of uniform boundedness* (*Theorem of Banach-Steinhaus*): let X_1 be a Banach space, X_2 a normed space and $\mathcal{F} \subset \mathcal{L}(X_1, X_2)$ a family of continuous linear maps from X_1 to X_2. If $\mathcal{F}$ is *pointwise bounded*, which means that

$$\sup_{f\in\mathcal{F}}\{\|f(x)\|_2\} = c_x < \infty$$

holds for every $x \in X_1$, then $\mathcal{F}$ is bounded in $\mathcal{L}(X_1, X_2)$, i.e. bounded with respect to the norm, which means explicitly that

$$\sup_{f\in\mathcal{F}} \|f\| = \sup_{f\in\mathcal{F}} \sup_{\substack{x\in X_1 \\ \|x\|_1\le 1}} \|f(x)\|_2 = c < \infty\,.$$

This result can also be formulated as follows: if $\sup_{f\in\mathcal{F}} \|f\| = +\infty$ holds, then there exists an $x \in X_1$ such that $\sup_{f\in\mathcal{F}} \|f(x)\|_2 = +\infty$.

Let us elaborate on the proof of this result which is used often. The fact that $\mathcal{F}$ is pointwise bounded shows that every point $x \in X_1$ lies in one of the sets

$$A_n = \{y \in X_1 \mid \|f(y)\|_2 \le n,\ \forall f \in \mathcal{F}\}$$

and thus $X_1 = \bigcup_{n=1}^{\infty} A_n$. The continuity of $f \in \mathcal{F}$ implies that all the sets A_n are closed. Therefore, the Banach space X_1 is the countable union of closed sets A_n. Thus, one of the sets A_n contains an open ball, e.g. $B_{r_{n_0}}(x) \subset A_{n_0}$ (*Baire's Theorem*). Consequently, one has $\|f(x)\|_2 \le n_0$ for all $x \in B_{r_{n_0}}(x_0)$ and all $f \in \mathcal{F}$, and therefore $\|f\| \le 2n_0/r$ for all $f \in \mathcal{F}$.

The principle of uniform boundedness is thus one of the important consequences of *Baire's Theorem*: if a Banach space X is a countable union of closed sets E_n, $X = \bigcup_{n=1}^{\infty} E_n$, then at least one of the sets E_n possesses a nonempty interior. (This theorem is valid in every complete metric space.)

The *proof* of this theorem can be given as follows: if the closed sets E_n all have empty interiors, then it follows that $\bigcup_{n=1}^{\infty} E_n \subsetneq X$. To prove this statement construct a sequence of open balls $B_{r_n}(x_n)$ in X such that

$$\overline{B_{r_n}}(x_n) \subset B_{r_{n-1}}(x_{n-1}) \quad \overline{B}_{r_n}(x_n) \cap E_n = \emptyset \quad \text{and} \quad r_n \underset{n\to\infty}{\longrightarrow} 0\,,$$

where the bar indicates closure. This construction can be done as follows: since $E_1^c = X \setminus E_1$ is open, there exists a closed ball $\overline{B}_{r_1}(x_1) \subset E_1^c$. If these balls up to the number $n-1$ have been constructed ($n > 1$), then we construct the ball $B_{r_n}(x_n)$ as follows: since E_n^c is an open, dense set in X, the intersection $B_{r_{n-1}}(x_{n-1}) \cap E_n^c$ is an open, nonempty set and therefore contains a closed ball $\overline{B}_{r_n}(x_n)$ where $0 < r_n < r_{n-1}/2$. By construction one has $x_m \in B_{r_n}(x_n)$ for all $m \geq n$; it follows therefore that $\|x_i - x_j\| \leq 2r_n$ for all $i, j \geq n$. Thus $(x_n)_{n\in\mathbb{N}}$ is a Cauchy sequence which converges to a point $x_0 \in X$ due to the completeness of X. It follows then that $x_0 \in \overline{B_{r_n}}(x_n)$ for all n and therefore by construction $x_0 \notin E_n$ for all $n \in \mathbb{N}$, or equivalently $x_0 \in X \setminus \bigcup_{n=1}^{\infty} E_n$.

As a simple but important application of the Theorem of Banach-Steinhaus we now prove that every weakly bounded subset M of a normed space X is strongly bounded. Regarding M as the subset $J(M)$ of $X'' = \mathcal{L}(X', \mathbb{R})$ by virtue of the canonical isometry $J : X \to X''$, M is pointwise bounded: for all $x \in M$ and every $f \in X'$ one has $|J(x)(f)| = |f(x)| \leq c_f < \infty$. Since $(X', \|\cdot\|')$ is a Banach space, one has $\sup_{x\in M} \|J(x)\|'' = c < \infty$ which follows from the isometry of J. Furthermore, the following often-used result also follows: every weakly convergent sequence in a normed space is bounded (with respect to the norm). This is due to the fact that a weakly convergent sequence is weakly bounded.

In general no useful criteria exist for controlling the relationship between strong and weak convergences. However, the Hilbert spaces are an exception in this regard. Kadez has proved that a sequence $(x_n)_{n\in\mathbb{N}}$ in a Hilbert space converges strongly to an $x \in \mathcal{H}$ if and only if it converges weakly to x and the sequence of norms converges, i.e. $\lim_{n\to\infty} \|x_n\| = \|x\|$. This is actually true in every uniformly convex space, and in particular also in every L^p space where $1 < p < +\infty$.

A property of *convex* sets of a normed space frequently used is the result that the closure with respect to the weak topology is identical to the closure with respect to the norm topology. This is in fact a simple consequence of the Hahn-Banach separation theorem.

Appendix B. Continuity and Semicontinuity

It was *Baire* who first introduced the concept of a semicontinuous real function, at the beginning of this century. This concept arises as a natural generalisation of the concept of a continuous function.

Let us recall that a real function f on a Hausdorff space X is *continuous* at a point $x_0 \in X$ if, and only if, for every $\varepsilon > 0$ there exists a neighborhood $U = U_\varepsilon(x_0) \subset X$ of x_0 such that for all $x \in U$ one has

$$f(x_0) - \varepsilon < f(x) < f(x_0) + \varepsilon\,,$$

which means

$$U_\varepsilon(x_0) \subset [f > f(x_0) - \varepsilon] \cap [f < f(x_0) + \varepsilon]$$

where we have set $[f > \lambda] = \{x \in X \mid f(x) > \lambda\}$ and, correspondingly, $[f < \lambda]$.

If one imposes this requirement only for one of the sets $[f > f(x_0) - \varepsilon]$ or $[f < f(x_0) + \varepsilon]$, then one arrives at the concept of a semicontinuous function: a function $f : X \to \mathbb{R} \cup \{+\infty\}$ is said to be *lower semicontinuous at a point* $x_0 \in X$ if for every $\varepsilon > 0$ there exists a neighborhood $U_\varepsilon(x_0)$ of x_0 such that

$$U_\varepsilon(x_0) \subset [f > f(x_0) - \varepsilon]\,.$$

Analogously $f : X \to \mathbb{R} \cup \{-\infty\}$ is said to be *upper semicontinuous at a point* $x_0 \in X$ if for every $\varepsilon > 0$ there exists a neighborhood $U_\varepsilon(x_0)$ such that

$$U_\varepsilon(x_0) \subset [f < f(x_0) + \varepsilon]\,.$$

Therefore, it follows that f is upper semicontinuous at a point x_0 if, and only if, $-f$ is lower semicontinuous at x_0. f is continuous at x_0 if, and only if, f is simultaneously upper and lower semicontinuous at x_0.

f is said to be *lower (upper) semicontinuous on* X if at all points of X, f is lower (upper) semicontinuous.

f is lower semicontinuous on X if, and only if, all the sets $[f > \lambda]$, $\lambda \in \mathbb{R}$ are open.

Semicontinuity, like continuity, is a local property.

In order to test the continuity of a function one often utilises the sequence criterion. A corresponding criterion for lower semicontinuity is also known.

Lemma B.1. *Let X be a Hausdorff space and $f : X \to \mathbb{R} \cup \{+\infty\}$ a function on X. Then the following is true:*

(a) *If f is lower semicontinuous at $x_0 \in X$, then for every sequence $(x_n)_{n\in\mathbb{N}} \subset X$ converging to x_0 one has*

$$f(x_0) \leq \liminf_{n\to\infty} f(x_n)\,.$$

(b) *If X satisfies the first axiom of countability, then the converse of the statement* (a) *is also true.*

Proof. Let us first recall the following characterisation of $\liminf_{n\to\infty} a_n$ of a sequence $(a_n)_{n\in\mathbb{N}} \subset \mathbb{R}$:

$$a = \liminf_{n\to\infty} a_n \iff \text{for every } \varepsilon > 0 \text{ one has}$$

(i) $a - \varepsilon < a_n$ for almost all n and
(ii) $a_n < a + \varepsilon$ for infinitely many n.

(a) Let f be lower semicontinuous at $x_0 \in X$ and $(x_n)_{n\in\mathbb{N}} \subset X$ be a sequence converging to x_0. Now set

$$A = \liminf_{n\to\infty} f(x_n)\,.$$

If $A < f(x_0)$, then there would exist an $\varepsilon > 0$ such that $A + \varepsilon < f(x_0)$. Since f is lower semicontinuous at x_0, there exists a neighborhood U of x_0 for which $f(x) > A + \varepsilon$ holds for all $x \in U$. It follows then that there exists an n_0 such that for all $n \geq n_0$ one has $x_n \in U$ and therefore $f(x_n) > A + \varepsilon$ holds, which is in contradiction to condition (ii) in the characterisation of A.

(b) Now let

$$f(x_0) \leq \liminf_{n\to\infty} f(x_n)$$

hold for all sequences $(x_n)_{n\in\mathbb{N}} \subset X$ converging to x_0. If f were not lower semicontinuous at x_0, then there would exist a $\lambda \in \mathbb{R}$, $\lambda < f(x_0)$, so that x_0 would not be an interior point of $[f > \lambda]$, and therefore for all neighborhoods U of x_0 there would exist a point $x_u \in U$ which did not belong to $[f > \lambda]$. x_0 would then be a point of accumulation of $[f \geq \lambda]$. Since X satisfies the first countability axiom, x_0 is the limit of a sequence $(x_n)_{n\in\mathbb{N}} \subset [f \leq \lambda]$. We would therefore have $f(x_n) \leq \lambda$ and thus

$$\liminf_{n\to\infty} f(x_n) \leq \lambda < f(x_0)$$

in contradiction to the assumption. Consequently x_0 is an interior point of all sets $[f > \lambda]$, $\lambda < f(x_0)$, which means that f is lower semicontinuous at x_0. □

If X does not satisfy the first axiom of countability, then a function $f : X \to \mathbb{R} \cup \{+\infty\}$ which satisfies the conditions of Lemma B.1 is said to be *sequentially lower semicontinuous.*

Appendix C. Compactness in Banach Spaces

The direct methods of the calculus of variations are based on the properties of the compact sets and the (semi)continuity of functions on such sets. We therefore give here an outline of the most important aspects of compactness.

(a) *The Most Important Notions of Compactness.* Intuitively, the simplest sets are those which consist of a finite number of elements. The sets consisting of a finite number of "simple" sets then appear as a natural generalisation of this. We obtain the definition of compact sets by making this concept more precise: a Hausdorff topological space X is said to be *compact* if every open covering

$$X = \bigcup_{\alpha \in A} O_\alpha \,, \quad O_\alpha \text{ open} \,,$$

of X admits of a finite subcovering

$$X = \bigcup_{j=1}^{N} O_{\alpha_j} \,, \quad \alpha_j \in A \,.$$

A subset M of X is called *compact* if M is a compact space with respect to the relative topology. $M \subset X$ is said to be *relatively compact* if the closure $\overline{M}$ of M is compact. Every compact set is closed.

The weak topology $\sigma(X, X')$ of a normed space X does not always satisfy the first axiom of countability (that is to say that every point of X possess a countable basis of neighborhoods) or the second axiom of countability (which means that $\sigma(X, X')$ possesses a countable basis). Therefore one often utilises also the following concepts of compactness which differ in general from that introduced above. A subset M of a Hausdorff space is said to be *countably compact* if every open covering of M by a countable family of sets admits of a finite subcovering. A subset M of a Hausdorff space X is countably compact if, and only if, every infinite subset of M possesses at least one accumulation point.

A subset M of a Hausdorff space X is said to be *sequentially compact* if every sequence in M possesses a convergent subsequence. The notion of *relative countable compactness* and *relative sequential compactness* are constructed in analogy to the notion of relative compactness.

(b) *Equivalence of the Various Notions of Compactness.* The following relations between these notions of compactness are obvious. Every compact set is countably compact. Every sequentially compact set is countably compact. In general the converse is not true for either of the cases. However, there are also many important examples where the three concepts of compactness come together. For this let us first consider the following.

If in a Hausdorff space X the first axiom of countability holds, then every cluster point of a subset $M \subset X$ is the limit of a convergent sequence in M. Therefore, if in a Hausdorff space X the first axiom of countability holds, then every countably compact set is sequentially compact. If, moreover, the second axiom of countability is valid, then every countably compact set is compact. If the first axiom of countability holds, then every compact set is sequentially compact.

The following diagram shows once again the relationship between the different notions of compactness (CA = countability axiom):

compact

2nd CA $\nearrow\!\swarrow$ $\qquad\qquad$ $\searrow\!\nwarrow$ 2nd CA

countably compact $\underset{\text{1st CA}}{\overset{\text{1st CA}}{\leftrightarrows}}$ sequentially compact

It follows that in a separable Banach space the notions of "compactness" and "sequential compactness" are the same, a fact which has been decisively utilised in the calculus of variations.

(c) *Examples of Compact Sets.* As is well known, a subset in $\mathbb{R}^n$ is compact if, and only if, it is bounded and closed (*Heine-Borel Covering Theorem*).

In infinite-dimensional normed spaces this characterisation of the compact sets is no longer valid. This is because here, the *Theorem of F. Riesz* holds: a normed space is finite dimensional if, and only if, its closed unit ball is compact. Thus, in an infinite-dimensional normed space every compact set possesses an empty interior.

Therefore, in order that a bounded closed set be compact, additional properties are required which depend on the space concerned. We illustrate this by known criteria of compactness for the Banach spaces $\mathcal{C}(K)$ and $L^p(\mathbb{R}^n)$, $1 \leq p < \infty$.

The *Theorem of Arzela-Ascoli* characterises the compact sets of the Banach space $\mathcal{C}(K)$ of continuous real functions on a compact space K. A bounded closed subset M of $\mathcal{C}(K)$ is compact if, and only if, it is equicontinuous, i.e. only when, for every $\varepsilon > 0$ and every $x_0 \in K$, there exists a neighborhood $U_\varepsilon(x_0)$ of x_0 in K such that $|f(x) - f(x_0)| < \varepsilon$ for all $x \in U_\varepsilon(x_0)$ and all $f \in M$ holds.

Similarly, the *Theorem of Kolmogoroff and M. Riesz* characterises the compact subsets of the separable Banach space $L^p(\mathbb{R}^n)$, $1 \leq p < \infty$: a closed bounded set $M \subset L^p(\mathbb{R}^n)$ is compact if, and only if, the following holds uniformly in $f \in M$:

(i) $$\lim_{y\to 0} \int_{\mathbb{R}^n} |f(x+y) - f(x)|^p \, dx = 0 \,,$$

(ii) $$\lim_{R\to+\infty} \int_{|x|>R} |f(x)| \, dx = 0 \,.$$

This criterion of compactness is also valid with obvious modifications for the Banach spaces $L^p(\Omega)$, $\Omega \subset \mathbb{R}^n$ measurable. If Ω is bounded, then the condition (ii) is automatically satisfied.

(d) *Compactness and Convexity.* Under the assumption of a special geometry, the compact sets are determined by a "very small" number of special subsets. For this purpose let us recall that a convex polyhedron in $\mathbb{R}^n$ is determined by its vertices and a sphere by its boundary. In infinite-dimensional Banach spaces the compact sets possess a corresponding property (*Theorem of Krein-Milman*): a convex compact set M of a Banach space X is the closed convex hull $\operatorname{ch} \mathcal{E}(M)$ of its extremal points $\mathcal{E}(M) = \{x \in M \mid x$ does not possess any representation of the form $x = tx_1 + (1-t)x_2,\ x_i \in M,\ x_1 \neq x_2,\ 0 < t < 1\}$.

(e) *Maps on Compact Sets.* The image of a compact set under a continuous map into a Hausdorff space is compact. Thus, a continuous real-valued function on a compact set assumes its maximum and minimum values. Every continuous mapping of a compact metric space into a metric space is uniformly continuous. A bijective continuous mapping of a compact space onto a Hausdorff space is a homeomorphism.

A *map* A, which is not necessarily linear, of a normed space X into a normed space Y is called *compact* if it maps bounded sets of X onto relatively compact sets of Y. A continuous compact mapping is called *completely continuous.*

A linear map of X into Y is compact only if it maps the unit ball of X onto a relatively compact set of Y. Every relatively compact set of a normed space is bounded. Thus, every compact linear map between two normed spaces is bounded and hence continuous. For linear maps between normed spaces the concepts of compactness and complete continuity are therefore the same.

A compact mapping of a normed space X into a normed space Y maps every weakly convergent sequence in X onto a strongly convergent sequence in Y.

For a Banach space the interplay between the concepts of compactness and complete continuity leads to a very important consequence: every completely continuous mapping A which maps a closed, bounded, convex set $K \neq \emptyset$ of a Banach space X into itself possesses a fixed point x in K, which means that $Ax = x$ (*Fixed-Point Theorem of Schauder*). This theorem is thus to be regarded as the infinite-dimensional analogue of the *Fixed-Point Theorem of*

Brouwer (1910): every continuous map f of a compact, convex, nonempty set K of $\mathbb{R}^n$ possesses a fixed point in K.

(f) *Weak Topologies and Compactness.* Let us first note a few further properties of the weak topologies. Compactness, boundedness, convergence, etc. with respect to the weak topology $\sigma(X, X')$ of a normed space X with the dual space X' are expressed as *weak compactness, weak boundedness, weak convergence*, etc. In finite-dimensional spaces the weak and strong (i.e. norm) topologies coincide. However, this is not the case for infinite-dimensional normed spaces. The weak topology is then coarser than the strong topology in the proper sense. Thus, for example, in an infinite-dimensional Banach space X there exist sequences $(x_n)_{n\in\mathbb{N}}$ with $\|x_n\| = 1$ for all n which converge weakly to null. This implies:

(i) The open unit ball $B_1 = \{x \in X \mid \|x\| < 1\}$ of an infinite-dimensional Banach space X is not weakly open.
(ii) The unit sphere $S_1(X) = \{x \in X \mid \|x\| = 1\}$ is not weakly closed.

In analogy with the weak topology $\sigma(X, X')$ on a normed space X, one introduce the *weak $\star$-topology* $\sigma(X', X)$ on the dual space X' and X. $\sigma(X', X)$ is the coarsest locally convex topology on X' with respect to which all linear functions $F : X' \to \mathbb{R}$, $F \in J(X) \subset X''$, are continuous. $\sigma(X', X)$ is a Hausdorff convex topology on X' which is generally coarser than the weak topology $\sigma(X', X'')$.

The weak $\star$-convergence of a sequence $\{f_n\}_{n\in\mathbb{N}} \subset X'$ to $f \in X'$ is equivalent to "$\lim_{n\to\infty} f_n(x) = f(x)$ for all $x \in X$".

The convergence of a sequence with respect to the norm topology of X' implies its weak $\star$-convergence with the same limit. The weakly $\star$-convergent sequences in X' are strongly bounded. Bounded sets of X and X' are weakly bounded and weakly $\star$-bounded, respectively, and conversely.

The closed unit ball $\overline{B}_1(X') = \{f \in X' \mid \|f\|' \leq 1\}$ in the dual space X' of a normed space is $\sigma(X', X)$-compact. This gives the following characterisation of the weakly $\star$-compact sets (*Theorem of Banach-Alaoglu-Bourbaki*) which corresponds to the characterisation of the compact sets in a finite-dimensional space: a set $M \subset X'$ is weakly $\star$-compact if, and only if, it is weakly $\star$-closed and bounded.

(g) *Weak Compactness in Reflexive Banach Spaces.* For a reflexive Banach space X one can identify X with the bidual space X'' by virtue of the canonical embedding $J : X \to X''$. It follows then that the weak topology and the weak $\star$-topology on X' coincide and therefore the results for the weak $\star$-topology can be carried over to the weak topology on X:

In a reflexive Banach space, a subset is weakly compact if, and only if, it is weakly closed and bounded.

Since a convex set in a Banach space is closed if, and only if, it is weakly closed, the following sharpening ensues:

A convex subset of a reflexive Banach space is weakly compact if, and only if, it is closed and bounded (with respect to the norm topology).

The reflexivity of a Banach space can be regarded as the compactness property of the closed unit ball $\overline{B}_1(X) = \{x \in X \mid \|x\| \leq 1\}$ since one has:

A Banach space X is reflexive if, and only if,

(i) $\overline{B_1}(X)$ *is weakly compact or*
(ii) *the weak topology and the weak $\star$-topology on X' coincide.*

(h) *Equivalence of the Different Concepts of Weak Compactness.* In general, the weak topology of a normed space does not satisfy the first countability axiom, so that a weakly compact set is not automatically weakly sequentially compact.

As the first step, the *Theorems of Šmulian and Kaplansky* lead to the following characterisation: a subset M of a normed space X is weakly compact if, and only if, it is relatively weakly compact and weakly sequentially closed. Finally, for the purposes of the calculus of variations, the following very deep result is of fundamental importance:

In a Banach space the weakly compact subsets are identical with those which are weakly countably compact and weakly sequentially compact.

The proof of this result can be found in [C.1, p. 321, § 24.3, Theorem 9]. With this we obtain immediately a new criterion for the reflexivity:

A Banach space is reflexive if, and only if, its closed unit ball is weakly sequentially compact (or weakly countably compact).

The Banach spaces in our applications are reflexive and we utilise as the crucial argument the fact that their closed unit balls are weakly sequentially compact. We therefore give a direct proof for this fundamental property by showing the following:

In a reflexive Banach space the bounded sets are relatively weakly sequentially compact.

Proof. We have to show that every bounded (by norms) sequence $(x_n)_{n \in \mathbb{N}}$ in a reflexive Banach space X possesses a weakly convergent subsequence. Let E denote, the subspace of X generated by this sequence. Then E is a separable Banach space and, as the subspace of a reflexive Banach space, is itself reflexive. Lemma C.1 shows that the dual space E' of E is also a separable reflexive Banach space. We can thus assume that E' is generated by a sequence $(f_n)_{n \in \mathbb{N}}$ of unit vectors f_n with $\|f_n\|' = 1$. The uniform boundedness of all sequences $(f_n(x_j))_{j \in \mathbb{N}}$, $n \in \mathbb{N}$, enables one to apply the trick of diagonal sequences (Cantor) in a standard way. One obtains a subsequence $(x_{j(i)})_{i \in \mathbb{N}}$ with the property that all real sequences $(f_n(x_{j(i)})_{i \in \mathbb{N}})$, $n \in \mathbb{N}$, possess a limit

$$\lim_{i\to\infty} f_n(x_{j(i)}) = a_n\,, \quad n \in \mathbb{N}\,.$$

Let $\varepsilon > 0$ and $f \in E'$ be arbitrarily given; then the exists an

$$f_\varepsilon \in \operatorname{lin}\{f_1, f_2, \ldots\} \quad \text{with} \quad \|f - f_\varepsilon\|' < \frac{\varepsilon}{4c}\,, \quad c = \sup_{i\in\mathbb{N}} \|x_{j(i)}\|\,.$$

We have then

$$\begin{aligned}|f(x_{j(i)}) - f(x_{j(k)}) &\le |(f - f_\varepsilon)(x_{j(i)} - x_{j(k)})| + |f_\varepsilon(x_{j(i)} - x_{j(k)})| \\ &\le \|f - f_\varepsilon\|'\,\|x_{j(i)} - x_{j(k)}\| + |f_\varepsilon(x_{j(i)}) - f_\varepsilon(x_{j(k)})| \\ &\le \frac{\varepsilon}{2} + |f_\varepsilon(x_{j(i)}) - f_\varepsilon(x_{j(k)})| < \varepsilon\end{aligned}$$

whenever $i, k \ge i_\varepsilon$. Defining

$$G(f) = \lim_{i\to\infty} f(x_{j(i)})\,, \quad f \in E'\,,$$

we obtain immediately the fact that G is a continuous linear function $E' \to \mathbb{R}$ (since $|G(f)| \le c\|f\|'$), and thus $G \in E''$. Since E is reflexive, G has the form $G = J(x)$ for a uniquely determined $x \in E$. This shows that

$$f(x_{j(i)})_{i\to\infty} \to J(x)(f) = f(x)\,, \quad f \in E'\,,$$

and hence the weak convergence of the sequence $(x_{j(i)})_{i\to\mathbb{N}}$ is obtained in E and therefore in X. □

Lemma C.1. *If* $(X, \|\cdot\|)$ *is a separable reflexive Banach space, then the dual space* $(X', \|\cdot\|')$ *is also a separable reflexive Banach space.*

Proof. First, $(X', \|\cdot\|')$ is a Banach space. Since X is reflexive, X' is also reflexive, since if X is isomorphic to X'', then X' is also isomorphic to $X''' = (X')''$ (under the canonical embedding). We can assume that X is generated by a sequence of unit vectors x_n, $\|x_n\| = 1$. The Hahn-Banach extension theorem then implies that for all $n \in \mathbb{N}$ there exists an $f_n \in X'$, $\|f_n\|' = 1$, $f_n(x_n) = 1$.

We show that these f_n, $n \in \mathbb{N}$, generate the dual space X'. To this end, let E' be the subspace of X' generated by f_n, $n \in \mathbb{N}$. Assume that there exists an $f' \in X'$, $f' \notin E'$. The separation theorem of Hahn-Banach gives an $F \in X''$ such that $F(f') = 1$ and $F(g) = 0$ for all $g \in E'$. Since X is reflexive, F has the form $F = J(x)$, $x \in X$. For all $n \in \mathbb{N}$ it follows then that

$$\|x - x_n\| = \|J(x - x_n)\|'' \ge |J(x - x_n)(f_n)| = |F(f_n) - f_n(x_n)| = 1$$

in contradiction to the assumption that $(x_n)_{n\in\mathbb{N}}$ generates X. Therefore one has $E' = X'$ and hence X' is separable. □

We conclude this appendix by *proving* the important Theorem of F. Riesz.

Proof. If X is finite dimensional, then X is isomorphic to $\mathbb{R}^{\dim X}$. The Heine-Borel theorem implies that the closed unit ball of X is compact. If, conversely, the closed unit ball $\overline{B}_1(0)$ of X is compact, then for every ε, $0 < \varepsilon < 1$, there exist points $a_1, \ldots, a_N \in X$ such that

$$\overline{B}_1(0) \subseteq \bigcup_{j=1}^{N} B_\varepsilon(a_j) = \bigcup_{j=1}^{N} \{a_j + \varepsilon B_1(0)\}$$

holds, since $B_\varepsilon(a) = a + \varepsilon B_1(0)$. If V is the finite-dimensional subspace of X generated by $a_1, \ldots, a_N$, then it follows that

$$\overline{B}_1(0) \subseteq V + \varepsilon B_1(0) \subseteq V + \varepsilon \overline{B}_1(0).$$

By iteration one obtains

$$B_1(0) \subseteq V + \varepsilon(V + \varepsilon B_1(0)) = V + \varepsilon^2 B_1(0),$$

and thus

$$\overline{B}_1(0) \subseteq V + \varepsilon^n B_1(0), \quad n \in \mathbb{N},$$

and further

$$\overline{B}_1(0) \subseteq \bigcap_n (V + \varepsilon^n B_1(0)) = \overline{V} = V.$$

Since the unit ball absorbs the points of X, we obtain the result

$$X = \bigcup_{n=1}^{\infty} n\overline{B}_1(0) \subseteq \bigcup_{n=1}^{\infty} nV = V. \qquad \square$$

Appendix D. The Sobolev Spaces $W^{m,p}(\Omega)$

D.1 Definition and Properties

Let us begin with the "classical" definition of the Sobolev spaces $W^{m,p}(\Omega)$, $m = 0, 1, 2, \ldots, 1 \le p < \infty$, for an arbitrary nonempty open subset $\Omega \subset \mathbb{R}^n$. A vector subspace $\mathcal{C}^m(\Omega)_p$ of the vector space of all $\mathcal{C}^m$ functions on Ω is defined by the condition

$$\int_\Omega |D^\alpha u(x)|^p \, dx < \infty \,, \quad |\alpha| = \sum_{i=1}^n \alpha_j \le m \,.$$

This subspace can be normed by setting

$$u \to \|u\|_{m,p} = \Big(\sum_{|\alpha| \le m} \int_\Omega |D^\alpha u(x)|^p \, dx \Big)^{1/p} = \Big(\sum_{|\alpha| \le m} \|D^\alpha u\|_p^p \Big)^{1/p} .$$

The normed space $(\mathcal{C}^m(\Omega)_p, \| \cdot \|_{m,p})$ is, however, not complete.

The *Sobolev space* $W^{m,p}(\Omega)$ is defined as the completion of the normed space $(\mathcal{C}^m(\Omega)_p, \|\cdot\|_{m,p})$. It follows immediately from the definition that the following relation holds between the Sobolev spaces $W^{m,p}(\Omega)$ for $m = 0, 1, 2, \ldots$:

$$W^{m+1,p}(\Omega) \subseteq W^{m,p}(\Omega) \subseteq \; \ldots \; \subseteq W^{0,p}(\Omega) = L^p(\Omega) \,.$$

The identical embeddings of these Banach spaces are continuous since for all $u \in W^{m+1,p}(\Omega)$ one has

$$\|u\|_{m,p} \le \|u\|_{m+1,p} \quad \text{for } m = 0, 1, 2, \ldots \,.$$

The separability and reflexivity of the Sobolev spaces follow from the corresponding properties of the Lebesgue spaces and the fact that closed subspaces of a separable (reflexive) Banach space are separable (reflexive). The direct sum $\dot{+}_{|\alpha| \le m} L^p(\Omega)$ of the Banach spaces equipped with the norm

$$\dot{+}_{|\alpha| \le m} L^p(\Omega) \ni \underline{v} = (v_\alpha)_{|\alpha| \le m} \to \|\underline{v}\|_{m,p} = \Big(\sum_{|\alpha| \le m} \|v_\alpha\|_p^p \Big)^{1/p}$$

is a Banach space. Because of

$$W^{m,p}(\Omega) \ni u \to \underline{u} = (D^\alpha u)_{|\alpha| \le m} \in \dot{+}_{|\alpha| \le m} L^p(\Omega)$$

one can identify $W^{m,p}(\Omega)$ with a closed subspace of this Banach space.

Now the Banach space $L^p(\Omega)$ for $1 \le p < \infty$ is separable. It follows therefore that $W^{m,p}(\Omega)$ is also separable for $1 \le p < \infty$ and $m = 0, 1, 2, \ldots$, since a finite direct sum of separable spaces is separable.

The Banach space $L^p(\Omega)$ is reflexive for $1 < p < \infty$, and, consequently, the finite direct sum $(\dot{+}_{|\alpha| \le m} L^p(\Omega), \| \cdot \|_{m,p})$ is also reflexive. Therefore the Sobolev spaces $W^{m,p}(\Omega)$, $1 < p < \infty$, $m = 0, 1, 2, \ldots$, are reflexive.

The space $\mathcal{D}(\Omega)$ of all $\mathcal{C}^\infty$ functions on Ω which possess a compact support in Ω is naturally contained in $W^{m,p}(\Omega)$; the closure of $\mathcal{D}(\Omega)$ in $W^{m,p}(\Omega)$ is denoted by $W_0^{m,p}(\Omega)$. In general, $W_0^{m,p}(\Omega)$ is a proper subspace of $W^{m,p}(\Omega)$. For $\Omega = \mathbb{R}^n$, however, equality holds. In any case the Sobolev spaces $W_0^{m,p}(\Omega)$, $1 < p < \infty$, $m = 0, 1, 2, \ldots$, are also separable reflexive Banach spaces and, like $W^{m,p}(\Omega)$, are uniformly convex.

In particular, the spaces $W^{m,2}(\Omega)$ and $W_0^{m,2}(\Omega)$ are Hilbert spaces where

$$\langle u, v \rangle_{m,2} = \sum_{|\alpha| \le m} \int_\Omega D^\alpha u(x) D^\alpha v(x)\, dx$$

is the scalar product. For these spaces the following notation is often used:

$$H^m(\Omega) = W^{m,2}(\Omega), \quad H_0^m(\Omega) = W_0^{m,2}(\Omega).$$

In the original definition of Sobolev (1938) the spaces $W^{m,p}(\Omega)$ were defined as the subspaces of the space of all regular distributions on Ω. This approach is of great advantage for the investigation of partial differential operators. Therefore we explain it briefly too. Let us first define what we mean by a distribution on Ω.

A *distribution* on Ω is a continuous linear functional on the space $\mathcal{D}(\Omega)$ of all $\mathcal{C}^\infty$ functions on Ω with compact support, whereby the following topology is used on $\mathcal{D}(\Omega)$. For every compact subset $K \subset \Omega$ the vector subspace

$$\mathcal{D}_K(\Omega) = \{\varphi \in \mathcal{D}(\Omega) \mid \operatorname{supp} \varphi \subseteq K\}$$

on $\mathcal{D}(\Omega)$ is topologised through the norms

$$\varphi \to q_{K,l}(\varphi) = \sup_{|\alpha| \le l} \sup_{x \in K} |D^\alpha \varphi(x)|, \quad l = 0, 1, 2, \ldots .$$

For

$$\mathcal{D}(\Omega) = \bigcup_{\substack{K \subset \Omega \\ K \text{ compact}}} \mathcal{D}_K(\Omega)$$

one subsequently introduces the topology of the "inductive limit" of the spaces $\mathcal{D}_K(\Omega)$. This means that a linear map A of $\mathcal{D}(\Omega)$ into a normed space X is continuous if, and only if, all $A \restriction \mathcal{D}_K(\Omega) : \mathcal{D}_K(\Omega) \to X$, $K \subset \Omega$, K compact, are continuous. Thus a linear function : $\mathcal{D}(\Omega) \to \mathbb{R}$ (or $\mathbb{C}$) is a distribution if, and only if, for every compact subset $K \subset \Omega$ there exist a constant $C = C_{T,K}$ and a number $l = l_{T,K} \in \mathbb{N}$ such that

$$|T(\varphi)| \le C q_{K,l}(\varphi), \quad \forall \varphi \in \mathcal{D}_K(\Omega),$$

holds. The set of all distributions on Ω is then exactly the topological dual space $\mathcal{D}'(\Omega)$ of $\mathcal{D}(\Omega)$. In accordance with this, every locally integrable function f on Ω (i.e. $f : \Omega \to \overline{\mathbb{R}}$ is measurable and $|f|$ is integrable on every compact subset K of Ω), for example, can be considered as a distribution I_f: for $\varphi \in \mathcal{D}(\Omega)$ let

$$I_f(\varphi) = \int_\Omega f(x)\varphi(x)\,dx$$

be defined. For all $\varphi \in \mathcal{D}_K(\Omega)$ one has then

$$|I_f(\varphi)| \le \|f\|_{1,K} q_{K,0}(\varphi), \quad \|f\|_{1,K} = \int_K |f(x)|\,dx < \infty.$$

Let $L^1_{\text{loc}}(\Omega)$ denote the vector space of all locally integrable functions on Ω. The linear map $L^1_{\text{loc}}(\Omega) \ni f \to I_f \in \mathcal{D}'(\Omega)$ is injective, which means that $I_f = 0$ in $\mathcal{D}'(\Omega)$ implies $f = 0 \in L^1_{\text{loc}}(\Omega)$. Thus I is an embedding of $L^1_{\text{loc}}(\Omega)$ in $\mathcal{D}'(\Omega)$ and

$$\mathcal{D}'_{\text{reg}}(\Omega) = I(L^1_{\text{loc}}(\Omega))$$

is known as the space of all *regular distributions* on Ω. By virtue of this embedding we can continuously embed all Lebesgue spaces $L^p(\Omega) \subset L^1_{\text{loc}}(\Omega)$, $1 \le p \le \infty$, in $\mathcal{D}'(\Omega)$. If the weak $*$-topology is used on $\mathcal{D}'(\Omega)$, the following holds for every fixed $\varphi \in \mathcal{D}_K(\Omega)$, using Hölder's inequality ($|K|$ = volume of the compact set $K \subset \Omega$, $1/p' + 1/p = 1$):

$$|\langle I_{f_n} - I_f, \varphi\rangle| \le q_{K,0}(\varphi) \int_K |f_n - f|\,dx \le q_{K,0}(\varphi)|K|^{1/p'} \|f - f_n\|_p \underset{n\to\infty}{\longrightarrow} 0.$$

As is well known, $\mathcal{D}(\Omega)$, for $1 \le p < \infty$, is dense in $L^p(\Omega)$. The identity map of $\mathcal{D}(\Omega)$ is again continuous in $L^p(\Omega)$ and thus defines a continuous embedding.

An important and, for applications, very useful property of distributions is that they can be differentiated arbitrarily many times according to the formula

$$(D^\alpha T)(\varphi) = (-1)^{|\alpha|} T(D^\alpha \varphi), \quad \varphi \in \mathcal{D}(\Omega).$$

It follows that all the functions f from $L^p(\Omega)$, $1 \le p < \infty$, are arbitrarily differentiable in $\mathcal{D}'(\Omega)$. But, in general, one does not have

$$f \in L^p(\Omega) \Longrightarrow D^\alpha I_f \in I(L^p(\Omega)).$$

The collection of those f from $L^p(\Omega)$ for which this statement is true for $|\alpha| \le m$ is exactly the space $W^{m,p}(\Omega)$ (Sobolev 1938),

$$W^{m,p}(\Omega) = \{f \in L^p(\Omega) \mid D^\alpha I_f = I_{f_\alpha},\ f_\alpha \in L^p(\Omega),\ |\alpha| \le m\}.$$

The equivalence of the two definitions was shown by Meyers and Serrin in 1964. Here $f_\alpha \in L^p(\Omega)$, $|\alpha| \leq m$, is called the αth *weak* or *distributional* derivative $\partial^\alpha f$ of $f \in L^p(\Omega)$, $f_\alpha \equiv \partial^\alpha f$. This means that for all $\varphi \in \mathcal{D}(\Omega)$,

$$(-1)^{|\alpha|} \int f D^\alpha \varphi \, dx = \int f_\alpha \varphi \, dx$$

holds. With this definition, the above equation for the two function spaces says that any function $u \in L^p(\Omega)$ which has weak derivatives $\partial^\alpha u$ in $L^p(\Omega)$ for $|\alpha| \leq m$ is, according to the classical definition of $W^{m,p}(\Omega)$, the limit of a sequence of functions f_j, $j \in \mathbb{N}$, in $(\mathcal{C}^m(\Omega)_p, \|\cdot\|_{m,p})$.

For $1 < p < \infty$, an important and very useful characterisation of the existence of weak derivatives in terms of L^p properties is available. We state this now for $m = 1$.

For $u \in L^p(\Omega)$, we introduce the *difference quotient* $\Delta_i^h u$ on $\Omega' \Subset \Omega$ for $0 < |h| < d(\Omega') \equiv \operatorname{dist}(\Omega', \partial\Omega)$:

$$\Delta_i^h u(x) = \frac{1}{h}(u(x + he_i) - u(x)) \, , \quad x \in \Omega' \, ,$$

where e_i is the unit vector in coordinate direction i. Clearly, $\Delta_i^h u \in L^p(\Omega')$ for $0 < |h| < d(\Omega')$ and $i = 1, \ldots, n$. Now, according to [Ref. D.1, Lemmas 7.23 and 7.24], we have for $i = 1 \ldots n$:

The weak derivative $\partial_i u$ of $u \in L^p(\Omega)$ exists and belongs to $L^p(\Omega)$ if and only if there is a constant C, $0 \leq C < \infty$, such that for all $\Omega' \subset \Omega$ and all $0 < |h| < d(\Omega')$

$$\|\Delta_i^h u\|_{L^p(\Omega')} \leq C \, .$$

and if this condition is satisfied then one has

$$\|\Delta_i^h u\|_{L^p(\Omega')} \leq \|\partial_i u\|_{L^p(\Omega)} \leq C \, .$$

The fact that $W_0^{m,p}(\Omega)$ is, in general, a proper subspace of $W^{m,p}(\Omega)$ plays a decisive role in the formulation of boundary value problems. We wish to determine the difference between the two spaces in the somewhat simpler case of $p = 2$.

We then have $H_0^m(\Omega) = W_0^{m,2}(\Omega)$ as a subspace of the Hilbert space $H^m(\Omega) = W^{m,2}(\Omega)$ with an orthogonal complement $\mathcal{K}^m(\Omega)$. $u \in \mathcal{K}^m(\Omega)$ is characterised by

$$0 = \langle u, v \rangle_{m,2} = \sum_{|\alpha| \leq m} \int_\Omega D^\alpha u(x) D^\alpha v(x) \, dx \, , \quad \forall v \in H_0^m(\Omega) \, ,$$

or, since $\mathcal{D}(\Omega)$ is dense in $H_0^m(\Omega)$, by

$$0 = \sum_{|\alpha| \leq m} \int_\Omega D^\alpha u(x) D^\alpha \varphi(x) \, dx \, , \quad \forall \varphi \in \mathcal{D}(\Omega) \, ,$$

and thus

$$0 = \sum_{|\alpha|\le m} (-1)^{|\alpha|} D^{2\alpha} u \quad \text{in } \mathcal{D}'(\Omega) .$$

This shows that

$$H^m(\Omega) = H_0^m(\Omega) \oplus \mathcal{K}^m(\Omega)$$

with

$$\begin{aligned}\mathcal{K}^m(\Omega) &= \left\{ u \in H^m(\Omega) \Big| \sum_{|\alpha|\le m} (-1)^{|\alpha|} D^{2\alpha} u = 0 \text{ in } \mathcal{D}'(\Omega) \right\} \\ &= \operatorname{Ker}\left(\sum_{|\alpha|\le m} (-1)^{|\alpha|} D^{2\alpha} \right),\end{aligned}$$

whereby the differential operator $\sum_{|\alpha|\le m} (-1)^{|\alpha|} D^{2\alpha}$ is regarded as a map of $H^m(\Omega)$ in $\mathcal{D}'(\Omega)$.

The difference between $H^m(\Omega)$ and $H_0^m(\Omega)$ could therefore be quite substantial in some circumstances, since, for example, for $m = 1$ and a bounded region Ω all the functions of the form

$$x \to \exp\left(\sum_{j=1}^{n} \alpha_j x_j \right), \quad \alpha_j \in \mathbb{R},$$

belong to $H^1(\Omega)$. Among all these functions, those with parameters satisfying the condition $\sum_{j=1}^{n} \alpha_j^2 = 1$ belong to $\mathcal{K}^m(\Omega)$.

From the definition it should be clear that the elements of $W^{m,p}(\Omega)$ and $W_0^{m,p}(\Omega)$ differ in their behaviour on the boundary. Intuitively, $W_0^{m,p}(\Omega)$ should contain just those u from $W^{m,p}(\Omega)$ which vanish on the boundary of Ω. The difficulty in making this definition precise is mainly due to the fact that a function from $W^{m,p}(\Omega)$ for $n \ge 2$ is not necessarily continuous, and in particular it is not continuous on $\overline{\Omega}$, so that its restriction to the boundary $\Gamma = \partial\Omega$ of Ω is not necessarity defined. A proper investigation of this restriction to the boundary Γ cannot be undertaken without making further assumptions about the smoothness of the boundary. For example, let

$$\Omega = \{x \in \mathbb{R}^2 \,|\, |x| < R\}, \quad R < 1, \quad m = 1 \quad \text{and} \quad p = 2 .$$

Then the function

$$x \to v(x) = (-\log |x|)^k \quad \text{for } 0 < k < 1/2$$

belongs to $H^1(\Omega)$ but has a singularity at $x = 0$.

If the boundary Γ of Ω is sufficiently smooth, the above conjecture can be confirmed. The following theorem gives a typical result.

Theorem D.1.1. *Let $\Omega \subset \mathbb{R}^n$ be a bounded open subset whose boundary $\Gamma = \partial\Omega$ is piecewise C^1. Then the following holds:*

(i) *every* $v \in H^1(\Omega)$ *possesses a restriction* $\gamma_0 v = v \restriction \Gamma$ *to the boundary.*
(ii) $H_0^1(\Omega) = \operatorname{Ker}\gamma_0 = \{v \in H^1(\Omega) \,|\, \gamma_0 v = 0\}$.

Remark D.1.1. Utilising the fact that the Fourier transformation $\mathcal{F}$ is a unitary map in $L^2(\mathbb{R}^n)$, the Sobolev spaces $H^m(\mathbb{R}^m)$ can be characterised in a convenient manner as

$$\begin{aligned} H^m(\mathbb{R}^m) &= \left\{ u \in L^2(\mathbb{R}^n) \Big| \int_{\mathbb{R}^n} |p^\alpha \mathcal{F}u(p)|^2 \, dp < \infty \,, \; |\alpha| \leq m \right\} \\ &= \left\{ u \in L^2(\mathbb{R}^n) \Big| \int_{\mathbb{R}^n} |\mathcal{F}u(p)|^2 (1+p^2)^m \, dp < \infty \right\}. \end{aligned}$$

From this form we recognise that the exponent m does not need to be an integer. For an arbitrary $s \in \mathbb{R}$ we can then introduce the Hilbert spaces

$$H^s(\mathbb{R}^n) = \left\{ u \in L^2(\mathbb{R}^n) \Big| \int_{\mathbb{R}^n} |\mathcal{F}u(p)|^2 (1+p^2)^s \, dp < \infty \right\}.$$

D.2 Poincaré's Inequality

This inequality establishes when and how to control the $L^2(\Omega)$ norm through the $L^2(\Omega)$ norm of the derivatives. If an open set $\Omega \subset \mathbb{R}^n$ is contained in a strip of finite width, the $L^2(\Omega)$ norm of all functions $f \in H_0^l(\Omega)$, $l = 1, 2, \ldots$, can be estimated through the $L^2(\Omega)$ norms of the derivatives.

Without loss of generality we can assume that Ω is contained in the strip $\{x \in \mathbb{R}^n \,|\, 0 < x_1 < b\}$. Since $\mathcal{D}(\Omega)$ is dense in $H_0^l(\Omega)$, it is sufficient to obtain the crucial estimate for $f \in \mathcal{D}(\Omega)$: a simple partial integration leads to

$$\|f\|_2^2 = \int_\Omega |f(x)|^2 \, dx = -\int_\Omega x_1 \frac{\partial}{\partial x_1}(|f(x)|^2) \, dx = -2\int_\Omega x_1 f(x) \frac{\partial f}{\partial x_1}(x) \, dx \,.$$

It follows then that

$$\|f\|_2^2 \leq 2b \int_\Omega |f(x)| \left| \frac{\partial f}{\partial x_1}(x) \right| dx \leq 2b \|f\|_2 \left\| \frac{\partial f}{\partial x_1} \right\|_2$$

and thus

$$\|f\|_2 \leq 2b \left\| \frac{\partial f}{\partial x_1} \right\|_2 \leq 2b \Big(\sum_{|\alpha|=1} \|D^\alpha f\|_2^2 \Big)^{1/2} = 2b \|\nabla f\|_2 \,.$$

Since all the norms appearing in this inequality are continuous with respect to the $\|\cdot\|_{2,l}$ norm, $l \leq 1$, this inequality can be extended to all $f \in H_0^l(\Omega)$, $l = 1, 2, \ldots$. By iteration of this inequality one obtains various versions of Poincaré's inequality. This inequality holds in particular for all bounded open sets.

D.3 Continuous Embeddings of Sobolev Spaces

The solution of boundary and eigenvalue problems by variational calculus are essentially based on the existence of suitable continuous and compact embeddings of Sobolev spaces. We wish to discuss here the most important embedding theorems, give in part the ideas behind the proofs, and finally comment on the meaning of these embeddings.

If X_1 and X_2 are topological spaces and X_1 is contained in X_2, we can consider the *identity embedding* i of X_1 in X_2 which assigns to $x \in X_1$ the same element x in X_2. This identity embedding is said to be *continuous (compact, completely continuous)* if the map $i\colon X_1 \to X_2$ is continuous (compact, completely continuous). For normed spaces X_1 and X_2 the identity embedding is linear, so that compact embeddings are automatically continuous, and thus the concept of compactness and complete continuity are the same.

We begin with two theorems of Sobolev on regularity which are to be considered as important, typical results on the properties of differentiability of functions from $H^l(\Omega)$ in the classical sense.

Theorem D.3.1 (Sobolev). *Let $k \in \mathbb{N}$ and $s > k + n/2$. Then $H^s(\mathbb{R}^n)$ is continuously embedded in the space*

$$C_b^k(\mathbb{R}^n) = \left\{ f \in C^k(\mathbb{R}^n) \,|\, \|f\|_{k,\infty} = \sup_{x\in\mathbb{R}^n} \sup_{|\alpha|\le k} |D^\alpha f(x)| < \infty \right\}$$

and for every $u \in H^s(\mathbb{R}^n)$ one has

$$\lim_{|x|\to\infty} |D^\alpha u(x)| = 0\,, \quad |\alpha| \le k\,,$$

$$\|u\|_{k,\infty} \le C\|u\|_{s,2}\,.$$

Proof. The Fourier transform of an $L^1(\mathbb{R}^n)$ function is continuous and vanishes at infinity (Lemma of Riemann-Lebesgue). For $|\alpha| \le k$ and $s > k + n/2$ one has

$$\int_{\mathbb{R}^n} \frac{|p^{2\alpha}|}{(1+p^2)^s}\, dp = C_\alpha^2 < \infty\,.$$

For $u \in H^s(\mathbb{R}^n)$ one therefore has

$$\int_{\mathbb{R}^n} |p^\alpha(\mathcal{F}u)(p)\,|\, dp \le C\left(\int_{\mathbb{R}^n} (1+p^2)^s |\mathcal{F}u(p)|^2\, dp\right)^{1/2} = C_\alpha\|u\|_{s,2}$$

and therefore for all $x \in \mathbb{R}^n$

$$|D^\alpha u(x)| = \left|\int_{\mathbb{R}^n} e^{ipx} p^\alpha(\mathcal{F}u)(p)\, dp\right| \le C_\alpha\|u\|_{s,2}\,,$$

and therefore $\|u\|_{k,\infty} \le C\|u\|_{s,2}$.

Counterexamples show that this estimate cannot be improved. □

Theorem D.3.2 (Sobolev). *Let $l > k + n/2$ and $\Omega \subset \mathbb{R}^n$ be arbitrary. Then one has $H^1(\Omega) \subseteq C_b^k(\Omega)$, and for $u \in H_0^l(\Omega)$*

$$\|u\|_{k,\infty} \leq C\|u\|_{l,2}\,.$$

These *regularity theorems of Sobolev* contain the following statements: in every equivalence class $u \in H^s(\Omega)$ there exists a representative $\overline{u}$ in $C_l^k(\Omega)$; and every sequence in $C_b^k(\Omega)$ which is Cauchy sequence in $H^s(\Omega)$ is also a Cauchy sequence in $C_l^k(\Omega)$.

We mention a further regularity result of this type which is easy to prove: let $n < p$ and $\alpha = 1 - n/p$. Then every element $u \in W^{1,p}(\mathbb{R}^n)$ is a Hölder continuous function with exponent α. Furthermore, the following holds:

$$|u(x) - u(y)| \leq C|x-y|^{\alpha}\|\nabla u\|_p\,.$$

Whether a continuous embedding of a Sobolev space $W^{m,p}(\Omega)$ exists in normed space $X(\Omega)$ of functions on Ω depends naturally on the region and the exponents m, p. Investigation shows that the smoothness of the boundary $\Gamma = \partial\Omega$ and the boundedness of the region Ω are the deciding factors for this problem.

Theorem D.3.3 (Sobolev's Embedding Theorem). *Let $\Omega \subset \mathbb{R}^n$ be an open but not necessarily bounded set and $1 \leq p < \infty$, $m = 0, 1, \ldots$.*

(a) *If the boundary of Ω is sufficiently smooth (which means, precisely, if Ω possesses the cone property* [D.2]), *then the following embeddings exist and are continuous:*

(i) $mp < n : W^{m,p}(\Omega) \hookrightarrow L^q(\Omega),\ p \leq q \leq q_0 = \dfrac{np}{n-mp},$

(ii) $mp = n : W^{m,p}(\Omega) \hookrightarrow L^q(\Omega),\ p \leq q < \infty,$

(iii) $mp > n : W^{m,p}(\Omega) \hookrightarrow C_b^0(\Omega).$

(b) *For the subspace $W_0^{m,p}(\Omega)$ the continuous embeddings* (i)–(iii) *exist for arbitrary open subsets Ω.*

(c) *If Ω is bounded (or, more precisely, if Ω has a finite volume), then the continuous embeddings* (i) *and* (ii) *in* (a) *and* (b) *also exist for the exponent $q, 1 \leq q < p$.*

Remark D.3.1. (i) As can be shown by examples, the embedding of $W^{m,p}(\Omega)$ in $L^q(p)$ for $q < p$ is only possible if Ω possesses a finite volume.

(ii) The continuity of the above embeddings implies that there are constants $K = K(\Omega, m, p, n, q)$ such that the estimates

$$\|u\|_q \leq K\|u\|_{m,p} \quad \text{and} \quad \|u\|_{0,\infty} \leq K\|u\|_{m,p}$$

hold for all $u \in W^{m,p}(\Omega)$.

(iii) As the central point of the proof, which is relatively elementary but rather lengthy, one has to prove inequalities of the type

$$\|u\|_q \leq K\|\nabla u\|_p \equiv K\left(\sum_{j=1}^{n}\left\|\frac{\partial u}{\partial x_j}\right\|_p^p\right)^{1/p}.$$

This very important inequality is due to *Sobolev*. (See Chapter 9.4.)

D.4 Compact Embeddings of Sobolev Spaces

The most important results on the compactness of embeddings of Sobolev spaces are due to Kondrachov and are based on a well-known lemma of Rellich which is referred to in the literature as the Rellich-Kondrachov Theorem. The compactness of embeddings can be proved for some of the cases for which we have previously established the continuity.

Theorem D.4.1 (Rellich-Kondrachov). *Let $\Omega \subset \mathbb{R}^n$ be a region and $\Omega_0 \subset \Omega$ a bounded subregion. Further, let $1 \leq p < \infty$ and $m = 1, 2, \ldots$.*

(a) *Let Ω possess the cone property. Then the following embeddings are compact:*

(i) $mp < n : W^{m,p}(\Omega) \hookrightarrow L^q(\Omega_0)$, $1 \leq q < q_0 = \dfrac{np}{n - mp}$,

(ii) $mp = n : W^{m,p}(\Omega) \hookrightarrow L^q(\Omega_0)$, $1 \leq q < \infty$,

(iii) $mp > n : W^{m,p}(\Omega) \hookrightarrow C_b^0(\Omega)$.

(b) *For the subspace $W_0^{m,p}(\Omega)$ the embeddings* (i)–(iii) *for arbitrary open sets Ω are compact.*

(c) *If Ω is bounded, we can set $\Omega_0 = \Omega$ in* (a) *and* (b) *for the embeddings* (i) *and* (ii).

Proof. We give the *proof* for the compactness of the embedding (i) where we assume the continuous embedding. For $u \in W^{m,p}(\Omega)$ a simple application of Hölder's inequality shows that

$$\|u\|_q \leq \|u\|_1^{\alpha}\|u\|_{q_0}^{1-\alpha}, \quad \alpha = \frac{1}{q}\frac{q_0 - q}{q_0 - 1}, \quad 1 - \alpha = \frac{q_0}{q}\frac{q - 1}{q_0 - 1}.$$

Since we have assumed the continuous embedding $W^{m,p}(\Omega) \hookrightarrow L^{q_0}(\Omega)$, it is known that $\|u\|_{q_0} \leq K\|u\|_{m,p}$.

To prove the compactness of the embedding (i), we use the criterion of Kolmogorov and M. Riesz and show that every closed bounded subset M of $W^{m,p}(\Omega)$ satisfies the condition (i) of this criterion. Because $1 \leq q < q_0$, we have $\alpha > 0$, so that for all $u \in M$ an estimate of the form $\|u\|_q \leq C_M\|u\|_1^{\alpha}$

is obtained. Thus it is sufficient to show condition (i) with respect to the L^1 norm. For $j = 1, 2, \ldots$, set

$$\Omega_j = \{x \in \Omega_0 \mid d(x, \partial\Omega) \mid > 2/j\},$$

where $d(x, \partial\Omega)$ denotes the distance of the point x from the boundary $\partial\Omega$. With Hölder's inequality, it follows for all $u \in M$ that

$$\begin{aligned}\int_{\Omega_0 \setminus \Omega_j} |u(x)|\, dx &\le \left(\int_{\Omega_0 \setminus \Omega_j} |u(x)|^{q_0}\right)^{1/q_0} \left(\int_{\Omega_0 \setminus \Omega_j} dx\right)^{1-1/q_0} \\ &\le \|u\|_{q_0} |\Omega_0 \setminus \Omega_j|^{1-1/q_0} \\ &\le K\|u\|_{m,p} |\Omega_0 \setminus \Omega_j|^{1-1/q_0} \le C'_M |\Omega_0 \setminus \Omega_j|^{1-1/q_0} .\end{aligned}$$

For a given $\varepsilon > 0$ we can therefore find a $j = j_\varepsilon$ such that

$$\int_{\Omega_0 \setminus \Omega_j} |u(x)|\, dx < \varepsilon/4$$

holds for all $u \in M$. Setting

$$\widehat{u}(x) = \begin{cases} u(x), & x \in \Omega_0, \\ 0, & \text{otherwise}, \end{cases}$$

we have

$$\begin{aligned}\int_{\Omega_0} |\widehat{u}(x+y) - \widehat{u}(x)|\, dx &= \int_{\Omega_j} |\widehat{u}(x+y) - \widehat{u}(x)|\, dx \\ &\quad + \int_{\Omega_0 \setminus \Omega_j} |\widehat{u}(x+y) - \widehat{u}(x)|\, dx \\ &\le \frac{\varepsilon}{2} + \int_{\Omega_j} |\widehat{u}(x+y) - \widehat{u}(x)|\, dx .\end{aligned}$$

Now let $u \in M \cap C^1(\Omega)$, $|y| < 1/j$; then it follows that $(1/p' + 1/p = 1)$

$$\begin{aligned}\int_{\Omega_j} |\widehat{u}(x+y) - \widehat{u}(x)|\, dx &\le \int_{\Omega_j} dx \int_0^1 dt \left|\frac{d}{dt} u(x+ty)\right| \\ &\le |y| \int_0^1 dt \int_{\Omega_{2j}} dx |\nabla u(x)| \\ &\le |y|\, |\Omega_{2j}|^{1/p'} \|\nabla u\|_{L^p(\Omega_{2j})} \\ &\le |y|\, |\Omega_0|^{1/p'} \|u\|_{m,p} \le |y|\, \Omega_0|^{1/p'} K' .\end{aligned}$$

Since $C^1(\Omega) \cap W^{m,p}(\Omega)$ is dense in $W^{m,p}(\Omega)$, the estimate

$$\int_{\Omega_j} |\widehat{u}(x+y) - \widehat{u}(x)|\, dx \le K_0 |y|$$

for all $u \in M$ and all $y \in \mathbb{R}^n$, $|y| < 1/j$, follows where $K_0 = |\Omega_0|^{1/p'} K'$ is independent of u.

Now, combining the two estimates, condition (i) of the criterion of Kolmogorov and Riesz is obtained easily. Since it has been assumed that Ω_0 is bounded, condition (ii) holds automatically, and hence the compactness of M in $L^q(\Omega)$, $1 \leq q < q_0$, is proved. □

Remark D.4.1. (i) Simple examples show that the theorem of Rellich-Kondrachov is wrong for unbounded regions.

(ii) A certain analogy can be made between the theorem of Rellich-Kondrachov and the theorem of Arzela-Ascoli: if, namely, for a closed bounded set $M \subset \mathcal{C}(K)$, K being a compact topological space, we know a uniform bound for the derivatives, then M is equicontinuous and hence compact since

$$\sup_{f \in M} \sup_{x \in M} \sup_{i=1 \cdots h} \left| \frac{\partial f}{\partial x_i}(x) \right| = C < \infty$$

implies for all $f \in M$ that

$$|f(x) - f(y)| = \left| \sum_{i=1}^{n} (x_i - y_i) \int_0^1 \frac{\partial f}{\partial x_i}(x + t(x-y))\, dt \right| \leq |x-y| C\,,$$

where K has been assumed to be convex.

The theorem of Rellich-Kondrachov is of similar type: if, say all the derivatives of a function of a bounded set $M \subset H^1(\Omega)$ in $L^1(\Omega)$ are bounded in $L^2(\Omega)$, then M is relatively compact in $L^2(\Omega)$.

Appendix E

E.1 Bessel Potentials

Some elementary facts about Bessel potentials which are used in Sects. 9.3 and 9.9 are collected here. Good references for these facts are [E.1, 2].

The *Bessel potential of order* α on $\mathbb{R}^d$ is by definition the kernel G_α of the operator $(1-\Delta)^{-\alpha/2}$, i.e. in the sense of (tempered) distributions G_α is the Fourier transform of the function $p \mapsto (1+p^2)^{-\alpha/2}$ on $\mathbb{R}^d$. This means

$$G_\alpha * f = (1-\Delta)^{-\alpha/2} f \quad \text{for all } f \in \mathcal{S}(\mathbb{R}^d) \tag{E.1}$$

where $\mathcal{S}(\mathbb{R}^d)$ is the Schwartz space of rapidly decreasing $\mathcal{C}^\infty$ functions on $\mathbb{R}^d$ and $*$ denotes convolution.

For $0 < \alpha < d$ the following integral representation is available:

$$G_\alpha(x) = (2\pi)^{d/2} \Gamma\left(\frac{\alpha}{2}\right)^{-1} \int_0^\infty t^{(\alpha-d)/2} \mathrm{e}^{-t-|x|^2/4t} \frac{dt}{t}, \quad x \in \mathbb{R}^d, \quad x \neq 0. \tag{E.2}$$

Since for all $|x| \geq 1$ and all $t > 0$ the following elementary inequality holds:

$$\mathrm{e}^{-t-|x|^2/4t} \leq \mathrm{e}^{-|x|/2} \mathrm{e}^{(t+1/4t)/2},$$

one obtains the following bound simply by inserting this inequality into (E.2) and by evaluating an integral:

$$G_\alpha(x) \leq C_{\alpha,d} \mathrm{e}^{-|x|/2} \quad \text{for all } |x| \geq 1. \tag{E.3}$$

In order to study the behaviour of G_α near its singular point $x = 0$ one substitutes $t = |x|^2\tau$, $\tau > 0$, $x \neq 0$, in (E.2) to obtain

$$G_\alpha(x) = |x|^{\alpha-d} g_\alpha(x) \tag{E.4}$$

and controls g_α near $x = 0$. The result is

$$G_\alpha(x) = |x|^{\alpha-d}\{g_\alpha(0) + \mathrm{O}(|x|^2)\} \quad \text{for } 0 < |x|,\ |x| \to 0. \tag{E.5}$$

Hence, with $s_0 = d/(d-\alpha)$, the above estimates imply the following integrability properties of the function G_α:

$$G_\alpha \in L^s(\mathbb{R}^d) \quad \text{for all } s \in [1, s_0)\,, \tag{E.6 a}$$

$$G_\alpha \in L_w^{s_0}(\mathbb{R}^d)\,. \tag{E.6 b}$$

Recall that (E.6 b) means

$$\sup_{t>0} t\|[G_\alpha \ge t]\|^{1/s_0} < \infty\,.$$

Notice that for $\alpha = 2$ the exponent s_0 is just half of the Sobolev exponent $2^* : s_0 = 2^*/2$.

In the two-dimensional case the integral representation (E.2) does not apply for the Bessel potential of order $\alpha = 2$. Instead, for $\alpha = d = 2$, the following representation can be used:

$$J(x) = c \int_0^\infty \mathrm{e}^{-|x|(1+t)}(t + t^2/2)^{-1/2}\, dt\,, \quad x \ne 0\,, \tag{E.7}$$

with $c = 2^{3/2}\pi$. We rewrite this as

$$J(x) = |x|^{-1/2}\mathrm{e}^{-|x|}g(x)\,, \quad x \in \mathbb{R}^2\,, \quad x \ne 0\,, \tag{E.8}$$

where now

$$g(x) = c \int_0^\infty \left[\frac{2|x|}{2|x| + s}\right]^{1/2} \frac{\mathrm{e}^{-s}}{s^{1/2}}\, ds$$

is a bounded continuous function on $\mathbb{R}^2$ with $g(0) = 0$. Thus the integrability properties of J can easily be derived:

$$J \in L^s(\mathbb{R}^2)\,, \quad s \in [1, 4)\,, \quad J \in L_w^4(\mathbb{R}^2)\,. \tag{E.9}$$

E.2 Some Properties of Weakly Differentiable Functions

In addition to the results on weakly differentiable functions which have been presented in Appendix D, we collect here some facts which are used in Sect. 9.9.

For a nonempty domain $\Omega \subset \mathbb{R}^d$, $d \ge 1$, denote by $W^1(\Omega)$ the set of all real L^1_{loc} functions on Ω whose first weak derivative also belongs to $L^1_{\text{loc}}(\Omega)$. Some of the most basic properties of the space $W^1(\Omega)$ are

(a) If $u, v \in W^1(\Omega)$ then $\max\{u, v\}$ and $\min\{u, v\} \in W^1(\Omega)$.
(b) If $u \in W^1(\Omega)$ then $\nabla u = 0$ almost everywhere on

$$[u = \text{const.}] = \{x \in \Omega \,|\, u(x) = c\}\,.$$

(c) If $u \in W^1(\Omega)$ and if for some $x, y \in \Omega$ we have $a = u(x) < u(y) = b$, then the set $[a < u < b] = \{x \in \Omega \,|\, a < u(x) < b\}$ is of positive Lebesgue measure: $\|[a < u < b]\| > 0$.

The proof of (a) and (b) is given in [Ref. E.3, Lemma 7.8, 7.7]. Property (c) follows from the following. Clearly the domain Ω has the following decomposition: $\Omega = [u \le a] \cup [a < u < b] \cup [b \le u]$. By property (a),

$$v := \max\{\min\{u, b\}, a\}$$

is known to belong to $W^1(\Omega)$. This function takes at least two different values:

$$[v = a] = [u \leq a] \quad \text{and} \quad [v = b] = [b \leq u] .$$

By property (b) it is known that $\nabla v = 0$ almost everywhere on $[u \leq a] \cup [b \leq u]$. If $\|[a < u < b]\| = 0$ is assumed then it follows that $\nabla v = 0$ almost everywhere on Ω, and hence $v = \text{const.}$ on Ω. This contradiction proves $\|[a < u < b]\| > 0$.

The intuitive meaning of property (c) is that weakly differentiable functions have 'no jumps'.

This property is used in Sect. 9.9 in the following way.

Lemma E.2.1. *Suppose $g : \mathbb{R}^n \to \mathbb{R}$ is a continuous function and satisfies*

$$g^{-1}(0) = \{y \in \mathbb{R}^n \mid g(y) = 0\} \subseteq \{0\} \cup \{y \in \mathbb{R}^n \mid |y| \geq \delta\}$$

for some $\delta > 0$. If now for some $u \in W^1(\Omega)$,

$$\widehat{g}(u)(x) = g(u(x)) = 0$$

holds almost everywhere on Ω, then it follows that either $u = 0$ or $|u(x)| \geq \delta$ almost everywhere on Ω.

Proof. On the one hand, we always have the following disjoint decomposition:

$$\Omega = [|u| = 0] \cup [0 < |u| < \delta] \cup [\delta \leq |u|] ,$$

while on the other hand, by assumption, the set

$$\Omega_0 = [\widehat{g}(u) = 0] \subseteq [|u| = 0] \cup [|u| \geq \delta]$$

differs from Ω only in a set of measure zero. Hence $[0 < |u| < \delta]$ is of measure zero and therefore by property (c), either $[\delta \leq |u|]$ or $[|u| = 0]$ are empty. □

If this lemma is applied in Sect. 9.9, the second possibility is excluded by an integrability condition ($u \in L^q(\mathbb{R}^2)$).

E.3 Proof of Theorem 9.2.3

As for a proof ob Sobolev's inequality, the starting point is the following *inequality of Gagliardo-Nirenberg* ([E.3–5] and references therein):

$$\|f\|_p \leq \frac{1}{d} \|\nabla f\|_1 , \quad \frac{1}{p} = 1 - \frac{1}{d}, \quad f \in C_0^1(\mathbb{R}^d) . \tag{E.10}$$

Denote by $D^{1,1}$ the completion of $\mathcal{D} = C_0^\infty(\mathbb{R}^d)$ with respect to the norm $f \mapsto \|\nabla f\|_1$. By continuous extension, inequality (E.10) still holds on all of $\mathcal{D}^{1,1}$. By induction on $k = 0, 1, 2, \ldots$, we will show that for any fixed $u \in E_q$,

$$v_k = |u|^{\alpha+k}, \quad \alpha = 1 + q/p, \tag{E.11}$$

belongs to $\mathcal{D}^{1,1}$ so that the above inequality applies.

Choose some functions $\chi, \eta \in \mathcal{D}$ such that $0 \le \eta \le 1$, $0 \in \chi$, $n(x) = 1$ for $|x| \le 1/2$, $\operatorname{supp} \chi, \operatorname{supp} \eta \subseteq \{x : |x| \le 1\}$, $\int \chi \, dx = 1$. Then define for $m = 1, 2, \ldots$, and $x \in \mathbb{R}^d$,

$$\chi_m(x) = m^{-d} \chi(mx) \quad \text{and} \quad \eta_m(x) = \eta(x/m).$$

Clearly, $\eta_m, \chi_m \in \mathcal{D}$, $\int \chi_m \, dx = 1$, and $\eta_m(x) = 1$ on $\{x : |x| \le m/2\}$. This implies that all the functions

$$\psi_m^k = \eta_m(v_k * \chi_m), \quad m = 1, 2, \ldots, \quad k = 0, 1, 2, \ldots,$$

belong to $\mathcal{D}$.

If v_{k-1} belongs to $L^P(\mathbb{R}^d)$ then $v_k \in L^{q_k}(\mathbb{R}^d)$ where $q_k = p(\alpha + k - 1)(\alpha + k)^{-1} \ge 1$ and, according to the rules for weak derivatives,

$$\nabla v_k = (\alpha + k) v_{k-1} \nabla |u|; \tag{E.12}$$

hence, $\nabla v_k \in L^1$, since $\|\nabla |u|\,\|_d \le \|\nabla u\|_d$. For the differences $\nabla v_k - \nabla \psi_m^k$, $m = 1, 2, \ldots$, $k = 0, 1, 2, \ldots$, the following estimates are available:

$$\|\nabla v_k - \nabla \psi_m^k\|_1 \le \|(1 - \eta_m) \nabla v_k\|_1 + \|\eta_m(\nabla v_k - \chi_m * \nabla v_k\|_1 + \|(\nabla \eta_m)(v_k * \chi_m)\|_1 .$$

Since $\nabla v_k \in L^1$, the first term tends to zero for $m \to \infty$ by dominated convergence. The limit $m \to \infty$ of the second term also vanishes since $\|\eta_m\|_\infty \le 1$ and $\nabla v_k \in L^1$ by the fact that $(\chi_m)_{m \in \mathbb{N}}$ is a "δ-sequence". Finally, the last term is dominated by Hölder's and Young's inequality by

$$\|\nabla \eta_m\|_{q_k'} \|v_k * \chi_m\|_{q_k} \le \|\nabla \eta_m\|_{q_k'} \|v_k\|_{q_k} \|\chi_m\|_1 .$$

By definition we have, for all $m \in \mathbb{N}$,

$$\|\chi_m\|_1 = 1 \quad \text{and} \quad \|\nabla \eta_m\|_{q_k'} = m^{d/q_k' - 1} \|\nabla \eta\|_{q_k'}$$

with $\frac{d}{q_k'} - 1 = -\frac{d}{r(k)} < 0$. Therefore, this term also vanishes in the limit $m \to \infty$ and thus proves $\|\nabla v_k - \nabla \psi_m^k\|_1 \underset{m \to \infty}{\longrightarrow} 0$. We conclude that $v_k \in \mathcal{D}^{1,1}$ and therefore, by (E.10), we also get $v_k \in L^p$ and thus by (E.12)

$$\|v_k\|_p \le \frac{1}{d} \|\nabla v_k\|_1 \le \frac{\alpha + k}{d} \|v_{k-1}\|_p \|\nabla u\|_d . \tag{E.13}$$

By choice of α our hypothesis $v_{k-1} \in L^p$ is satisfied for $k = 0$. Hence by induction on k it follows that estimate (E.13) holds for all $k = 0, 1, 2, \ldots$. Taking into account definition (E.11), a simple iteration implies the inequalities of part (b) of Theorem 9.2.3. The inequalities in part (a) follow from this using the interpolation inequality, where the constant C_r is explicitly known.

References

Chapter 1

1.1 Lebesgue, H.: Sur le problème de Dirichlet. Rend. Circ. mat. Palermo bf 24 (1905) 371–402
1.2 Schäfer, H.: Topological vector spaces. Springer, Berlin Heidelberg 1971
1.3 Vainberg, M.M.: Variational methods for the study of nonlinear operators. Holden Day, London 1964
1.4 Carrol, R.W.: Abstract methods in partial differential equations. Harper and Row, New York 1969
1.5 Berger, M.S.: Non-linearity and functional analysis. Academic Press, New York 1977
1.6 Choquet, G.: Lectures on analysis II. Benjamin, New York 1969
1.7 Ritz, W.: Über eine neue Methode zur Lösung gewisser Variationsprobleme der mathematischen Physik. J. reine angew. Math. **135** (1908) 1–61
1.8 Courant, R., Hilbert, D.: Methods of mathematical physics. Wiley-Interscience, New York 1966
1.9 Cea, J.: Optimisation, théorie et algorithmes. Dunod, Paris 1971. Optimisation techniques, Proc. 7th IFIP Conf., Nice 1975. Springer, Berlin Heidelberg 1976

Chapter 2

2.1 Dieudonné, J.: Foundations of modern analysis. Academic Press, New York 1969
2.2 Averbukh, V., Smolanov, O.: Uspekhi mat. Nauk **22**, 6 (1967); **23**, 4 (1968)
2.3 Zeidler, E.: Nonlinear functional analysis and its applications. I Fixed points. II Monotome operators. III Variational methods and optimization. IV Applications to mathematical physics. Springer, Berlin Heidelberg 1985, 1988

Chapter 3

3.1 Vainberg, M.M.: Variational methods for the study of nonlinear operators. Holden Day, London 1964
3.2 Dieudonné, J.: Foundations of modern analysis. Academic Press, New York 1969

Further Reading

Courant, R., Hilbert, D.: Methods of mathematical physics. Wiley-Interscience, New York 1966
Velte, W.: Direkte Methoden der Variationsrechnung. Teubner, Stuttgart 1976

Young, L.C.: Calculus of variation and optimal control theory. Saunders, Philadelphia 1969
Gelfand, I.M., Fomin, S.V.: Calculus of variations. Prentice-Hall, Englewood Cliffs 1963
Funk, P.: Variationsrechnung und ihre Anwendung in Physik und Technik. Springer, Berlin Heidelberg 1962
Morse, M.: The calculus of variations in the large. Am. Math. Soc., Providence 1934
Fučik, S., Nečas, J., Souček, V.: Einführung in die Variationsrechnung. Teubner, Leipzig 1977
Jaffe, A., Taubes, C.: Vortices and monopoles. Birkhäuser, Boston 1980

Chapter 4

4.1 Ljusternik, L.A.: On conditional extrema of functions. Mat. Sbornik **41**, 3 (1934)
4.2 Robertson, A.P., Robertson, W.J.: Topological vector spaces. Cambridge University Press, Cambridge 1973
4.3 Dieudonné, J.: Foundations of modern analysis. Academic Press, New York 1969
4.4 Hirzebruch, F., Scharlau, W.: Einführung in die Funktionalanalysis. BI Taschenbücher **296**, BI, Mannheim 1971
4.5 Gelfand, I.M., Fomin, S.V.: Calculus of variations. Prentice-Hall, Englewood Cliffs 1963

Further Reading

Schwartz, J.T.: Nonlinear functional analysis. Gordon and Breach, New York 1969
Maurin, K.: Calculus of variations and classical field theory, Part 1. Lecture Notes Series **34**, Matematisk Institut, Aarhus University 1976

Chapter 5

5.1 Noether, E.: Invariante Variationsprobleme (F. Klein zum 50jährigen Doktorjubiläum). Nachr. kgl. Ges. Wiss. Göttingen math. phys. Kl. S. (1918) 235–257
5.2 Rektory, K.: Variational methods in mathematics, science and engineering. Reidel, Dordrecht 1980
5.3 Landau, L. D., Lifschitz, E. M.: The classical theory of fields. Course in theoretical physics, vol. 2. Pergamon, Oxford 1971
5.4 Ashby, N., Brittin, W. E., Love, W. F., Wyss, W.: Am. J. Phys. **43** (1975) 902–906
5.5 Soper, D. E.: Classical field theory. Wiley, New York 1976
5.6 Itzykson C., Zuber, J. B.: Quantum field theory. McGraw-Hill, New York 1980
5.7 Bogoliubov, N. N., Shirkov, D. V.: Introduction to the theory of quantized fields. Wiley, New York 1959
5.8 Sexl, R. U., Urbantke, H. K.: Relativität, Gruppen, Teilchen. 2nd revised edition. Springer, Vienna, New York 1982
5.9 Pauli, W.: Theory of relativity. Pergamon, Oxford 1958
5.10 Weyl, H.: Space-time-matter. Dover, New York 1951
5.11 Coleman, S.: Classical lumps and their quantum descendants. In: New phenomena in subnuclear physics; International School of Subnuclear Physics, Erice 1975. A. Zichichi (ed.), Plenum Press 1977
5.12 Palais, R. S.: The principle of symmetric criticality. Commun. Math. Phys. **69** (1979) 19–30

5.13 Hewitt, E., Ross, K. A.: Abstract harmonic analysis I. GMW Bd. 115. Springer, Berlin Göttingen Heidelberg 1963
5.14 Lang, S.: Introduction to differentiable manifolds. Interscience, New York 1966

Chapter 6

6.1 Achieser, H. T., Glasmann, T. M.: Theory of linear operators in Hilbert spaces. Pitman, Boston 1981
6.2 Hirzebruch, F., Scharlau, W.: Einführung in die Funktionalanalysis. BI Taschenbücher **296**, BI, Mannheim 1971
6.3 Adams, R. A.: Sobolev spaces. Academic Press, New York 1975
6.4 Courant, R.: Dirichlet's principle, conformal mapping and minimal surfaces. Springer, Berlin Heidelberg 1977
6.5 Courant, R., Hilbert, D.: Methods of mathematical physics. Wiley-Interscience, New York 1966
6.6 Aubin, J. P.: Applied functional analysis. Wiley, New York 1979
6.7 Lions, J. L., Magenes, E.: Non-homogeneous boundary value problems and applications. Springer, Berlin Heidelberg 1972
6.8 Temam, R.: Navier-Stokes equations. North-Holland, Amsterdam 1977

Further Reading

Nečas, J.: Les méthodes directes en théorie des équations elliptiques. Academia, Prague 1967
Rektory, K.: Variational methods in mathematics, science and engineering. Reidel, Dordrecht 1980

Chapter 7

7.1 Browder, F. E.: Nonlinear eigenvalue problems and group invariance. In: Functional analysis and related fields, F. E. Browder (ed.). Springer, Berlin Heidelberg 1970
7.2 Oden, J. T., Reddy, J. N.: An introduction to the mathematical theory of finite elements. Wiley, London 1976
7.3 Oden, J. T., Reddy, J. N.: Variational methods in theoretical mechanics. Springer, Berlin Heidelberg 1976
7.4 Carrol, R. W.: Abstract methods in partial differential equations. Harper and Row, New York 1969
7.5 Browder, F. E.: Pseudomonotone operators and the direct method of the calculus of variations. Arch. Rat. Mech. Anal. **38** (1970) 268–277
7.6 Vainberg, M. M.: Variational methods for the study of nonlinear operators. Holden Day, London 1964
7.7 Browder, F. E.: Non linear eigenvalue problems and Galerkin-approximation. Bull. Amer. Math. Soc. **74** (1968) 651–656
7.8 Dunford, N., Schwartz, J. T.: Linear operators. Wiley-Interscience, New York 1963

Further Reading

Browder, F. E.: Problèmes non-linéaires. Séminaire de Mathématiques Supérieures, Montréal 1966
Minty, G.: On a "monotonicity" method for the solution of nonlinear equations in Banach spaces. Proc. Nat. Acad. Sci. USA **50** (1963) 1038–1041

Lions, J. L.: Quelques méthodes de résolution des problèmes aux limites non linéaires. Dunod, Paris 1969

Chapter 8

8.1 Browder, F. E.: Nonlinear eigenvalue problems and group invariance. In: Functional analysis and related fields, F. E. Browder (ed.). Springer, Berlin Heidelberg 1970

8.2 Browder, F. E.: Existence theorems for nonlinear partial differential equations. Global Analysis, Proc. Symp. Pure Math. **16** (1970) 1–62

8.3 Reed, M., Simon, B.: Methods of modern mathematical physics IV. Academic Press, New York 1975

8.4 Ljusternik, L.: Sur quelques méthodes topologiques en géometrie différentielle. Atti dei Congresso Internationale dei Matematici Bologna **4** (1928) 291–296

8.5 Ljusternik, L., Schnirelman, T.: Méthodes topologiques dans les problèmes variationels. Hermann, Paris 1934

8.6 Ljusternik, L. A., Schnirelman, L. G.: Topological methods in variational problems. Trudy Inst. Math. Mech., Moscow State University 1936, pp. 1–68

8.7 Palais, R. S.: Ljusternik-Schnirelman theory on Banach manifolds. Topology **5** (1967) 115–132

8.8 Rabinowitz, P. H.: Variational methods for nonlinear elliptic eigenvalue problems. Indiana Univ. Math. J. **23** (1974) 729–754

8.9 Browder, F. E.: Non linear eigenvalue problems and Galerkin-approximation. Bull. Amer. Math. Soc. **74** (1968) 651–656

8.10 Bröcker, T., Jänisch, K.: Einführung in die Differentialtopologie. Heidelberger Taschenbücher **143**. Springer, Berlin Heidelberg 1973

8.11 Lang, S.: Differential manifolds. Addison-Wesley, Reading, MA 1972

8.12 Palais, R. S., Smale, S.: A generalized Morse theory. Bull. Amer. Math. Soc. **70** (1964) 165–171

8.13 Krasnoselskij, M.: Topological methods in the theory of nonlinear integral equations. Pergamon, New York 1964

8.14 Coffman, C. V.: A minimum-maximum principle for a class of nonlinear integral equations. Analyse Mathématique **22** (1969) 391–419

8.15 Dieudonné, J.: Foundations of modern analysis. Academic Press, New York 1969

8.16 Robertson, A. P., Robertson, W. J.: Topological vector spaces. Cambridge University Press, Cambridge 1973

8.17 Schwartz, J. T.: Nonlinear functional analysis. Gordon and Breach, New York 1969

8.18 Rabinowitz, P. H.: Some aspects of nonlinear eigenvalue problems. Rocky Mountain J. Math. **3** (1973) 161–202

8.19 Vainberg, M. M.: Variational methods for the study of nonlinear operators. Holden Day, London 1964

8.20 Berger, M. S.: Non linearity and functional analysis. Academic Press, New York 1977

Further Reading

Palais, R. S.: Critical point theory and the minimax principle. Proc. Am. Math. Soc. Summer Institute on Global Analysis, S. S. Chen and S. Smale (eds.), 1968

Rabinowitz, P. H.: Pairs of positive solutions for nonlinear elliptic partial differential equations. Indiana Univ. Math. J. **23** (1974) 173–186

Alber, S.I.: The topology of functional manifolds and the calculus of variations in the large. Russian Mathematical Surveys **25** (1970) 4

Amann, H.: Ljusternik-Schnirelman theory and nonlinear eigenvalue problems. Math. Ann. **199** (1972) 55–72

Chapter 9

9.1 Berestycki, H., Lions, P.L.: Nonlinear scalar field equations. I: Existence of a ground state, pp. 313–345; II: Existence of infinitely many solutions, pp. 347–375. Arch. Rat. Mech. Anal. **84** (1983)

9.2 Bossavit, A., Damiamian, A., Fremond, M. (eds.): Free boundary problems: applications and theory, vols. III and IV. Research Notes in Mathematics **120, 121**. Pitman, Boston 1985

9.3 Fife, P.C.: Mathematical aspects of reacting and diffusing systems. Lecture Notes in Biomathematics **28**. Springer, Berlin Heidelberg 1979

9.4 Fitzgibbon (III), W.E., Walker, H.F. (eds.): Nonlinear diffusion. Research Notes in Mathematics **14**. Pitman, Boston 1977

9.5 Friedman, A.: Variational principles and free boundary problems. Wiley, New York 1982

9.6 Lions, P.L.: The concentration-compactness principle in the calculus of variations. The locally compact case: Ann. I. H. P. Anal. non linéaire **1** (1984); part 1, pp. 109–145; part 2, pp. 223–283. The limit case. Revista Matematica Ibero Americana (1985): part 1, **1**(1), pp. 145–201; part 2, **1**(2), pp. 45–121

9.7 Zeidler, E.: Nonlinear functional analysis and its applications III. Variational methods and optimization. Springer, Berlin Heidelberg 1985

9.8 Aubin, T.: Nonlinear analysis on manifolds. Monge-Ampère equations. Springer, Berlin Heidelberg 1982

9.9 Hörmander, L., Lions, J.L.: Sur la complétion par rapport à une intégrale de Dirichlet. Math. Scand. **4** (1956) 259–270

9.10 Talenti, G.: Best constant in Sobolev inequality. Ann. di Matem. Pura et Appl. **110** (1976) 353–372

9.11 Strauss, W.A.: Existence of solitary waves in higher dimensions. Commun. Math. Phy. **55** (1977) 149–162

9.12 Coleman, S., Glaser, V., Martin, A.: Action minimum among solutions to a class of Euclidean scalar field equations. Commun. Math. Phys. **58** (1978) 211-221

9.13 Brezis, H., Lieb, E.H.: Minimum action solutions of some vector field equations. Commun. Math. Phys. **96** (1984) 97–113

9.14 Brüning, E.: On the variational approach to semilinear elliptic equations with scale-covariance. J. Diff. Eq. **83** (1990) 109–144

9.15 Strauss, W.A., Vàzquez, L.: Existence of localized solutions for certain model field theories. J. Math. Phys. **22** (1981) 1005–1009

9.16 Lions, P.L.: On the existence of positive solutions of semilinear elliptic equations. SIAM Rev. **24** (1982) 441–467

9.17 Berestycki, H., Gallouët, Th., Kavian, O.: Equations de champs scalaires Euclidiens non linéaires dans le plan. Compt. Rend. Acad. Sci. **297** (1983) 307–310

9.18 Struwe, M.: Multiple solutions of differential equations without the Palais-Smale condition. Math. Ann. **261** (1982) 399–412

9.19 Struwe, M.: A generalized Palais-Smale condition and application. Proc. Symposia in Pure Mathematics **45**, part 2 (1986) 229–241. American Mathematical Society, Providence, RI

9.20 Brezis, H., Nirenberg, L.: Positive solutions of nonlinear elliptic equations involving critical Sobolev exponents. Commun. Pure Appl. Math. **36** (1983) 437–477
9.21 Brezis, H., Lieb, E. H.: A relation between pointwise convergence of functions and convergence of functionals. Proc. Am. Math. Soc. **88** (1983) 486–490
9.22 Lieb, E. H.: Existence and uniqueness of the minimizing solution of Choquard's nonlinear equation. Studies in Appl. Math. **57** (1977) 93–105
9.23 Brascamp, H. J., Lieb, E. H., Luttinger, J. M.: A general rearrangement inequality for multiple integrals. J. Funct. Anal. **17** (1974) 227–237
9.24 Lieb, E. H.: On the lowest eigenvalue of the Laplacian for the intersection of two domains. Invent. Math. **74** (1983) 441–448
9.25 Gilbarg, D., Trudinger, N. S.: Elliptic partial differential equations of second order. Grundlehren der mathematischen Wissenschaften **224**. Springer, Berlin Heidelberg 1977
9.26 Kato, T.: Schrödinger operators with singular potentials. Israel J. Math. **13** (1972) 135–148
9.27 Agmon, S.: Lectures on exponential decay of solutions of second order elliptic equations: bounds on eigenfunctions of n-body Schrödinger operators. Princeton University Press, Princeton, NJ 1982
9.28 Pohozaev, S. I.: Eigenfunctions of the equations $\Delta u + \lambda f(u) = 0$. Sov. Math. Dokl. **6** (1965) 1408–1411
9.29 Brüning, E.: A note on the existence of solutions for two-dimensional vector field equations with strong nonlinearity. Proc. Conf. on Nonlinear Classical Fields, Bielefeld, July 1987 (Lecture Notes in Physics, vol. 347, ed. Ph. Blanchard et al.). Springer, Berlin Heidelberg 1989, pp. 37–52
9.30 Gidas, B.: Euclidean Yang Mills and related equations. In: Bifurcation phenomena in mathematical physics and related topics, C. Bardos and D. Bessis (eds.). Reidel, Dordrecht 1980
9.31 Kazdan, J., Warner, F.: Remarks on some quasilinear elliptic equations. Commun. Pure App. Math. **28** (1975) 567–697
9.32 Gidas, B., Ni, W. M., Nirenberg, L.: Symmetry and related properties via the maximum principle. Commun. Math. Phys. **68** (1979) 209–243
9.33 Protter, M. H., Morrey, C. B.: A first course in real analysis. Springer, Berlin Heidelberg 1977
9.34 Volevich, L. R., Paneyakh, B. P.: Certain spaces of generalized functions and embedding theorems. Russian Mathematical Surveys **20** (1965) 1–73
9.35 Moser, J.: A sharp form of an inequality of N. Trudinger. Ind. Univ. Math. J. **20** (1971) 1077–1092
9.36 Ding, W. Y., Ni, W. M.: On the existence of positive entire solutions of a semilinear elliptic equation. Arch. Rat. Mech. Anal. **91** (1986) 283–308
9.37 Kufner, A., John, O., Fučik, S.: Function spaces. Noordhoff, Leyden 1977
9.39 Vainberg, M. M.: Variational methods and the method of monotone operators in the theory of nonlinear equations. Wiley, New York 1972
9.40 Brezis, H.: Some variational problems with lack of compactness. Proc. Symposia in Pure Mathematics **45**, part 1 (1986) 165–201. American Mathematical Society, Providence, RI
9.40 Ambrosetti, A., Rabinowitz, P.: Dual variational methods in critical point theory and applications. J. Funct. Anal. **14** (1973) 349–381
9.41 Rabinowitz, P. H.: Variational methods for nonlinear elliptic eigenvalue problems. Indiana Univ. Math. J. **23** (1974) 729–754
9.42 Rabinowitz, P.: Minimax methods for indefinite functionals. Proc. Symposia in Pure Mathematics **45**, part 2 (1986) 287–306. American Mathematical Society, Providence, RI

9.43 Peletier, L. A., Serrin, J.: Uniqueness of positive solutions of semilinear equations in $\mathbb{R}^n$. Arch. Rat. Mech. Anal. **81** (1983) 181–197

9.44 Ni, W. M.: Uniqueness, nonuniqueness and related questions of nonlinear elliptic and parabolic equations. Proc. Symposia in Pure Mathematics **45**, part 2 (1986) 229–241. American Mathematical Society, Providence, RI

9.45 Cerami, G., Fortunato, D., Struwe, W.: Bifurcation and multiplicity results for nonlinear elliptic problems involving critical Sobolev exponents. Ann. Inst. H. Poincaré, Analyse Non Linéaire **1** (1984) 341–350

9.46 Struwe, M.: A global compactness result for elliptic boundary value problems involving limiting nonlinearities. Math. Z. **187** (1984) 511–517

9.47 Capozzi, A., Fortunato, D., Palmieri, G.: An existence result for nonlinear elliptic problems involving critical Sobolev exponents. Ann. Inst. H. Poincaré, Analyse Non Linéaire **2**, no. 6 (1985) 463–470

9.48 Cerami, G., Solimini, S., Struwe, M.: Some existence results for superlinear elliptic boundary value problems involving critical exponents. J. Funct. Anal. **69** (3) (1986) 289–306

9.49 Ambrosetti, A., Struwe, M.: A note on the problem $-\Delta u = \lambda u + u|u|^{2^*-2}$. Manuscripta Math. **54** (1986) 373–379

Further Reading

Friedman, A., McLeod, B.: Strict inequalities for integrals of decreasingly rearranged functions. Proc. Roy. Soc. Edinburgh **102A** (1986) 277–289

Kato, T.: Growth properties of solutions for the reduced wave equation with a variable coefficient. Commun. Pure Appl. Math. **42** (1959) 403–425

Lieb, E. H.: Sharp constants in the Hardy-Littlewood-Sobolev and related inequalities. Ann. Math. **118** (1983) 349–374

Nirenberg, L.: On elliptic partial differential equations. Ann. Scuola Norm. Sup. Pisa Cl. Sci. (3) **13** (1959) 55–72

Chapter 10

10.1 Thirring, W.: A course in mathematical physics, vol. 4. Springer, Berlin Heidelberg 1986

10.2 Lieb, E. H., and Simon, B.: The Thomas-Fermi theory of atoms, molecules and solids. Adv. in Math. **23** (1977) 22–116; Thomas-Fermi theory revisited. Phys. Rev. Lett. **31** (1973) 681–683

10.3 Lieb, E. H.: The stability of matter. Rev. Mod. Phys. **48** (1970) 553–569

10.4 Lieb, E. H.: Thomas-Fermi and related theories of atoms and molecules. Rev. Mod. Phys. **53** (1981) 603–641; Errata, ibid **54** (1982) 311

10.5 Reed, M., Simon, B.: Methods of modern mathematical physics II. Fourier analysis, self adjointness. Academic Press, New York 1975

10.6 Brezis, H.: Nonlinear problems related to the Thomas-Fermi equation. In: Contemporary developments in continuum mechanics and partial differential equations, G. M. de la Penha and L. A. Medreiros (eds.). North-Holland, Amsterdam 1978. Some variational problems of the Thomas-Fermi type. In: Variational inequalities and complementary problems: theory and applications, R. W. Cottle, F. Gianessi and J. L. Lions (eds.). Wiley, New York 1980, pp. 55–73

10.7 Thomas, L. H.: The calculation of atomic fields. Proc. Camb. Phil. Soc. **23** (1927) 542–548

10.8 Fermi, E.: Un metodo statistico per la determinazione di alcune prioretà dell'atomo. Rend. Acad. Naz. Lincei **6** (1927) 602–607

10.9 Messer, J.: Temperature dependent Thomas-Fermi theory. Lecture Notes in Physics **147**. Springer, Berlin Heidelberg 1981

10.10 Baumgartner, B., Narnhofer, H., Thirring, W.: Thomas-Fermi limit of Bose-jellium. Ann. Phys. **150** (1983) 373–391

10.11 Benguria, R., Brezis, H., Lieb, E.: The Thomas-Fermi-von Weizsäcker theory of atoms and molecules. Commun. Math. Phys. **79** (1981) 167–180

10.12 Daubechies, I., Lieb, E.: One electron relativistic molecules with Coulomb interaction. Commun. Math. Phys. **90** (1983) 497–510

10.13 Lieb, E., Thirring, W.: Gravitational collapse in quantum mechanics with relativistic kinetic energy. Ann. Phys. **155** (1984) 494–512

10.14 Fröhlich, J., Lieb, E., Loss, M.: Stability of Coulomb systems with magnetic fields I. The one electron atom. Commun. Math. Phys. **104** (1986) 251–270

10.15 Lieb, E., Loss, M.: Stability of Coulomb systems with magnetic fields II. The many electron atom and the one electron molecule. Commun. Math. Phys. **104** (1986) 271–282

10.16 Loss, M., Yau, H. T.: Stability of Coulomb systems with magnetic fields III. Zero energy bound states of the Pauli operator. Commun. Math. Phys. **104** (1986) 283–290

10.17 Lieb, E., Thirring, W.: Bounds for the kinetic energy of fermions which prove the stability of matter. Phys. Rev. Lett. **35** (1976) 687–689; Errata, ibid **35** (1976) 1116

10.18 Lieb, E. H.: The stability of matter: from atoms to stars. American Mathematical Society Josiah Willard Gibbs Lecture, Phoenix, AZ 1989. Bulletin of the AMS, vol. 22 no. 1, Jan. 1990, pp. 1–49

Further Reading

Thirring, W.: A lower bound with the best possible constant for Coulomb Hamiltonians. Commun. Math. Phys. **79** (1981) 1–7

Dyson, F. J.: In: Brandeis University Summer Institute in Theoretical Physics 1966, vol. 1. M. Chretien, E. P. Gross and S. Deser (eds.). Gordon and Breach, New York 1978

Lenard, A.: In: Statistical mechanics and mathematical problems. Lecture Notes in Physics **20**. Springer, Berlin Heidelberg 1973

Hille, E.: On the Thomas-Fermi equation. Proc. Nat. Acad. Sci. (USA) **62** (1969) 7–10

Appendix A

Further Reading

Hirzebruch, F., Scharlau, W.: Einführung in die Funktionalanalysis. BI Taschenbücher **296**, BI, Mannheim 1971

Ljusternik, L. A., Sobolev, W. I.: Elements of functional analysis. Ungar, New York 1965

Wloka, J.: Funktionalanalysis und Anwendungen. de Gryyter, Berlin 1971

Richtmyer, R. D.: Principles of advanced mathematical physics, vol. 1. Springer, Berlin Heidelberg 1978

Reed, M., Simon, B.: Methods of modern mathematical physics, vol. 1, functional analysis. Revised and enlarged edition. Academic Press, New York 1980

Kelley, J. L., Namioka, I.: Linear topological spaces. Springer, Berlin Heidelberg 1974

Köthe, G.: Topological vector spaces, vol. 1. Springer, Berlin Heidelberg 1969

Hille, E.: Methods in classical and functional analysis. Addison-Wesley, Reading, MA 1972

Appendix C

C.1 Köthe, G.: Topological vector spaces I. Springer, Berlin Heidelberg 1983

Further Reading

Dieudonné, J.: Foundations of modern analysis. Academic Press, New York 1969
Hirzebruch, F., Scharlau, W.: Einführung in die Funktionalanalysis. BI Taschenbücher **296**, BI, Mannheim 1971
Dunford, N., Schwartz, J.T.: Linear operators. Part I: general theory. Wiley-Interscience, New York 1963
Heuser, H.: Funktionalanalysis. Teubner, Stuttgart 1975
Schaefer, H.H.: Topological vector spaces. Springer, Berlin Heidelberg 1971
Yosida, K.: Functional analysis. Springer, Berlin Heidelberg 1965

Appendix D

D.1 Gilbarg, D., Trudinger, N.S.: Elliptic partial differential equations of second order. Grundlehren der mathematischen Wissenschaften **224**. Springer, Berlin Heidelberg 1977
D.2 Adams, R.A.: Sobolev spaces. Academic Press, New York 1975

Further Reading

Lions, J.L., Magenes, E.: Problèmes aux limites non homogènes et applications I. Dunod, Paris 1968
Sobolev, S.L.: Applications of functional analysis in mathematical physics. American Mathematical Society. Providence, RI 1963

Appendix E

E.1 Aronszajn, N., Smith, K.T.: Theory of Bessel potentials I. Ann. Inst. Fourier **11** (1961) 385–475
E.2 Stein, E.M.: Singular integrals and differentiability properties of functions. Princeton University Press, Princeton, NJ 1970
E.3 Gilbarg, D., Trudinger, N.S.: Elliptic partial differential equations of second order. Grundlehren der mathematischen Wissenschaften **224**. Springer, Berlin Heidelberg 1977
E.4 Aubin, T.: Nonlinear analysis on manifolds. Monge-Ampère equations. Springer, Berlin Heidelberg 1982
E.5 Nirenberg, L.: On elliptic partial differential equations. Ann. Scuola Norm. Sup. Pisa Cl. Sci. (3) **13** (1959) 55–72

Index of Names

Baire, R.L. 10, 373
Berestycki, H. 322, 337
Bernoulli, Jakob 3, 4
Bernoulli, Johann 3
Bolzano, B. 12
Brezis, H. 245, 247, 288, 320, 336, 361
Browder, F.E. 10, 171, 173, 176, 195, 207
Brüning, E. 245

Carathéodory, C. 10

Dirichlet, G. 7
Du Bois-Reymond, P. 6

Euler, L. 4

Fermat, P. de 7
Fréchet, M. 6

Gâteaux, R. 6
Gauß, C.F. 4
Glaser, V. 245

Hamilton, W.R. 4
Hilbert, D. 6, 9

Jacobi, C. 7

Krasnoselskij, M.A. 211

Lagrange, L. de 6
Legendre, A.M. 7
Lebesgue, H. 9
Lieb, E.H. 245, 257, 261, 283, 320, 341, 361
Lions, J.L. 10
Lions, P.L. 337

Maupertuis, P.L. 4
Minty, G. 173, 176

Nirenberg, L. 247, 288
Noether, E. 11

Palais, R.S. 195
Pohozaev, S.I. 278

Riemann, B. 7

Schrödinger, E. 11
Schwartz, J.T. 194
Simon, B. 341, 361
Stampacchia, G. 10
Strauss, W. 244, 246, 259
Struwe, M. 337

Talenti, G. 245, 257, 283
Tonelli, L. 10

Weierstraß, K. 8

Subject Index

Action functional 80
Algebraic dual space 367
Approximation property 368

Bauer's maximum principle 30
Bessel potential 269
- of order α 391
Best Sobolev constant 283
Bidual space 367
Bounded linear map 367
Bounded set 364
Boundary conditions 149
natural – 172
Brachystochrone problem 3, 77, 113 ff.
- with Coulomb friction 116

Canonical momenta 107
Carathéodory-(C–) function 179
Cauchy-Schwarz inequality 364
Cauchy sequence 363
Chain rule 45
Charge 124
Chemical potential 357
Classical Dirichlet problem 159
Classical minimax principle 144, 145
Classical Poisson problem 159
Compact 373
countably – 373
relatively – 373
sequentially – 12, 16, 373
Compact map 375
Compactness principle 245
Complete 363
Completely continuous map 375
Concentration functional 306
Concentration of a minimizing sequence 246
Condition C of Palais and Smale 210
Condition of
- concentration 253, 304, 306
- incomplete uniform decay 253, 304, 306
Coercive 24, 174
strictly – 28
Coerciveness 173, 182
Conjugate point 92, 99
Continuously differentiable 37
Constrained minimisation (maximisation) problem 63, 244
Convergent sequence 363
Convex set 368
Covariance condition 295, 296
- relations 310
Critical point 1, 35, 55, 208
- subject to a constraint 74
Critical Sobolev exponent 241, 288
Current 124
Cycloid 115

Deformation 205
- theorem 209
Derivative 37
nth – 44
Differentiable
- at a point 37
- on a set 37
Direct orthogonal sum 364
Dirichlet
- boundary conditions 150
- form 7
- –, generalised 171, 193
- principle 29
- problem 7, 159
Distribution 381
regular – 382
Dual exponent 365
Duality 367
Dual space (topological) 367

Eigenvalue problem 149
general – 152

solution of – 193
Ellipticity condition 151
Elliptic regularity theory 246, 266
Energy norm 243
Euler condition 5
– equation 85
Euler-Lagrange equation 97, 104, 118
Equivalent norms 363
Euclidean space 364
Existence of a minimising point 17
Extremal point 30
Extreme value problem 15

Field 116
Field equations (nonlinear elliptic) 122
Field theory 116
local classical – 117
Finite codimension 364
Fixed point of a flow 208
Fréchet derivative 39
Fredholm operator, – mapping 69
Functional analysis 6
Functionals 40

Galerkin approximation 32
Gâteaux derivative 46
– differentiable 47
– differential 46
Genus 211
– of spheres 213
Global field equations 244
Global minimum 54
Green's function 159
Ground state solution 195

Hamiltonian 107
Hamilton's equation of motion 108
– principle 80, 116
Harmonic function 134
Hilbert space 364
Hölder's inequality 365, 366
Homeomorphic 367
Homeomorphism 367

Identity embedding 386
Indefinite functionals 336
Inequality of Gagliardo-Nirenberg 393
Integral of motion 102, 109
Isoperimetric problem 2, 7, 9

Jacobi's condition 14, 90
– equation 99, 120

Kato's inequality for vector fields 272

Kinetic energy 242
Klein-Gordon equation, nonlinear 123

Lack of compactness 335, 337
Lagrangian 80, 107
Lebesgue spaces 366
Legendre's condition 85
– conjecture 89
– necessary condition 97, 104, 119
Legendre transformation 107
Lemma of
– concentration by translation 245, 261
– Du Bois-Reymond 6, 83
Limit case 287
Linearisable 139
Ljusternik-Schnirelman category 210
Local extremal point 54

Mass zero case 244
Maximising point 12, 15
– subset 206
Mayer determinant 100
Minimal action solution 299
Minimax principle
– classical 145
– generalised 206
Minimax value 205
Minimising point 12, 15
– subset 309
– sequence 31
Monotone 52
strictly – 52
Monotonicity 154, 173, 182
Moser-Trudinger inequality
– global version 326
Mountain-pass lemma 336

Navier-Stokes equation 168
Necessary condition of Euler-Lagrange 55
Niemytski operator 179
Nontranslation-invariant case 244, 277
Norm, normed space 363
Norm (strong) topology 368

Palais-Smale condition, generalised 337
Partial derivative 39
Phase space 108
Poincaré inequality 151
Pointwise bounded 369
Poisson bracket 109
Pohozaev identities 282, 289

Polynomial behaviour at zero (infinity) 310
Positive mass case 244
Potential 57
- energy 243
- operator (gradient operator) 57
Principle of least action 80, 88, 104 116
- symmetric criticality 131
- uniform boundedness 369
Projection theorem 364

Quadratic functionals 28

Radial retraction 175
Reflexive Banach space, reflexivity 367, 377
Regular point 55, 208
Regularity of a local minimum 86, 87
Riccati's differential equation 89

Scalar field equation 244
Scalar product 364
Semicontinuous at a point
lower - 12, 16, 371
upper - 16, 371
Semiconvex 26
Seminorm 363
Separable 363
Sequentially lower semicontinuous 372
Sequence space 365
Smale condition 177
Smallest accumulation point 16
Snell's law of refraction 116
Sobolev's inequality 242, 388
Sobolev's spaces 382
- classical definition 380
Spaces of continuous functions 365, 366
Spaces of differentiable functions 365
Sphere like
- (sub-) manifold (214), 215, 217
- constraint 218, 219
Spherically symmetric rearrangement 257
Stable quantum mechanical N-body system 341
Stokes equation 168
Strict local minimum 54
Strong nonlinearity 322
Sturm-Liouville boundary value problem 159
Symmetry 107 ff., 124 ff., 303
Symmetry group 107 ff., 124 ff., 303
Symmetric points 131

Theorem of
Arzela-Ascoli 374
Baire 370
Banach-Alaoglu-Bourbaki 376
Banach-Steinhaus 369
Browder-Minty 176
Clarkson 368
Eberlein 21
Heine-Borel 374
Jordan-von Neumann 364
Kolmogoroff-M. Riesz 374
Krein-Milman 375
Milman 368
Noether 109, 112, 117
Rellich-Kondrachov 388
Riesz 20
Riesz-Fischer 366
Riesz-Fréchet 367
Sobolev (embedding) 387
- (regularity) 387
Šmulian-Kaplansky 377
the inverse operator 367
(fixed point) theorem of
Brouwer 376
Schauder 375

Thomas-Fermi energy functional 340
- equations 340, 341
Translation-invariant case 244, 277
Twice differentiable 43
Two dimensional mass zero vector field equation 323

Uniformly convex space 368
Uniqueness of a minimising point 17

Variation, nth. 49
Variational boundary value problem 172
Variational derivative 41
Variational form of eigenvalue problems 150
Variational problem with subsidiary condition or constraint minimisation (maximisation) 63 ff., 196, 200, 204, 225, 237, 244, 294 ff., 304, 320, 328
Vector field equation 244
Virial equation 282

Weak compactness 376
Weak convergence 368
Weak (distributional) derivative 383
Weakly bounded 369
- complete 369

Weakly countably compact 21
- sequentially compact 13
- sequentially lower (upper) semicontinuous 22, 23
Weak neighborhood 368
- topology 13, 20, 368
- * topology 376
Wronskian 100

Yamabe problem 242
Young's inequality 268, 342
generalised - 268